STUDY GUIDE to accompany

Precalculus

Third Edition

Mustafa A. Munem

William Tschirhart

James P. Yizze

MACOMB COUNTY COMMUNITY COLLEGE

Worth Publishers, Inc.

STUDY GUIDE

to accompany

PRECALCULUS, Third Edition

Printed in the United States of America
ISBN: 0-87901-092-4
First printing, March 1978

WORTH PUBLISHERS, INC.
444 Park Avenue South
New York, New York 10016

PREFACE

PURPOSE This study guide has been revised to conform with the changes made in the third edition of *Precalculus*, by Mustafa Munem and James Yizze. It is not intended as an independent textbook, but rather should be used as a workbook to help the student master the material presented in *Precalculus.*

ORGANIZATION The study guide is written in a semiprogrammed format, in that the correct answer to one question leads to answering the following question. Difficult concepts and techniques presented in the textbook are broken down into a sequence of steps that make them easier to understand. Questions that are shaded emphasize important definitions, properties, and theorems. The chapters in this workbook conform section by section to the ideas and material developed in the textbook.

The tests at the end of each chapter were compiled from test questions used by the authors with their own students in class-testing the third edition of *Precalculus.* Each test is followed by answers to all the test problems. The test problems are similar in scope and difficulty to those appearing throughout each chapter and will be useful in preparing for class examinations.

HOW TO USE We recommend that the workbook be used in the following way:

1 Place a strip of paper over the left column to cover the answers.

2 Read the statement in the right column carefully and complete the statement by writing the answers in the blanks provided.

3 After all of the answers for that problem are filled in, lower the strip of paper and compare your answers with the correct answers. Review the material in the textbook whenever you do not fully understand the answers given.

4 Continue to the next problem and repeat the above procedure.

The use of this workbook obviously depends on the needs of the student. However, here are some suggestions:

As an auxiliary learning resource to be used for each section covered in the parent textbook.

As a supplementary learning resource to be used only when a student is having difficulty mastering the material in a particular section of the textbook.

As an auxiliary learning resource to be used in reviewing and studying for examinations.

As a learning resource to be used to cover material missed because of absences from class.

MUSTAFA A. MUNEM
WILLIAM TSCHIRHART
JAMES P. YIZZE

CONTENTS

Chapter 2 FUNCTIONS AND GRAPHS 55

Chapter 3 POLYNOMIAL AND RATIONAL FUNCTIONS 95

Chapter 4
EXPONENTIAL AND LOGARITHMIC FUNCTIONS 144

Chapter 5
TRIGONOMETRIC FUNCTIONS 174

APPENDIX

Chapter 1 FUNDAMENTALS OF ALGEBRA

The main objective of this chapter is to review some of the fundamentals of algebra needed throughout the remainder of this study guide and the text that it accompanies. After working through appropriate problems in the chapter, the student will be able to:

1 Locate real numbers on the number line.
2 Perform operations on polynomials, factor polynomials, and solve linear equations.
3 Perform operations on fractions.
4 Simplify exponential and radical expressions.
5 Find the solution sets of linear inequalities and show each solution on a number line.
6 Find the solution sets of absolute-value equations and inequalities.

1 Sets and Real Numbers

Answer	No.	Frame
set	1	A collection of objects is called a _____.
elements (or members)	2	The objects of sets are called the _____ of the set.
$A = \{2, 3, 5, 7\}$	3	The set A, whose elements are 2, 3, 5, and 7, is written as _____.
element (or member)	4	If $x \in A$, we say that x is an _____ of set A.
element	5	If $x \notin A$, we say that x is *not* an _____ of set A.
empty, null $\emptyset$	6	The set with no elements is called the _____ set or _____ set and is represented symbolically by _____.
5	7	The set of counting numbers between 3 and 6 is represented symbolically by $\{4, \_\_\_\}$.
$\{8, 9, 10, 11, 12\}$	8	The set of counting numbers between 7 and 13 is _____.

In Problems 9–19, let A be the set of all counting numbers less than or equal to 14. Indicate which of the statements are true and which are false.

Answer	No.	Frame
True	9	$8 \in A$ _____
True	10	$15 \notin A$ _____
False	11	$0 \in A$ _____
False	12	$-3 \in A$ _____
True	13	$6 \in A$ _____
False	14	$\frac{1}{3} \in A$ _____
False	15	$-7 \in A$ _____
True	16	$5 \in A$ _____
False	17	$\emptyset \in A$ _____
False	18	$\{2\} \in A$ _____
False	19	$\{3, 4\} \in A$ _____
5	20	The set of fingers on one hand has _____ elements.
12	21	The set of months in a year has _____ elements.
$\{8, 9, 10, 11, 12, 13, 14, 15\}$	22	The set of counting numbers between 7 and 16 is described by enumeration as _____.
empty	23	The set of women vice-presidents of the United States is the _____ set.
$\in$, $\notin$	24	Let $A = \{1, 2, 3, 4, 5\}$; then 2 _____ A and 6 _____ A.
empty	25	The set of cats that can fly is the _____ set.
empty	26	The set of squared odd counting numbers that are even is the _____ set.

27 The set of counting numbers between 5 and 10 is described by enumeration as ______.

$\{6, 7, 8, 9\}$

28 A set A whose elements can be counted, in the usual way, using the counting numbers 1, 2, 3, 4, . . . , that eventually terminates, resulting in a specific number of elements in the set, is called a ______ set.

finite

29 The set $\{1, 2, 3, 5, 6\}$ is a ______ set.

finite

30 The set of cities in Canada is a ______ set.

finite

31 The set of counting numbers less than 15 is a ______ set.

finite

32 A set that is neither finite nor empty is called an ______ set.

infinite

33 The set of odd counting numbers is an ______ set.

infinite

34 The set of counting numbers whose elements are divisible by 3 is an ______ set.

infinite

35 The set of people in the world is a ______ set.

finite

36 The set of letters in the word Michigan is a ______ set.

finite

37 The set of women who model for an art class is a ______ set.

finite

38 The set of students in a mathematics class is a ______ set.

finite

39 The set of even counting numbers is an ______ set.

infinite

1.1 Set Relations and Operations

40 If every element of the set A is also an element of the set B, then we say that A is a ______ of the set B. This is represented symbolically by ______.

subset

$A \subseteq B$

41 The set $\{3, 4\}$ is a ______ of the set $\{3, 4, 5\}$.

subset

In Problems 42–48, indicate which of the statements are true and which are false. Consider the sets $A = \{1, 2, 3, 4\}$, $B = \{1, 2\}$, $C = \{3, 4\}$, and $D = \{3, 4, 5\}$.

42 $B \subseteq A$ ______ — True

43 $C \subseteq D$ ______ — True

44 $\{3\} \subseteq C$ ______ — True

45 $4 \in D$ ______ — True

46 $\emptyset \subseteq C$ ______ — True

47 $B \subseteq \emptyset$ ______ — False

48 $D \subseteq B$ ______ — False

49 If set A is a subset of set B, and set B is a subset of set A, then we say that set A is ______ to set B. It is represented symbolically by A ______ B.

equal

=

50 $\{1, 2, 5, 7\}$ ______ $\{5, 2, 1, 7\}$

=

Answer	No.	Frame
=	51	$\{3, x, y\}$ ____ $\{y, 3, x\}$
=	52	$\{3, 4, 5, 6, 7\}$ ____ $\{6, 5, 3, 4, 7\}$
proper; $\subset$	53	A set A is a ________ subset of a set B if all members of A are in B and B has at least one member not in A. It is represented symbolically by A ____ B.
proper	54	$\{1, 2\}$ is a ________ subset of $\{1, 2, 3\}$.
$\emptyset$, $\{2\}$, $\{3\}$, $\{2, 3\}$	55	The possible subsets of $\{2, 3\}$ are ________________.
$\not\subset$	56	The symbol $\not\subset$ means " is not a subset of." If $A = \{1, 2\}$ and $B = \{2, 3, 4\}$, then A ____ B.
3	57	The set $\{x, y\}$ has ____ proper subsets.
$\{3\}$, $\{4\}$, $\{5\}$, $\{3, 4\}$ $\{3, 5\}$, $\{4, 5\}$, $\emptyset$	58	List all the proper subsets of the set $\{3, 4, 5\}$. ________________________
2^n	59	If set A has n elements, the number of subsets in set A is ____.
No	60	If A is any set, is $A \subset \emptyset$? ____
union; $A \cup B$	61	The set of all elements in set A or in set B, or in both sets A and B, is called the ____________ of A and B, and it is written as ____________.
$\{1, 3, 5, 7\}$	62	$\{3, 1\} \cup \{5, 7, 3\} =$ __________
$\{1, 3, 5, 7, 9\}$	63	$\{5, 7, 9\} \cup \{9, 3, 1\} =$ __________
$\{x, y, z, r, s, t\}$	64	$\{x, y, z, t\} \cup \{x, y, r, s\} =$ __________
$\{a, b\}$	65	$\{a, b\} \cup \emptyset =$ ________
$\{12, 4, 8, 1, 6, 5\}$	66	$\{5, 4, 12\} \cup \{4, 12, 6, 8, 1\} =$ __________
$\{1, 2, 3, 4, 5, \ldots\}$	67	$\{1, 3, 5, 7, \ldots\} \cup \{2, 4, 6, \ldots\} =$ __________
seven	68	If set A has four members and set B has five members, as shown in Figure 1, then $A \cup B$ has ________ members.

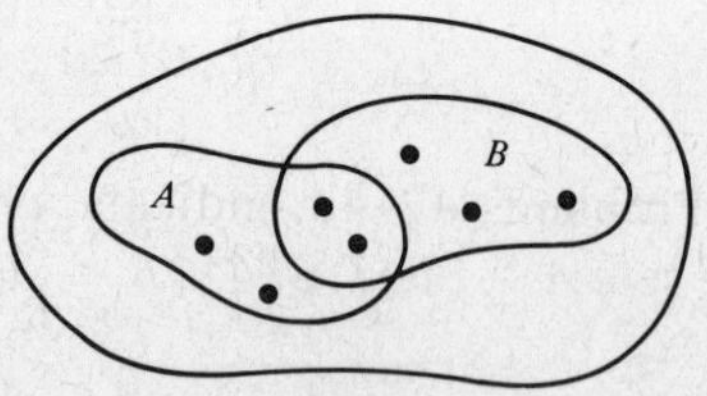

Figure 1

Answer	No.	Frame
B	69	If $A \subseteq B$, then $A \cup B =$ ____.
finite	70	Let A and B be finite sets. Then $A \cup B$ is a ________ set.
commutative	71	$A \cup B = B \cup A$ is a statement of the ____________ property of the union of two sets.
intersection, $A \cap B$	72	The set of all elements common to both sets A and B is called the ____________ of sets A and B and it is written as ________.
$\{3, 4\}$	73	$\{2, 3, 4\} \cap \{3, 4, 7\} =$ ________
$\{a, b, c, d\}$	74	$\{a, b, c, d, e, f\} \cap \{a, b, c, d\} =$ __________

Answer	Frame
$\{x, y\}$	75 $\{x, y, r, s\} \cap \{x, y, z, t\} =$ ________
$\{2, 6\}$	76 $\{2, 4, 6, 8, 10\} \cap \{2, 6, 5, 7, 9\} =$ ________
$\{d, e, f\}$	77 $\{a, b, c, d, e, f\} \cap \{d, e, f, g, h, i\} =$ ________
disjoint	78 Sets A and B are said to be ________ sets if $A \cap B = \emptyset$.
$\emptyset$	79 $\{2, 4, 6, 8\} \cap \{1, 3, 5, 7\} =$ ____
$\subseteq$	80 $A \cap B = B$ if B ____ A
$\subseteq$	81 $A \cap B = A$ if A ____ B
A	82 $A \cap A =$ ____
$\emptyset$	83 $\emptyset \cap A =$ ____
$=$	84 If $A = \{1, 2, 3, 4, 5\}$, $B = \{4, 5, 6, 7\}$, and $C = \{2, 3, 4, 5, 6\}$, then $(A \cap B) \cap C$ ____ $A \cap (B \cap C)$.
$\cap$	85 The shaded region in Figure 2 represents A ____ B.

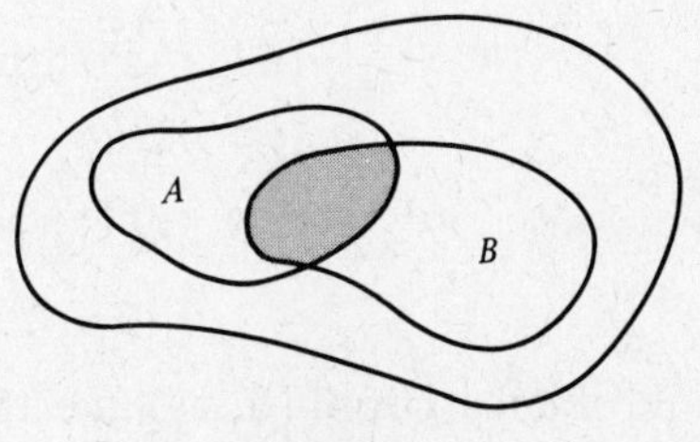

Figure 2

Answer	Frame
$\emptyset$	86 $\{2, 3, 4, 5, 7\} \cap \emptyset =$ ____
$\emptyset$; $\{1, 4, 5, 6\}$	87 If $A = \{1, 5\}$ and $B = \{4, 6\}$, then $A \cap B =$ ____ and $A \cup B =$ ________.

1.2 Real Number Sets

Answer	Frame
positive integers, counting; I_p	88 The set $\{1, 2, 3, \ldots\}$ is called the set of ________ or the set of ________ numbers. Symbolically, it is denoted by ____.
nonnegative integers	89 The union of the set of positive integers and the set $\{0\}$ is called the set of ________.
integers, I	90 The set $\{\ldots, -3, -2, -1, 0, 1, 2, 3, \ldots\}$ is called the set of ________. Symbolically, it is denoted by ____.
rational	91 The set $Q = \left\{q \mid q = \dfrac{x}{y}, \text{ where } y \neq 0 \text{ and } x, y \in I\right\}$ is called the set of ________ numbers.
irrational	92 Such numbers as $\sqrt{2}$ and $3 + \sqrt{5}$ are called ________ numbers.
$\emptyset$	93 {Rational numbers} $\cap$ {irrational numbers} = ____.
{real numbers}; R	94 {Rational numbers} $\cup$ {irrational numbers} = ________. Symbolically, it is denoted by ____.

In Problems 95–99, change each of the rational numbers to fractional form.

Answer	Problem
$\frac{7}{9}$	**95** $0.\overline{7} =$ ____
$\frac{29}{99}$	**96** $0.\overline{29} =$ ____
$\frac{8}{9}$	**97** $0.\overline{8} =$ ____
$\frac{1}{2}$	**98** $0.4\overline{9} =$ ____
$\frac{325}{999}$	**99** $0.\overline{325} =$ ____
proper	**100** The set of integers is a ________ subset of the set of rational numbers.

In Problems 101–103, indicate which of the statements are true and which are false.

Answer	Problem
False	**101** {Rational numbers} $\subseteq$ {integers} ______
False	**102** {Counting numbers} $\subseteq$ {irrational numbers} ______
True	**103** {Positive integers} $\subseteq$ {integers} ______
I	**104** $Q \cap I =$ ______
Q	**105** $Q \cup I =$ ______

In Problems 106–116, assume that Q = the set of rational numbers, I = the set of integers, and L = the set of irrational numbers. Indicate the set to which each of the given elements belongs.

Answer	Problem
L	**106** $\sqrt{2} \in$ ____
Q	**107** $-\frac{11}{7} \in$ ____
L	**108** $\sqrt{2} - 1 \in$ ____
L	**109** $\frac{\pi}{2} \in$ ____
Q	**110** $0.\overline{15} \in$ ____
$I \cup Q$	**111** $-8 \in$ ______
$I \cup Q$	**112** $1 \in$ ______
Q	**113** $2.25 \in$ ____
Q	**114** $\frac{25}{6} \in$ ____
Q	**115** $-\frac{4}{3} \in$ ____
$I \cup Q$	**116** $5 - 7 \in$ ______

In Problems 117–130, express each of the given rational numbers in decimal representation.

Answer	Problem
3.8	**117** $3\frac{4}{5} =$ ____
0.15	**118** $\frac{3}{20} =$ ______
5.125	**119** $5\frac{1}{8} =$ ______
$27.\overline{6}$	**120** $27\frac{2}{3} =$ ______
7.625	**121** $7\frac{5}{8} =$ ______
9.3125	**122** $9\frac{5}{16} =$ ________

Answer	Frame	
$5.\overline{18}$	123	$5\frac{2}{11} =$ ____
$0.\overline{714285}$	124	$\frac{5}{7} =$ ________
$0.52\overline{3}$	125	$\frac{157}{300} =$ ______
$1.8\overline{5}$	126	$\frac{167}{90} =$ ____
$3.\overline{27}$	127	$3\frac{3}{11} =$ ____
$2.\overline{2}$	128	$2\frac{2}{9} =$ ____
0.31	129	$\frac{31}{100} =$ _____
$17.\overline{3}$	130	$17\frac{1}{3} =$ ____

1.3 Real Number Properties

Answer	Frame	
R	131	The closure property for addition states that if $a, b \in R$, then $a + b \in$ ____.
R	132	$\frac{3}{5} + \frac{7}{12} \in$ ____
R	133	$\sqrt{2} + 3 \in$ ____
$b + a$	134	The commutative property for addition states that if $a, b \in R$, then $a + b =$ ________.
7	135	$7 + 6 = 6 +$ ____
9	136	$13 + 9 =$ ____ $+ 13$
4	137	If $x + 8 = 8 + 4$, then $x =$ ____.
4	138	If $x + y + 3 = 4 + 3$, then $x + y =$ ____.
18	139	If $3x + 5 = 5 + 18$, then $3x =$ ____.
c	140	The associative property for addition states that if $a, b, c \in R$, then $a + (b + c) = (a + b) +$ ____.
2	141	$5 + (2 + 7) = (5 +$ ____ $) + 7$
4	142	$(3 + 1) + 4 = 3 + (1 +$ ____ $)$
5	143	If $x + (2 + 3) = (5 + 2) + 3$, then $x =$ ____.
12	144	If $3x + (7 + 2) = (12 + 7) + 2$, then $3x =$ ____.
K	145	$K + (2m + n) = ($ ____ $+ 2m) + n$
5	146	If $x + (2 + 7) = (5 + 2) + 7$, then $x =$ ____.
9	147	If $5 + (3x + 2) = (5 + 9) + 2$, then $3x =$ ____.
14	148	If $7x + (2 + 3) = (14 + 2) + 3$, then $7x =$ ____.
R	149	The closure property for multiplication states that if $a, b \in R$, then $a \cdot b \in$ ____.
R	150	$3 \cdot 2 \in$ ____
R	151	If $x, y \in R$ and $x \cdot y = z$, then $z \in$ ____.
R	152	If $\frac{1}{3} \cdot \frac{5}{6} = x$, then $x \in$ ____.
$b \cdot a$	153	The commutative property for multiplication states that if $a, b \in R$, then $a \cdot b =$ ____.
5	154	$5 \cdot 7 = 7 \cdot$ ____
5	155	$5(3 + 2) = (3 + 2) \cdot$ ____

Answer	Frame	
12	**156**	If $x(5+7) = (5+7)12$, then $x =$ ____.
15	**157**	If $3 \cdot (5x) = 15 \cdot 3$, then $5x =$ ____.
−2	**158**	$-2[5+(-8)] = [5+(-8)] \cdot$ (____)
x	**159**	$x(2y+z) = (2y+z) \cdot$ ____
$a \cdot (b \cdot c)$	**160**	The associative property for multiplication states that if $a, b, c \in R$, then $(a \cdot b) \cdot c =$ ________.
5	**161**	$2 \cdot (3 \cdot 5) = (2 \cdot 3) \cdot$ ____
3	**162**	If $x \cdot (4 \cdot 5) = (3 \cdot 4) \cdot 5$, then $x =$ ____.
8	**163**	If $2x \cdot (3 \cdot 4) = (8 \cdot 3) \cdot 4$, then $2x =$ ____.
$z+w$	**164**	$x[y(z+w)] = x \cdot y$ (____)
23	**165**	$(5 \cdot 16) \cdot 23 = 5 \cdot (16 \cdot$ ____)
$c+d$	**166**	$(a \cdot b) \cdot (c+d) = a \cdot [b \cdot ($______$)]$
$3x$	**167**	$(3x \cdot y) \cdot 5z =$ ____ $\cdot (y \cdot 5z)$
$a \cdot b + a \cdot c$ $a \cdot c + b \cdot c$	**168**	The distributive property states that if $a, b \in R$, then $a \cdot (b+c) =$ __________ and $(a+b) \cdot c =$ ________.
7	**169**	$5 \cdot (4+7) = 5 \cdot 4 + 5 \cdot$ ____
$x \cdot 2$	**170**	$(3+x) \cdot 2 = 3 \cdot 2 +$ ____
$z,\ w$ yw	**171**	$(x+y)(z+w) = (x+y)($____$) + (x+y)($____$)$ $= xz + yz + xw +$ ____
$y \cdot c$	**172**	$(2x+y) \cdot c = 2x \cdot c +$ ____
$3xz$	**173**	$3x(y+z) = 3xy +$ ____
$8 \cdot 11$	**174**	$8 \cdot (15+11) = (8 \cdot 15) + ($______$)$
$0+a,\ a$	**175**	The identity property for addition assumes the existence of a unique real number 0 with the property $a+0 =$ ______ $=$ ____.
4	**176**	$4+0 =$ ____
0	**177**	$5 +$ ____ $= 5$
7	**178**	____ $+ 0 = 7$
$2x+y$	**179**	$(2x+y)+0 =$ ______
$3a$	**180**	____ $+ 0 = 3a$
$a,\ a$	**181**	The identity property for multiplication assumes the existence of a unique real number 1 such that $1 \cdot a =$ ____ and $a \cdot 1 =$ ____.
$11+2$	**182**	$(11+2) \cdot 1 =$ ______
1	**183**	____ $\cdot (2+7) = 2+7$
$x+1$	**184**	$(x+1) \cdot 1 =$ ____
1	**185**	$16 \cdot$ ____ $= 16$
$3x+y$	**186**	$(3x+y) \cdot 1 =$ ______
1	**187**	____ $\cdot (5x+7) = 5x+7$
$2x+1$	**188**	$1 \cdot ($______$) = 2x+1$
$4x+3y$	**189**	$1 \cdot (4x+3y) =$ ______

Answer	Frame
	190 If $a \in R$, then there exists a unique real number $-a$, called the
0	*additive inverse*, such that $a + (-a) = (-a) + a =$ ____.
0	**191** $3 + (-3) =$ ____
-7	**192** $7 + ($____$) = 0$
$-3x$	**193** $3x + ($____$) = 0$
12	**194** If $3x + (-12) = 0$, then $3x =$ ____.
15	**195** If $15 + (-5x) = 0$, then $5x =$ ____.
0	**196** $(x + y) + [-(x + y)] =$ ____
$-(3x + 2)$	**197** $(3x + 2) + [$________$] = 0$
	198 If $a \neq 0$ and $a \in R$, then there exists a unique real number $\frac{1}{a}$,
reciprocal	called the *multiplicative inverse* or ________, such that
1	$a \cdot \frac{1}{a} = \frac{1}{a} \cdot a =$ ____
1	**199** $\sqrt{2}\left(\frac{1}{\sqrt{2}}\right) = ($____$)$
$\frac{1}{3}$	**200** $3 \cdot$ ____ $= 1$
$\frac{1}{5}$	**201** $5 \cdot$ ____ $= 1$
2	**202** $\frac{1}{2} \cdot$ ____ $= 1$
$\frac{1}{x + y}$	**203** $(x + y) \cdot \frac{}{\rule{2cm}{0.4pt}} = 1$
$a + 5$	**204** $($____$) \cdot \frac{1}{a + 5} = 1$
1	**205** $-3 \cdot (-\frac{1}{3}) =$ ____
	206 The cancellation property for addition states that if $a, b, c \in R$
b	and $a + c = b + c$, then $a =$ ____.
$r + s$	**207** If $(r + s) + 2 = 5 + 2$, then ________ $= 5$.
15	**208** If $5x + 3 = 15 + 3$, then $5x =$ ____.
18	**209** If $3x + y + z = 18 + y + z$, then $3x =$ ____.
$p + q$	**210** If $4 + (p + q) = 4 + 7$, then ________ $= 7$.
	211 The cancellation property for multiplication states that if
b	$a, b, c \in R$, $c \neq 0$, and $ac = bc$, then $a =$ ____.
5	**212** If $3x = 3 \cdot 5$, then $x =$ ____.
7	**213** If $4(p + q) = 4 \cdot 7$, then $p + q =$ ____.
8	**214** If $3xy = 24y$, then $x =$ ____.
5	**215** If $(a + x)(c + d) = 5(c + d)$, then $a + x =$ ____.
$a + b$	**216** If $(3x + y)z = (a + b)z$, then $3x + y =$ ________.
$2x - 3y$	**217** If $(2x - 3y)(-2) = (20)(-2)$, then ________ $= 20$.
0	**218** For every $a \in R$, $a \cdot 0 = 0 \cdot a =$ ____.
0	**219** $3 \cdot 0 =$ ____
0	**220** $4 \cdot$ ____ $= 0$

Answer	Frame	
0	221	$(3x + y) \cdot$ ____ $= 0$
0	222	$0 \cdot (5x + 2) =$ ____
0, 0	223	If $a, b \in R$ and $a \cdot b = 0$, then $a =$ ____, $b =$ ____, or both.
0	224	If $3x = 0$, then $x =$ ____.
0	225	If $x(5 + z) = 0$, then either $x = 0$ or $5 + z =$ ____.
0	226	If $(5x + y) \cdot (3x - 2y) = 0$, then either $5x + y =$ ____ or $3x - 2y = 0$.
$a + b$	227	If $a, b \in R$, then $(-a) + (-b) = -($ ________ $)$.
a	228	If $a \in R$, then $-(-a) =$ ____.
3	229	$-(-3) =$ ____
5	230	$-(-5) =$ ____
8	231	$-(-$ ____ $) = 8$
$-ab$	232	If $a, b \in R$, then $(-a)b =$ ____.
$-3x$	233	$(-3)x =$ ____
$-5x$	234	$x(-5) =$ ____
ab	235	If $a, b \in R$, then $(-a)(-b) =$ ____.
6	236	$(-2)(-3) =$ ____
$6xy$	237	$(-3x)(-2y) =$ ____
$x(x + y)$	238	$(-x)[-(x + y)] =$ ____________
$(x + 1)(x + 2)$	239	$[-(x + 1)][-(x + 2)] =$ ____________________
$3x(5x + 13)$	240	$(-3x)[-(5x + 13)] =$ ________________
$-(x + y)$	241	$(-a) \cdot [$ ____________ $] = a(x + y)$
$-b$	242	The difference of two real numbers, a and b, is defined by $a - b = a + ($ ____ $)$
-3, 2	243	$5 - 3 = 5 + ($ ____ $) =$ ____
-5, 3	244	$8 - 5 = 8 + ($ ____ $) =$ ____
$-3y$	245	$x - 3y = x + ($ ____ $)$
$a \cdot \frac{1}{b}$	246	The quotient of real numbers a and b, for $b \neq 0$, denoted by $a \div b$ or $\frac{a}{b}$, is defined by $\frac{a}{b} =$ ________.
$\frac{1}{3}$, 5	247	$15 \div 3 = 15 \cdot$ ____ $=$ ____
5, 5	248	$1 \div \frac{1}{5} = 1 \cdot$ ____ $=$ ____
$\frac{1}{a}$, $\frac{1}{b}$	249	$\frac{a}{b} \div a = \frac{a}{b} \cdot$ ____ $=$ ____
$2 + y$, x	250	$\frac{x}{2 + y} \div \frac{1}{2 + y} = \frac{x}{2 + y} \cdot ($ ________ $) =$ ____

2 Polynomials—Operations and Factoring

2.1 Positive Integer Exponents

$\overbrace{x \cdot x \cdot x \cdot \cdots \cdot x}^{n \text{ times}}$	**251** Let $x \in R$ and $n \in I_P$; then $x^n =$ ____________
base	The number x is called the ____, and the product x^n is called the
nth power	________ of x.
5^4	**252** $5 \cdot 5 \cdot 5 \cdot 5 =$ ____
x^5	**253** $x \cdot x \cdot x \cdot x \cdot x =$ ____
y^3	**254** $y \cdot y \cdot y =$ ____
$4x^2y^3$	**255** $4 \cdot x \cdot x \cdot y \cdot y \cdot y =$ ______
$7x^3 + 2y^2z^4$	**256** $7 \cdot x \cdot x \cdot x + 2 \cdot y \cdot y \cdot z \cdot z \cdot z \cdot z =$ __________
$(x + y)^2$	**257** $(x + y)(x + y) =$ ______
3, 9	**258** $3^2 = 3 \times$ ____ $=$ ____
$2 \times 2 \times 2$, 8	**259** $2^3 =$ __________ $=$ ____
$3 \times 3 \times 3$	**260** $2^4 \cdot 3^3 = (2 \times 2 \times 2 \times 2)($________$)$
27, 432	$= (16)($____$) =$ ____
5, 5	**261** $5^2 + 2(5) = (5 \times 5) + ($____ $+$ ____$)$
10, 35	$= 25 +$ ____ $=$ ____
a^{m+n}	**262** Let $a \in R$ and $m, n \in I_p$; then $a^m \cdot a^n =$ ____.
^{3}, 4^5	**263** $4^2 \cdot 4^3 = 4^{2+\_\_\_} =$ ____
^{7}, $(-5)^{11}$	**264** $(-5)^7 \cdot (-5)^4 = (-5)^{\_\_\_+4} =$ ______
a^9	**265** $a^4 \cdot a^5 = a^{4+5} =$ ____
^{7}, x^9	**266** $x^2 \cdot x^7 = x^{2+\_\_\_} =$ ____
a^9	**267** $a^3 \cdot a^2 \cdot a^4 = a^{3+2+4} =$ ____
^{1}, $(-2)^6$	**268** $(-2)^3 \cdot (-2)^2 \cdot (-2) = (-2)^{3+2+\_\_\_} =$ ____
^{1}, x^7	**269** $x^4 \cdot x \cdot x^2 = x^{4+\_\_\_+2} =$ ____
y^8	**270** $y^3 \cdot y^5 =$ ____
x^{21}	**271** $x^{10} \cdot x^{11} =$ ____
b^{11}	**272** $b^2 \cdot b^4 \cdot b^5 =$ ____

2.2 Polynomial Terminology

	273 Such expressions as $5x$, $3x - 2$, $y^2 + 3y + 1$, and $\dfrac{5x + 1}{3x + 2}$
algebraic expressions	are called ______________.
	274 Each of the expressions $5x + 7$, $3x^2 - 7x + 9$, and
polynomial	$4x^2 + 5xy + 3y^2 - 1$ is an example of a ________.

Answer	Frame	
coefficient constant	275	The polynomial $5x + 7$ in one variable is the sum of the two terms $5x$ and 7. The number 5 in the term $5x$ is called its ____________, and the term 7, which contains no variable, is called the ____________ term.
5, -7, 11	276	The polynomial $5x^2 - 7x + 11$ in one variable is the sum of the three terms $5x^2$, $-7x$, and 11, whose coefficients are, respectively, ____, ____, and ____.
number integer power	277	A term may be defined as either a ____________ or a number times a positive ____________ of one or more variables.
monomial	278	A polynomial that contains one term is called a ____________.
binomial	279	A polynomial that contains two terms is called a ____________.
trinomial	280	A polynomial that contains three terms is called a ____________.
binomial	281	$3x^2y + 2b$ is a ____________.
monomial	282	$6xyz^2$ is a ____________.
trinomial	283	$3x^2 + 2ab + c^3$ is a ____________.
binomial	284	$3y - 7x$ is a ____________.
trinomial	285	$3ax + 2b + c$ is a ____________.
binomial	286	$x^2y^3 + 5xy^2$ is a ____________.
degree polynomial of degree zero	287	The highest power of the variable in a polynomial in one variable is called the ________ of the polynomial. A nonzero real number, such as 5, is called a ______________________.

In Problems 288–294, find the degree of each polynomial.

Answer	Frame	
2	288	$x^2 - 3x + 6$ ____
1	289	$3a - 5$ ____
5	290	$7x^5$ ____
0	291	9 ____
4	292	$x^4 - 3x^2 + 2$ ____
3	293	$a^3 - 27$ ____
5	294	$-x^5 - 3x^2 + 17$ ____

2.3 Addition, Subtraction, and Multiplication of Polynomials

Answer	Frame	
similar	295	Terms such as $4x$ and $8x$ are called ________ terms.
$9x^2$	296	$7x^2 + 2x^2 = (7 + 2)x^2 =$ ____
$10x^4$	297	$5x^4 + 2x^4 + 3x^4 = (5 + 2 + 3)x^4 =$ ____
$4xy$	298	$7xy + (-3xy) = [7 + (-3)]xy =$ ____
$11x^2$	299	$6x^2 + 8x^2 + (-3x^2) = [6 + 8 + (-3)]x^2 =$ ____
$21x^4$	300	$12x^4 + 9x^4 =$ ____

$13x^2y$

301 $15x^2y + 5x^2y + (-7x^2y) =$ ____

302 $(3x + 2) + (5x + 6) = (3 + 5)x + (2 + 6)$

$8x + 8$

$=$ ____

303 $(2x^2 + 1) + (3x^2 - 7) = (2 + 3)x^2 + [1 + (-7)]$

$5x^2 - 6$

$=$ ____

304 $(x^2 + 3x + 5) + (2x^2 + x + 3)$

$= (1 + 2)x^2 + (3 + 1)x + (5 + 3)$

$3x^2 + 4x + 8$

$=$ ____

305 $(2x^2 - xy + y^2) + (3x^2 + 2xy - 3y^2)$

$= (2 + 3)x^2 + (-1 + 2)xy + [1 + (-3)]y^2$

$5x^2 + xy - 2y^2$

$=$ ____

$7x + 7y$

306 $(2x + 3y) + (5x + 4y) =$ ____

$3xy + 5$

307 $(5xy - 3) + (-2xy + 8) =$ ____

$5x^2 - 2x + 1$

308 $(2x^2 - 3x + 5) + (3x^2 + x - 4) =$ ____

$7x^2 + 7xy + 2y^2$

309 $(3x^2 + 8xy - y^2) + (4x^2 - xy + 3y^2) =$ ____

310 $(x^3 - 4x^2 + 6x + 3) + (2x^3 + x^2 - 2x + 4)$

$3x^3 - 3x^2 + 4x + 7$

$=$ ____

311 $(2x^2y + 3xy - 5xy^2) + (-3x^2y + 7xy^2 - 2xy)$

$-x^2y + xy + 2xy^2$

$=$ ____

$5x$

312 $8x - 3x = (8 - 3)x =$ ____

$5y^2$

313 $12y^2 - 7y^2 = (12 - 7)y^2 =$ ____

$23xy$

314 $20xy - (-3xy) = [20 - (-3)]xy =$ ____

$6x^2y^3$

315 $17x^2y^3 - 11x^2y^3 =$ ____

$-7y^3$

316 $-5y^3 - 2y^3 =$ ____

$-5xy$

317 $-12xy - (-7xy) =$ ____

$x + 5$

318 $(3x + 6) - (2x + 1) = (3 - 2)x + (6 - 1) =$ ____

$3x^2 - 10$

319 $(5x^2 - 7) - (2x^2 + 3) = (5 - 2)x^2 + (-7 - 3) =$ ____

320 $(3x^2 + 6x + 5) - (x^2 + 2x + 3)$

$= (3 - 1)x^2 + (6 - 2)x + (5 - 3)$

$2x^2 + 4x + 2$

$=$ ____

321 $(5x^2 + 3xy - 4y^2) - (2x^2 - 5xy + 6y^2)$

$= (5 - 2)x^2 + [3 - (-5)]xy + (-4 - 6)y^2$

$3x^2 + 8xy - 10y^2$

$=$ ____

$x^3 + 4y$

322 $(2x^3 + 7y) - (x^3 + 3y) =$ ____

$10x^2y^2 - 14$

323 $(8x^2y^2 - 11) - (-2x^2y^2 + 3) =$ ____

$2x^2 - 5x + 3$

324 $(5x^2 - 3x + 2) - (3x^2 + 2x - 1) =$ ____

325 $(10x^3 - 7x^2 + x - 3) - (8x^3 - 5x^2 - 3x + 6)$

$2x^3 - 2x^2 + 4x - 9$

$=$ ____

326 $(3x^3y - 4x^2y^2 - 7xy^3) - (2x^2y^2 + x^3y - 4xy^3)$

$2x^3y - 6x^2y^2 - 3xy^3$

$=$ ____

327 $(2x+3)+(4x+5)-(3x+2)$

$= (2+4-3)x+(3+5-2)$

$=$ ________

$3x+6$

328 $(x^2+2x+1)+(3x^2-4x+6)-(2x^2+x+2)$

$= (1+3-2)x^2+(2-4-1)x+(1+6-2)$

$=$ ________________

$2x^2-3x+5$

329 $(3x^2+x-2)+(x^2+4x+7)-(3x^2+x+2) =$ ________________

x^2+4x+3

330 $(5x^2-7x+8)-(3x^2+x-2)+(-x^2+3x-4)$

$=$ ____________________

x^2-5x+6

331 $(12x^2y-3xy)-(8x^2y+4xy)-(-3x^2y+2xy) =$ ______________

$7x^2y-9xy$

332 If P and Q are polynomial expressions, then PQ is called the ____________ of P and Q.

product

333 $(2x^3)(4x^2) = (2 \times 4)(x^3 \cdot x^2) =$ ____

$8x^5$

334 $(-3x^2)(5x) = (-3 \times 5)(x^2 \cdot x) =$ ________

$-15x^3$

335 $(4x^2y)(3x^3y^2) = (4 \times 3)(x^2 \cdot x^3)(y \cdot y^2) =$ ________

$12x^5y^3$

336 $(6x^4)(3x^2y) =$ ________

$18x^6y$

337 $(-4x^2y^3)(7x^3y) =$ ________

$-28x^5y^4$

338 $(3x^2y)(2xy^2)(-4x^2y^3) =$ ________

$-24x^5y^6$

339 $3x(2x+4) = (3x)(2x)+(3x)(4) =$ ________

$6x^2+12x$

340 $-2x^2(x^2-3) = (-2x^2)(x^2)+(-2x^2)(-3)$

$=$ ________

$-2x^4+6x^2$

341 $5xy(x^2+2xy-y^2) = (5xy)(x^2)+(5xy)(2xy)+(5xy)(-y^2)$

$=$ ________________________

$5x^3y+10x^2y^2-5xy^3$

342 $-2x^2y(-x^3y+3x^2y^2-4xy^3)$

$= (-2x^2y)(-x^3y)+(-2x^2y)(3x^2y^2)+(-2x^2y)(-4xy^3)$

$=$ __________________________

$2x^5y^2-6x^4y^3+8x^3y^4$

343 $(x+3)(x+1) = x(x+1)+3(x+1)$

$= x^2+x+$ ________

$3x+3$

$=$ ____________

x^2+4x+3

344 $(2x-1)(x^2+x+1) = (2x-1)x^2+(2x-1)x+(2x-1)1$

$=$ ____________ $+2x^2-x+2x-1$

$2x^3-x^2$

$=$ ____________________

$2x^3+x^2+x-1$

345 $(x+y)(x^2+xy+y^2)$

$= (x+y)x^2+(x+y)xy+$ ___________

$(x+y)y^2$

$= x^3+x^2y+x^2y+xy^2+$ _________

xy^2+y^3

$=$ __________________________

$x^3+2x^2y+2xy^2+y^3$

346 $(x-2)(x^2+2x+3)$

$= x(x^2+2x+3)-2($ ______________ $)$

x^2+2x+3

$= x^3+2x^2+3x+($ ______________ $)$

$-2x^2-4x-6$

$=$ _____________

x^3-x-6

347 $(x + 3)(x^2 - 3x + 9)$
$= x(x^2 - 3x + 9) + 3(\underline{\qquad\qquad})$
$= x^3 - 3x^2 + 9x + \underline{\qquad\qquad}$
$= \underline{\qquad}$

$x^2 - 3x + 9$
$3x^2 - 9x + 27$
$x^3 + 27$

348 $(a + 2b - 3c)(2a - 4b + 5c)$
$= a(2a - 4b + 5c) + 2b(\underline{\qquad\qquad})$
$- 3c(\underline{\qquad\qquad})$
$= 2a^2 - 4ab + 5ac + (\underline{\qquad\qquad})$
$+ (\underline{\qquad\qquad})$
$= \underline{\qquad\qquad\qquad}$

$2a - 4b + 5c$
$2a - 4b + 5c$
$4ab - 8b^2 + 10bc$
$-6ac + 12bc - 15c^2$
$2a^2 - ac + 22bc - 8b^2 - 15c^2$

349 $(a + 3b)(2a^2 - 3ab + b^2)$
$= \underline{\qquad\qquad\qquad}$

$2a^3 + 3a^2b - 8ab^2 + 3b^3$

350 $(x - 2)(x + 5) = x^2 + 5x - 2x - 2(\underline{\quad})$
$= x^2 + \underline{\quad}x - 10$

5
3

351 $(x + 3)(x + 4) = x^2 + 4x + 3x + \underline{\quad}$
$= x^2 + \underline{\quad}x + 12$

12
7

352 $(x - 3)(x - 5) = x^2 \underline{\quad} x - 3x + 15$
$= x^2 - \underline{\quad}x + 15$

-5
8

353 $(x - 5)(2x - 7) = 2x^2 - 7x - \underline{\quad}x + 35$
$= 2x^2 - \underline{\quad}x + 35$

10
17

354 $(x + 2y)(3x - 4y) = 3x^2 - 4\underline{\quad} + 6xy - 8y^2$
$= 3x^2 + \underline{\quad}xy - 8y^2$

xy
2

355 $(5x - 6y)(9x - 14y) = \underline{\quad}x^2 - 70xy - 54xy + 84y^2$
$= 45x^2 - \underline{\quad}xy + 84y^2$

45
124

356 $(3x + y)(5x - 2y) = 15x^2 - 6xy + \underline{\qquad\qquad}$
$= \underline{\qquad\qquad}$

$5xy - 2y^2$
$15x^2 - xy - 2y^2$

357 $(4x - 3)(2x + 1) = \underline{\qquad\qquad}$

$8x^2 - 2x - 3$

358 $(2x - 3y)(3x + y) = \underline{\qquad\qquad}$

$6x^2 - 7xy - 3y^2$

359 $(5y + 2z)(3y - 2z) = \underline{\qquad\qquad}$

$15y^2 - 4yz - 4z^2$

Special Products

360 $(a + b)^2 = a^2 + \underline{\quad} + b^2$

$2ab$

361 $(x + 4)^2 = x^2 + 2(\underline{\quad}) + 4^2$
$= \underline{\qquad\qquad}$

$4x$
$x^2 + 8x + 16$

362 $(x + 7)^2 = x^2 + 2(\underline{\quad}) + 7^2$
$= \underline{\qquad\qquad}$

$7x$
$x^2 + 14x + 49$

363 $(2x + 3)^2 = (2x)^2 + 2(\underline{\quad}) + (3)^2$
$= \underline{\qquad\qquad}$

$6x$
$4x^2 + 12x + 9$

364 $(3x + 5y)^2 = (\underline{\quad})^2 + 2(3x)(5y) + (\underline{\quad})^2$
$= \underline{\qquad\qquad}$

$3x$, $5y$
$9x^2 + 30xy + 25y^2$

$4y,\ 4y,\ 9z,\ 9z$	**365** $(4y + 9z)^2 = (\_\_\_\_)^2 + 2(\_\_\_\_)(\_\_\_\_) + (\_\_\_\_)^2$
$16y^2 + 72yz + 81z^2$	$= \_\_\_\_\_\_\_\_$
$4x^2 + 4x + 1$	**366** $(2x + 1)^2 = \_\_\_\_\_\_\_\_$
$25x^2 + 30x + 9$	**367** $(5x + 3)^2 = \_\_\_\_\_\_\_\_$
$9x^2 + 42xy + 49y^2$	**368** $(3x + 7y)^2 = \_\_\_\_\_\_\_\_$
ab	**369** $(a - b)^2 = a^2 - 2(\_\_\_\_) + b^2$
$2x$	**370** $(x - 2)^2 = x^2 - 2(\_\_\_\_) + 2^2$
$x^2 - 4x + 4$	$= \_\_\_\_\_\_\_\_$
$6y$	**371** $(y - 6)^2 = y^2 - 2(\_\_\_\_) + 6^2$
$y^2 - 12y + 36$	$= \_\_\_\_\_\_\_\_$
$12xy$	**372** $(3x - 4y)^2 = (3x)^2 - 2(\_\_\_\_) + (4y)^2$
$9x^2 - 24xy + 16y^2$	$= \_\_\_\_\_\_\_\_$
$2x,\ 5y$	**373** $(2x - 5y)^2 = (\_\_\_\_)^2 - 2(2x)(5y) + (\_\_\_\_)^2$
$4x^2 - 20xy + 25y^2$	$= \_\_\_\_\_\_\_\_$
$2x,\ 2x,\ 7y,\ 7y$	**374** $(2x - 7y)^2 = (\_\_\_\_)^2 - 2(\_\_\_\_)(\_\_\_\_) + (\_\_\_\_)^2$
$4x^2 - 28xy + 49y^2$	$= \_\_\_\_\_\_\_\_$
$9x^2 - 6x + 1$	**375** $(3x - 1)^2 = \_\_\_\_\_\_\_\_$
$25x^2 - 30xy + 9y^2$	**376** $(5x - 3y)^2 = \_\_\_\_\_\_\_\_$
$49z^2 - 42z + 9$	**377** $(7z - 3)^2 = \_\_\_\_\_\_\_\_$
$a^2 - b^2$	**378** $(a - b)(a + b) = \_\_\_\_\_\_\_\_$
$x,\ 7y$	**379** $(x - 7y)(x + 7y) = (\_\_\_\_)^2 - (\_\_\_\_)^2$
$x^2 - 49y^2$	$= \_\_\_\_\_\_\_\_$
$x,\ 5$	**380** $(x + 5)(x - 5) = (\_\_\_\_)^2 - (\_\_\_\_)^2$
$x^2 - 25$	$= \_\_\_\_\_\_\_\_$
$3x,\ 4$	**381** $(3x - 4)(3x + 4) = (\_\_\_\_)^2 - (\_\_\_\_)^2$
$9x^2 - 16$	$= \_\_\_\_\_\_\_\_$
$5,\ 7x$	**382** $(5 - 7x)(5 + 7x) = (\_\_\_\_)^2 - (\_\_\_\_)^2$
$25 - 49x^2$	$= \_\_\_\_\_\_\_\_$
$a,\ b,\ a^3 + b^3$	**383** $(a + b)(a^2 - ab + b^2) = (\_\_\_\_)^3 + (\_\_\_\_)^3 = \_\_\_\_\_\_\_\_$
$4,\ a,\ 64 + a^3$	**384** $(4 + a)(16 - 4a + a^2) = (\_\_\_\_)^3 + (\_\_\_\_)^3 = \_\_\_\_\_\_\_\_$
$1,\ 10x$	**385** $(1 + 10x)(1 - 10x + 100x^2) = (\_\_\_\_)^3 + (\_\_\_\_)^3$
$1 + 1000x^3$	$= \_\_\_\_\_\_\_\_$
$5a^2,\ 2b$	**386** $(5a^2 + 2b)(25a^4 - 10a^2b + 4b^2) = (\_\_\_\_)^3 + (\_\_\_\_)^3$
$125a^6 + 8b^3$	$= \_\_\_\_\_\_\_\_$
$x^3 + z^3$	**387** $(x + z)(x^2 - xz + z^2) = \_\_\_\_\_\_\_\_$
$8x^3 + y^3$	**388** $(2x + y)(4x^2 - 2xy + y^2) = \_\_\_\_\_\_\_\_$
$a,\ b,\ a^3 - b^3$	**389** $(a - b)(a^2 + ab + b^2) = (\_\_\_\_)^3 - (\_\_\_\_)^3 = \_\_\_\_\_\_\_\_$
$x,\ 3a$	**390** $(x - 3a)(x^2 + 3xa + 9a^2) = (\_\_\_\_)^3 - (\_\_\_\_)^3$
$x^3 - 27a^3$	$= \_\_\_\_\_\_\_\_$
$5x,\ y$	**391** $(5x - y)(25x^2 + 5xy + y^2) = (\_\_\_\_)^3 - (\_\_\_\_)^3$
$125x^3 - y^3$	$= \_\_\_\_\_\_\_\_$

$3m$, 1	**392** $(3m - 1)(9m^2 + 3m + 1) = (\underline{\qquad})^3 - (\underline{\qquad})^3$
$27m^3 - 1$	$= \underline{\qquad\qquad}$
$y^3 - z^3$	**393** $(y - z)(y^2 + yz + z^2) = \underline{\qquad\qquad}$
$8x^3 - 27y^3$	**394** $(2x - 3y)(4x^2 + 6xy + 9y^2) = \underline{\qquad\qquad}$

2.4 Factoring Polynomials

	395 The process of writing a given polynomial as the product of two
factoring	or more polynomials is called ____________.
	396 A polynomial that has no factors other than itself and 1, or its
prime	negative and -1, is called ________.
	To factor a polynomial, we express it as the product of
prime polynomials	____________________.
	397 The monomial factor of highest degree and largest numerical
	coefficient common to every term of a polynomial is called the
highest common factor	____________________ of the polynomial.

In Problems 398–408, express each polynomial in completely factored form.

3	**398** $5x + 15 = 5 \cdot x + 5 \cdot 3 = 5(x + \underline{\quad})$
a	**399** $121 - 11a = 11 \cdot 11 - 11 \cdot a = 11(11 - \underline{\quad})$
1	**400** $ab^2 - ab = ab \cdot b - ab \cdot 1 = ab(b - \underline{\quad})$
x	**401** $5x + 3x^2 = 5 \cdot x + 3x \cdot x = (5 + 3x)\underline{\quad}$
4	**402** $15xy^2 + 20x = (5x)(3y^2) + (5x)(4) = 5x(3y^2 + \underline{\quad})$
	403 $x^3 + 3x^2y - x^2 = x^2 \cdot x + x^2 \cdot 3y - x^2 \cdot 1$
$x + 3y - 1$	$= x^2(\underline{\qquad\qquad})$
	404 $5xy + 4x^2y^2 + x^3y^3 = xy \cdot 5 + xy \cdot 4xy + xy \cdot x^2y^2$
$5 + 4xy + x^2y^2$	$= xy(\underline{\qquad\qquad\qquad})$
$my + ny$	**405** $mx + nx + my + ny = (mx + nx) + (\underline{\qquad\qquad})$
$(m + n)y$	$= (m + n)x + \underline{\qquad\qquad}$
$x + y$	$= (m + n)(\underline{\qquad})$
$ax - ay$	**406** $ax - ay + bx - by = (\underline{\qquad\qquad}) + (bx - by)$
$a(x - y)$	$= \underline{\qquad} + b(x - y)$
$a + b$	$= (\underline{\qquad})(x - y)$
$1 - x$	**407** $mx + 1 - m - x = (mx - m) + (\underline{\qquad})$
-1	$= m(x - 1) + (\underline{\qquad})(x - 1)$
$(m - 1)(x - 1)$	$= \underline{\qquad\qquad\qquad}$
$-2a - 2b$	**408** $y^2a + y^2b - 2a - 2b = (y^2a + y^2b) + (\underline{\qquad\qquad})$
-2	$= y^2(a + b) + (\underline{\qquad})(a + b)$
$(y^2 - 2)(a + b)$	$= \underline{\qquad\qquad\qquad}$

Answer	Frame
$a - b$	**409** $a^2 - b^2 = (a + b)(\_\_\_\_\_\_\_\_)$
$5,\ x - 5$	**410** $x^2 - 25 = x^2 - (\_\_\_\_)^2 = (x + 5)(\_\_\_\_\_\_\_\_)$
$ab,\ 3,\ ab,\ 3,\ ab,\ 3$	**411** $a^2b^2 - 9 = (\_\_\_\_)^2 - (\_\_\_\_)^2 = (\_\_\_\_ + \_\_\_\_)(\_\_\_\_ - \_\_\_\_)$
$ab,\ 7,\ ab - 7,\ ab + 7$	**412** $a^2b^2 - 49 = (\_\_\_\_)^2 - (\_\_\_\_)^2 = (\_\_\_\_\_\_\_\_)(\_\_\_\_\_\_\_\_)$
$6,\ ab,\ 6 - ab,\ 6 + ab$	**413** $36 - a^2b^2 = (\_\_\_\_)^2 - (\_\_\_\_)^2 = (\_\_\_\_\_\_\_\_)(\_\_\_\_\_\_\_\_)$
$xyz,\ 8$	**414** $x^2y^2z^2 - 64 = (\_\_\_\_)^2 - (\_\_\_\_)^2$
$xyz - 8,\ xyz + 8$	$= (\_\_\_\_\_\_\_\_)(\_\_\_\_\_\_\_\_)$
$2xy,\ 3m$	**415** $4x^2y^2 - 9m^2 = (\_\_\_\_)^2 - (\_\_\_\_)^2$
$2xy - 3m,\ 2xy + 3m$	$= (\_\_\_\_\_\_\_\_)(\_\_\_\_\_\_\_\_)$
$3ab,\ 4$	**416** $9a^2b^2 - 16 = (\_\_\_\_)^2 - (\_\_\_\_)^2$
$3ab - 4,\ 3ab + 4$	$= (\_\_\_\_\_\_\_\_)(\_\_\_\_\_\_\_\_)$
	417 $4(a - b)^2 - 49c^2$
$2(a - b),\ 7c$	$= [\_\_\_\_\_\_\_\_]^2 - (\_\_\_\_)^2$
$2(a - b) - 7c,\ 2(a - b) + 7c$	$= [\_\_\_\_\_\_\_\_][\_\_\_\_\_\_\_\_]$
b^2	**418** $a^3 + b^3 = (a + b)(a^2 - ab + \_\_\_\_)$
a^2	**419** $a^3 - b^3 = (a - b)(\_\_\_\_ + ab + b^2)$
	420 $27a^3 + b^3 = (3a)^3 + (b)^3$
$-,\ +$	$= (3a + b)(9a^2\ \_\_\_\_\ 3ab\ \_\_\_\_\ b^2)$
	421 $27a^3 - b^3 = (3a)^3 - (b)^3$
$+,\ +$	$= (3a - b)(9a^2\ \_\_\_\_\ 3ab\ \_\_\_\_\ b^2)$
$2y$	**422** $x^3 + 8y^3 = (x)^3 + (\_\_\_\_)^3$
$x^2 - 2xy + 4y^2$	$= (x + 2y)(\_\_\_\_\_\_\_\_)$
4	**423** $27m^3 + 64 = (3m)^3 + (\_\_\_\_)^3$
$9m^2 - 12m + 16$	$= (3m + 4)(\_\_\_\_\_\_\_\_)$
b	**424** $125a^3 + b^3 = (5a)^3 + (\_\_\_\_)^3$
$-,\ +$	$= (5a + b)(25a^2\ \_\_\_\_\ 5ab\ \_\_\_\_\ b^2)$
$10x$	**425** $1 - 1000x^3 = (1)^3 - (\_\_\_\_)^3$
$+,\ +$	$= (1 - 10x)(1\ \_\_\_\_\ 10x\ \_\_\_\_\ 100x^2)$
$3b^2$	**426** $64a^6 + 27b^6 = (4a^2)^3 + (\_\_\_\_)^3$
$12a^2b^2$	$= (4a^2 + 3b^2)(16a^4 - \_\_\_\_\_\_\_\_ + 9b^4)$
$5a$	**427** $216 + 125a^3 = (6)^3 + (\_\_\_\_)^3$
36	$= (6 + 5a)(\_\_\_\_ - 30a + 25a^2)$
	428 $(x - 5)^3 + 8y^3$
$2y$	$= (x - 5)^3 + (\_\_\_\_)^3$
$-(x - 5)(2y)$	$= [(x - 5) + 2y][(x - 5)^2\ \_\_\_\_\_\_\_\_ + 4y^2]$
$(x - 5)^2 - 2y(x - 5) + 4y^2$	$= (x - 5 + 2y)[\_\_\_\_\_\_\_\_]$
	429 If $ax^2 + bx + c = (sx + p)(rx + q) = rsx^2 + (sq + pr)x + pq$,
$rs,\ sq + pr,\ pq$	then $a = \_\_\_\_$, $b = \_\_\_\_\_\_\_\_$, and $c = \_\_\_\_$.
	430 $x^2 - 6x + 9 = x^2 - (3 + 3)x + 3 \cdot 3$
3	$= (x - 3)(x - \_\_\_\_)$

Answer	No.	Problem
	431	$x^2 + 13x + 42 = x^2 + (6 + 7)x + 7 \cdot 6$
6, 7		$= (x + \_\_\_\_)(x + \_\_\_\_)$
$a + 2$	432	$6a^2 + 5a - 14 = (6a - 7)(\_\_\_\_\_\_\_\_)$
$x + 6$	433	$x^2 + 10x + 24 = (x + 4)(\_\_\_\_\_\_\_\_)$
$b + 5$	434	$b^2 - 6b - 55 = (b - 11)(\_\_\_\_\_\_\_\_)$
$1 + y$	435	$2 + y - y^2 = (2 - y)(\_\_\_\_\_\_\_\_)$
$2 + 7a$	436	$6 + 11a - 35a^2 = (3 - 5a)(\_\_\_\_\_\_\_\_)$
8, 9	437	$a^2 + 17a + 72 = (a + \_\_\_\_)(a + \_\_\_\_)$
$5y$, $3y$	438	$15y^2 - 8y + 1 = (\_\_\_\_ - 1)(\_\_\_\_ - 1)$
$3x + 2y$	439	$6x^2 - 5xy - 6y^2 = (2x - 3y)(\_\_\_\_\_\_\_\_)$

2.5 Linear Equations

440 Equations such as $3x + 1 = 5$, $2(x - 1) = 3 - 4x$, and $7x + 6 = 0$ are called ________________ equations.

Answer: first-degree (*or* linear)

441 Any value of the variable x that makes the equation a true statement is called a ____________. To solve an equation means to find *all* values of the variable for which the equation is a ________________.

Answers: solution; true statement

442 All of the solutions of an equation form a set, and this set is called the ______________ of the equation.

Answer: solution set

443 Two equations are ______________ if all values, and no other, that satisfy one of those equations also satisfy the second equation.

Answer: equivalent

444 The addition of the same number to both sides of an equation produces an ______________ equation. This property is called the ____________ property. In symbols, the property is: If $P = Q$, then $P + R =$ ________.

Answers: equivalent; addition; $Q + R$

445 The multiplication of both sides of an equation by the same nonzero number produces an ______________ equation. This is called the ________________ property. In symbols, the property is: If $P = Q$ and $R \neq 0$, then $PR =$ ____.

Answers: equivalent; multiplication; QR

In Problems 446–452, find the solution set of each first-degree equation.

Answer	No.	Problem
	446	$3x - 1 = 14$
1		$3x - 1 + 1 = 14 + \_\_\_\_$
15		$3x = \_\_\_\_$
5		$x = \_\_\_\_$
{5}		The solution set is ____.

Answer	Frame
	447 $4x + 3 = 19$
3, 3	$4x + 3 -$ ____ $= 19 -$ ____
16	$4x =$ ____
4	$x =$ ____
{4}	The solution set is ____.
	448 $\frac{1}{5}x - 3 = 2$
3, 3	$\frac{1}{5}x - 3 +$ ____ $= 2 +$ ____
5	$\frac{1}{5}x =$ ____
5	$5(\frac{1}{5}x) = 5($____$)$
25	$x =$ ____
{25}	The solution set is ____.
	449 $\frac{1}{3}x + 4 = 8$
4, 4	$\frac{1}{3}x + 4 -$ ____ $= 8 -$ ____
4	$\frac{1}{3}x =$ ____
12	$x =$ ____
{12}	The solution set is ______.
	450 $\frac{7}{4}x + 5 = \frac{3}{4}x + 1$
5	$\frac{7}{4}x - \frac{3}{4}x = 1 -$ ____
-4	$\frac{4}{4}x =$ ____
-4	$x =$ ____
{-4}	The solution set is ______.
	451 $8x + 14 = 5x + 44$
5*x*	$8x -$ ____ $= 44 - 14$
30	$3x =$ ____
10	$x =$ ____
{10}	The solution set is ____.
	452 $7x - 5 + 3x = 5x + 10$
(-5*x*)	$7x + 3x +$ ________ $= 10 + 5$
15	$5x =$ ____
3	$x =$ ____
{3}	The solution set is ____.
	453 John is 28 years younger than his father. In 15 years, his father's age will be twice John's age. How old is each now?
x + 28	Let *x* years be John's age now. Then, ______ years is his father's
x + 15	age now, and ______ years is John's age 15 years from now.
15 years	The expression $(x + 28 + 15)$ is his father's age ______ from now.
x + 28 + 15	The equation is ________ $= 2(x + 15)$. Solving the equation,
13, 13, 41	$x =$ ____. Therefore, ____ years is John's age now and ____ years is his father's age now.

454 A man invested \$30,000, part of it at 4 percent interest and the remainder at 6 percent. The interest for one year is \$1600. How much did he invest at 4 percent and how much at 6 percent? Let x be the amount invested at 4 percent. Then $(30{,}000 - x)$ is the amount invested at 6 percent. The equation is formed as follows:

1600 — $4\%x + 6\%\,(30{,}000 - x) =$ ______

1600 — $\frac{4}{100}x + \frac{6}{100}\,(30{,}000 - x) =$ ______

$$4x + 6\,(30{,}000 - x) = 160{,}000$$

10,000, \$10,000 — Solving the equation, $x =$ ________. Therefore, _________ is

\$20,000 — invested at 4 percent and _________ is invested at 6 percent.

455 A woman has \$5000 invested at 6 percent and \$3000 invested at 7 percent. How much must she invest at 9 percent to make an average investment of 8 percent?

Let x represent the amount invested at 9 percent. The equation is formed as follows:

8% — $6\%\,(5000) + 7\%\,(3000) + 9\%x = ($____$)(5000 + 3000 + x)$

$\frac{8}{100}$ — $\frac{6}{100}\,(5000) + \frac{7}{100}\,(3000) + \frac{9}{100}x = ($____$)(8000 + x)$

640 — $300 + 210 + 0.09x =$ ____ $+ 0.08x$

130 — $0.01x =$ ____

$$x = 13{,}000$$

\$13,000 — Therefore, _________ must be invested at 9 percent.

456 A druggist has a 10 percent solution and an 18 percent solution of a medicine. How much of each must he use to obtain 40 ounces of a 12 percent solution?

If x represents the amount of the 10 percent solution used,

$40 - x$ — then ________ represents the amount of the 18 percent solution needed. The equation is formed as follows:

$40 - x$ — $10\%x + 18\%\,($________$) = 12\%(40)$

$40 - x$ — $0.10x + 0.18\,($________$) = 0.12(40)$

$40 - x$ — $10x + 18\,($________$) = 12(40)$

$720 - 18x$ — $10x +$ ____________ $= 480$

30, 30 — Solving this equation, $x =$ _____. Therefore, _____ ounces of the

10 — 10 percent solution and _____ ounces of the 18 percent solution are required.

3 Fractions

rational — **457** The expression $\dfrac{3x^2 + x}{5x + 1}$ is called a ____________ expression.

458 The fraction is not defined for any value of the variable that makes

denominator — the ______________ equal to zero.

In Problems 459–463, determine all values for which the rational expressions are not defined.

459 $\frac{5}{x}$

The denominator is zero for $x =$ ____.

Answer: 0

460 $\frac{7x - 1}{x^2}$

The denominator is zero for $x =$ ____.

Answer: 0

461 $\frac{x - 1}{x - 2}$

The denominator is zero for $x =$ ____.

Answer: 2

462 $\frac{3x}{(x - 1)(2x + 1)}$

The denominator is zero for $x =$ ____ or $x =$ ____.

Answer: $1, -\frac{1}{2}$

463 $\frac{2x + 7}{(2x + 1)(3x - 2)(x + 11)}$

The denominator is zero for $x =$ ____ or $x =$ ____ or $x =$ ____.

Answer: $-\frac{1}{2}, \frac{2}{3}, -11$

464 Two fractions are ________ if they give equal real numbers for each assignment of values to their variables, except when such an assignment gives an undefined value to either (or both) fractions.

Answer: equivalent

465 The fundamental principle of rational expressions states that if $\frac{P}{Q}$ is a rational expression and $K \neq 0$ is another rational expression, then $\frac{PK}{QK} = \frac{\_\_\_}{Q}$.

Answer: P

466 A fraction is said to be in ________ if the numerator and the denominator have no common factor other than 1 or −1.

Answer: lowest terms

467 The procedure of expressing a given fraction as an equivalent fraction in lowest terms is called ________.

Answer: reducing the fraction

In Problems 468–477, reduce each fraction.

468 $\frac{x^2 - x}{x^2 - 1} = \frac{x(\_\_\_\_)}{(x - 1)(x + 1)} =$ ____

Answer: $x - 1, \frac{x}{x + 1}$

469 $\frac{x + y}{x^2 - y^2} = \frac{x + y}{(x - y)(x + y)} =$ ____

Answer: $\frac{1}{x - y}$

470 $\frac{x^2 - 4}{x^2 + 3x + 2} = \frac{\_\_\_\_}{(x + 1)(x + 2)} =$ ____

Answer: $(x - 2)(x + 2), \frac{x - 2}{x + 1}$

471 $\frac{x^2 - 25}{x^2 + 8x + 15} = \frac{\_\_\_\_}{(x + 3)(x + 5)} =$ ____

Answer: $(x - 5)(x + 5), \frac{x - 5}{x + 3}$

$(x-6)(x-4),\ \dfrac{x-4}{x+6}$

472 $\dfrac{x^2-10x+24}{x^2-36}=\dfrac{\_\_\_\_\_\_}{(x-6)(x+6)}=$ ______

$\dfrac{a-4}{a+1}$

473 $\dfrac{a^2-16}{a^2+5a+4}=\dfrac{(a-4)(a+4)}{(a+4)(a+1)}=$ ______

$x-2,\ \dfrac{x(x+2)}{x-1}$

474 $\dfrac{x^3-4x}{x^2-3x+2}=\dfrac{x(\_\_\_)(x+2)}{(x-1)(x-2)}=$ ______

$b-3,\ \dfrac{b-3}{b+8}$

475 $\dfrac{b^2+4b-21}{b^2+15b+56}=\dfrac{(b+7)(\_\_\_\_\_)}{(b+7)(b+8)}=$ ______

$(3-x)(3+x),\ \dfrac{3-x}{x-5}$

476 $\dfrac{9-x^2}{x^2-2x-15}=\dfrac{\_\_\_\_\_\_}{(x-5)(x+3)}=$ ______

$\dfrac{x+4}{x-2}$

477 $\dfrac{x^2-4x-32}{x^2-10x+16}=\dfrac{(x-8)(x+4)}{(x-8)(x-2)}=$ ______

3.1 Multiplication and Division of Fractions

$\dfrac{P\cdot R}{Q\cdot S}$

478 Let $\dfrac{P}{Q}$ and $\dfrac{R}{S}$ be rational expressions, with $Q \neq 0$ and $S \neq 0$. Then $\dfrac{P}{Q}\cdot\dfrac{R}{S}=$ ______.

In Problems 479–488, find the products and reduce them.

$\dfrac{2}{3}$

479 $\dfrac{3}{4}\times\dfrac{8}{9}=\dfrac{3\times 8}{4\times 9}=\dfrac{24}{36}=$ ____

$\dfrac{6}{55}$

480 $\dfrac{21}{44}\times\dfrac{8}{35}=\dfrac{21\times 8}{44\times 35}=\dfrac{3\times 7\times 2\times 4}{4\times 11\times 5\times 7}=$ ____

$\dfrac{8x^2}{3}$

481 $\dfrac{16}{5x}\cdot\dfrac{20x^3}{24}=\dfrac{(16)(20x^3)}{(5x)(24)}=\dfrac{(40x)(8x^2)}{(40x)(3)}=$ ____

$15w^2$

482 $\dfrac{8x^2}{5wz}\cdot\dfrac{15w^2}{14x^3}=\dfrac{8x^2\cdot\_\_\_\_}{5wz\cdot 14x^3}$

$=\dfrac{10x^2w\cdot 12w}{10x^2w\cdot\_\_\_\_}$

$7xz$

$\dfrac{12w}{7xz}$

$=$ ____

483 $\dfrac{24}{3x-6}\cdot\dfrac{x^2-4}{x+2}=\dfrac{24(x^2-4)}{(3x-6)(x+2)}$

$x+2$

$=\dfrac{3(8)(x-2)(\_\_\_\_\_)}{3(x-2)(x+2)}$

8

$=$ ____

xy

484 $\dfrac{x^2-y^2}{x^2}\cdot\dfrac{xy}{x+y}=\dfrac{(x^2-y^2)\cdot(\_\_\_)}{x^2(x+y)}$

$y(x-y),\ \dfrac{xy-y^2}{x}$

$=\dfrac{\_\_\_\_\_\_}{x}=$ ______

$x^2 - 25y^2$

$(x - 5y)(x + 5y)$

$(x + 3y)(x + 5y)$

485 $\dfrac{x(x+3y)^2}{x-5y} \cdot \dfrac{x^2 - 25y^2}{x^2 + 3xy} = \dfrac{x(x+3y)^2(\underline{\qquad\qquad})}{(x-5y)(x^2+3xy)}$

$= \dfrac{x(x+3y)^2 \;\underline{\qquad\qquad}}{(x-5y)(x)(x+3y)}$

$= \underline{\qquad\qquad}$

$x^2 - 16$

$(x - 4)(x + 4)$

$(x + 12)(x - 4)$

486 $\dfrac{x^2 - 144}{x + 4} \cdot \dfrac{x^2 - 16}{x - 12}$

$= \dfrac{(x^2 - 144)(\underline{\qquad\qquad})}{(x+4)(x-12)}$

$= \dfrac{(x-12)(x+12)\;\underline{\qquad\qquad}}{(x+4)(x-12)}$

$= \underline{\qquad\qquad}$

$x^2 + 12x + 20$

$(x - 2)(x - 5)$

$\dfrac{(x-5)(x+2)}{(x-10)(x-2)}$

487 $\dfrac{x^2 - 10x + 25}{x^2 - 100} \cdot \dfrac{x^2 + 12x + 20}{x^2 - 7x + 10}$

$= \dfrac{(x^2 - 10x + 25)(\underline{\qquad\qquad})}{(x^2 - 100)(x^2 - 7x + 10)}$

$= \dfrac{(x-5)(x-5)(x+10)(x+2)}{(x-10)(x+10)\;\underline{\qquad\qquad}}$

$= \underline{\qquad\qquad}$

$a^2(a - 2b)$

$(2a + b)(a - 2b)a$

$\dfrac{a}{2a - b}$

488 $\dfrac{2a+b}{a^2 - 2ab} \cdot \dfrac{a^3 - 2a^2 b}{4a^2 - b^2} = \dfrac{(2a+b)\;\underline{\qquad\qquad}}{a(a-2b)(2a-b)(2a+b)}$

$= \dfrac{[\underline{\qquad\qquad}]\,a}{[(2a+b)(a-2b)a]\,(2a-b)}$

$= \underline{\qquad\qquad}$

$\dfrac{P}{Q} \cdot \dfrac{S}{R}$

489 Let P, Q, R, and S be polynomials with $Q \neq 0$ and $\dfrac{R}{S} \neq 0$; then the rule of dividing two fractions is given by

$\dfrac{P}{Q} \div \dfrac{R}{S} = \underline{\qquad\qquad} = \dfrac{PS}{QR}$

In Problems 490–500, find the quotients and reduce them.

$\dfrac{21x}{15}$, $63x$

$3x$, $\dfrac{21}{25}$

490 $\dfrac{3}{5x} \div \dfrac{15}{21x} = \dfrac{3}{5x} \cdot \left(\underline{\qquad}\right) = \dfrac{\underline{\qquad}}{75x}$

$= \dfrac{21(\underline{\qquad})}{25(3x)} = \underline{\qquad}$

44

$99x^2$, $\dfrac{4}{7}$

491 $\dfrac{9x^2}{11} \div \dfrac{63x^2}{44} = \dfrac{9x^2}{11} \cdot \dfrac{\underline{\qquad}}{63x^2}$

$= \dfrac{4(\underline{\qquad})}{7(99x^2)} = \underline{\qquad}$

$2x^3$, $\dfrac{x^5}{5y^3}$

492 $\dfrac{3x^2}{5y} \div \dfrac{6y^2}{2x^3} = \dfrac{3x^2}{5y} \cdot \dfrac{\underline{\qquad}}{6y^2} = \underline{\qquad}$

ac

ac

a

$\dfrac{a}{b^2c}$

493 $\dfrac{a^2bc}{abc^2} \div \dfrac{ab^2c}{ac} = \dfrac{a^2bc}{abc^2} \cdot \dfrac{\underline{\qquad}}{ab^2c}$

$$= \frac{(a^2bc)\,(\underline{\qquad})}{(abc^2)\,(ab^2c)}$$

$$= \frac{(a^2bc^2)\,(\underline{\qquad})}{(a^2bc^2)\,(b^2c)}$$

$$= \underline{\qquad}$$

$6x + 3$

$2x + 1$

$\dfrac{3(x-1)}{4(2x-1)}$

494 $\dfrac{x^2 - 2x + 1}{4x^2 - 1} \div \dfrac{4x - 4}{6x + 3} = \dfrac{x^2 - 2x + 1}{4x^2 - 1} \cdot \dfrac{\underline{\qquad\qquad}}{4x - 4}$

$$= \frac{(x-1)^2(3)\,(\underline{\qquad\qquad})}{(2x-1)\,(2x+1)\,(4)\,(x-1)}$$

$$= \underline{\qquad\qquad\qquad}$$

$(x + 2)^2$

$x^2 - 25$

$x - 5$

$\dfrac{(x+5)\,(x+2)}{(x-5)\,(x-2)}$

495 $\dfrac{(x+5)^2}{x^2 - 4} \div \dfrac{x^2 - 25}{(x+2)^2} = \dfrac{(x+5)^2}{x^2 - 4} \cdot \dfrac{\underline{\qquad\qquad}}{x^2 - 25}$

$$= \frac{(x+5)^2(x+2)^2}{(x^2-4)\,(\underline{\qquad\qquad})}$$

$$= \frac{(x+5)\,(x+5)\,(x+2)\,(x+2)}{(x-2)\,(x+2)\,(x+5)\,(\underline{\qquad\qquad})}$$

$$= \underline{\qquad\qquad\qquad\qquad}$$

496 $\dfrac{9a^2 - 4}{a^2 + 4a - 12} \div \dfrac{12a + 8}{a^2 + a - 30}$

$a^2 + a - 30$

$a^2 + a - 30$

$a + 6$

$\dfrac{(3a-2)\,(a-5)}{4(a-2)}$

$$= \frac{9a^2 - 4}{a^2 + 4a - 12} \cdot \frac{\underline{\qquad\qquad\qquad}}{12a + 8}$$

$$= \frac{(9a^2 - 4)\,(\underline{\qquad\qquad\qquad})}{(a^2 + 4a - 12)\,(12a + 8)}$$

$$= \frac{(3a-2)\,(3a+2)\,(a-5)\,(\underline{\qquad\qquad})}{(a+6)\,(a-2)\,(4)\,(3a+2)}$$

$$= \underline{\qquad\qquad\qquad\qquad}$$

497 $\dfrac{x^2 + 5x + 6}{2x^2 + x - 1} \div \dfrac{x^2 - 9}{(x+1)^2}$

$(x + 1)^2$

$(x + 1)^2$

$2x - 1$

$\dfrac{(x+2)\,(x+1)}{(2x-1)\,(x-3)}$

$$= \frac{x^2 + 5x + 6}{2x^2 + x - 1} \cdot \frac{\underline{\qquad\qquad}}{x^2 - 9}$$

$$= \frac{(x^2 + 5x + 6)\,\underline{\qquad\qquad}}{(2x^2 + x - 1)\,(x^2 - 9)}$$

$$= \frac{(x+2)\,(x+3)\,(x+1)\,(x+1)}{(\underline{\qquad\qquad})\,(x+1)\,(x-3)\,(x+3)}$$

$$= \underline{\qquad\qquad\qquad\qquad}$$

498 $\dfrac{4x^2 - 36}{x^2 + 4x + 3} \div \dfrac{x^2 + x - 12}{72 - 6x - 6x^2}$

$= \dfrac{4x^2 - 36}{x^2 + 4x + 3} \cdot \dfrac{\_\_\_\_\_\_\_\_}{x^2 + x - 12}$ — $72 - 6x - 6x^2$

$= \dfrac{(4)(x + 3)(x - 3)(-6)(\_\_\_\_\_\_)(x - 3)}{(x + 3)(x + 1)(x - 3)(x + 4)}$ — $x + 4$

$=$ ________ — $\dfrac{-24(x - 3)}{x + 1}$

499 $\dfrac{2a}{15} \cdot \dfrac{8x}{9} \div \dfrac{a^2x^2}{90} = \dfrac{2a}{15} \cdot \dfrac{8x}{9} \cdot \dfrac{\_\_\_}{a^2x^2}$ — 90

$= \dfrac{(2a)(8x)(\_\_\_)}{(15)(9)(a^2x^2)} =$ ____ — 90, $\dfrac{32}{3ax}$

500 $\dfrac{2x^2 + x - 1}{3x^4y^2} \cdot \dfrac{2x^6y^3}{x^2 + 2x + 1} \div \dfrac{4x^2 - 1}{x^2 - 1}$

$= \dfrac{2x^2 + x - 1}{3x^4y^2} \cdot \dfrac{2x^6y^3}{x^2 + 2x + 1} \cdot \dfrac{\_\_\_\_\_\_}{4x^2 - 1}$ — $x^2 - 1$

$= \dfrac{(2x - 1)(x + 1)(2x^6y^3)(x - 1)(\_\_\_\_\_\_)}{(3x^4y^2)(x + 1)^2(2x - 1)(2x + 1)}$ — $x + 1$

$=$ ________ — $\dfrac{2x^2y(x - 1)}{3(2x + 1)}$

3.2 Addition and Subtraction of Fractions

501 To add like fractions, add the ________ to find the numerator of the sum and retain the common denominator as the denominator of the sum. That is, if $\dfrac{P}{Q}$ and $\dfrac{R}{Q}$ are fractions, with $Q \neq 0$, then $\dfrac{P}{Q} + \dfrac{R}{Q} = \dfrac{\_\_\_}{Q}$. — numerators; $P + R$

502 To subtract like fractions, subtract the ________ to find the numerator of the difference and retain the common denominator as the denominator of the difference. That is, if $\dfrac{P}{Q}$ and $\dfrac{R}{Q}$ are fractions, with $Q \neq 0$, then $\dfrac{P}{Q} - \dfrac{R}{Q} = \dfrac{\_\_\_}{Q}$. — numerators; $P - R$

In Problems 503–506, perform the indicated operations.

503 $\dfrac{3x + 1}{y} + \dfrac{5x - 2}{y} = \dfrac{3x + 1 + \_\_\_\_\_\_}{y} =$ ______ — $5x - 2$, $\dfrac{8x - 1}{y}$

504 $\dfrac{4x + 3}{4y} - \dfrac{2x + 1}{4y} = \dfrac{4x + 3 - (\_\_\_\_\_\_)}{4y}$ — $2x + 1$

$= \dfrac{\_\_\_}{4y} =$ ______ — $2x + 2$, $\dfrac{x + 1}{2y}$

Answers	Frames
$-x^2 + 36$	**505** $\dfrac{-x^2}{6-x} + \dfrac{36}{6-x} = \dfrac{\_\_\_\_\_\_}{6-x}$
$(-x+6)$	$= \dfrac{\_\_\_\_\_\_(x+6)}{6-x}$
$(6-x)(6+x)$	$= \dfrac{\_\_\_\_\_\_}{6-x}$
$6+x$	$=$ ______
$16 - x^2,\ 4+x$	**506** $\dfrac{16}{4-x} - \dfrac{x^2}{4-x} = \dfrac{\_\_\_\_\_\_}{4-x} = \dfrac{(4-x)(\_\_\_\_\_\_)}{4-x}$
$4+x$	$=$ ______
least common denominator L.C.D.	**507** The common denominator with the smallest number of factors that all denominators divide into evenly is called the ________________. It is denoted by ______.
prime factors product	**508** To find the L.C.D. of a set of fractions, we use the following steps: *Step 1* Factor each denominator into a product of __________ or into a product of prime factors and -1. *Step 2* In any factored denominator, list all the prime factors and consider the one with the largest exponent; then find the ______ of the factors so listed.

In Problems 509–516, perform the indicated operations and simplify.

Answers	Frames
	509 $\dfrac{5}{8} + \dfrac{7}{12}$
24	The L.C.D. of the given fractions is ____, so that
14, 14	$\dfrac{5}{8} + \dfrac{7}{12} = \dfrac{15}{24} + \dfrac{\_\_\_}{24} = \dfrac{15 + \_\_\_}{24}$
$\dfrac{29}{24}$	$= \dfrac{}{\_\_\_}$
	510 $\dfrac{5}{14x^2y} - \dfrac{3}{21xy^2}$
$42x^2y^2$	The L.C.D. of the given fractions is ______, so that
$6x$	$\dfrac{5}{14x^2y} - \dfrac{3}{21xy^2} = \dfrac{15y}{42x^2y^2} - \dfrac{\_\_\_}{42x^2y^2}$
$15y - 6x$	$= \dfrac{\_\_\_\_\_\_}{42x^2y^2}$

511 $\frac{2x-3}{18}+\frac{x+2}{12}$

36 | The L.C.D. of the given fractions is ____, so that

$$\frac{2x-3}{18}+\frac{x+2}{12}=\frac{2(2x-3)}{36}+\frac{3(x+2)}{36}$$

$3x+6$, $\frac{7x}{36}$ | $$=\frac{4x-6+\underline{\qquad}}{36}=\underline{\qquad}$$

512 $\frac{2x+1}{24x}-\frac{3x^2+2x}{36x^2}$

$72x^2$ | The L.C.D. is ____, so that

$3x^2+2x$ | $$\frac{2x+1}{24x}-\frac{3x^2+2x}{36x^2}=\frac{3x(2x+1)}{72x^2}-\frac{2(\underline{\qquad})}{72x^2}$$

$4x$ | $$=\frac{6x^2+3x-6x^2-\underline{\qquad}}{72x^2}$$

$-x$, $-\frac{1}{72x}$ | $$=\frac{\underline{\qquad}}{72x^2}=\underline{\qquad}$$

513 $\frac{-7}{2x^2-5x-3}+\frac{6}{x^2-9}$

To find the L.C.D., factor each denominator, so that

$(2x+1)(x-3)$ | $2x^2-5x-3=$ ____

$(x-3)(x+3)$ | $x^2-9=$ ____

$(2x+1)(x-3)(x+3)$ | The L.C.D. is ____, so that

$$\frac{-7}{2x^2-5x-3}+\frac{6}{x^2-9}$$

$$=\frac{-7}{(2x+1)(x-3)}+\frac{6}{(x-3)(x+3)}$$

$x+3$, $2x+1$ | $$=\frac{-7(\underline{\qquad})}{(2x+1)(x-3)(x+3)}+\frac{6(\underline{\qquad})}{(x-3)(x+3)(2x+1)}$$

$5x-15$, 5 | $$=\frac{\underline{\qquad}}{(2x+1)(x-3)(x+3)}=\frac{\underline{\qquad}}{(2x+1)(x+3)}$$

514 $\frac{6}{x^2-2x-8}+\frac{x}{x+2}$

x^2-2x-8 | The L.C.D. is ____, so that

$(x-4)(x+2)$ | $$\frac{6}{x^2-2x-8}+\frac{x}{x+2}=\frac{6}{\underline{\qquad}}+\frac{x}{x+2}$$

$x(x-4)$ | $$=\frac{6}{(x-4)(x+2)}+\frac{\underline{\qquad}}{(x-4)(x+2)}$$

x^2-4x+6 | $$=\frac{\underline{\qquad}}{(x-4)(x+2)}$$

515 $\dfrac{x}{x^2+4x-5}+\dfrac{2}{x+5}$

x^2+4x-5 The L. C. D. is ______________, so that

$2(x-1)$ $\dfrac{x}{x^2+4x-5}+\dfrac{2}{x+5}=\dfrac{x}{(x+5)(x-1)}+\dfrac{\_\_\_\_\_\_}{(x+5)(x-1)}$

$3x-2$ $=\dfrac{\_\_\_\_\_\_}{(x+5)(x-1)}$

516 $\dfrac{1}{x-2}-\dfrac{1}{x-1}+\dfrac{x}{x^2+2x-3}$

$(x-1)(x-2)(x+3)$ The L.C.D. is ______________________, so that

$\dfrac{1}{x-2}-\dfrac{1}{x-1}+\dfrac{x}{x^2+2x-3}$

$(x+3)(x-1)$ $=\dfrac{1}{x-2}-\dfrac{1}{x-1}+\dfrac{x}{\_\_\_\_\_\_\_\_\_\_}$

$x-2$ $=\dfrac{(x+3)(x-1)-(x-2)(x+3)+x(\_\_\_\_\_\_)}{(x-1)(x-2)(x+3)}$

x^2-2x $=\dfrac{x^2+2x-3-x^2-x+6+(\_\_\_\_\_\_)}{(x-1)(x-2)(x+3)}$

x^2-x+3 $=\dfrac{\_\_\_\_\_\_}{(x-1)(x-2)(x+3)}$

517 A fraction that is of a form such that its numerator or denominator or both are fractions is called a __________ fraction.

complex

518 One method for simplifying a complex fraction is first to multiply its numerator and its denominator by the __________ of all the fractions occurring in both its numerator and its denominator, and then to reduce the fraction obtained.

L.C.D.

519 Another method for simplifying complex fractions is to combine the fractions in the numerator and denominator separately and then divide the numerator by the ______________.

denominator

In Problems 520–524, express each fraction as a single fraction in lowest terms.

520 $\dfrac{x+\dfrac{1}{y}}{y+\dfrac{1}{x}}$

xy^2+y $\dfrac{x+\dfrac{1}{y}}{y+\dfrac{1}{x}}=\dfrac{\left(x+\dfrac{1}{y}\right)(xy)}{\left(y+\dfrac{1}{x}\right)(xy)}=\dfrac{x^2y+x}{\_\_\_\_\_\_}$

$xy+1,\ \dfrac{x}{y}$ $=\dfrac{x(xy+1)}{y(\_\_\_\_\_\_)}=\_\_\_\_$

521 $$\frac{1 - \dfrac{3y}{x+y}}{1 - \dfrac{y}{x-y}}$$

$(x - y)(x + y)$

$$\frac{1 - \dfrac{3y}{x+y}}{1 - \dfrac{y}{x-y}} = \frac{\left(1 - \dfrac{3y}{x+y}\right)\underline{\hspace{3cm}}}{\left(1 - \dfrac{y}{x-y}\right)(x-y)(x+y)}$$

$x - y$

$$= \frac{(x-y)(x+y) - 3y(\underline{\hspace{2cm}})}{(x-y)(x+y) - y(x+y)}$$

$$= \frac{x^2 - y^2 - 3xy + 3y^2}{x^2 - y^2 - xy - y^2}$$

$x^2 - 3xy + 2y^2$

$$= \frac{\underline{\hspace{3cm}}}{x^2 - xy - 2y^2}$$

$x - 2y,\ x - y$

$$= \frac{(x-y)(\underline{\hspace{2cm}})}{(x+y)(x-2y)} = \frac{\underline{\hspace{2cm}}}{x+y}$$

522 $$\frac{\dfrac{1}{x} - \dfrac{1}{y}}{\dfrac{1}{x} + \dfrac{1}{y}}$$

xy

$$\frac{\dfrac{1}{x} - \dfrac{1}{y}}{\dfrac{1}{x} + \dfrac{1}{y}} = \frac{\dfrac{y-x}{xy}}{\dfrac{y+x}{xy}} = \frac{y-x}{xy} \cdot \frac{\underline{\hspace{1cm}}}{y+x}$$

$\dfrac{y-x}{y+x}$

$$= \underline{\hspace{3cm}}$$

523 $$\frac{\dfrac{x}{y} - 1}{\dfrac{y}{x} - 1}$$

x

$$\frac{\dfrac{x}{y} - 1}{\dfrac{y}{x} - 1} = \frac{\dfrac{x-y}{y}}{\dfrac{y-x}{x}} = \frac{x-y}{y} \cdot \frac{\underline{\hspace{1cm}}}{y-x}$$

$\dfrac{x(x-y)}{y(y-x)},\ -\dfrac{x}{y}$

$$= \underline{\hspace{4cm}} = \underline{\hspace{1.5cm}}$$

524 $$\frac{a^2 - \dfrac{1}{a}}{a + \dfrac{1}{a} - 1}$$

$a^3 - 1,\ a - 1,\ a - 1$

$$\frac{a^2 - \dfrac{1}{a}}{a + \dfrac{1}{a} + 1} = \frac{\dfrac{\underline{\hspace{2cm}}}{a}}{\dfrac{a^2 + 1 + a}{a}} = \frac{(\underline{\hspace{1.5cm}})(a^2 + a + 1)}{a^2 + a + 1} = \underline{\hspace{2cm}}$$

3.3 Equations Involving Fractions

Answer	Frame
fractional	**525** Equations of the form $\frac{x+2}{5} = 7$ and $\frac{x+1}{3x} = 2$ are called __________ equations.
L.C.D.	**526** Fractional equations can be solved by first multiplying both sides of the equation by the ______ of the fractions in the equation.

In Problems 527–534, find the solution set of each equation.

Answer	Frame
	527 $\frac{x}{2} + \frac{x}{3} = 10$
6	$6\left(\frac{x}{2} + \frac{x}{3}\right) = (\_\_\_)\ 10$
60	$3x + 2x = \_\_\_$
60	$5x = \_\_\_$
12	$x = \_\_\_$
{12}	The solution set is ____.
	528 $\frac{x-10}{8} + \frac{13}{4} = \frac{4x+6}{3}$
24	$24\left(\frac{x-10}{8} + \frac{13}{4}\right) = \_\_\_\ \left(\frac{4x+6}{3}\right)$
8	$3(x-10) + 6(13) = \_\_\_\ (4x+6)$
$32x + 48$	$3x - 30 + 78 = \_\_\_\_\_\_$
−29, 0	$\_\_\_ x = \_\_\_$
0	$x = \_\_\_$
{0}	The solution set is ____.
	529 $\frac{6}{x} + \frac{x-3}{2x} = 2$
$2x$	$2x\left(\frac{6}{x} + \frac{x-3}{2x}\right) = (\_\_\_)\ 2$
$4x$	$12 + x - 3 = \_\_\_$
12	$x - 4x = 3 - \_\_\_$
−9	$-3x = \_\_\_$
3	$x = \_\_\_$
{3}	The solution set is ____.
	530 $\frac{2}{3x} + \frac{1}{6x} = \frac{1}{4}$
$12x$, $12x$	$(\_\_\_)\left(\frac{2}{3x} + \frac{1}{6x}\right) = (\_\_\_)\ \frac{1}{4}$
8 + 2	$\_\_\_\_ = 3x$
10	$\_\_\_ = 3x$
$\frac{10}{3}$	$\_\_\_ = x$
$\{\frac{10}{3}\}$	The solution set is ____.

Answers	Frames
	531 $\dfrac{8}{x-3} = 2$
$x - 3$	$(x-3)\dfrac{8}{x-3} = (\_\_\_\_)2$
6	$8 = 2x - \_\_\_\_$
14	$\_\_\_\_ = 2x$
7	$x = \_\_\_\_$
$\{7\}$	The solution set is ____.
	532 $\dfrac{4}{x-8} = \dfrac{3}{x-9}$
$(x-8)(x-9)$	$(x-8)(x-9)\left(\dfrac{4}{x-8}\right) = \_\_\_\_\_\_\_\_\left(\dfrac{3}{x-9}\right)$
$x - 8$	$4(x-9) = 3(\_\_\_\_)$
24	$4x - 36 = 3x - \_\_\_\_$
$3x$	$4x - \_\_\_\_ = -24 + 36$
12	$x = \_\_\_\_$
$\{12\}$	The solution set is ____.
	533 $\dfrac{x+2}{x-2} - \dfrac{x-2}{x+2} = 1 - \dfrac{x^2}{x^2-4}$
$(x-2)(x+2)$	$(x-2)(x+2)\left(\dfrac{x+2}{x-2} - \dfrac{x-2}{x+2}\right) = \_\_\_\_\_\_\_\_\left(1 - \dfrac{x^2}{x^2-4}\right)$
$x - 2$	$(x+2)^2 - (\_\_\_\_)^2 = (x-2)(x+2) - x^2$
$x^2 - 4$	$x^2 + 4x + 4 - (x^2 - 4x + 4) = (\_\_\_\_) - x^2$
-4	$x^2 + 4x + 4 - x^2 + 4x - 4 = \_\_\_\_$
-4	$8x = \_\_\_\_$
$-\frac{1}{2}$	$x = \_\_\_\_$
$\{-\frac{1}{2}\}$	The solution set is ____.
	534 $\dfrac{2x}{x-2} = \dfrac{4}{x-2} - 1$
$x - 2$	$(x-2)\left(\dfrac{2x}{x-2}\right) = (\_\_\_\_)\left(\dfrac{4}{x-2} - 1\right)$
$x - 2$	$2x = 4 - (\_\_\_\_)$
6	$2x + x = \_\_\_\_$
6, 2	$3x = \_\_\_\_$ or $x = \_\_\_\_$
$\emptyset$	Since $x = 2$ is an extraneous solution, the solution set is ____.
	535 A large truck can haul the gravel needed for a certain concrete job in 24 hours and a small truck can haul the gravel in 56 hours. How long should it take if the two trucks work together?
	Let x be the number of hours it will take both trucks to haul the
$\frac{1}{24}$ of the work	gravel. The large truck will do ________ in 1 hour,
$\frac{1}{56}$ of the work	and the small truck will do ________ in 1 hour.

Answers	
$\frac{1}{x}$ of the work	Then both trucks will do ______ in 1 hour. The
$\frac{1}{24}+\frac{1}{56}=\frac{1}{x}$, $16\frac{4}{5}$	equation is ______, so that $x =$ ____.
16 hours and 48 minutes	Therefore, it will take ______ for both trucks to do the work together.

4 Exponents and Radicals

Answers	No.	Problem
	536	Let $a, b \in R$ and m and n be integers; then
a^{m+n}		$a^m a^n =$ ____
a^{mn}		$(a^m)^n =$ ____
$a^n b^n$		$(ab)^n =$ ____
$\frac{a^n}{b^n}$		$\left(\frac{a}{b}\right)^n = \frac{}{\_\_\_\_}$, for $b \neq 0$
a^{m-n}		$\frac{a^m}{a^n} =$ ____, for $a \neq 0$
1	537	$a^0 =$ ____, for $a \neq 0$
1	538	$(\frac{3}{8})^0 =$ ____
1	539	$(\frac{5}{6})^0 =$ ____
1	540	$\frac{4^5}{4^5} = 4^{5-5} = 4^0 =$ ____
10, $\frac{5}{2}$	541	$\frac{3^2+3^0}{3+7^0} = \frac{\_\_\_\_}{3+1} =$ ____
0, 0	542	$\frac{x^0-y^0}{x^0+y^0} = \frac{\_\_\_\_}{1+1} =$ ____
1	543	$a^{m-m} = a^0 =$ ____
a^n	544	$a^{-n} = \frac{1}{\_\_\_\_}$, for $a \neq 0$
a^{-n}	545	$\frac{1}{a^n} =$ ____, for $a \neq 0$

In Problems 546–565, eliminate the negative exponents and simplify.

Answers	No.	Problem
$\frac{1}{5^3}$, $\frac{1}{125}$	546	$5^{-3} = \frac{}{\_\_\_\_} = \frac{}{\_\_\_\_}$
$(\frac{3}{5})^2$, $\frac{25}{9}$	547	$\left(\frac{3}{5}\right)^{-2} = \frac{1}{\_\_\_\_} =$ ____
$\frac{4a}{b^2}$, b^2	548	$\frac{4ab^{-2}}{c^0 d^{-2}} = \frac{\_\_\_\_}{\frac{1}{d^2}} = \frac{4ad^2}{\_\_\_\_}$
$\frac{1}{x^5}$, x^2, x^3	549	$\frac{x^{-2}}{x^{-5}} = \frac{\frac{1}{x^2}}{\_\_\_\_} = \frac{x^5}{\_\_\_\_} =$ ____

12	**550**	$\frac{3}{2^{-2}} = \frac{3}{\frac{1}{2^2}} = \underline{\qquad}$
$x^3,\ \frac{y^5}{x^3}$	**551**	$x^{-3}y^5 = \frac{1}{\underline{\qquad}}\, y^5 = \underline{\qquad}$
$\frac{1}{z^3},\ 5y^2$	**552**	$\frac{3x^3y^{-2}}{5z^{-3}} = \frac{3x^3\,\frac{1}{y^2}}{5\left(\underline{\qquad}\right)} = \frac{\frac{3x^3}{y^2}}{\frac{5}{z^3}} = \frac{3x^3z^3}{\underline{\qquad}}$
$\frac{1}{3^2}\cdot\frac{1}{y^2},\ 9y^2$	**553**	$\frac{3^{-2}x^0y^{-2}}{z^{-4}} = \frac{\underline{\qquad}}{\frac{1}{z^4}} = \frac{\frac{1}{9y^2}}{\frac{1}{z^4}} = \frac{z^4}{\underline{\qquad}}$
y^6	**554**	$(x^2y^{-3})^2 = x^4y^{-6} = \frac{x^4}{\underline{\qquad}}$
$(a^2b)^3,\ (ab^2)^2,\ a^2b^4,\ \frac{a^4}{b}$	**555**	$\frac{(ab^2)^{-2}}{(a^2b)^{-3}} = \frac{\frac{1}{(ab^2)^2}}{\frac{1}{\underline{\qquad}}} = \frac{(a^2b)^3}{\underline{\qquad}} = \frac{a^6b^3}{\underline{\qquad}} = \underline{\qquad}$
$\frac{1}{b},\ \frac{b}{a},\ \frac{2b}{a}$	**556**	$\frac{a^{-1}}{b^{-1}} + \frac{b}{a} = \frac{\frac{1}{a}}{\underline{\qquad}} + \frac{b}{a} = \underline{\qquad} + \frac{b}{a} = \underline{\qquad}$
$\frac{1}{y},\ y+x,\ \frac{1}{xy}$	**557**	$\frac{x^{-1}+y^{-1}}{x+y} = \frac{\frac{1}{x}+\underline{\qquad}}{x+y} = \frac{\frac{\underline{\qquad}}{xy}}{x+y} = \underline{\qquad}$
$\frac{1}{xy},\ y+x$	**558**	$\frac{x^{-1}+y^{-1}}{(xy)^{-1}} = \frac{\frac{1}{x}+\frac{1}{y}}{\underline{\qquad}} = \frac{\frac{y+x}{xy}}{\frac{1}{xy}} = \underline{\qquad}$
$\frac{1}{(xy)(x+y)^3}$	**559**	$\frac{(x+y)^{-3}}{xy} = \underline{\qquad}$
$\frac{1}{y},\ y-x,\ \frac{y+x}{y-x}$	**560**	$\frac{x^{-1}+y^{-1}}{x^{-1}-y^{-1}} = \frac{\frac{1}{x}+\frac{1}{y}}{\frac{1}{x}-\underline{\qquad}} = \frac{\frac{y+x}{xy}}{\frac{\underline{\qquad}}{xy}} = \underline{\qquad}$
$y^2,\ y^2+x^2$	**561**	$x^{-2}+y^{-2} = \frac{1}{x^2} + \frac{1}{\underline{\qquad}} = \frac{\underline{\qquad}}{x^2y^2}$
$y,\ y-x$	**562**	$\frac{x^{-1}-y^{-1}}{x-y} = \frac{\frac{1}{x}-\frac{1}{\underline{\qquad}}}{x-y} = \frac{\frac{\underline{\qquad}}{xy}}{x-y}$
$x-y,\ -\frac{1}{xy}$		$= \frac{-(\underline{\qquad})}{xy(x-y)} = \underline{\qquad}$

y^3, $y^3 + x^3$

563 $\dfrac{x^{-3} + y^{-3}}{x + y} = \dfrac{\frac{1}{x^3} + \frac{1}{\underline{\quad}}}{x + y} = \dfrac{\frac{\underline{\qquad}}{x^3y^3}}{x + y}$

$y^2 - yx + x^2$

$= \dfrac{(y + x)(\underline{\qquad\qquad})}{x^3y^3(x + y)}$

$\dfrac{y^2 - yx + x^2}{x^3y^3}$

$= \underline{\qquad\qquad}$

$\dfrac{1}{y}$, $y - x$, x^2y^2

564 $\dfrac{x^{-2} - y^{-2}}{x^{-1} - y^{-1}} = \dfrac{\frac{1}{x^2} - \frac{1}{y^2}}{\frac{1}{x} - \underline{\quad}} = \dfrac{\frac{y^2 - x^2}{x^2y^2}}{\frac{\underline{\quad}}{xy}} = \dfrac{(y^2 - x^2)xy}{(y - x)\underline{\quad}}$

$y - x$, $\dfrac{y + x}{xy}$

$= \dfrac{(\underline{\qquad})(y + x)(xy)}{(y - x)x^2y^2} = \underline{\qquad}$

y^2, $y^2 - x^2$

565 $\dfrac{x^{-2} - y^{-2}}{x + y} = \dfrac{\frac{1}{x^2} - \frac{1}{\underline{\quad}}}{x + y} = \dfrac{\frac{\underline{\qquad}}{x^2y^2}}{x + y}$

$y^2 - x^2$, $(y - x)(y + x)$, $\dfrac{y - x}{x^2y^2}$

$= \dfrac{\underline{\qquad}}{x^2y^2(x + y)} = \dfrac{\underline{\qquad\qquad}}{x^2y^2(x + y)} = \underline{\qquad}$

In Problems 566–577, use the properties of exponents to simplify each expression.

x

566 $x^{-2}x^3 = x^{-2+3} = \underline{\quad}$

x^{2n+1}, x^{4n+2}

567 $(x^{3n}x^{1-n})^2 = (\underline{\qquad})^2 = \underline{\qquad}$

x^{3n+3}, x^{2n+3}

568 $\dfrac{(x^{n+1})^3}{x^n} = \dfrac{\underline{\qquad}}{x^n} = x^{3n+3-n} = \underline{\qquad}$

x^{2n-1}, $2n-1-n+2$, x^{n+1}

569 $\dfrac{x^nx^{n-1}}{x^{n-2}} = \dfrac{\underline{\qquad}}{x^{n-2}} = x^{\underline{\qquad}} = \underline{\quad}$

x^{-n}, x^{2n}

570 $\left(\dfrac{x^{-n}}{x^n}\right)^{-1} = \dfrac{x^n}{\underline{\quad}} = x^{n+n} = \underline{\quad}$

$(x^{n+1})^n$, x^{-n^2-2n}

571 $\dfrac{(x^{n+1})^{-n}}{x^n} = \dfrac{\frac{1}{\underline{\qquad}}}{x^n} = \dfrac{\frac{1}{x^{n^2+n}}}{x^n} = \dfrac{1}{x^nx^{n^2+n}} = \underline{\qquad}$

$x^{2n-2-n+2}$, x^n

572 $\dfrac{(x^{n-1})^2}{x^{n-2}} = \dfrac{x^{2n-2}}{x^{n-2}} = \underline{\qquad\qquad} = \underline{\quad}$

$\dfrac{1}{x}$, $x^{2n-2}y^{2n}$, $x^{2n}y^{2n}$

573 $\left(\dfrac{x^{n-1}y^n}{x^{-1}}\right)^2 = \dfrac{(x^{n-1}y^n)^2}{(\underline{\quad})^2} = x^2 \cdot \underline{\qquad\qquad} = \underline{\qquad}$

y^{4-4n}, y^{-4-4n}

574 $(y^{-2}y^{1-n})^4 = y^{-8} \cdot \underline{\qquad} = \underline{\qquad}$

y^{-2n-3n}

575 $\left(\dfrac{x^{2n-3}y^{-2n}}{x^{-2}y^{3n}}\right)^2 = (x^{2n-3-(-2)}\underline{\qquad})^2$

y^{-5n}

$= (x^{2n-1}\underline{\quad})^2$

$x^{4n-2}y^{-10n}$

$= \underline{\qquad\qquad}$

Answer	Frame
$y^{n-(-2n)}$	**576** $\left(\dfrac{x^{3n-2}y^{n}}{x^{-3}y^{-2n}}\right)^{3} = (x^{3n-2-(-3)}$ ____________ $)^3$
y^{3n}	$= (x^{3n+1}$ ______ $)^3$
$x^{9n+3}y^{9n}$	= ______________
$y^{-3n+2-(-2n)}$	**577** $\left(\dfrac{x^{-2n}y^{-3n+2}}{y^{-2n}}\right)^{-1} = (x^{-2n}$ ____________________ $)^{-1}$
y^{-n+2}	$= (x^{-2n}$ ________ $)^{-1}$
$x^{2n}y^{n-2}$	= ____________
$(a^{1/q})^p$	**578** If $a \in R$, then $a^{p/q} =$ ___________, provided that $a^{1/q}$ exists.
root	The number $a^{1/q}$ is called the principal qth ______ of a.
zero	**579** If a is zero, then $a^{1/q}$ is also ______.
	580 If a is a negative number and q is an odd positive integer, then
negative	$a^{1/q}$ is also ______________.
3	**581** $9^{1/2} = (3^2)^{1/2} =$ _____
3	**582** $(27)^{1/3} = (3^3)^{1/3} =$ _____
-5	**583** $-25^{1/2} = -(5^2)^{1/2} =$ _____
	584 If a is a negative number and q is an even positive number, the
real number	number $a^{1/q}$ is not a __________________, because an even power
	of either a positive or negative real number is positive.
	585 $(-9)^{1/2}$ is not a real number, since there exists no real number
-9	whose square is _____.
real	**586** $(-81)^{1/4}$ is meaningless in the set of ______ numbers.
	587 If a is a negative real number and q is an odd positive number,
	then there exists a real number $a^{1/q}$, since an odd power of a
negative	negative number is ______________.
-32	**588** $(-32)^{1/5} = -2$, since $(-2)(-2)(-2)(-2)(-2) =$ _____.
-2	**589** $(-8)^{1/3} =$ _____
-4	**590** $(-64)^{1/3} =$ _____
a^p	**591** If $a^{1/q}$ exists, then $a^{p/q} = (a^{1/q})^p = ($_____$)^{1/q}$.

In Problems 592–597, compute the value of the given numbers.

Answer	Frame
^{4}, 81	**592** $(27)^{4/3} = (27^{1/3})$——— $= (3)^4 =$ _____
$64^{1/6}$, 32	**593** $(64)^{5/6} = ($_______$)^5 = (2)^5 =$ _____
-2, 4	**594** $(-8)^{2/3} = [(-8)^{1/3}]^2 = ($_____$)^2 =$ _____
-2, -8	**595** $(-32)^{3/5} = [(-32)^{1/5}]^3 = ($_____$)^3 =$ _____
$(-27)^{1/3}$, 9	**596** $(-27)^{2/3} = [$____________$]^2 = (-3)^2 =$ _____
2, 8	**597** $(16)^{3/4} = ($_____$)^3 =$ _____

In Problems 598–607, compute the value of each expression. Assume that the bases are positive.

Answers	Problems
x^4, x^8	598 $(x^{12})^{2/3} = (\_\_\_\_)^2 = \_\_\_\_$
$-x^2$, x^4	599 $(-x^6)^{2/3} = (\_\_\_\_)^2 = \_\_\_\_$
$3x^9$, $9x^{18}$	600 $(27x^{27})^{2/3} = (\_\_\_\_)^2 = \_\_\_\_$
$2p^2q^3$, $4p^4q^6$	601 $(8p^6q^9)^{2/3} = (\_\_\_\_)^2 = \_\_\_\_$
$2x^2y^4$, $8x^6y^{12}$	602 $(4x^4y^8)^{3/2} = (\_\_\_\_)^3 = \_\_\_\_$
$-4x^4$, $16x^8$	603 $(-64x^{12})^{2/3} = (\_\_\_\_)^2 = \_\_\_\_$
$-3x^3$, $9x^6$	604 $(-27x^9)^{2/3} = (\_\_\_\_)^2 = \_\_\_\_$
$3/4$, $\frac{2x}{y^2}$, $\frac{8x^3}{y^6}$, $\frac{y^6}{8x^3}$	605 $\left(\frac{16x^4}{y^8}\right)^{-3/4} = \frac{1}{\left(\frac{16x^4}{y^8}\right)^{\_\_\_\_}} = \frac{1}{\left(\frac{\_\_\_\_}{\_\_\_\_}\right)^3} = \frac{1}{\_\_\_\_} = \_\_\_\_$
$\frac{125x^3}{64y^6}$, $\frac{5x}{4y^2}$, $\frac{25x^2}{16y^4}$, $\frac{16y^4}{25x^2}$	606 $\left(\frac{125x^3}{64y^6}\right)^{-2/3} = \frac{1}{\left(\frac{\_\_\_\_}{\_\_\_\_}\right)^{2/3}} = \frac{1}{\left(\frac{\_\_\_\_}{\_\_\_\_}\right)^2} = \frac{1}{\_\_\_\_} = \_\_\_\_$
$a^{-6}x^{-27}$, $a^{-2}x^{-9}$, $a^{-4}x^{-18}$ a^4x^{18}	607 $(a^{-6}x^{-27})^{-2/3} = \frac{1}{(\_\_\_\_)^{2/3}} = \frac{1}{(\_\_\_\_)^2} = \frac{1}{\_\_\_\_}$ $= \_\_\_\_$

In Problems 608–612, assume that a and b are real numbers and that r and s are reduced rational numbers. If the individual factors exist, then

Answers	Problems
a^{r+s}	608 $a^r \cdot a^s = \_\_\_\_$
a^{rs}	609 $(a^r)^s = \_\_\_\_$
a^rb^r	610 $(ab)^r = \_\_\_\_$
$\frac{a^r}{b^r}$	611 $\left(\frac{a}{b}\right)^r = \frac{\_\_\_\_}{\_\_\_\_}$, for $b \neq 0$
a^{r-s}	612 $\frac{a^r}{a^s} = \_\_\_\_$, for $a \neq 0$

In Problems 613–622, simplify each expression. Assume that all the bases are positive.

Answers	Problems
$x^{13/4}$	613 $x^{3/2} \cdot x^{7/4} = x^{3/2+7/4} = \_\_\_\_$
$x^{1/6}$	614 $x^{-2/3} \cdot x^{5/6} = x^{-2/3+5/6} = \_\_\_\_$
$x^{7/4}$	615 $x^{5/2} \cdot x^{-3/4} = x^{5/2-3/4} = \_\_\_\_$
$x^{1/8}$	616 $\frac{x^{-3/4}}{x^{-7/8}} = x^{-3/4-(-7/8)} = \_\_\_\_$
$x^{14/15}$	617 $\frac{x^{4/3}}{x^{2/5}} = x^{4/3-2/5} = \_\_\_\_$
$x^{16/3}$	618 $(x^8)^{2/3} = x^{(8)(2/3)} = \_\_\_\_$
x^2	619 $(x^{-1/6})^{-12} = x^{(-1/6)(-12)} = \_\_\_\_$
$y^{1/3}$	620 $(y^{-1/4})^{-4/3} = y^{(-1/4)(-4/3)} = \_\_\_\_$
$y^{1/4}$, $x^{4/3}y^{1/2}$	621 $(x^{2/3}y^{1/4})^2 = (x^{2/3})^2(\_\_\_\_)^2 = \_\_\_\_$
x^9y^8	622 $(x^{-3/4}y^{-2/3})^{-12} = (x^{-3/4})^{-12}(y^{-2/3})^{-12} = \_\_\_\_$

Answer	No.	Problem
	623	The principal nth root of a, where n is a positive integer and $n \geqslant 2$,
$\sqrt[n]{a}$, radical		is defined by $a^{1/n}$ = ____. The symbol $\sqrt{}$ is called the ________,
index, radicand		n is called the ________, and a is called the ________.
positive	624	If $a > 0$ and n is any positive integer, then $\sqrt[n]{a}$ is ________.
negative	625	If $a < 0$ and n is any positive odd integer, then $\sqrt[n]{a}$ is ________.
	626	If $a < 0$ and n is any positive even integer, then $\sqrt[n]{a}$ is
undefined		________ in the real number system.
undefined	627	$\sqrt{-4}$ is ________ in the real number system.

In Problems 628–631, write each expression in radical form.

Answer	No.	Problem
$\sqrt[5]{(-32)^3}$	628	$(-32)^{3/5}$ = ________
$\sqrt[5]{(x-y)^2}$	629	$(x-y)^{2/5}$ = ________
$\sqrt[4]{(2x+y)^3}$	630	$(2x+y)^{3/4}$ = ________
$\sqrt[11]{(3x+2y)^2}$	631	$(3x+2y)^{2/11}$ = ________

In Problems 632–634, write each expression in rational exponent form.

Answer	No.	Problem
$5^{1/2}$	632	$\sqrt{5}$ = ____
$11^{1/4}$	633	$\sqrt[4]{11}$ = ____
$y^{5/4}$	634	$\sqrt[4]{y^5}$ = ____

In Problems 635–639, assume that each root exists for the positive integers m, n, and c.

Answer	No.	Problem
$\sqrt[n]{b}$	635	$\sqrt[n]{ab} = \sqrt[n]{a} \cdot$ ____
$\sqrt[n]{b}$	636	$\sqrt[n]{\dfrac{a}{b}} = \dfrac{\sqrt[n]{a}}{\_\_\_\_}$, for $b \neq 0$
$a^{1/n}$	637	$\sqrt[n]{a^m} = (\sqrt[n]{a})^m = (\_\_\_\_)^m$
$\sqrt[mn]{a}$	638	$\sqrt[m]{\sqrt[n]{a}} = \sqrt[n]{\sqrt[m]{a}}$ = ____
$\sqrt[m]{a^n}$	639	$\sqrt[cm]{a^{cn}}$ = ____
$\sqrt[4]{3}$, $2\sqrt[4]{3}$	640	$\sqrt[4]{48} = \sqrt[4]{16} \cdot$ ____ = ____
$\sqrt{5}$, $2\sqrt{5}$	641	$\sqrt{20} = \sqrt{4} \cdot$ ____ = ____
$\sqrt{9}$, $3\sqrt{3}$	642	$\sqrt{27}$ = ____ $\cdot \sqrt{3}$ = ____
$\sqrt{9}$, $3\sqrt{7}$	643	$\sqrt{63}$ = ____ $\cdot \sqrt{7}$ = ____
$\sqrt{5}$, $10\sqrt{5}$	644	$\sqrt{500} = \sqrt{100} \cdot$ ____ = ____

In Problems 645–655, assume that all bases are positive.

Answer	No.	Problem
$\sqrt{z}$, $(x+y)\sqrt{z}$	645	$\sqrt{(x+y)^2 z} = \sqrt{(x+y)^2} \cdot$ ____ = ________
$\sqrt{3(x+y)}$	646	$\sqrt{27(x+y)^3} = \sqrt{9(x+y)^2} \cdot$ ________
$3(x+y)\sqrt{3(x+y)}$		= ________
$\sqrt{64}$, $\dfrac{5}{8}$	647	$\sqrt{\dfrac{25}{64}} = \dfrac{\sqrt{25}}{\_\_\_\_}$ = ____

$\sqrt{9}$, $\dfrac{\sqrt{11}}{3}$

648 $\sqrt{\dfrac{11}{9}} = \dfrac{\sqrt{11}}{\_\_\_\_} = \_\_\_\_$

$\dfrac{5}{2}$

649 $\sqrt[3]{\dfrac{125}{8}} = \dfrac{\sqrt[3]{125}}{\sqrt[3]{8}} = \_\_\_\_$

$\sqrt[3]{5}$, $\dfrac{\sqrt[3]{5}}{3}$

650 $\sqrt[3]{\dfrac{5}{27}} = \dfrac{\_\_\_\_}{\sqrt[3]{27}} = \_\_\_\_$

$\sqrt[4]{16}$, $\dfrac{5}{2}$

651 $\sqrt[4]{\dfrac{625}{16}} = \dfrac{\sqrt[4]{625}}{\_\_\_\_} = \_\_\_\_$

$\sqrt{y^2}$, $\dfrac{\sqrt{x}}{y}$

652 $\sqrt{\dfrac{x}{y^2}} = \dfrac{\sqrt{x}}{\_\_\_\_} = \_\_\_\_$

$\sqrt[3]{a}$, $\dfrac{\sqrt[3]{a}}{a+b}$

653 $\sqrt[3]{\dfrac{a}{(a+b)^3}} = \dfrac{\_\_\_\_}{\sqrt[3]{(a+b)^3}} = \_\_\_\_$

$\sqrt[5]{3x}$, $\dfrac{\sqrt[5]{3x}}{2}$

654 $\sqrt[5]{\dfrac{3x}{32}} = \dfrac{\_\_\_\_}{\sqrt[5]{32}} = \_\_\_\_$

$\sqrt[5]{(a+b)^5}$, $\dfrac{\sqrt[5]{27}}{a+b}$

655 $\sqrt[5]{\dfrac{27}{(a+b)^5}} = \dfrac{\sqrt[5]{27}}{\_\_\_\_} = \_\_\_\_$

rationalizing

656 If a radical occurs in the denominator of a fraction, the process of eliminating the radical is called ________ the denominator.

rationalizing

657 If the product of two radical expressions is free of radicals, each factor is called a ________ factor of the other.

$\sqrt{16}$, 4

658 $\sqrt{2}$ is a rationalizing factor of $\sqrt{8}$, since $\sqrt{2} \cdot \sqrt{8} = \_\_\_\_ = \_\_\_\_$.

$a^2x - b^2y$

659 $a\sqrt{x} + b\sqrt{y}$ is a rationalizing factor of $a\sqrt{x} - b\sqrt{y}$, since $(a\sqrt{x} + b\sqrt{y})(a\sqrt{x} - b\sqrt{y}) = \_\_\_\_\_\_$.

2, 3

660 $(\sqrt{5} - \sqrt{2})(\sqrt{5} + \sqrt{2}) = (\sqrt{5})^2 - (\sqrt{2})^2 = 5 - \_\_\_\_ = \_\_\_\_$

$\sqrt{3}$, 4

661 $(\sqrt{7} - \sqrt{3})(\sqrt{7} + \sqrt{3}) = (\sqrt{7})^2 - (\_\_\_\_)^2 = \_\_\_\_$

8, 28

662 $(6 - 2\sqrt{2})(6 + 2\sqrt{2}) = (6)^2 - (2\sqrt{2})^2 = 36 - \_\_\_\_ = \_\_\_\_$

12, 2, 10

663 $(2\sqrt{3} - \sqrt{2})(2\sqrt{3} + \sqrt{2}) = (2\sqrt{3})^2 - (\sqrt{2})^2$
$= \_\_\_\_ - \_\_\_\_ = \_\_\_\_$

$\sqrt{21}$, $\sqrt{5}$, 5, 16

664 $(\sqrt{21} - \sqrt{5})(\sqrt{21} + \sqrt{5}) = (\_\_\_\_)^2 - (\_\_\_\_)^2 = 21 - \_\_\_\_ = \_\_\_\_$

$\sqrt{33}$, 6, 27

665 $(\sqrt{33} - \sqrt{6})(\sqrt{33} + \sqrt{6}) = (\_\_\_\_)^2 - (\sqrt{6})^2 = 33 - \_\_\_\_ = \_\_\_\_$

rationalizing

666 To rationalize the denominator of a fraction, multiply the numerator and the denominator of the fraction by a ________ factor of the denominator.

In Problems 667–679, write the fractions as equivalent fractions in which the denominator is rationalized. Assume that all bases are positive real numbers.

$\sqrt{3}$, 3

667 $\dfrac{2}{\sqrt{3}} = \dfrac{2 \cdot \sqrt{3}}{\sqrt{3} \cdot \_\_\_\_} = \dfrac{2\sqrt{3}}{\_\_\_\_}$

$\sqrt{3}$, 6, 2

668 $\dfrac{3\sqrt{2}}{2\sqrt{3}} = \dfrac{3\sqrt{2} \cdot \sqrt{3}}{2\sqrt{3} \cdot \_\_\_\_} = \dfrac{3\sqrt{6}}{\_\_\_\_} = \dfrac{\sqrt{6}}{\_\_\_\_}$

$\sqrt{12}$, $6\sqrt{2}$, $\sqrt{2}$

669 $\dfrac{2\sqrt{6}}{\sqrt{12}} = \dfrac{2\sqrt{6}\cdot\sqrt{12}}{\sqrt{12}\cdot\_\_\_\_} = \dfrac{2\sqrt{72}}{12} = \dfrac{\sqrt{36\cdot 2}}{6} = \dfrac{\_\_\_\_}{6} = \_\_\_\_$

$\sqrt[3]{6^2}$, $3\sqrt[3]{36}$, $\dfrac{\sqrt[3]{36}}{2}$

670 $\dfrac{3}{\sqrt[3]{6}} = \dfrac{3}{\sqrt[3]{6}}\cdot\dfrac{\sqrt[3]{6^2}}{\_\_\_\_} = \dfrac{\_\_\_\_}{6} = \_\_\_\_$

$\sqrt[3]{5}$, 5

671 $\dfrac{1}{\sqrt[3]{25}} = \dfrac{1}{\sqrt[3]{5^2}}\cdot\dfrac{\sqrt[3]{5}}{\_\_\_\_} = \dfrac{\sqrt[3]{5}}{\_\_\_\_}$

$\sqrt{x}$, x

672 $\dfrac{5}{\sqrt{x}} = \dfrac{5(\_\_\_\_)}{\sqrt{x}\cdot\sqrt{x}} = \dfrac{5\sqrt{x}}{\_\_\_\_}$

$\dfrac{3-\sqrt{2}}{3-\sqrt{2}}$, 2, $\dfrac{3-\sqrt{2}}{7}$

673 $\dfrac{1}{3+\sqrt{2}} = \dfrac{1}{3+\sqrt{2}}\cdot\dfrac{\qquad}{\_\_\_\_} = \dfrac{3-\sqrt{2}}{9-\_\_\_\_} = \_\_\_\_$

$\sqrt{5}+\sqrt{3}$

674 $\dfrac{3}{\sqrt{5}-\sqrt{3}} = \dfrac{3(\_\_\_\_)}{(\sqrt{5}-\sqrt{3})(\sqrt{5}+\sqrt{3})}$

$\sqrt{3}$, $\dfrac{3(\sqrt{5}+\sqrt{3})}{2}$

$= \dfrac{3(\sqrt{5}+\sqrt{3})}{(\sqrt{5})^2-(\_\_\_\_)^2} = \_\_\_\_$

$2\sqrt{2}+\sqrt{3}$

675 $\dfrac{\sqrt{2}-3\sqrt{3}}{2\sqrt{2}-\sqrt{3}} = \dfrac{(\sqrt{2}-3\sqrt{3})(\_\_\_\_)}{(2\sqrt{2}-\sqrt{3})(2\sqrt{2}+\sqrt{3})}$

$2\sqrt{2}$, $-5-5\sqrt{6}$, $-1-\sqrt{6}$

$= \dfrac{4-5\sqrt{6}-9}{(\_\_\_\_)^2-(\sqrt{3})^2} = \dfrac{\_\_\_\_}{5} = \_\_\_\_$

$x-\sqrt{y}$, $\sqrt{y}$, $\dfrac{(x-\sqrt{y})^2}{x^2-y}$

676 $\dfrac{x-\sqrt{y}}{x+\sqrt{y}} = \dfrac{x-\sqrt{y}}{x+\sqrt{y}}\cdot\dfrac{\_\_\_\_}{x-\sqrt{y}} = \dfrac{(x-\sqrt{y})^2}{x^2-(\_\_\_\_)^2} = \_\_\_\_$

$\sqrt{x}+\sqrt{y}$, $\dfrac{x(\sqrt{x}+\sqrt{y})}{x-y}$

677 $\dfrac{x}{\sqrt{x}-\sqrt{y}} = \dfrac{x(\sqrt{x}+\sqrt{y})}{(\sqrt{x}-\sqrt{y})(\_\_\_\_)} = \_\_\_\_$

$\sqrt{x}-1$, $\sqrt{x}$, $\dfrac{\sqrt{x}}{x}$

678 $\dfrac{1-\dfrac{1}{\sqrt{x}}}{\sqrt{x}-1} = \dfrac{\dfrac{\sqrt{x}-1}{\sqrt{x}}}{\sqrt{x}-1} = \dfrac{\_\_\_\_}{\sqrt{x}}\cdot\dfrac{1}{\sqrt{x}-1} = \dfrac{1\cdot\sqrt{x}}{\sqrt{x}\cdot\_\_\_\_} = \_\_\_\_$

$(1+\sqrt{2})+\sqrt{3}$

679 $\dfrac{2}{1+\sqrt{2}-\sqrt{3}} = \dfrac{2}{(1+\sqrt{2})-\sqrt{3}}\cdot\dfrac{(1+\sqrt{2})+\sqrt{3}}{\_\_\_\_}$

$1+\sqrt{2}$

$= \dfrac{2[(1+\sqrt{2})+\sqrt{3}]}{(\_\_\_\_)^2-(\sqrt{3})^2} = \dfrac{2[(1+\sqrt{2})+\sqrt{3}]}{1+2\sqrt{2}+2-3}$

$\sqrt{2}$, $\sqrt{2}+2+\sqrt{6}$

$= \dfrac{2[(1+\sqrt{2})+\sqrt{3}]}{2\sqrt{2}}\cdot\dfrac{\sqrt{2}}{\_\_\_\_} = \dfrac{\_\_\_\_}{2}$

4.1 Equations Involving Radicals

original

extraneous

680 To solve equations involving radicals, assume that when each side of an equation is raised to the same power, the solution set of the resulting equation will contain all the solutions of the ______ equation.

original

681 The process of raising both sides of an equation to the same power may introduce an apparent solution that is invalid. This occurs because the result of raising both sides of an equation to a power is not necessarily equivalent to the ______ equation. Care must be taken, therefore, to check the validity of the solutions obtained. These invalid solutions, if any, are called ______ solutions.

In Problems 682–690, find the solution set of the given equation.

Answers	Problems
	682 $\sqrt{x} = 3$
9, 9	$(\sqrt{x})^2 =$ ____, so $x =$ ____.
3, {9}	*Check:* $\sqrt{9} \stackrel{?}{=}$ ____. Yes. Therefore, the solution set is ____.
	683 $\sqrt{7y} = 5$
25, 25, $\frac{25}{7}$	$(\sqrt{7y})^2 =$ ____, so $7y =$ ____ or $y =$ ____.
5, $\{\frac{25}{7}\}$	*Check:* $\sqrt{7(\frac{25}{7})} \stackrel{?}{=}$ ____. Yes. Therefore, the solution set is ____.
	684 $\sqrt{x - 3} = 2$
2	$(\sqrt{x - 3})^2 = ($____$)^2$
$x - 3$	____ $= 4$
7	$x =$ ____
2, {7}	*Check:* $\sqrt{7 - 3} \stackrel{?}{=}$ ____. Yes. Therefore, the solution set is ____.
	685 $\sqrt{x - 2} = 4$
16, 16, 18	$(\sqrt{x - 2})^2 =$ ____, so $x - 2 =$ ____ or $x =$ ____.
4, {18}	*Check:* $\sqrt{18 - 2} \stackrel{?}{=}$ ____. Yes. Thus, the solution set is ____.
	686 $\sqrt{2x + 1} - 4 = 2$
6	$\sqrt{2x + 1} =$ ____
6	$(\sqrt{2x + 1})^2 = ($____$)^2$
36	$2x + 1 =$ ____
35	$2x =$ ____
	$x = \frac{35}{2}$
	Check:
2, $\{\frac{35}{2}\}$	$\sqrt{2(\frac{35}{2}) + 1} - 4 \stackrel{?}{=}$ ____. Yes. Therefore, the solution set is ____.
	687 $\sqrt{x^2 - 5} = x + 1$
$x + 1$	$(\sqrt{x^2 - 5})^2 = ($_______$)^2$
$x^2 + 2x + 1$	$x^2 - 5 =$ ____________
6	$-2x =$ ____
-3	$x =$ ____
	Check:
	$\sqrt{(-3)^2 - 5} \stackrel{?}{=} -3 + 1$
4	$\sqrt{\_\_\_\_} \stackrel{?}{=} -2$
$\emptyset$	$2 \stackrel{?}{=} -2$. No. Therefore, the solution set is ____.
	688 $\sqrt{x^2 - 8} = 4 - x$
	$(\sqrt{x^2 - 8})^2 = (4 - x)^2$
$16 - 8x + x^2$	$x^2 - 8 =$ ____________
-24	$-8x =$ ____
3	$x =$ ____
3, {3}	*Check:* $\sqrt{9 - 8} \stackrel{?}{=} 4 -$ ____. Yes. Thus, the solution set is ____.

Answers	Frame
	689 $\sqrt{x+5} = 1 + \sqrt{x}$
2	$(\sqrt{x+5})^2 = (1 + \sqrt{x})$—
$2\sqrt{x}$	$x + 5 = 1 +$ ____ $+ x$
4	$2\sqrt{x} =$ ____
4	$x =$ ____
2, {4}	*Check:* $\sqrt{4+5} \stackrel{?}{=} 1 +$ ____. Yes. Thus, the solution set is ____.
	690 $\sqrt{x} = \sqrt{x+7} - 1$
$2\sqrt{x+7}$	$(\sqrt{x})^2 = x + 7 -$ ________ $+ 1$
$\sqrt{x+7}$	$-8 = -2($________$)$
$\sqrt{x+7}$	$4 =$ ________
9	$x =$ ____
$\sqrt{9+7}$, {9}	*Check:* $\sqrt{9} \stackrel{?}{=}$ ________ $- 1$. Yes. Thus, the solution set is ____.

5 Inequalities

Answers	Frame
	691 The set of real numbers R can be represented geometrically on a line by associating each real number with a point on that line.
number line	Such a representation is called a ____________.
	692 The distance between the two points labeled 0 and 1 is called the
unit length, scale unit	____________ or the ____________ of the number line.
	693 The point associated with a number on the number line is called
graph	the ______ of that number, and the number is called the
coordinate	__________ of that point.

In Problems 694–698, express the points on each number line as a set.

Answers	Frame
{-2, 2}	**694** Number line −5 −4 −3 −2 −1 0 1 2 3 4 5 ________
{-3, 1, 4}	**695** Number line −5 −4 −3 −2 −1 0 1 2 3 4 5 ________
{-3, -1, 2, 3}	**696** Number line −5 −4 −3 −2 −1 0 1 2 3 4 5 ________
{-4, 5}	**697** Number line −5 −4 −3 −2 −1 0 1 2 3 4 5 ________
∅	**698** Number line −5 −4 −3 −2 −1 0 1 2 3 4 5 ________

5.1 Positive and Negative Numbers

Answers	Frame
	699 Any number whose corresponding point on the number line lies to the right of the corresponding point of a second number is said to
greater than, >	be ____________ (denoted by ____) the second number.
less than	The second number is also said to be __________ (denoted by
<	____) the first number.

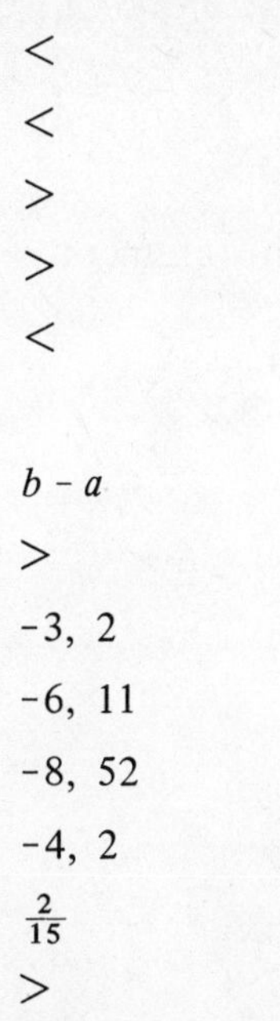

Answer	Frame
$<$	**700** -7 ___ 2
$<$	**701** -3 ___ -1
$>$	**702** 5 ___ 3
$>$	**703** 7 ___ 2
$<$	**704** If ——a——b——, then a ___ b.
$b - a$	**705** For any two real numbers a and b, a is less than b ($a < b$) or, equivalently, b is greater than a ($b > a$) if ___ is a positive number.
$>$	**706** $6 < 9$, since $9 - 6 = 3$ ___ 0.
$-3,\ 2$	**707** $-5 < -3$, since ___ $- (-5) =$ ___ > 0.
$-6,\ 11$	**708** $-6 < 5$, since $5 - ($___$) =$ ___ > 0.
$-8,\ 52$	**709** $44 > -8$, since $44 - ($___$) =$ ___ > 0.
$-4,\ 2$	**710** $-2 > -4$, since $-2 - ($___$) =$ ___ > 0.
$\frac{2}{15}$	**711** $\frac{2}{3} < \frac{4}{5}$, since $\frac{4}{5} - (\frac{2}{3}) =$ ___ > 0.
$>$	**712** If ——a——b——, then b ___ a.
$<$	**713** If ——a——0——b——, then $b > 0$ and a ___ 0.

5.2 Properties of Inequalities

Answer	Frame
$a = 0$	**714** The trichotomy principle states that if $a \in R$, then one and only one of the following can hold: $a < 0$, $a > 0$, or ______.
$a < c$	**715** The transitive property of inequalities states that if a, b, and $c \in R$ such that $a < b$ and $b < c$, then ______.
$2 < 5$	**716** If $2 < 3$ and $3 < 5$, then ______.
$-2 < 0$	**717** If $-2 < -1$ and $-1 < 0$, then ______.
$b + c$	**718** The addition property of inequalities states that if a, b, and $c \in R$ and $a < b$, then $a + c <$ ______.
$b - c$	**719** If $a, b, c \in R$ and $a < b$, then $a - c <$ ______.
$x + 1$	**720** If $x < y$, then ______ $< y + 1$.
2	**721** If $x + 1 < 2 + 1$, then $x <$ ___.
5	**722** If $x - 13 < 5 - 13$, then $x <$ ___.
$ac < bc$ $ac > bc$	**723** The multiplication property of inequalities states that for $a, b, c \in R$: (i) If $a < b$ and $c > 0$, then ______. (ii) If $a < b$ and $c < 0$, then ______.
$<$	**724** If $3 < 5$, then $2 \cdot 3$ ___ $2 \cdot 5$.
$>$	**725** If $3 < 7$, then $-2 \cdot 3$ ___ $-2 \cdot 7$.
$\leqslant$ $\leqslant$ $\geqslant$	**726** The symbol $\leqslant$ means less than or equal; that is, if a, b, and $c \in R$, then $a \leqslant b$ means either $a < b$ or $a = b$. Thus, (i) If $a \leqslant b$ and c is any number, then $a + c$ ___ $b + c$. (ii) If $a \leqslant b$ and $c > 0$, then ac ___ bc. (iii) If $a \leqslant b$ and $c < 0$, then ac ___ bc.
$\frac{y}{c}$	**727** If $x < y$ and $c > 0$, then $\frac{x}{c} <$ ___.

$\frac{y}{c}$	**728** If $x < y$ and $c < 0$, then $\frac{x}{c} >$ ____

In Problems 729–751, use the above properties of inequalities to complete each statement.

$b + 2$	**729** If $a < b$, then $a + 2 <$ ________.
$b - 11$	**730** If $a < b$, then $a - 11 <$ ________.
3	**731** If $a + 5 < 3 + 5$, then $a <$ ____.
6	**732** If $a - 7 < 6 - 7$, then $a <$ ____.
7	**733** If $x < 2$ and $y < 5$, then $x + y <$ ____.
11	**734** If $x > 3$ and $y > 8$, then $x + y >$ ____.
$7y$	**735** If $x < y$, then $7x <$ ____.
y	**736** If $9x < 9y$, then $x <$ ____.
$-5y$	**737** If $x < y$, then $-5x >$ ____.
y	**738** If $-3x < -3y$, then $x >$ ____.
$\frac{y}{3}$	**739** If $x < y$, then $\frac{x}{3} <$ ____.
$x < y$	**740** If $\frac{x}{7} < \frac{y}{7}$, then ________.
$\frac{y}{-2}$	**741** If $x < y$, then $\frac{x}{-2} >$ ____.
y	**742** If $\frac{x}{-7} < \frac{y}{-7}$, then $x >$ ____.
10	**743** If $x < 5$, then $2x <$ ____.
$\frac{1}{x}$	**744** If $3x > 0$ and $x > 3$, then $\frac{1}{3} >$ ____.
$\frac{1}{x}$	**745** If $7x > 0$ and $x < 7$, then $\frac{1}{7} <$ ____.
4	**746** If $x > 2$, then $x^2 >$ ____.
3	**747** If $x^2 > 9$, then $x >$ ____.
$\frac{1}{y}$	**748** If $x > 0$ and $y > 0$ and $x < y$, then $\frac{1}{x} >$ ____.
y	**749** If $x > 0$ and $y > 0$ and $x^2 < y^2$, then $x <$ ____.
0	**750** If $x < 0$, then $x^2 >$ ____.
x	**751** If $0 < x < 1$, then $x^2 <$ ____.

5.3 Interval Notation

interval	**752** An ____________ is defined to be a set of real numbers such that, whenever a number x lies between two numbers that belong to the set, then x also ____________ to the set.
belongs	
$a < x < b$	**753** The open interval from a to b, denoted by (a, b), is defined to be the set of all real numbers x such that ______________.

Answer	No.	Frame
$a \leqslant x \leqslant b$	754	The closed interval from a to b, denoted by $[a, b]$, is defined to be the set of all real numbers x such that ________.
$a \leqslant x < b$	755	The half-open interval on the right from a to b, denoted by $[a, b)$, is defined to be the set of all real numbers x such that ________.
$a < x \leqslant b$	756	The half-open interval on the left from a to b, denoted by $(a, b]$, is defined to be the set of all real numbers x such that ________.
$x > a$	757	The open interval from a to ∞, denoted by (a, ∞), is defined to be the set of all real numbers x such that ________.
$x < a$	758	The open interval from $-\infty$ to a, denoted by $(-\infty, a)$, is defined to be the set of all real numbers x such that ________.
$x \geqslant a$	759	The closed interval from a to ∞, denoted by $[a, \infty)$, is defined to be the set of all real numbers x such that ________.
$x \leqslant a$	760	The closed interval from $-\infty$ to a, denoted by $(-\infty, a]$, is defined to be the set of all real numbers x such that ________.
real numbers	761	The interval notation $(-\infty, \infty)$ will be used to denote the set R of all ________.
$[1, 2]$	762	The interval notation of the set $\{x \mid 1 \leqslant x \leqslant 2\}$ is ________.
$(3, 5]$	763	The interval notation of $\{x \mid 3 < x \leqslant 5\}$ is ________.
$(-\infty, 3) \cup (5, \infty)$	764	The interval notation of $\{x \mid x < 3 \text{ or } x > 5\}$ is ________.
$(-\infty, -7] \cup [2, \infty)$	765	The interval notation of $\{x \mid x \geqslant 2 \text{ or } x \leqslant -7\}$ is ________.
$\{x \mid 3 < x < 8\}$	766	The set notation of $(3, 8)$ is ________.
$\{x \mid 1 \leqslant x \leqslant 4\}$	767	The set notation of $[1, 4]$ is ________.
$\{x \mid x < 1 \text{ or } x \geqslant 3\}$	768	The set notation of $(-\infty, 1) \cup [3, \infty)$ is ________.

5.4 Linear Inequalities in One Variable

Answer	No.	Frame
solution set	769	The set of all numbers that make a first-degree inequality a true statement is called the ________ of the inequality.
equivalent	770	Inequalities that have the same solution set are called ________ inequalities.
conditional	771	Inequalities that are true only for certain values of the variables involved are called ________ inequalities.
R	772	The solution set of the inequality $x + 3 < x + 5$ is ________.

In Problems 773–780, solve each of the given first-degree inequalities and represent its solution set on the real line.

773 $3x - 5 < 7$

$3x < 7 +$ ____

$3x <$ ____

$x <$ ____

The solution set is __________.

The graph is __________.

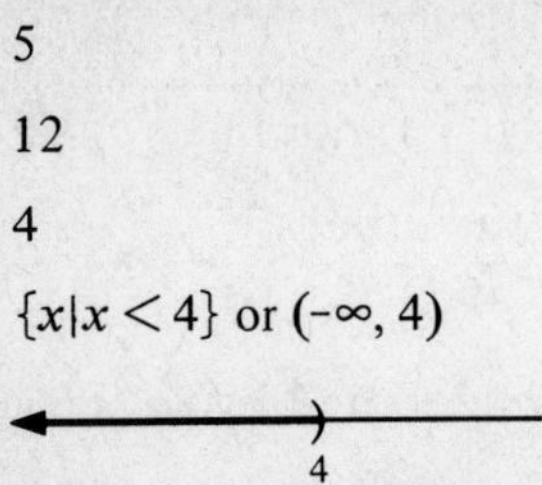

5

12

4

$\{x|x < 4\}$ or $(-\infty, 4)$

774 $3x - 2 < -7x + 1$

$3x < -7x + 1 +$ ____

____ < 3

$x <$ ____

The solution set is __________.

The graph is __________.

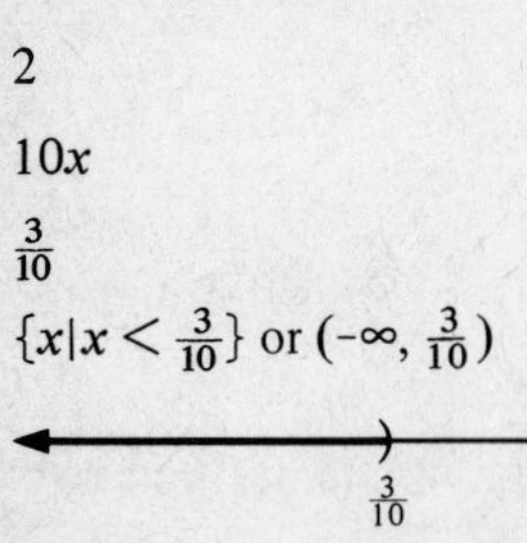

2

$10x$

$\frac{3}{10}$

$\{x|x < \frac{3}{10}\}$ or $(-\infty, \frac{3}{10})$

775 $\frac{1}{3}x - \frac{2}{5} < \frac{3}{2}x + 1$

$30(\frac{1}{3}x - \frac{2}{5}) <$ ____ $(\frac{3}{2}x + 1)$

$10x - 12 < 45x +$ ____

$10x -$ ____ $< 30 + 12$

$-35x <$ ____

x __ $-\frac{6}{5}$

The solution set is __________.

The graph is __________.

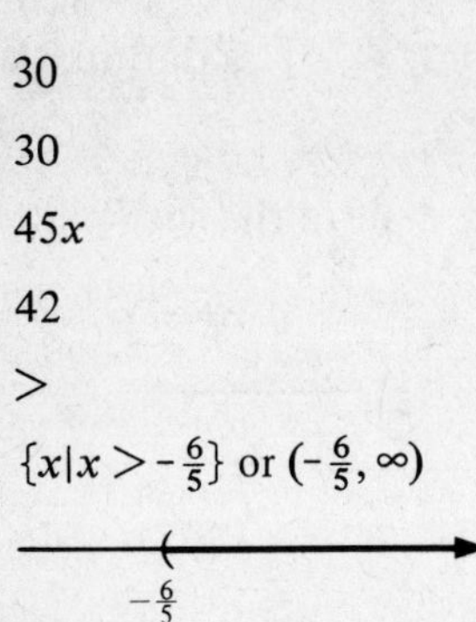

30

30

$45x$

42

$>$

$\{x|x > -\frac{6}{5}\}$ or $(-\frac{6}{5}, \infty)$

776 $x + 6 \geqslant 4 - 3x$

$x +$ ____ $\geqslant 4 - 6$

$4x \geqslant$ ____

$x \geqslant$ ____

The solution set is __________.

The graph is __________.

$3x$

-2

$-\frac{1}{2}$

$\{x|x \geqslant -\frac{1}{2}\}$ or $[-\frac{1}{2}, \infty)$

$-\frac{1}{2}$

777 $-3 < 1 - 2x < 4$

$-1 - 3 < -1 + 1 - 2x <$ ______

$-4 <$ ____ < 3

$2 > x$ ____ $-\frac{3}{2}$

The solution set is __________.

The graph is __________.

$-1 + 4$

$-2x$

$>$

$\{x|-\frac{3}{2} < x < 2\}$ or $(-\frac{3}{2}, 2)$

$-\frac{3}{2}$ 2

778 $\frac{2}{3}(2x - 1) - \frac{1}{2}(3x + 2) \leqslant 3$

$6[\frac{2}{3}(2x - 1) - \frac{1}{2}(3x + 2)] \leqslant$ ____ (3)

$4(2x - 1) - 3(3x + 2) \leqslant$ ____

$8x - 4 - 9x -$ ____ $\leqslant 18$

____ $- 10 \leqslant 18$

$-x \leqslant$ ____

x __ -28

6

18

6

$-x$

28

$\geqslant$

$\{x \mid x \geqslant -28\}$ or $[-28, \infty)$

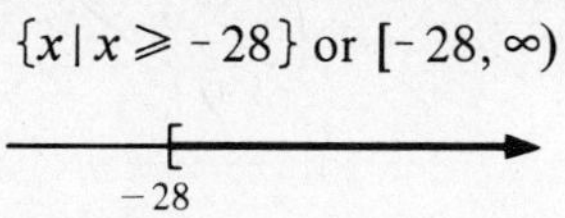

The solution set is ______.

The graph is ______.

779 $-0.2 \leqslant \dfrac{2x+3}{5} \leqslant 0.2$

$5(-0.2) \leqslant 5\left(\dfrac{2x+3}{5}\right) \leqslant$ ___ (0.2)

$-1 \leqslant$ ______ $\leqslant 1$

$-1 - 3 \leqslant 2x \leqslant$ ______

$-4 \leqslant 2x \leqslant$ ___

$-2 \leqslant x \leqslant$ ___

The solution set is ______.

The graph is ______.

5

$2x + 3$

$1 - 3$

-2

-1

$\{x \mid -2 \leqslant x \leqslant -1\}$ or $[-2, -1]$

780 $\dfrac{3x-7}{5} \geqslant 2x + 1$

$5\left(\dfrac{3x-7}{5}\right) \geqslant$ ___ $(2x+1)$

______ $\geqslant 10x + 5$

$3x - 10x \geqslant 5 +$ ___

$-7x \geqslant$ ___

x ___ $-\frac{12}{7}$

The solution set is ______.

The graph is ______.

5

$3x - 7$

7

12

$\leqslant$

$\{x \mid x \leqslant -\frac{12}{7}\}$ or $(-\infty, -\frac{12}{7}]$

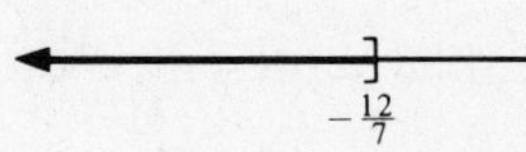

6 Absolute-Value Equations and Inequalities

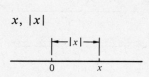

781 The absolute value of x, denoted by $|x|$, is used to represent the distance between ___ and 0 on the real line. Thus, $|-x| =$ ___.

The graph for $x > 0$ is ______.

x, $|x|$

782 Let x be a real number; then we define

$$|x| = \begin{cases} \_\_\_ & \text{if } x \geqslant 0 \\ \_\_\_ & \text{if } x < 0 \end{cases}$$

x

$-x$

783 $|7| =$ ___

7

784 $|-9| = -($___$) =$ ___

-9, 9

785 $|0| =$ ___

0

6.1 Absolute-Value Equations

786 If a and b are real numbers such that $a \leqslant b$, then the distance d between a and b is the nonnegative number ______, so that $d =$ ______, for $a \leqslant b$.

$b - a$

$b - a$

$|a - b|$ or $|b - a|$

787 Using absolute values, we can express the distance d between two real numbers a and b as $d =$ ______.

In Problems 788–790, use the formula $d = |a - b|$ to find the distance between the given numbers.

11
-6
6

788 5 and 11
$d = |5 - ___|$
$= |___|$
$= ___$

14
11
11

789 14 and 3
$d = |___ - 3|$
$= |___|$
$= ___$

8
-15
15

790 -7 and 8
$d = |-7 - ___|$
$= |___|$
$= ___$

$x - a$
$-(x - a)$

791 From the definition of the absolute value of a real number, we have

$$|x - a| = \begin{cases} ______, & \text{if } x \geqslant a \\ ______, & \text{if } x < a \end{cases}$$

In Problems 792–799, find the solution set of the given equation and then represent it on the real line.

-2
{-2, 2}

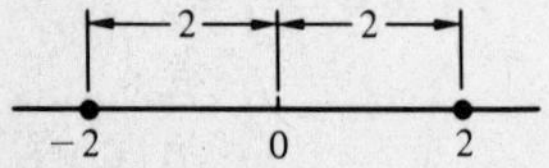

792 $|x| = 2$
$x = 2$ or $x = ___$
The solution set is ______.

The graph is ______.

7
{-7, 7}

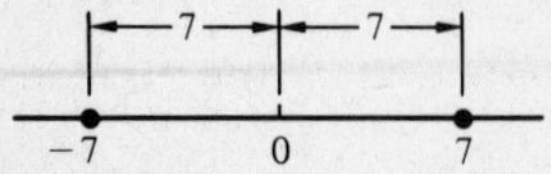

793 $|x| = 7$
$x = -7$ or $x = ___$
The solution set is ______.

The graph is ______.

15
5
{-5, 5}

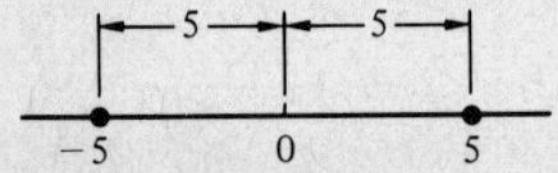

794 $|3x| = 15$
$3x = -15$ or $3x = ___$
$x = -5$ $\quad$ $x = ___$
The solution set is ______.

The graph is ______.

795 $|x| = -3$

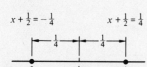

There is no value of x to satisfy this equation, since the absolute value is always ______________. The solution set is ____.

nonnegative, $\emptyset$

796 $|x - 2| = 3$

$x - 2 = 3$ or $x - 2 =$ ____

Solving each equation, $x = 5$ or $x =$ ____. The solution set is __________.

The graph is ____.

-3

-1

$\{-1, 5\}$

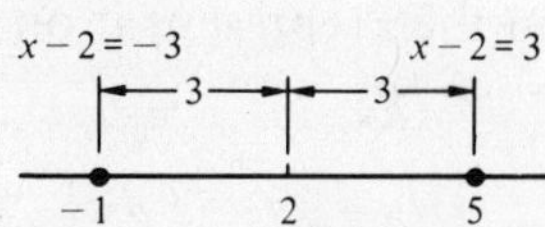

797 $|x + \frac{1}{2}| = \frac{1}{4}$

$x + \frac{1}{2} = \frac{1}{4}$ or $x + \frac{1}{2} =$ ____

Solving each equation, $x = -\frac{1}{4}$ or $x =$ _____. The solution set is ____________.

The graph is ____.

$-\frac{1}{4}$

$-\frac{3}{4}$

$\{-\frac{3}{4}, -\frac{1}{4}\}$.

$x + \frac{1}{2} = -\frac{1}{4}$ $x + \frac{1}{2} = \frac{1}{4}$

$\frac{1}{4}$ $\frac{1}{4}$

$-\frac{3}{4}$ $-\frac{1}{2}$ $-\frac{1}{4}$

798 $|2x - 5| = 9$

$2x - 5 = 9$ or $2x - 5 =$ ____

Solving each equation for x, we have

$2x = 14$ or $2x =$ ____

$x =$ __ $x = -2$

The solution set is __________.

The graph is ____.

-9

-4

7

$\{-2, 7\}$

$2x - 5 = -9$ $2x - 5 = 9$

$\frac{9}{2}$ $\frac{9}{2}$

-2 $\frac{5}{2}$ 7

799 $|2 - 3x| = 6$

$2 - 3x = 6$ or $2 - 3x =$ ____

Solving each equation for x, we have

$-3x = 4$ or $-3x =$ ____

$x = -\frac{4}{3}$ $x =$ ____

The solution set is __________.

The graph is ____.

-6

-8

$\frac{8}{3}$

$\{-\frac{4}{3}, \frac{8}{3}\}$

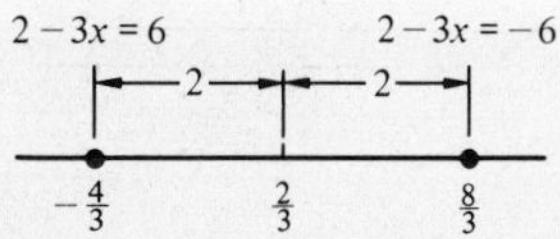

6.2 Absolute-Value Inequalities

800 From the definition of the absolute value of a real number, if $a > 0$, the following absolute-value inequalities will hold:

a
$-a$
a
$p-a$
$-a$
$p+a,\ p-a$

(i) If $|x| < a$, then $-a < x <$ ____.
(ii) If $|x| > a$, then $x > a$ or $x <$ ____.
(iii) If $|x-p| < a$, then $-a < x-p <$ ____, so that
____ $< x < p+a$
(iv) If $|x-p| > a$, then $x-p > a$ or $x-p <$ ____, so that
$x >$ ____ or $x <$ ____

In Problems 801–811, find the solution set of the given inequality and then represent it on the real line.

-2
$-2 < x < 2$

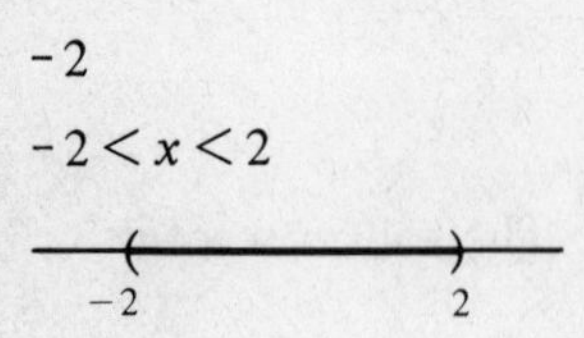

801 $|x| < 2$
____ $< x < 2$
The solution set is $\{x \mid$ ____ $\}$.
The graph is ____.

-4
$\{x \mid x > 4 \text{ or } x < -4\}$

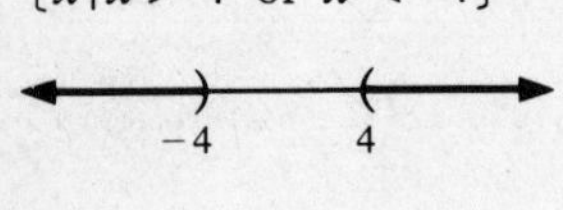

802 $|x| > 4$
$x > 4$ or $x <$ ____
The solution set is ____.
The graph is ____.

-5
$\{x \mid -5 \leqslant x \leqslant 5\}$

-5 5

803 $|x| \leqslant 5$
____ $\leqslant x \leqslant 5$
The solution set is ____.
The graph is ____.

$-7,\ 7$
$\{x \mid x \leqslant -7 \text{ or } x \geqslant 7\}$

-7 7

804 $|x| \geqslant 7$
$x \leqslant$ ____ or $x \geqslant$ ____
The solution set is ____.
The graph is ____.

$|3x|$
6
2
$\{x \mid -2 < x < 2\}$

-2 2

805 $|-3x| < 6$
$|-3x| =$ ____
$-6 < 3x <$ ____
$-2 < x <$ ____
The solution set is ____.
The graph is ____.

-6
-4
$\{x \mid -4 < x < 8\}$

-4 8

806 $|x-2| < 6$
____ $< x-2 < 6$
____ $< x < 8$
The solution set is ____.
The graph is ____.

-2

$-5, -1$

$\{x \mid x < -5 \text{ or } x > -1\}$

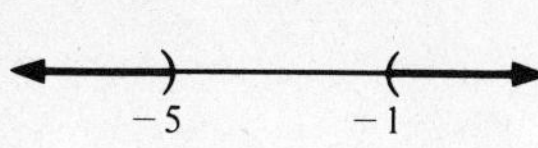

807 $|x + 3| > 2$

$x + 3 <$ ____ or $x + 3 > 2$

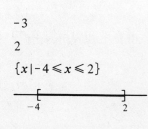

$x <$ ____ $\quad x >$ ____

The solution set is ________________.

The graph is ________________.

-3

2

$\{x \mid -4 \leqslant x \leqslant 2\}$

808 $|x + 1| \leqslant 3$

____ $\leqslant x + 1 \leqslant 3$

$-4 \leqslant x \leqslant$ ____

The solution set is ________________.

The graph is ________________.

-5

12

1

$\{x \mid x \geqslant 6 \text{ or } x \leqslant 1\}$

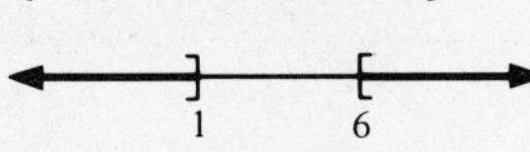

809 $|2x - 7| \geqslant 5$

$2x - 7 \geqslant 5$ or $2x - 7 \leqslant$ ____

Solving for x, we have

$2x \geqslant$ ____ or $2x \leqslant 2$

$x \geqslant 6 \qquad x \leqslant$ ____

The solution set is ________________.

The graph is ________________.

-12

-16

$-\frac{8}{3}$

$\{x \mid -\frac{8}{3} \leqslant x \leqslant \frac{16}{3}\}$

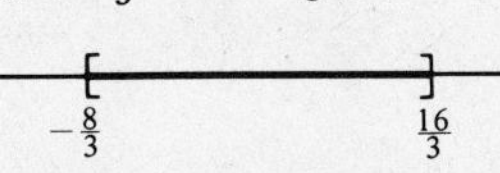

810 $|4 - 3x| \leqslant 12$

____ $\leqslant 4 - 3x \leqslant 12$

____ $\leqslant -3x \leqslant 8$

____ $\leqslant x \leqslant \frac{16}{3}$

The solution set is ________________.

The graph is ________________.

$3 - 2x$

4

$\{x \mid x \geqslant 4 \text{ or } x \leqslant -1\}$

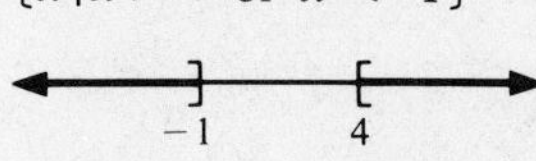

811 $|3 - 2x| \geqslant 5$

$3 - 2x \geqslant 5$ or ________ $\leqslant -5$

Solving for x, we have

$x \leqslant -1$ or $x \geqslant$ ____

The solution set is ________________.

The graph is ________________.

Chapter Test

1 Let $A = \{1, 2, 3, 4, 5\}$, $B = \{1, 2, 3, 6, 7\}$, and $C = \{3, 6, 7, 8\}$. Find

(a) $A \cup B$ (b) $A \cap B$ (c) $A \cup C$

(d) $A \cap C$ (e) $B \cap C$ (f) $B \cup C$

(g) $A \cup (B \cup C)$ (h) $A \cap (B \cap C)$ (i) $A \cap (B \cup C)$

(j) $A \cup (B \cap C)$

2 If I_p is the set of positive integers, I is the set of integers, and Q is the set of rational numbers, indicate which of the following are true and which are false.

(a) $I_p \subset Q$ ________ (b) $I_p \subset (I \cup Q)$ ________ (c) $I_p \subseteq I$ ________

(d) $I \cap Q = I_p$ ________ (e) $I_p \cap Q = Q \cap I$ ________ (f) $I_p \cup Q = Q \cup I$ ________

(g) $I_p = Q \cap I_p \cap I$ ________

3 (a) Draw a real line and locate each of the following points:

$\frac{1}{2}, \frac{-3}{2}, \frac{7}{2}, \frac{-7}{-4}, \frac{-17}{8}, \frac{19}{3}, \frac{-24}{8}$

(b) Express each of the numbers in part (a) in decimal form.

4 Justify the following statements by giving the appropriate property. Assume that all letters represent real numbers.

(a) $a + (x + 3) = (a + x) + 3$

(b) $ay = ya$

(c) $x(y + z) = xy + xz$

(d) $3 + (-3) = 0$

(e) If $x + z = y + z$, then $x = y$

5 Perform the indicated operation.

(a) $(x^2 - 3x + 6) + (2x^2 + 5x - 12)$

(b) $(3x^3 + 4x^2 - 5x + 7) + (-2x^3 + x^2 - 3)$

(c) $(x^4 - 3x^3 + 4x^2 - x + 2) - (2x^3 - x - 3)$

(d) $(2x^5 + 3x^3 - 7x - 1) - (-x^4 + x^2 + 3) + (x^5 + 2x^4 - 3x + 5)$

6 Determine the product.

(a) $(2x + 3)(x - 1)$ (b) $(4x + 3)(4x - 3)$ (c) $(x - 1)(x^2 + x + 1)$

(d) $(2x + 1)(4x^2 - 2x + 1)$ (e) $(x - y)^3$ (f) $(3x - y + 1)(3x + y - 1)$

7 Factor each expression completely.

(a) $xy^2 - 2x^2y - x^3y^3$ (b) $4x^2 - 9y^2$ (c) $x^2 + 5x + 6$

(d) $2x^2 + x - 1$ (e) $8x^3 + 1$ (f) $x^3 + 3x^2y + 3xy^2 + y^3$

(g) $x^2 + 4x + 4 - y^2$ (h) $mx^2 - my^2 + mx - my$

8 Perform the indicated operation and simplify the result.

(a) $\frac{24}{3x - 6} \cdot \frac{x^2 - 4}{x + 2}$ (b) $\frac{2x - 3}{x^2 - 1} \cdot \frac{2x^2 + x - 3}{4x^2 - 9}$

(c) $\frac{a^2bc}{abc^2} \div \frac{ab^2c}{ac}$ (d) $\frac{9x^2 - 4}{x^2 + 4x - 12} \div \frac{12x + 8}{x^2 + x - 30}$

(e) $\frac{xy}{1} \cdot \frac{y^2 - 4xy}{y - x} \div \frac{16x^2y^2 - y^4}{4x^2 - 3xy - y^2}$

9 Perform the indicated operation and simplify the result.

(a) $\frac{3}{x - 1} - \frac{3}{x} - \frac{2}{x^2} - \frac{1}{x^3}$ (b) $\frac{2x}{x^2 - 4xy + 4y^2} - \frac{y}{x^2 - 4y^2}$

10 Simplify each expression; express the answers in a form containing positive integer exponents.

(a) $3^{-4} \cdot 3^{-2}$ (b) $(5^{-2})^3$ (c) $(3x^{-1})^{-2}$

(d) $(\frac{3}{7})^{-5}$ (e) $\frac{5^{-7}}{5^{-13}}$ (f) $\frac{3ab^{-2}}{c^3d^{-4}}$

11 Use the properties of rational exponents to simplify each expression. Assume that all bases are positive.

(a) $5^{-1/2} \cdot 5^{5/2}$ (b) $(3^{-1/3})^{-15}$ (c) $(8x^3)^{1/3}$

(d) $(\frac{9}{4})^{-3/2}$ (e) $\frac{64^{2/3}}{64^{-1/2}}$ (f) $\left(\frac{8x^3}{27y^6}\right)^{-1/3}$

12 Use the properties of radicals to perform each operation and simplify.

(a) $\sqrt{3} \div \sqrt{27}$ (b) $\sqrt[3]{\dfrac{x^6}{8}}$ (c) $\sqrt[5]{x^{10}}$

(d) $\sqrt[3]{\sqrt{64}}$ (e) $\sqrt[3]{-64x^8y^{10}}$ (f) $\sqrt[4]{\dfrac{5y^5}{16x^8}}$

13 Rationalize each denominator.

(a) $\dfrac{5}{\sqrt{7}}$ (b) $\dfrac{3}{\sqrt[3]{16}}$ (c) $\dfrac{x}{\sqrt{x+y}}$

(d) $\dfrac{\sqrt{3}-\sqrt{2}}{\sqrt{3}+\sqrt{2}}$ (e) $\dfrac{3\sqrt{5}-7\sqrt{2}}{6\sqrt{5}-3\sqrt{2}}$

14 Find the solution set of each equation or inequality.

(a) $3x - 1 = 2x + 1$ (b) $\dfrac{1}{2x} - \dfrac{3}{x} = \dfrac{-25}{22}$ (c) $2x - 1 < 3$

(d) $x + 1 \geqslant 3 - 2x$ (e) $|2x - 1| = 5$ (f) $|3 - 4x| = 7$

(g) $|1 - 2x| \leqslant 2$ (h) $|2x - 3| \geqslant 4$ (i) $\sqrt{x+2} = 3$

15 In a football game, one team scored 8 points more than the other. Together, they scored a total of 34 points. How many points did each team score?

16 How many pounds of peanuts at 21 cents per pound should be mixed with 20 pounds of walnuts at 30 cents per pound to give a mixture worth 26 cents per pound?

Answers

1 (a) {1, 2, 3, 4, 5, 6, 7} (b) {1, 2, 3} (c) {1, 2, 3, 4, 5, 6, 7, 8}

(d) {3} (e) {3, 6, 7} (f) {1, 2, 3, 6, 7, 8}

(g) {1, 2, 3, 4, 5, 6, 7, 8} (h) {3} (i) {1, 2, 3}

(j) {1, 2, 3, 4, 5, 6, 7}

2 (a) True (b) True (c) True (d) False (e) False (f) True (g) True

3 (a)

$-\frac{17}{8}$

$-\frac{24}{8}$ $-\frac{3}{2}$ 0 $\frac{1}{2}$ $\frac{-7}{-4}$ $\frac{7}{2}$ $\frac{19}{3}$

(b) 0.5, −1.5, 3.5, 1.75, −2.125, $6.\overline{3}$, −3

4 (a) Associative property for addition (b) Commutative property for multiplication

(c) Distributive property (d) Additive inverse

(e) Cancellation property for addition

5 (a) $3x^2 + 2x - 6$ (b) $x^3 + 5x^2 - 5x + 4$

(c) $x^4 - 5x^3 + 4x^2 + 5$ (d) $3x^5 + 3x^4 + 3x^3 - x^2 - 10x + 1$

6 (a) $2x^2 + x - 3$ (b) $16x^2 - 9$ (c) $x^3 - 1$

(d) $8x^3 + 1$ (e) $x^3 - 3x^2y + 3xy^2 - y^3$ (f) $9x^2 - y^2 + 2y - 1$

7 (a) $xy(y - 2x - x^2y^2)$ (b) $(2x - 3y)(2x + 3y)$ (c) $(x + 2)(x + 3)$

(d) $(2x - 1)(x + 1)$ (e) $(2x + 1)(4x^2 - 2x + 1)$ (f) $(x + y)^3$

(g) $(x + y + 2)(x - y + 2)$ (h) $m(x - y)(x + y + 1)$

8 (a) 8 (b) $\dfrac{1}{x+1}$ (c) $\dfrac{a}{b^2c}$ (d) $\dfrac{(3x-2)(x-5)}{4(x-2)}$ (e) x

9 (a) $\dfrac{x^2 + x + 1}{x^3(x-1)}$ (b) $\dfrac{2x^2 + 3xy + 2y^2}{(x-2y)^2(x+2y)}$

10 (a) $\frac{1}{3^6}$ (b) $\frac{1}{5^6}$ (c) $\frac{x^2}{9}$ (d) $\frac{7^5}{3^5}$ (e) 5^6 (f) $\frac{3ad^4}{c^3b^2}$

11 (a) 25 (b) 3^5 (c) $2x$ (d) $\frac{8}{27}$ (e) 2^7 (f) $\frac{3y^2}{2x}$

12 (a) $\frac{1}{3}$ (b) $\frac{x^2}{2}$ (c) x^2 (d) 2 (e) $-4x^2y^3\sqrt[3]{x^2y}$
(f) $\frac{y\sqrt[4]{5y}}{2x^2}$

13 (a) $\frac{5\sqrt{7}}{7}$ (b) $\frac{3\sqrt[3]{4}}{4}$ (c) $\frac{x\sqrt{x}-xy}{x-y^2}$ (d) $5-2\sqrt{6}$ (e) $\frac{16-11\sqrt{10}}{54}$

14 (a) $\{2\}$ (b) $\{\frac{11}{5}\}$ (c) $\{x|x<2\}$ (d) $\{x|x\geqslant\frac{2}{3}\}$ (e) $\{-2, 3\}$ (f) $\{-1, \frac{5}{2}\}$
(g) $\{x|-\frac{1}{2}\leqslant x\leqslant\frac{3}{2}\}$ (h) $\{x|x\leqslant-\frac{1}{2}$ or $x\geqslant\frac{7}{2}\}$ (i) $\{7\}$

15 13 and 21

16 16 pounds

Chapter 2 FUNCTIONS AND GRAPHS

One of the most important concepts in mathematics is that of a function. Hence this chapter is devoted to the study of functions, with particular attention to special functions. The objectives of the chapter are that the student will be able to:

1 Locate points in the cartesian plane and find the distance between two points in the plane.
2 Define a relation and graph relations.
3 Define a function and determine its values.
4 Investigate the properties of functions and sketch their graphs.
5 Find the sum, difference, product, quotient, and composition of functions.
6 Find the inverse of a function.

1 Cartesian Coordinate System and Distance Formula

a b	1 The ordered pair (a, b) consists of the first element ____ and the second element ____.
4, 8	2 $(8, x) = (y, 4)$ if and only if $x =$ ____ and $y =$ ____.
3, 5	3 $(a, 5) = (3, c)$ if and only if $a =$ ____ and $c =$ ____.
4, $2y - 1$ 5, 1	4 $(x - 1, y) = (4, 2y - 1)$ if and only if $x - 1 =$ ____ and $y =$ ______ or, equivalently, $x =$ ____ and $y =$ ____.
8, 6	5 $(x - 3, y) = (5, 6)$ if and only if $x =$ ____ and $y =$ ____.

1.1 Cartesian Coordinate System

plane cartesian coordinate system	6 We can represent the set of all ordered pairs of real numbers as the set of points in a ________, using a two-dimensional indexing system called the ______________________________.
origin	7 The cartesian coordinate system is constructed as follows: First, two perpendicular lines L_1 and L_2 are constructed. The point of intersection of the two lines is called the ________ (Figure 1).
horizontal, vertical	8 The coordinate axes, L_1 and L_2, are often referred to as the ______________ axis, or x axis, and the ______________ axis, or y axis, respectively (Figure 1).

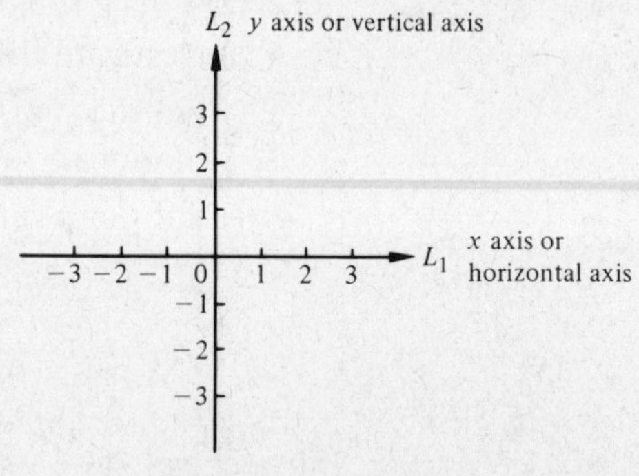

Figure 1

abscissa ordinate, coordinates	9 The first member, x, of the ordered pair (x, y) is called the ______________; the second member, y, of the pair is called the ______________; x and y are called the ______________.
graph	10 The set of all points in the plane whose coordinates correspond to the ordered pairs of a given set is called the ________ of the set of the ordered pairs.
graphing	11 Locating points in the plane by using the cartesian coordinate system is called ____________ the set of ordered pairs.

12 The coordinate axes divide the plane into four disjoint regions called ____________. The first quadrant includes all points (x, y) such that x is ____________ and, simultaneously, y is ____________; the second quadrant includes all points (x, y) such that x is ____________ and, simultaneously, y is ____________; the third quadrant includes all points (x, y) such that x is ____________ and, simultaneously, y is ____________; the fourth quadrant includes all points (x, y) such that x is ____________ and, simultaneously, y is ____________.

quadrants
positive
positive
negative, positive
negative, negative
positive, negative

In Problems 13–16, graph each given set of ordered pairs.

13 $\{(0, 2), (-1, -3), (-2, 1), (3, -2)\}$

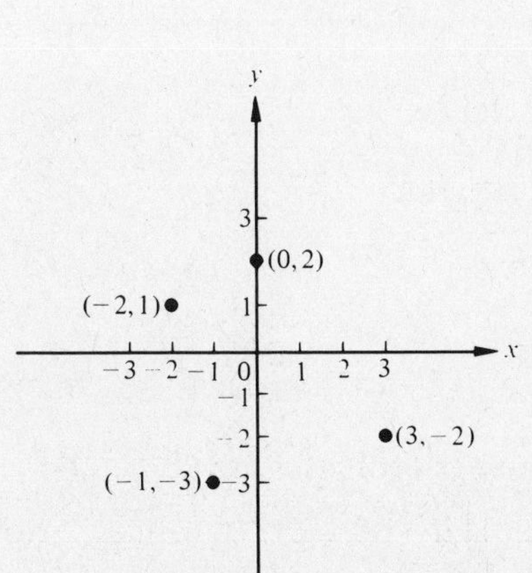

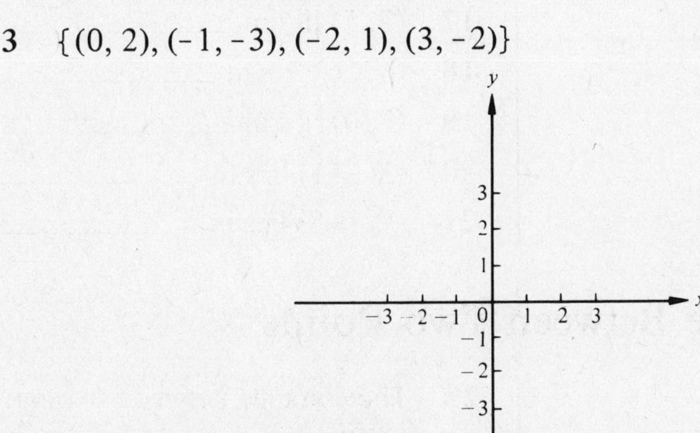

14 $\{(-2, 1), (-1, 2), (3, 3), (0, 4)\}$

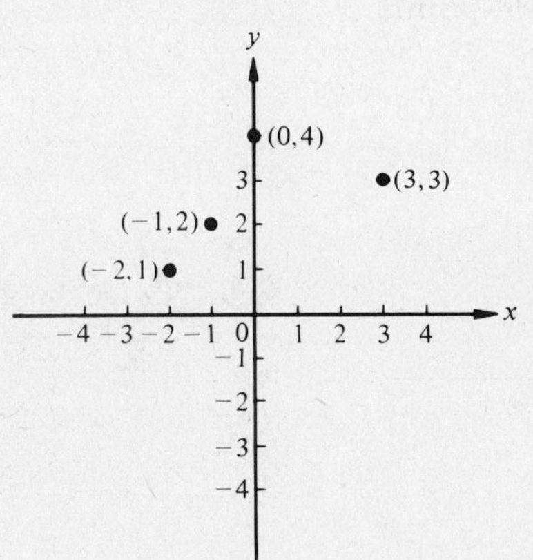

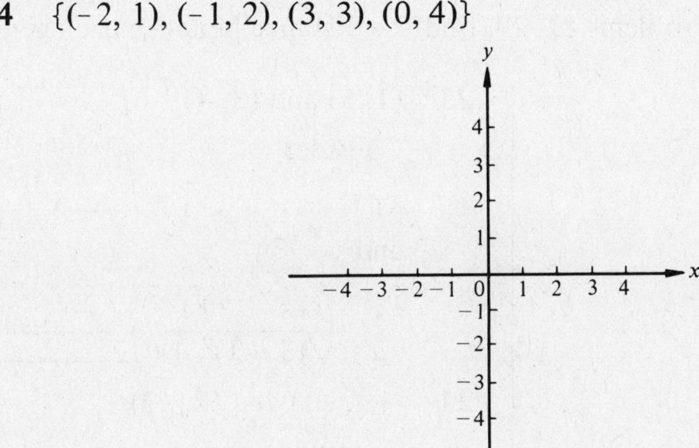

15 $\{(1, -2), (1, 0), (1, 1), (2, -2), (2, 0), (2, 1), (3, -2), (3, 0), (3, 1)\}$

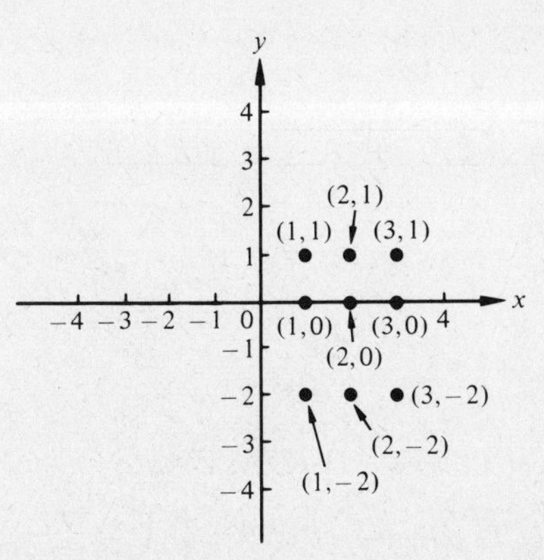

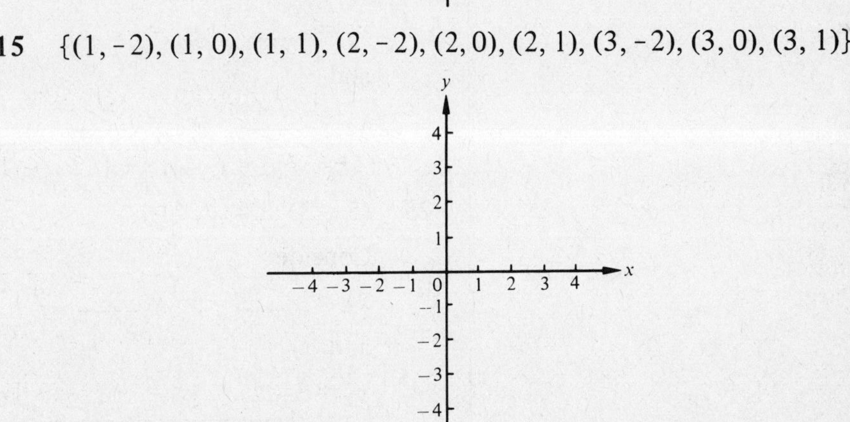

16 {(2, -3), (2, 1), (4, -3), (4, 1)}

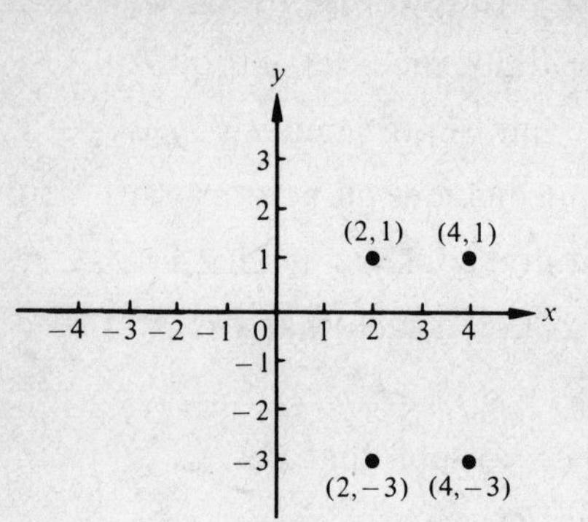

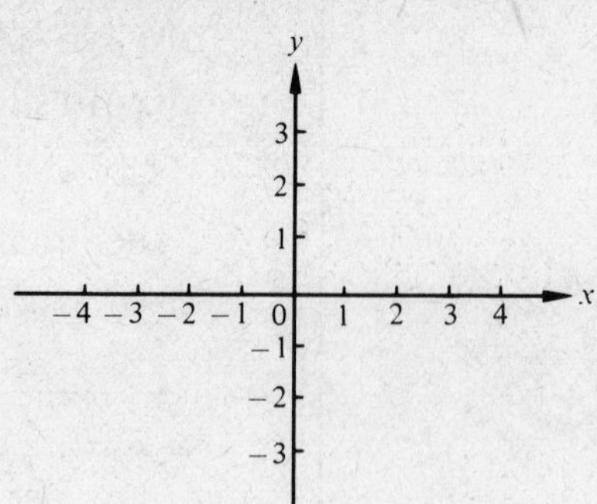

In Problems 17–21, indicate in which quadrant, if any, the point is found.

Answer	Problem
quadrant I	**17** (2, 5) lies in ________________.
quadrant II	**18** (-3, 6) lies in ________________.
no quadrant	**19** (7, 0) lies in ________________.
quadrant IV	**20** (3, -11) lies in ________________.
quadrant III	**21** (-7, -8) lies in ________________.

1.2 Distance Between Two Points

Answer	Problem
$\sqrt{(x_2 - x_1)^2 + (y_2 - y_1)^2}$	**22** The distance between two points whose coordinates are (x_1, y_1) and (x_2, y_2) is expressed by the formula $d =$ ________________.

In Problems 23–27, find the distance between each given pair of points.

Answer	Problem
	23 (1, 5) and (3, 4)
	Consider
1, 5	$x_1 =$ ____ $y_1 =$ ____
	and
3, 4	$x_2 =$ ____ $y_2 =$ ____
4 - 5, 1, $\sqrt{5}$	$d = \sqrt{(3-1)^2 + (\_\_\_\_\_\_)^2} = \sqrt{4 + \_\_\_\_} =$ ______
	24 (-4, -1) and (2, -2)
	Consider
-4, -1	$x_1 =$ ____ $y_1 =$ ____
	and
2, -2	$x_2 =$ ____ $y_2 =$ ____
36, 1, $\sqrt{37}$	$d = \sqrt{[2-(-4)]^2 + [-2-(-1)]^2} = \sqrt{\_\_\_\_ + \_\_\_\_} =$ ______
	25 (5, -3) and (7, 7)
	Consider
5, -3	$x_1 =$ ____ $y_1 =$ ____
	and
7, 7	$x_2 =$ ____ $y_2 =$ ____
2, 10, 104, $2\sqrt{26}$	$d = \sqrt{(\_\_\_\_)^2 + (\_\_\_\_)^2} = \sqrt{\_\_\_\_} =$ ______

26 (1, -7) and (5, 3)

Consider

1, -7 — $x_1 =$ ____ $y_1 =$ ____

and

5, 3 — $x_2 =$ ____ $y_2 =$ ____

4, 10, 116, $2\sqrt{29}$ — $d = \sqrt{(\_\_\_)^2 + (\_\_\_)^2} = \sqrt{\_\_\_} =$ ____

27 (-5, 12) and (0, 0)

Consider

-5, 12 — $x_1 =$ ____ $y_1 =$ ____

and

0, 0 — $x_2 =$ ____ $y_2 =$ ____

5, -12, 169, 13 — $d = \sqrt{(\_\_\_)^2 + (\_\_\_)^2} = \sqrt{\_\_\_} =$ ____

28 If two points (x_1, y_1) and (x_2, y_2) lie on the same vertical line, that is, $x_1 = x_2$, then $d =$ ____ $=$ ____.

$|y_2 - y_1|$, $|y_1 - y_2|$

29 If two points (x_1, y_1) and (x_2, y_2) lie on the same horizontal line, that is, $y_1 = y_2$, then $d =$ ____ $=$ ____.

$|x_2 - x_1|$, $|x_1 - x_2|$

In Problems 30–33, find the distance between each given pair of points.

30 (3, 5) and (-7, 5)

Consider

3, 5 — $x_1 =$ ____ $y_1 =$ ____

and

-7, 5 — $x_2 =$ ____ $y_2 =$ ____

3, -7, 10 — $d = |\_\_\_ - (\_\_\_)| =$ ____

31 (4, -2) and (-9, -2)

Consider

4, -2 — $x_1 =$ ____ $y_1 =$ ____

and

-9, -2 — $x_2 =$ ____ $y_2 =$ ____

4, -9, 13 — $d = |\_\_\_ - (\_\_\_)| =$ ____

32 (-3, 6) and (-3, 12)

Consider

-3, 6 — $x_1 =$ ____ $y_1 =$ ____

and

-3, 12 — $x_2 =$ ____ $y_2 =$ ____

6, 12, 6 — $d = |\_\_\_ - \_\_\_| =$ ____

33 (4, -2) and (4, 7)

Consider

x_1 = ____ y_1 = ____ — 4, -2

and

x_2 = ____ y_2 = ____ — 4, 7

d = |____ - (____)| = ____ — 7, -2, 9

2 Relations and Their Graphs

34 Any set of ordered pairs is called a ____________. — relation

35 The set of all first members of the ordered pairs is the ________ of the relation, and the set of all second members of the ordered pairs is the ________ of the relation. — domain; range

In Problems 36–39, find the domain and range of each relation.

36 R_1 = {(1, 3), (-2, 1), (4, 1), (0, 2), (3, 1)}

The domain is the set of ______ members and is ____________. — first, {1, -2, 4, 0, 3}

The range is the set of ________ members and is __________. — second, {3, 1, 2}

37 R_1 = {(0, 0), (-1, 2), (3, 2), (1, 5), (1, 3)}

The domain of the relation R_1 is ______________. The range of the relation R_1 is ____________. — {0, -1, 3, 1}; {0, 2, 5, 3}

38 R_1 = {(2, 0), (-3, 1), (7, 3), (-5, 0)}

The domain of the relation R_1 is ______________. The range of the relation R_1 is __________. — {2, -3, 7, -5}; {0, 1, 3}

39 R_1 = {(-1, 1), (-1, 2), (-1, 3), (-1, 4), (-1, -1)}

The domain of the relation R_1 is ________. The range of the relation R_1 is ______________. — {-1}; {1, 2, 3, 4, -1}

40 The graph of a relation, which is defined by an equation or inequality in the plane, is the graph of the solution set of the ______________. — equation or inequality

In Problems 41–53, graph the given relation, and indicate its domain and range.

41 R_1 = {(1, 1), (1, 2), (3, 5), (2, 4)}

The domain of the relation R_1 is ____________. The range of the relation R_1 is ____________. The graph is — {1, 3, 2}; {1, 2, 5, 4}

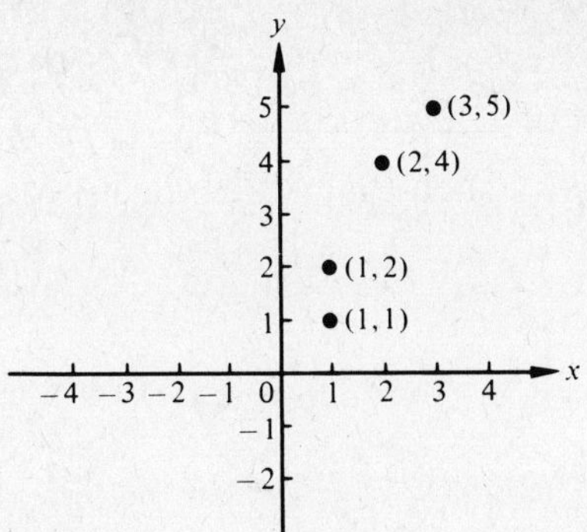

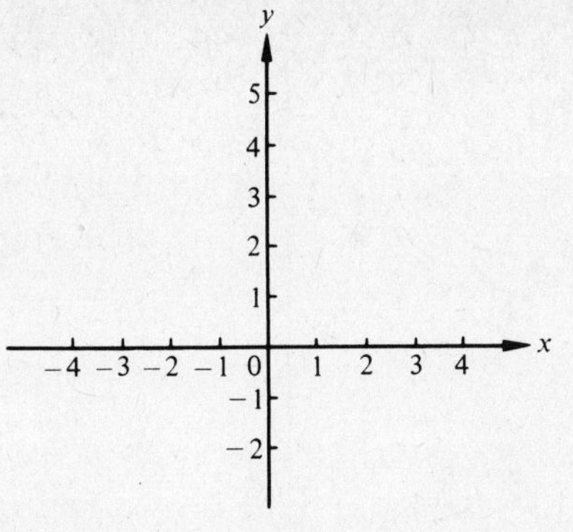

{-1, 2}

{2, 4, 3}

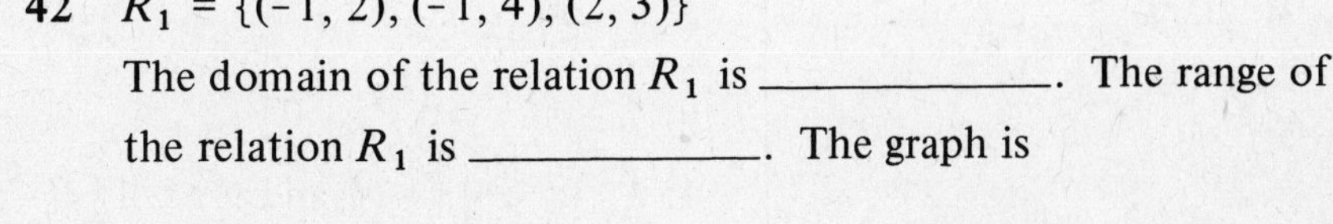

42 $R_1 = \{(-1, 2), (-1, 4), (2, 3)\}$

The domain of the relation R_1 is ____________. The range of the relation R_1 is ____________. The graph is

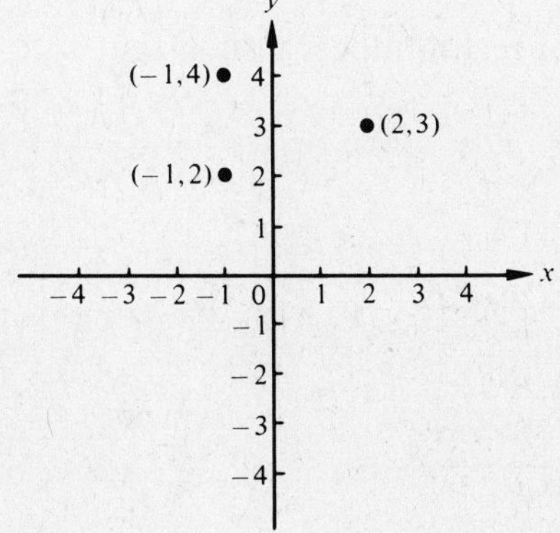

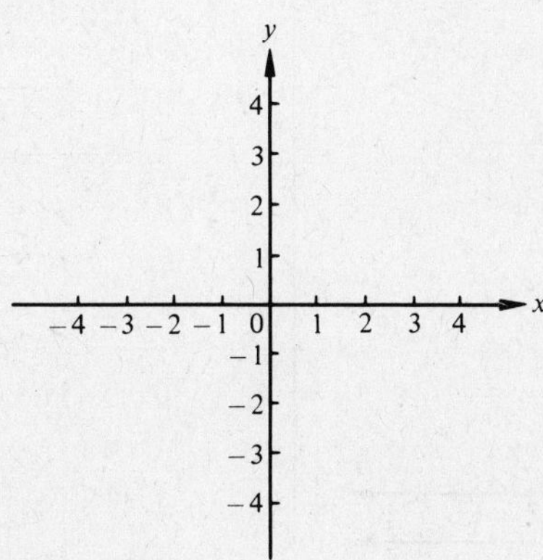

real numbers

real numbers

0

-2

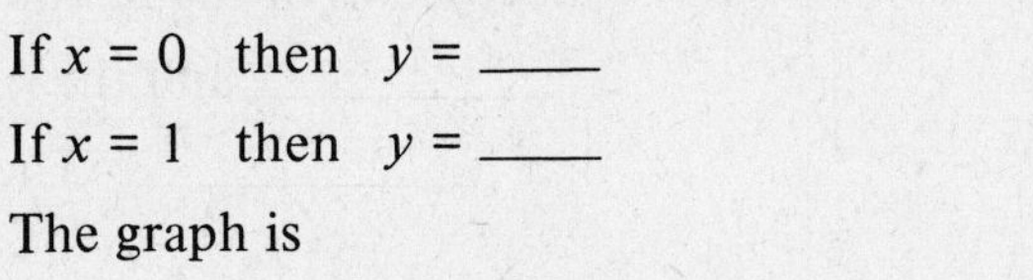

43 $\{(x, y) \mid y = -2x, x \in R\}$

The domain of the relation is the set of ____________________.

The range of the relation is the set of ____________________.

If $x = 0$ then $y =$ ____

If $x = 1$ then $y =$ ____

The graph is

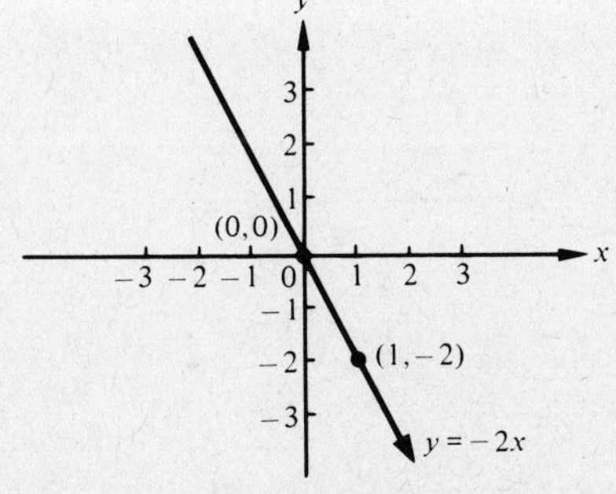

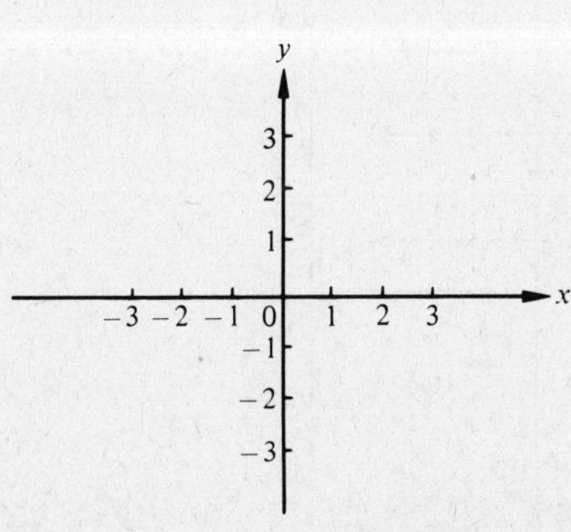

real numbers

$y < 0$

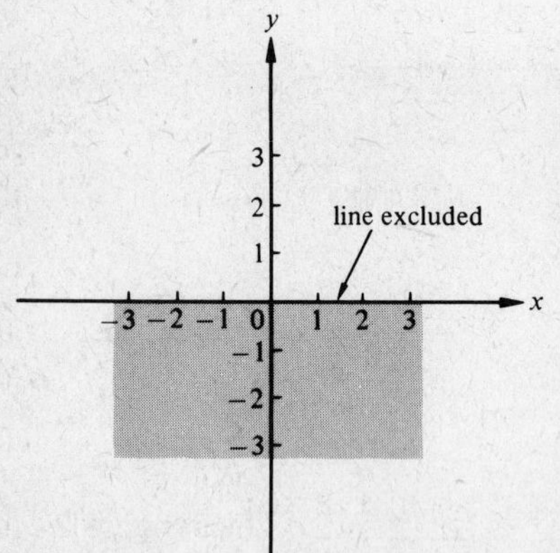

44 $\{(x, y) \mid y < 0\}$

The domain of the relation is the set of ________________ and the range is the set of all real numbers y such that ________.
The graph is

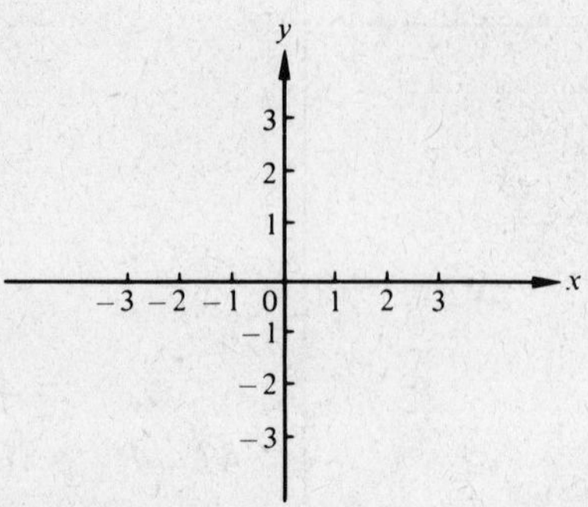

$x \leqslant 1$

$y \leqslant -1$

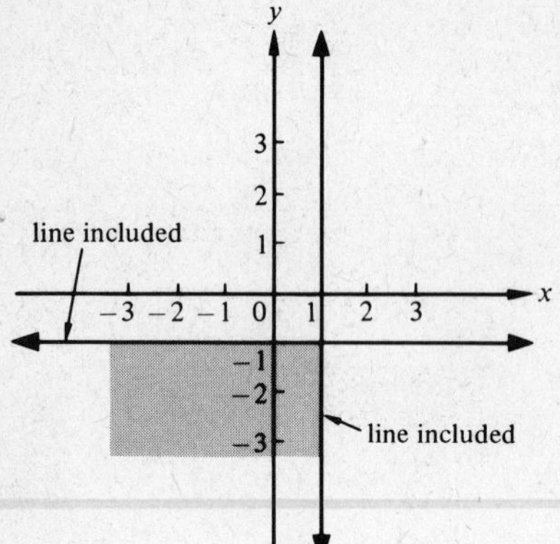

45 $\{(x, y) \mid x \leqslant 1 \text{ and } y \leqslant -1\}$

The domain of the relation is the set of real numbers x such that ________ and the range is the set of real numbers y such that ________. The graph is

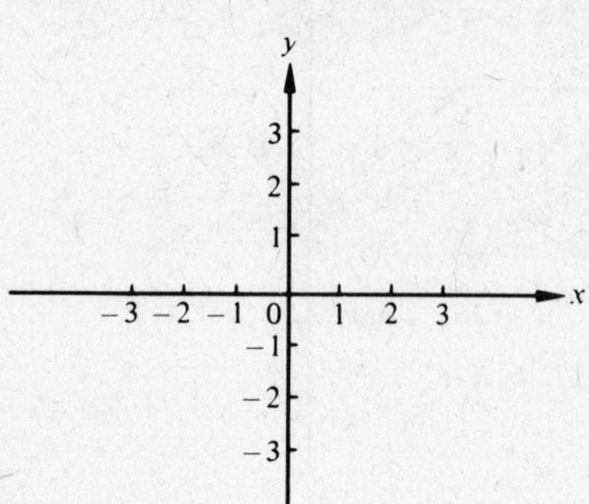

$0 \leqslant x \leqslant 2$

$-1 \leqslant y \leqslant 2$

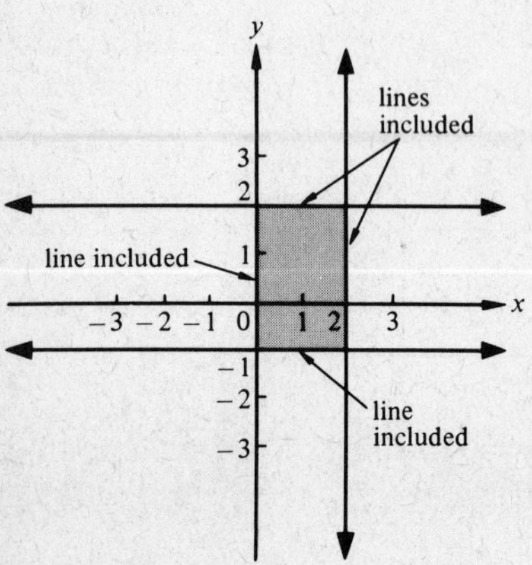

46 $\{(x, y) \mid 0 \leqslant x \leqslant 2 \text{ and } -1 \leqslant y \leqslant 2\}$

The domain of the relation is the set of real numbers x such that ____________ and the range is the set of real numbers y such that ____________. The graph is

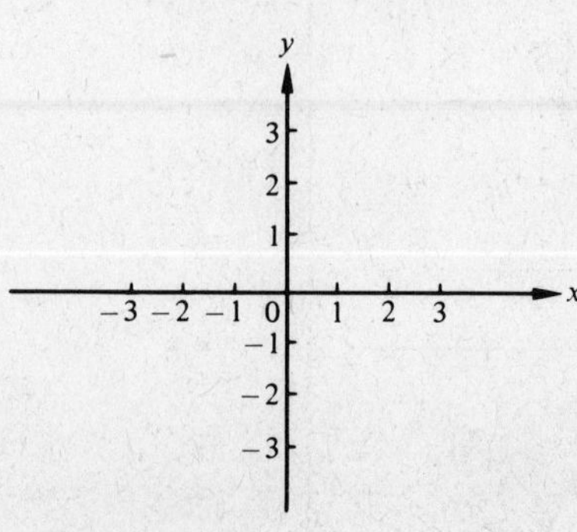

real numbers
real numbers
below

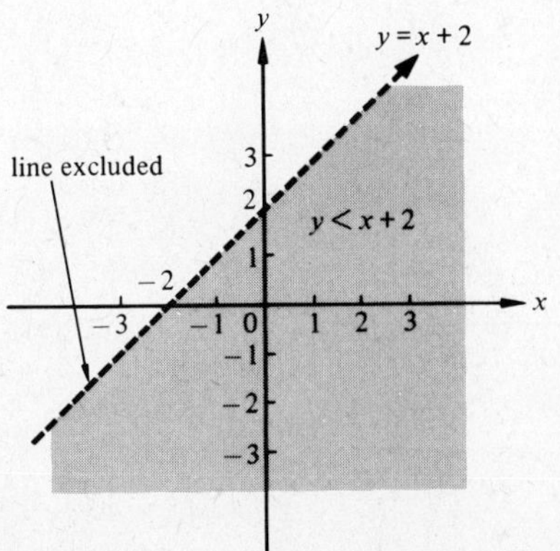

47 $y < x + 2$

The domain of the relation is the set of ________. The range is the set of ________. The graph of $y < x + 2$ consists of all points (x, y) that lie ________ the line $y = x + 2$. The graph is

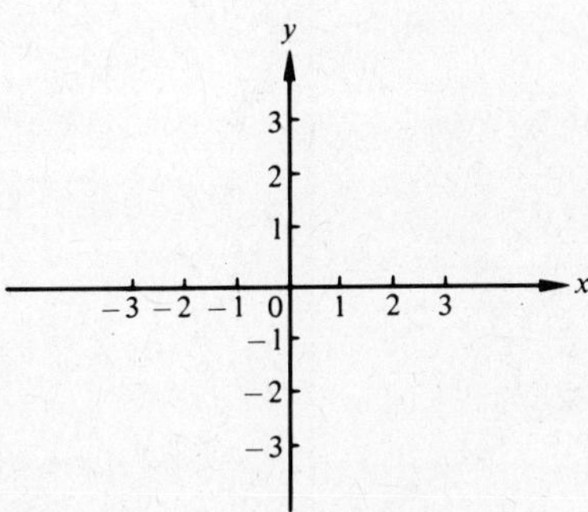

real numbers
real numbers
below or coincide with

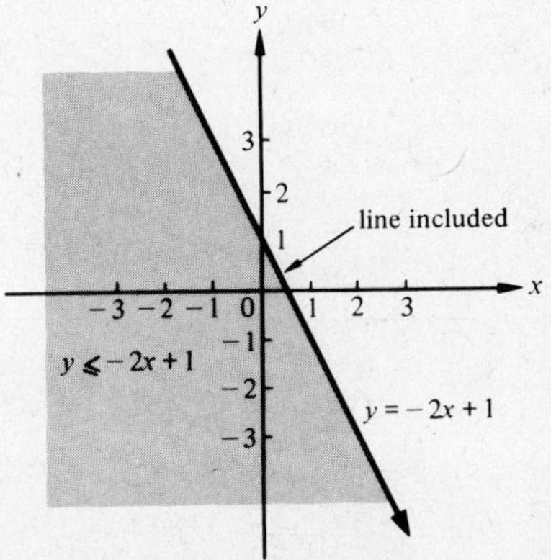

48 $y \leqslant -2x + 1$

The domain of the relation is the set of ________. The range is the set ________. The graph of $y \leqslant -2x + 1$ consists of all points (x, y) that lie ________ the line $y = -2x + 1$. The graph is

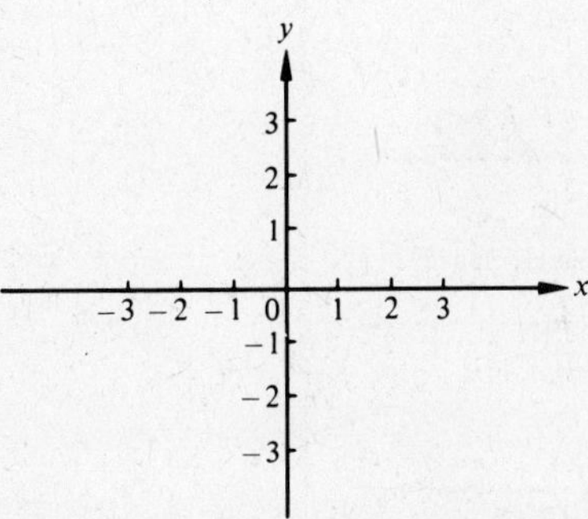

real numbers
real numbers
above

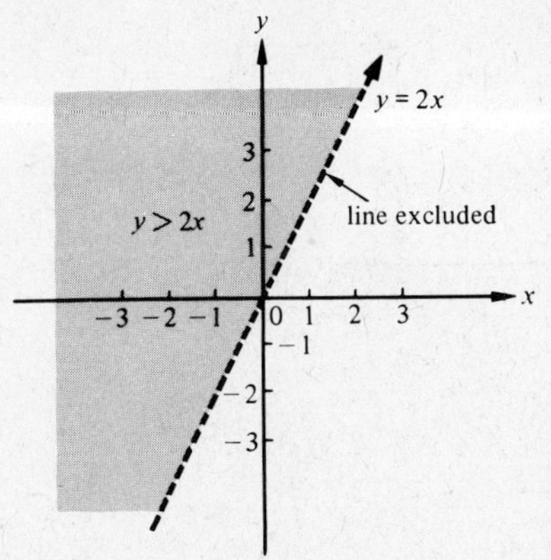

49 $y > 2x$

The domain of the relation is the set of ________. The range is the set of ________. The graph of $y > 2x$ consists of all points (x, y) that lie ________ the line $y = 2x$. The graph is

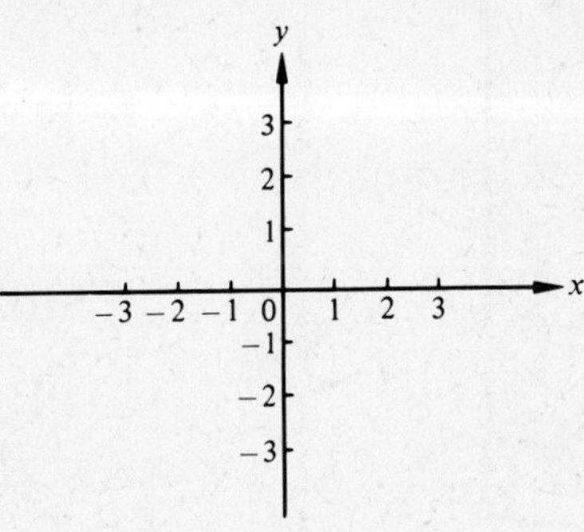

50 $y \geqslant -\frac{1}{2}x - 1$

The domain of the relation is the set of ________________. The range is the set of ________________. The graph of $y \geqslant -\frac{1}{2}x - 1$ consists of all points (x, y) that lie ________________ the graph of $y = -\frac{1}{2}x - 1$. The graph is

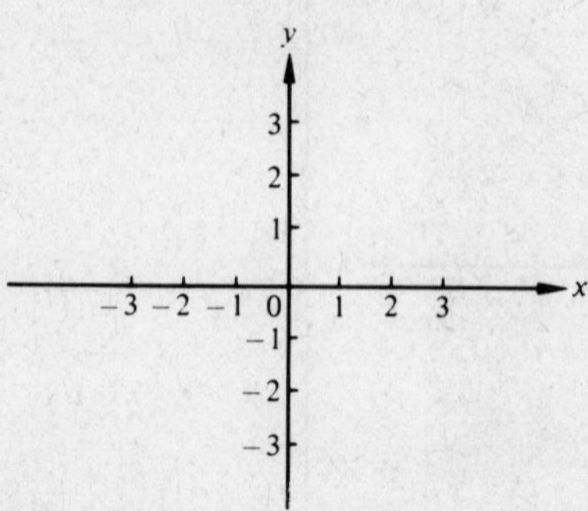

real numbers

real numbers

above or coincide with

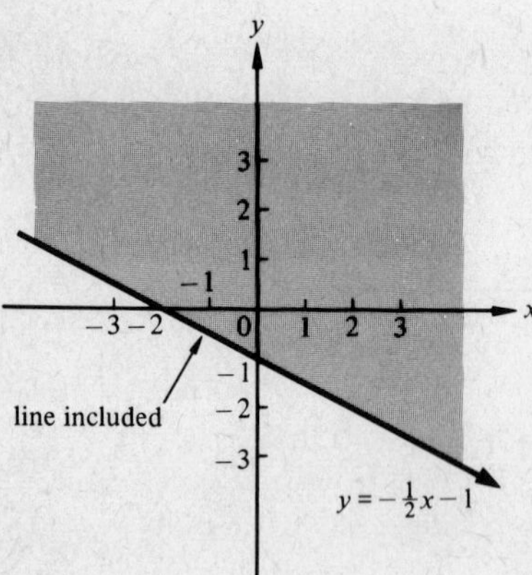

51 $|y| \geqslant 3$

The domain of the relation is the set of ________________. The range is the set of real numbers y such that $y \geqslant 3$ or ________.
The graph is

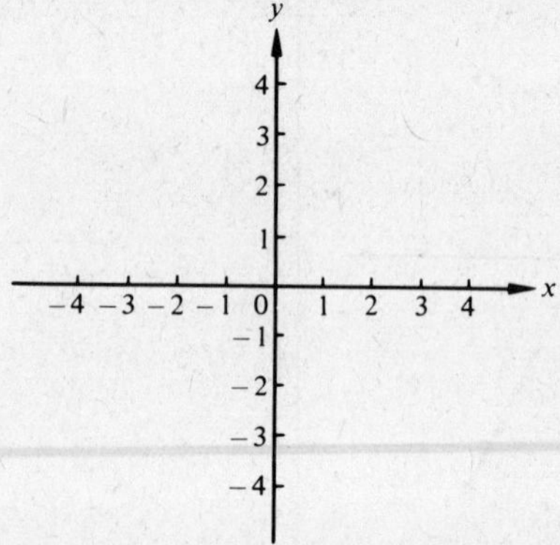

real numbers

$y \leqslant -3$

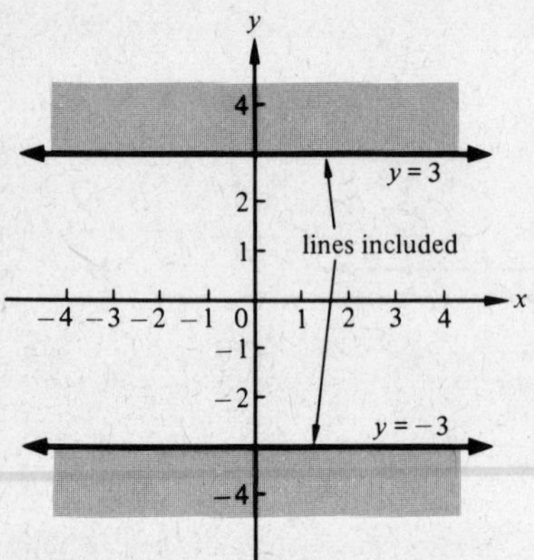

52 $|x - 2| \leqslant 4$

The domain of the relation is the set $\{x \mid$ ____________$\}$. The range is the set of ________________. The graph is

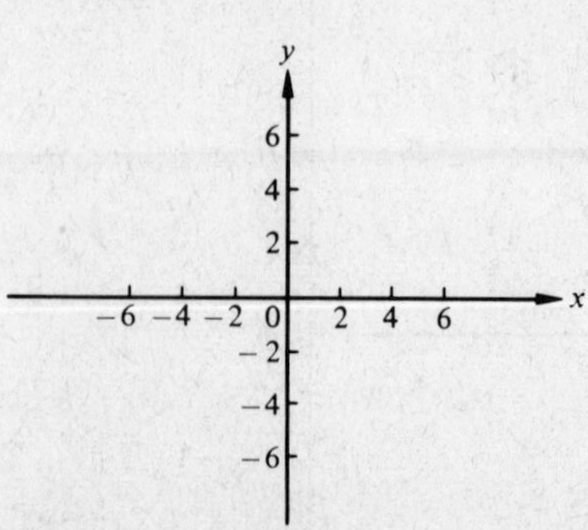

$-2 \leqslant x \leqslant 6$

real numbers

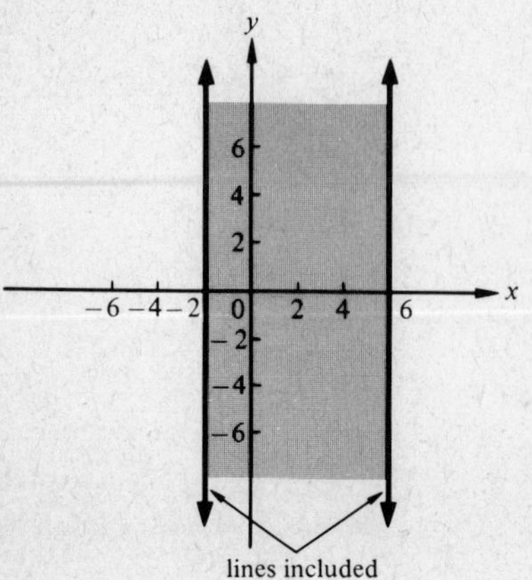

real numbers

$y \geqslant 0$

0

2

2

8

8

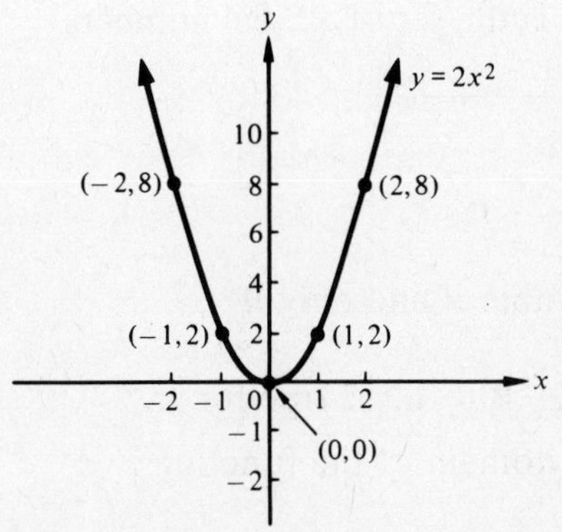

53 $y = 2x^2$

The domain of the relation is the set of ________________. The range is the set of real numbers y such that ________.

If $x = 0$ then $y =$ ____

If $x = -1$ then $y =$ ____

If $x = 1$ then $y =$ ____

If $x = -2$ then $y =$ ____

If $x = 2$ then $y =$ ____

The graph is

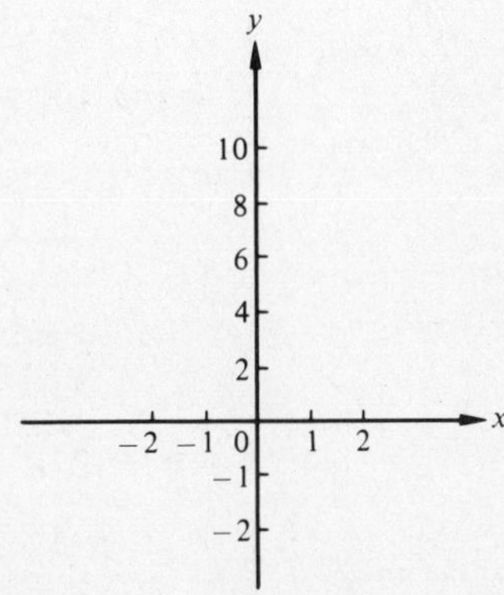

3 Functions

domain

range

54 A function f is a correspondence that assigns to each member in a certain set, called the __________ of f, exactly one member in a second set, called the ________ of f.

independent

dependent

55 If a variable quantity y depends in a definite way on another variable quantity x, we refer to x as the ____________ variable and y as the ____________ variable.

domain

range

56 If y is a function of x, then the independent variable x can take on any value in the ____________ of the function. The set of all possible corresponding dependent variables y is the ____________ of the function.

In Problems 57–60, find the domain of each function defined by the given equation.

$1 - x \geqslant 0$

$1 - x \geqslant 0,\ x \leqslant 1$

57 $y = \sqrt{1 - x}$

$\sqrt{1 - x}$ is defined and is a real number if and only if __________, so that the domain of the function is the set of all real numbers for which ____________, that is, for ____________.

58 $y = \frac{1}{x - 2}$

$x - 2 \neq 0$ | $\frac{1}{x - 2}$ is defined if and only if ________, so that the domain of the function is the set of all real numbers for which

$x - 2 \neq 0$, $x \neq 2$ | ________, that is, for ________.

59 $y = \frac{1}{\sqrt{x - 5}}$

$x - 5 > 0$ | $\frac{1}{\sqrt{x - 5}}$ is defined and is a real number if and only if ________, so that the domain of the function is the set of all real numbers

$x - 5 > 0$, $x > 5$ | for which ________ or ________.

60 $y = \frac{5}{\sqrt{x^2 + 9}}$

$\frac{5}{\sqrt{x^2 + 9}}$ is defined and is a real number if and only if

0, $x^2 + 9 \neq 0$ | $\sqrt{x^2 + 9} \neq$ ____ or ________. But this is true for any

real number | ________ x, so that the domain of the function is

real numbers | the set of all ________.

3.1 Function Notation

61 Letters such as f, g, and h, as well as F, G, and H, are used to denote functions. The symbol $f(x)$ is read "f of x," meaning

value | "the ________ of the function f at x."

In Problems 62–63, assume that $y = f(x)$ is defined by the given equation. Find the expression for $f(x)$.

62 $3x + y = 11$

$-3x + 11$ | Solving the equation for y in terms of x, we have $y =$ ________,

$-3x + 11$ | so that $y = f(x) =$ ________.

63 $5x - 3y = 15$

$5x - 15$ | Solving the equation for y in terms of x, we have $3y =$ ________

$\frac{5}{3}x - 5$, $\frac{5}{3}x - 5$ | or $y =$ ________, so that $y = f(x) =$ ________.

In Problems 64–77, find the values indicated for the given function. Indicate the domain of the function.

64 $f(2), f(3)$, and $f(0)$ if $f(x) = -3x + 1$

real numbers | The domain of f is the set of ________.

2, -5 | $f(2) = -3($____$) + 1 =$ ____

3, -8 | $f(3) = -3($____$) + 1 =$ ____

0, 1 | $f(0) = -3($____$) + 1 =$ ____

65 $f\left(\frac{1}{9}\right), f\left(\frac{1}{3}\right), f(3)$, and $f(15)$ if $f(x) = \frac{3}{x}$

real numbers — The domain of f is the set of ____________, except

0 — $x =$ ____.

$\frac{1}{9}$, 27 — $f\left(\frac{1}{9}\right) = \frac{3}{\_\_\_\_} = \_\_\_\_$

$\frac{1}{3}$, 9 — $f\left(\frac{1}{3}\right) = \frac{3}{\_\_\_\_} = \_\_\_\_$

3, 1 — $f(3) = \frac{3}{\_\_\_\_} = \_\_\_\_$

15, $\frac{1}{5}$ — $f(15) = \frac{3}{\_\_\_\_} = \_\_\_\_$

66 $f(-4), f(4)$, and $f(0)$ if $f(x) = \sqrt{16 - x^2}$

$-4 \leqslant x \leqslant 4$ — The domain is the set $\{x \mid \_\_\_\_\_\_\_\_\}$.

-4, 0 — $f(-4) = \sqrt{16 - (\_\_\_\_)^2} = \_\_\_\_$

4, 0 — $f(4) = \sqrt{16 - (\_\_\_\_)^2} = \_\_\_\_$

0, 4 — $f(0) = \sqrt{16 - (\_\_\_\_)^2} = \_\_\_\_$

67 $f(3), f(a)$, and $f(3 + a)$ if $f(x) = 2 - x$

real numbers — The domain is the set of ____________.

3, -1 — $f(3) = 2 -$ ____ $=$ ____

a, $2 - a$ — $f(a) = 2 -$ ____ $=$ ______

$3 + a$, $-1 - a$ — $f(3 + a) = 2 - (\_\_\_\_\_\_) =$ ______

68 $f(10), f(5)$, and $f(2)$ if $f(x) = \frac{3}{\sqrt{x - 1}}$

$\{x \mid x > 1\}$ — The domain is the set __________.

10, 1 — $f(10) = \frac{3}{\sqrt{\_\_\_\_ - 1}} = \_\_\_\_$

5, $\frac{3}{2}$ — $f(5) = \frac{3}{\sqrt{\_\_\_\_ - 1}} = \_\_\_\_$

2, 3 — $f(2) = \frac{3}{\sqrt{\_\_\_\_ - 1}} = \_\_\_\_$

69 $g(25), g(16)$, and $g(4)$ if $g(x) = \sqrt{x}$

$\{x \mid x \geqslant 0\}$ or $[0, \infty)$ — The domain is the set ____________.

25, 5 — $g(25) = \sqrt{\_\_\_\_} = \_\_\_\_$

16, 4 — $g(16) = \sqrt{\_\_\_\_} = \_\_\_\_$

4, 2 — $g(4) = \sqrt{\_\_\_\_} = \_\_\_\_$

70 $f(3), f(2), f(0)$, and $f(-2)$ if $f(x) = \frac{1}{2} - x^2$

real numbers — The domain is the set of ____________.

3, $-\frac{17}{2}$ — $f(3) = \frac{1}{2} - (\_\_\_\_)^2 = \_\_\_\_$

2, $-\frac{7}{2}$ — $f(2) = \frac{1}{2} - (\_\_\_\_)^2 = \_\_\_\_$

0, $\frac{1}{2}$ — $f(0) = \frac{1}{2} - (\_\_\_\_)^2 = \_\_\_\_$

-2, $-\frac{7}{2}$ — $f(-2) = \frac{1}{2} - (\_\_\_\_)^2 = \_\_\_\_$

Answers	Frames
	71 $\dfrac{f(a+h)-f(a)}{h}$ if $f(x) = 5x - 2$
real numbers	The domain is the set of ____________.
$a + h,\ 5a + 5h - 2$	$f(a+h) = 5(\_\_\_\_) - 2 = \_\_\_\_\_\_$
$a,\ 5a - 2$	$f(a) = 5(\_\_\_) - 2 = \_\_\_\_$
$5a - 2,\ 5$	$\dfrac{f(a+h)-f(a)}{h} = \dfrac{5a + 5h - 2 - (\_\_\_\_)}{h} = \_\_\_$
	72 $f(8) - f(6)$ if $f(x) = \sqrt{100 - x^2}$
$-10 \leqslant x \leqslant 10$	The domain is the set $\{x \mid \_\_\_\_\_\_\}$.
8, 6	$f(8) = \sqrt{100 - (\_\_\_)^2} = \_\_\_$
6, 8	$f(6) = \sqrt{100 - (\_\_\_)^2} = \_\_\_$
6, 8, -2	$f(8) - f(6) = \_\_\_ - \_\_\_ = \_\_\_$.
	73 $\dfrac{f(a+h)-f(a)}{h}$ if $f(x) = x^2$
real numbers	The domain is the set of ____________.
$a^2 + 2ah + h^2$	$f(a+h) = (a+h)^2 = \_\_\_\_\_\_$
$a,\ a^2$	$f(a) = (\_\_\_)^2 = \_\_\_$
a^2	$\dfrac{f(a+h)-f(a)}{h} = \dfrac{a^2 + 2ah + h^2 - \_\_\_}{h}$
$2a + h$	$= \_\_\_\_\_\_$
	74 $f(2), f(a^2)$, and $[f(a)]^2$ if $f(x) = \dfrac{3x - 1}{1 + 2x}$
$x \neq -\frac{1}{2}$	The domain is the set $\{x \mid \_\_\_\_\_\_\}$.
2, 1	$f(2) = \dfrac{3(\_\_\_) - 1}{1 + 2(2)} = \_\_\_$
$a^2,\ \dfrac{3a^2 - 1}{1 + 2a^2}$	$f(a^2) = \dfrac{3(\_\_\_) - 1}{1 + 2a^2} = \_\_\_\_\_\_$
$\dfrac{3a - 1}{1 + 2a},\ \dfrac{9a^2 - 6a + 1}{1 + 4a + 4a^2}$	$[f(a)]^2 = \left(\dfrac{\quad}{\_\_\_\_}\right)^2 = \_\_\_\_\_\_$
	75 $g(a) - g(-a)$ if $g(x) = \dfrac{2 + x}{2 - x}$
$x \neq 2$	The domain is the set $\{x \mid \_\_\_\_\}$.
$a,\ \dfrac{2 + a}{2 - a}$	$g(a) = \dfrac{2 + \_\_\_}{2 - a} = \_\_\_\_$
$-a,\ \dfrac{2 - a}{2 + a}$	$g(-a) = \dfrac{2 + (\_\_\_)}{2 + a} = \_\_\_\_$
$\dfrac{2 - a}{2 + a},\ \dfrac{8a}{4 - a^2}$	$g(a) - g(-a) = \dfrac{2 + a}{2 - a} - \_\_\_\_ = \_\_\_\_$
	76 $f(1)$ and $f(2)$ if $f(x) = (x - 1)(x - 2)(x - 3)$
$\in R$	The domain is the set $\{x \mid x \_\_\_\}$
0	$f(1) = (1 - 1)(1 - 2)(1 - 3) = \_\_\_$
0	$f(2) = (2 - 1)(2 - 2)(2 - 3) = \_\_\_$

77 $f(-1)$ and $f(-4)$ if $f(x) = \sqrt{-x}$

The domain is the set $\{x \mid \_\_\_\_\}$ — $x \leqslant 0$

$f(-1) = \sqrt{-\_\_\_\_} = \_\_\_\_$ — (-1), 1

$f(-4) = \sqrt{-\_\_\_\_} = \_\_\_\_$ — (-4), 2

78 If $y = f(x)$, then the expression

$$\frac{f(x+h) - f(x)}{h}, \quad \text{for } h \neq 0$$

is called the ________ of f at x. — difference quotient

In Problems 79–82, find the difference quotient of f at x.

79 $f(x) = 7x + 11$

$$\frac{f(x+h) - f(x)}{h} = \frac{[7(x+h) + 11] - (\_\_\_\_)}{h}$$ — $7x + 11$

$$= \frac{\_\_\_\_}{h} = \_\_\_\_$$ — $7h$, 7

80 $f(x) = -x^2$

$$\frac{f(x+h) - f(x)}{h} = \frac{-(x+h)^2 - (\_\_\_\_)}{h}$$ — $-x^2$

$$= \frac{-x^2 - 2xh - h^2 + \_\_\_\_}{h}$$ — x^2

$$= \frac{\_\_\_\_}{h} = \_\_\_\_$$ — $-2xh - h^2$, $-2x - h$

81 $f(x) = \sqrt{x+2}$

$$\frac{f(x+h) - f(x)}{h} = \frac{\sqrt{x+h+2} - \sqrt{\_\_\_\_}}{h}$$ — $x + 2$

$$= \frac{(\sqrt{x+h+2} - \sqrt{x+2})(\_\_\_\_)}{h(\sqrt{x+h+2} + \sqrt{x+2})}$$ — $\sqrt{x+h+2} + \sqrt{x+2}$

$$= \frac{\_\_\_\_}{h(\sqrt{x+h+2} + \sqrt{x+2})}$$ — h

$$= \frac{1}{\_\_\_\_}$$ — $\sqrt{x+h+2} + \sqrt{x+2}$

82 $f(x) = \dfrac{3}{x}$

$$\frac{f(x+h) - f(x)}{h} = \frac{\dfrac{3}{x+h} - \_\_\_\_}{h}$$ — $\dfrac{3}{x}$

$$= \frac{3x - [\_\_\_\_]}{hx(x+h)}$$ — $3(x+h)$

$$= \frac{\_\_\_\_}{hx(x+h)} = \_\_\_\_$$ — $-3h$, $\dfrac{-3}{x(x+h)}$

$\frac{\sqrt{3}}{2}x$

$\frac{\sqrt{3}}{2}x,\ \frac{\sqrt{3}}{4}x^2$

83 Express the area A of an equilateral triangle as a function of its side of length x.
Let x be the base of the triangle. Then its height is ________,
so that $A = \frac{1}{2}(x)\left(______\right) = $ ________.

$x + 3,\ 2x^2 + 6x$

84 Consider a rectangle of dimensions $2x$ by $x + 3$.
Express the area A of the rectangle as a function of x.
$A = (2x)(_____) = $ ____________

3.2 Functions as Relations

relation

85 Using relations, a function is defined to be a ____________ in which no two different ordered pairs have the same first member.

In Problems 86-91, indicate whether or not the relation is a function. What is the domain and the range in each case?

$\{1, 4, -2, -1\},\ \{3, 5\}$

is

86 $\{(1, 3), (4, 5), (-2, 5), (-1, 3)\}$
The domain is ________________. The range is ________.
Since the first members of the ordered pairs are different, the relation _____ a function.

$\{0, 2, -2, -1\},\ \{1, 0\}$

is

87 $\{(0, 1), (2, 0), (-2, 0), (-1, 0)\}$
The domain is ________________. The range is ________.
Since the first members of the ordered pairs are different, the relation _____ a function.

$\{-1, -2, 3\},\ \{1, 5, 2\}$

is not

88 $\{(-1, 1), (-2, 1), (3, 5), (3, 2)\}$
The domain is ________________. The range is ________.
The relation __________ a function, since all first members of the ordered pairs are not different.

real numbers

real numbers, is not

89 $\{(x, y) \mid y \leqslant -x\}$
The domain is the set of ________________. The range is the set of ________________. The relation ________ a function, since all first members of ordered pairs are not different. For example, $(2, -4)$ and $(2, -3)$ are members of the set.

real numbers

2, is

90 $\{(x, y \mid y = 2\}$
The domain is the set of ________________. The range is the set $\{\_\_\_\_\}$. The relation _____ a function, because no two different ordered pairs have the same first members.

-2

real numbers, is not

91 $\{(x, y) \mid x = -2\}$
The domain is $\{\_\_\_\_\}$. The range is the set of ________________. The relation ________ a function.
For example, $(-2, 3)$ and $(-2, 5)$ are members of the set.

92 If $y = f(x)$, we say that f maps x to y, or x is mapped to y by f, or y is the image of x under f. This mapping is symbolized as

$x \xrightarrow{f} y$ — follows: $f: x \longrightarrow y$ or __________ or $x \longrightarrow f(x)$

93 Let $f(x) = 3x^2 + 1$. Find the values of $f(-2), f(-1), f(0), f(1)$, and $f(2)$. Also describe these values using the mapping notation.

13 — $f(-2) = 3(-2)^2 + 1 =$ ____

4 — $f(-1) = 3(-1)^2 + 1 =$ ____

1 — $f(0) = 3(0)^2 + 1 =$ ____

4 — $f(1) = 3(1)^2 + 1 =$ ____

13 — $f(2) = 3(2)^2 + 1 =$ ____

Using the mapping notation, we have

13 — $f: -2 \longrightarrow$ ____

4 — $f: -1 \longrightarrow$ ____

1 — $f: 0 \longrightarrow$ ____

4 — $f: 1 \longrightarrow$ ____

13 — $f: 2 \longrightarrow$ ____

4 Graphs and Properties of Functions

94 To graph a function we simply locate all points in the cartesian

function — plane whose coordinates (x, y) belong to the __________.

4.1 Even and Odd Functions: Symmetry

95 A graph of a function is symmetric with respect to the y axis if for

$(-x, y)$ — each point (x, y) on the graph the point ________ is also on the graph.

96 A graph of a function is symmetric with respect to the origin if for

$(-x, -y)$ — each point (x, y) on the graph the point ________ is also on the graph.

97 A function f such that $f(x) = f(-x)$ for all x in the domain of f is

even — said to be an ________ function.

98 A function f such that $f(-x) = -f(x)$ for all x in the domain of f is

odd — said to be an ____ function.

In Problems 99–104, determine whether each given function is even or odd or neither.

99 $f(x) = 16x^4$

$16x^4$ — $f(-x) = 16(-x)^4 =$ ____ $= f(x)$

even — Therefore, f is an ________ function.

100 $f(x) = x^2 + 6$

$x^2 + 6$ — $f(-x) = (-x)^2 + 6 =$ ________ $= f(x)$

even — Therefore, f is an ________ function.

$-5x^3$

odd

101 $f(x) = 5x^3$

$f(-x) = 5(-x)^3 =$ ____ $= -f(x)$

Therefore, f is an ____ function.

$-\left(x^3 + \frac{1}{x}\right)$

odd

102 $f(x) = x^3 + \dfrac{1}{x}$

$f(-x) = (-x)^3 + \dfrac{1}{(-x)} =$ ________________ $= -f(x)$

Therefore, f is an ____ function.

$x^4 + 5x^2 + 3$

even

103 $f(x) = x^4 + 5x^2 + 3$

$f(-x) = (-x)^4 + 5(-x)^2 + 3$

$=$ ________________

$= f(x)$

Therefore, f is an ________ function.

$x^2 - 3x$

even

$-f(x)$

odd

104 $f(x) = x^2 + 3x$

$f(-x) = (-x)^2 + 3(-x)$

$=$ ________

$\neq f(x)$

Therefore, f is not an ________ function.

Also, $f(-x) \neq$ ________.

Hence, f is not an ____ function.

In Problems 105–110, discuss the symmetry of each of the given functions by determining if the function is even or odd. Use the results to sketch the graph.

x^2, y, even

105 $y = 2x^2$

The graph of $y = 2x^2$ is symmetric with respect to the y axis, since $2(-x)^2 = 2$____ $=$ ____. Therefore the function is ________, and it is only necessary to graph the function for the nonnegative abscissas. The remainder of the graph is determined by reflection across the y axis. The graph is

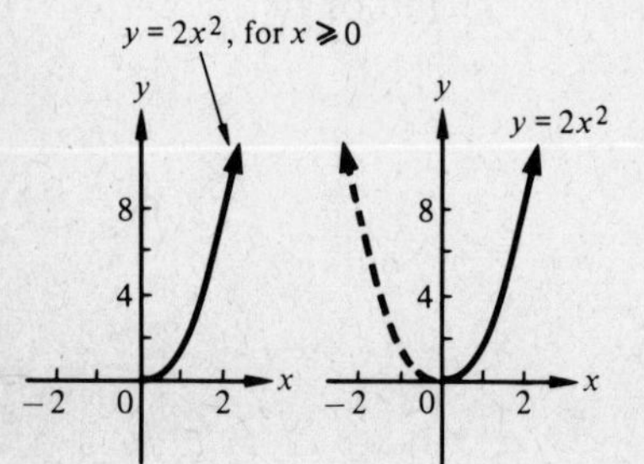

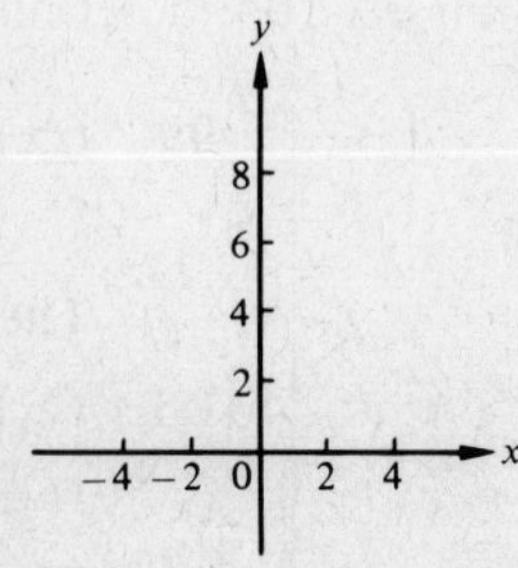

even

y axis

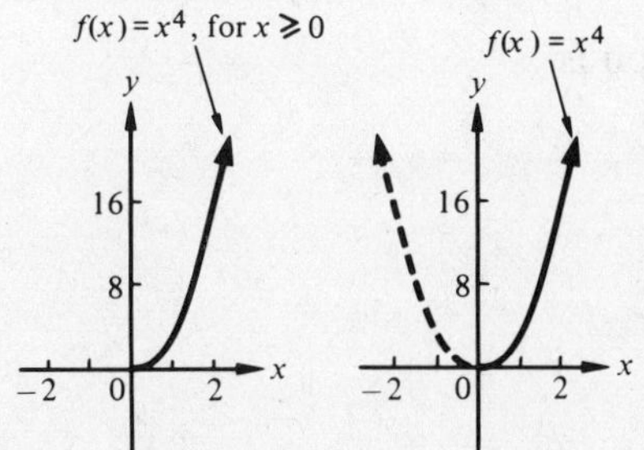

106 $f(x) = x^4$

Since $f(-x) = (-x)^4 = x^4 = f(x)$, the function f is ________ and the graph of f is symmetric with respect to the ________. The graph is

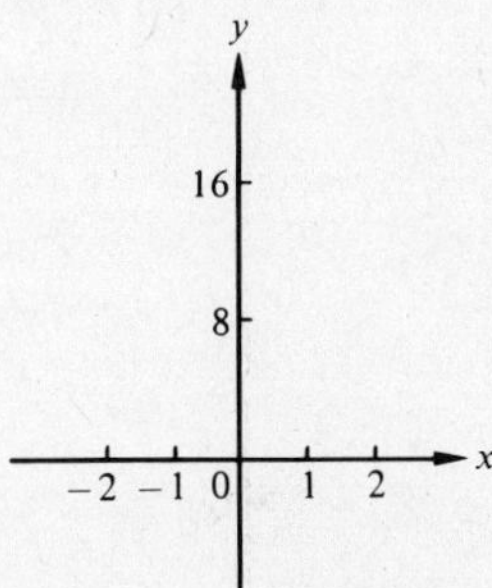

even

y axis

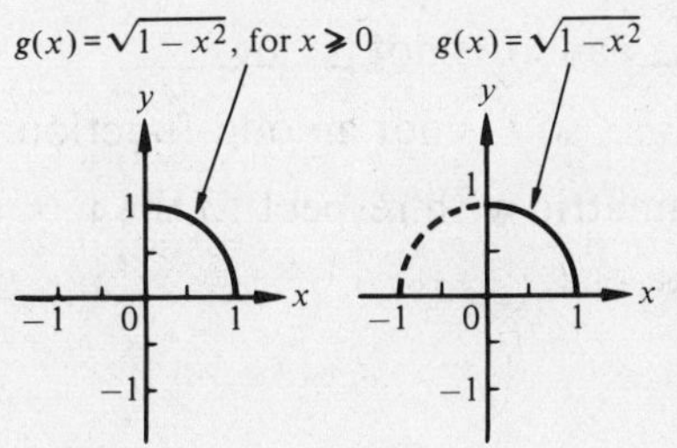

107 $g(x) = \sqrt{1 - x^2}$

Since $g(-x) = \sqrt{1 - (-x)^2} = \sqrt{1 - x^2} = g(x)$, the function g is ________ and the graph of g is symmetric with respect to the ________. The graph is

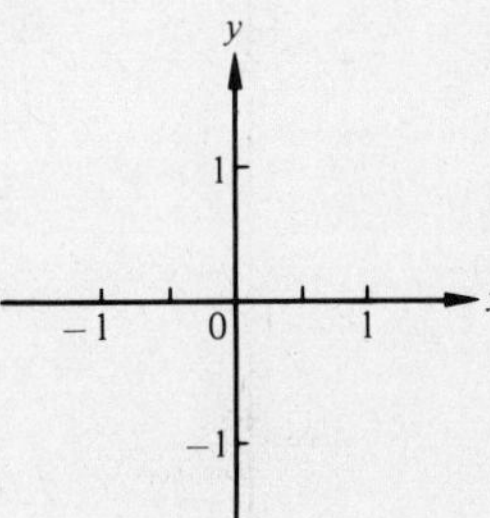

y axis

odd

origin

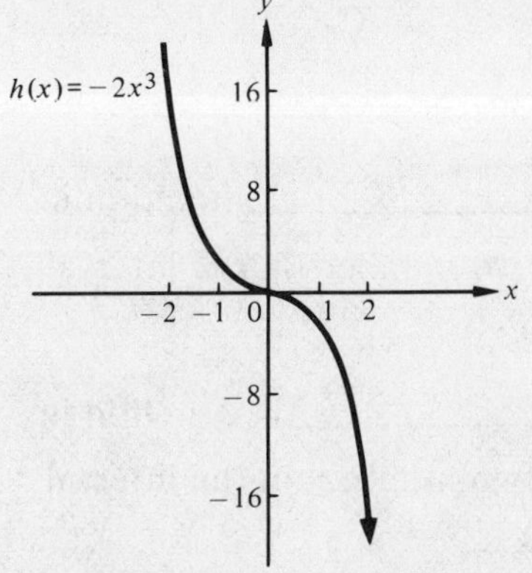

108 $y = h(x) = -2x^3$

Since $h(-x) = -2(-x)^3 \neq -2x^3 = h(x)$, the graph of h is not symmetric with respect to the ________. But if we replace x by $-x$, we obtain $h(-x) = -2(-x)^3 = -h(x)$, so that h is an ____ function. Thus, the graph of h is symmetric with respect to the ________. The graph is

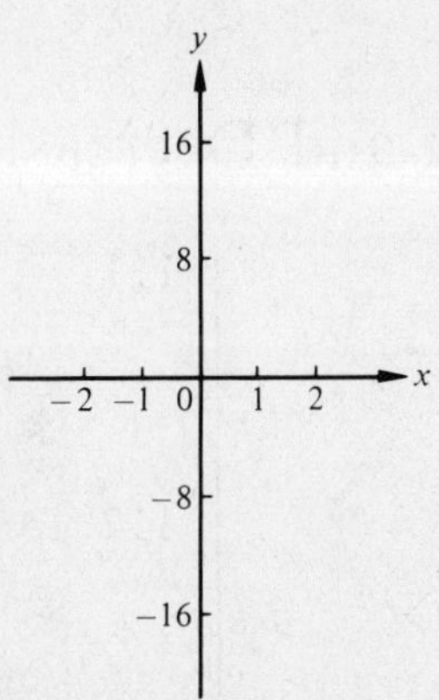

109 $y = F(x) = x^3$

Since $F(-x) = (-x)^3 \neq x^3 = F(x)$, the graph of the function is not symmetric with respect to the ________. But if we replace x by $-x$, we obtain $F(-x) = (-x)^3 = -F(x)$, so that F is an ______ function. Therefore, the graph of F is symmetric with respect to the ________. The graph is

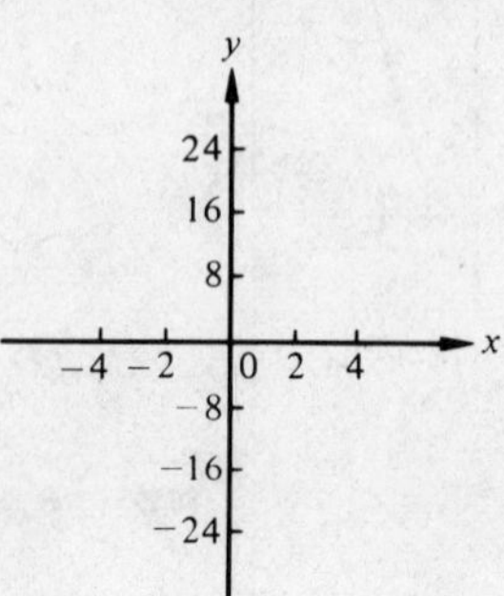

y axis

odd

origin

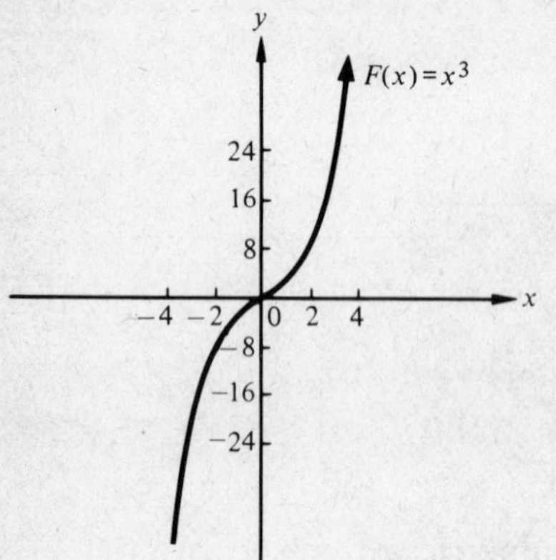

110 $f(x) = |x| - x$

$f(-x) = |-x| - (-x) = |x| + x \neq$ ______ and f is not ________.
Also, $f(-x) = |-x| - (-x) \neq$ ________, so f is not an odd function.
Therefore, the graph of f is not symmetric with respect to the y axis or the ________. The graph is

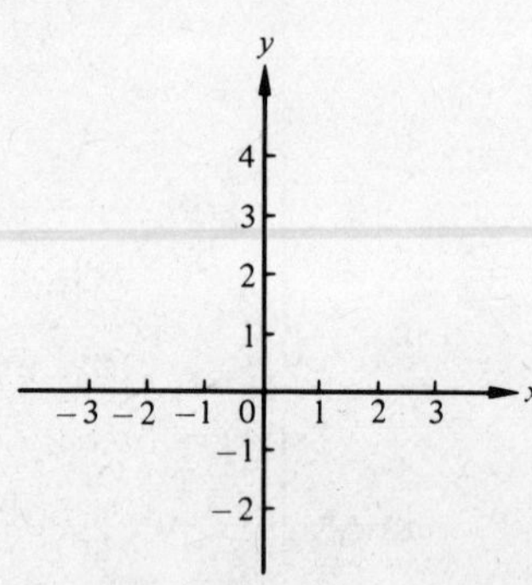

$f(x)$, even

$-f(x)$

origin

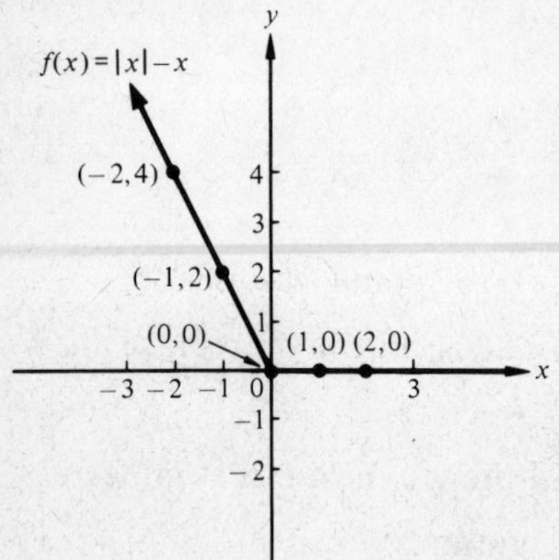

4.2 Increasing and Decreasing Functions

111 A function f is called a (strictly) ________________ function in an interval I if, whenever a and b are two numbers in the interval such that if $a < b$, we have $f(a) < f(b)$.

increasing

112 A function f is called a (strictly) ________________ function in an interval I if, whenever a and b are two numbers in the interval such that if $a < b$, we have $f(a) > f(b)$.

decreasing

In Problems 113–114, indicate the intervals for which the function is increasing or decreasing.

113

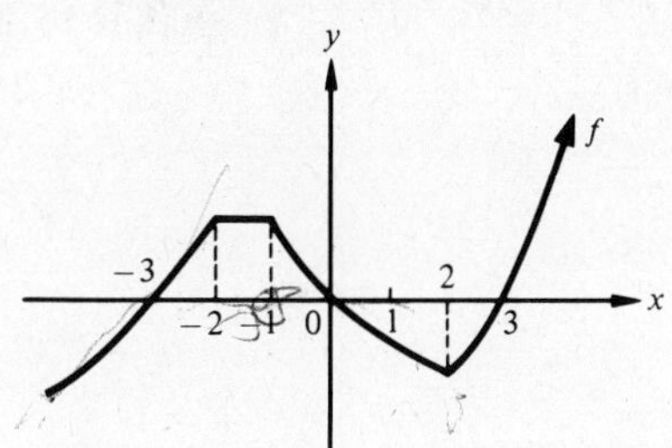

The graph indicates that f is increasing in the intervals ________ and ________; f is decreasing in the interval ________; and f is neither increasing nor decreasing in the interval ________.

$(-\infty, -2]$
$[2, \infty)$, $[-1, 2]$
$[-2, -1]$

114

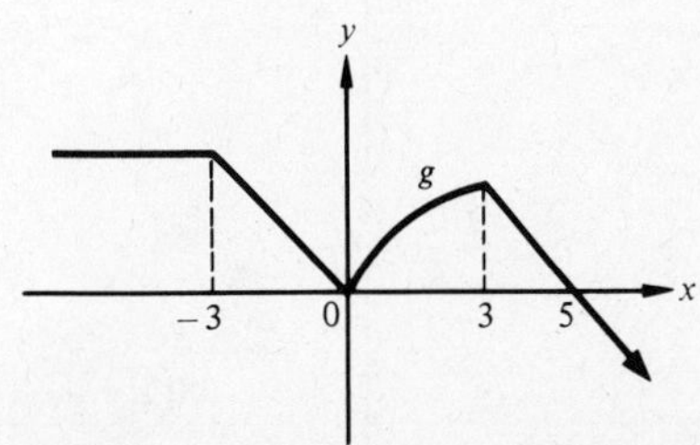

The graph indicates that g is increasing in the interval ________; g is decreasing in the intervals ________ and ________; and g is neither increasing nor decreasing in the interval ________.

$[0, 3]$
$[-3, 0]$, $[3, \infty]$
$(-\infty, -3]$

4.3 Graphs of Functions

115 The function f, defined by the equation $f(x) = x$, is called the ________ function. The domain is the set of ________ and the range is the set of ________.

identity, real numbers
real numbers

116 A function f, whose rule of correspondence is $f(x) = |x|$, is called the ________ function. The domain is the set of real numbers and the range is the set of ________.

absolute value
nonnegative real numbers

117 A function f, whose rule of correspondence is $f(x) = [\![x]\!]$, is called the ________ function. The domain is the set of real numbers and the range is the set of ________.

greatest integer
integers

In Problems 118–127, find the domain of each function and sketch its graph. Discuss the symmetry of the function and indicate if the function is even or odd and increasing or decreasing. Also, find the range of the function.

118 $f(x) = 5x$

The domain of f is the set of ________________. Since $f(-x) = -5x = -f(x)$, the graph is symmetric with respect to the ____________ and f is _____. The graph is

real numbers

origin, odd

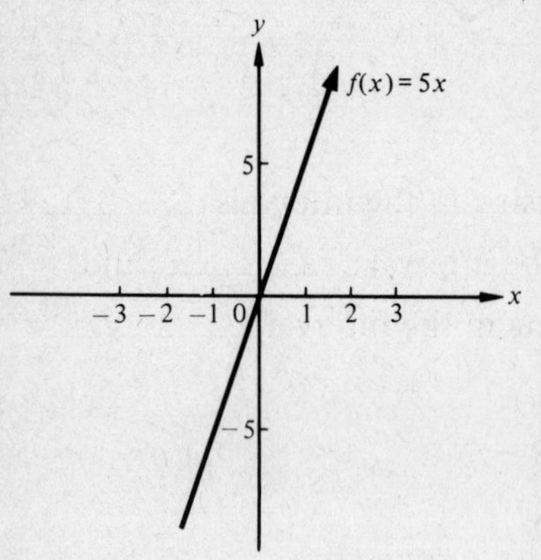

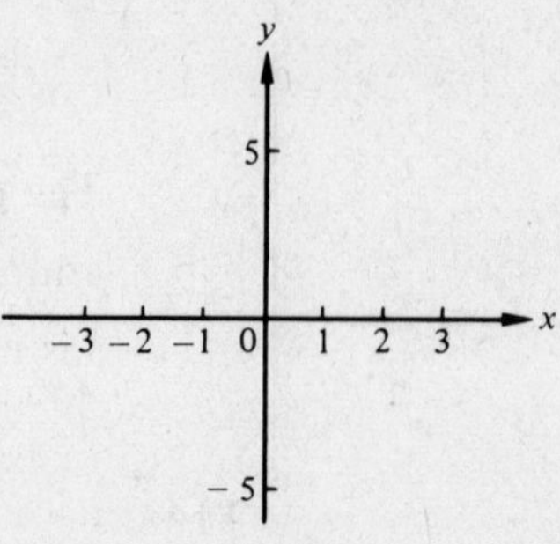

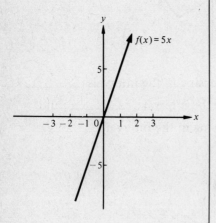

The graph shows that the range of f is the set of all ________________ and f is increasing for ________________.

real numbers, any real number

119 $g(x) = |x + 1|$

The domain of g is the set of all ________________. Since $g(-x) = |-x + 1| \neq$ _____, the graph is not symmetric with respect to the _________ and g is not _________. Also, $g(-x) \neq$ ______, so g is not an odd function. The graph is

real numbers

$g(x)$

y axis, even, $-g(x)$

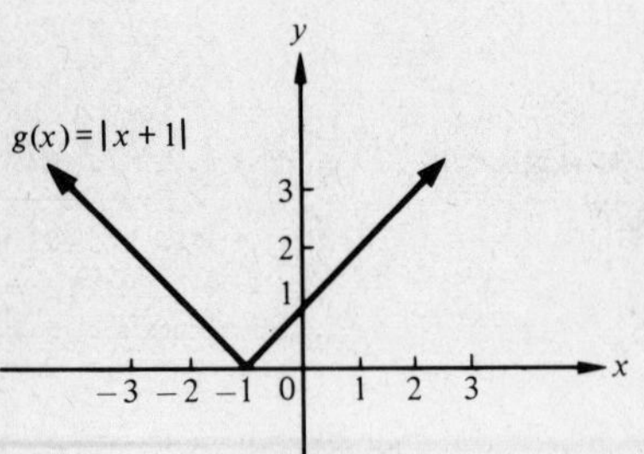

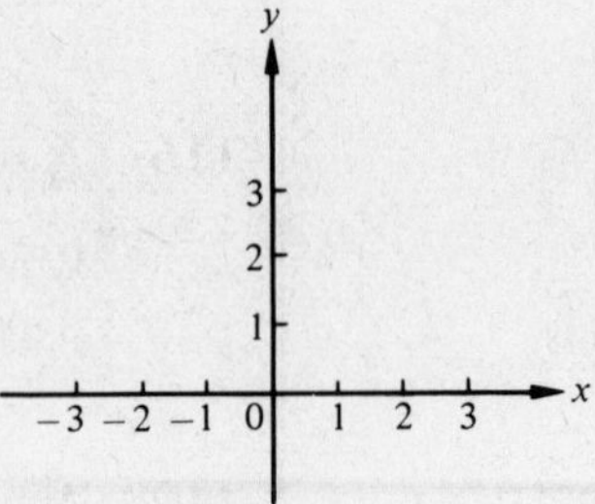

The graph shows that the range of g is the set of ______________ real numbers, that is $\{y \mid y \geqslant 0\}$, and it shows that g is increasing in ___________ and decreasing in ___________.

nonnegative

$[-1, \infty)$, $(-\infty, -1]$

120 $h(x) = |-x|$

$|-x| =$ _____. The domain of h is the set of ___________________.
Since $h(-x) = |-(-x)| =$ _____, the graph is symmetric with respect to the _________ and h is an _________ function. The graph is

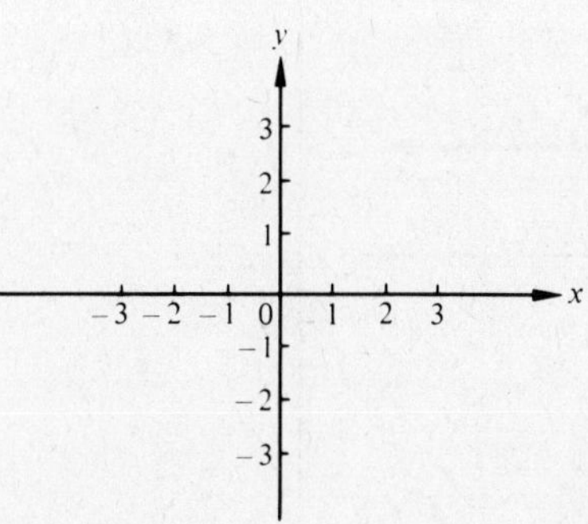

$|x|$, real numbers

$h(x)$

y axis, even

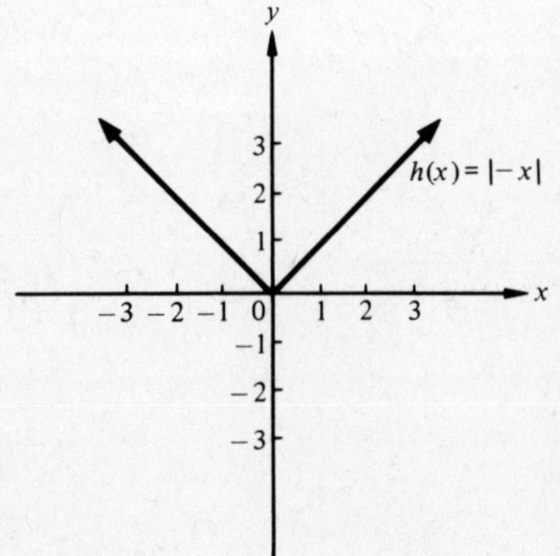

The graph reveals that the range of h is the set of _____________________________________ and h is increasing in ___________ and decreasing in ___________.

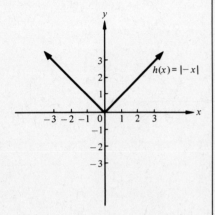

nonnegative real numbers

$[0, \infty)$, $(-\infty, 0]$

121 $f(x) = -\frac{1}{2}x^2$

The domain of f is the set of ___________________. Since $f(-x) = -\frac{1}{2}(-x)^2 = -\frac{1}{2}x^2 =$ _____, the graph is symmetric with respect to the _________ and f if an _________ function. The graph is

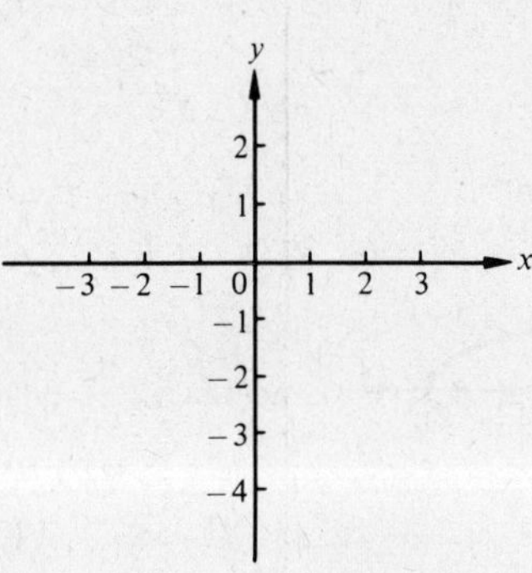

real numbers

$f(x)$

y axis, even

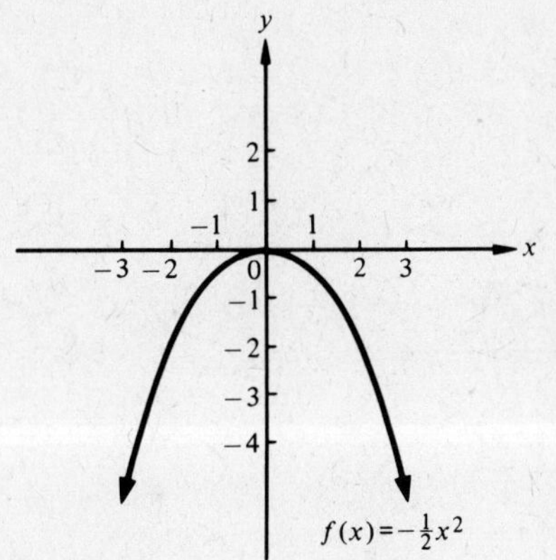

The graph shows that the range of f is the set of _____________________________________, that is, $\{y \mid y \leqslant 0\}$, and f is increasing in __________ and decreasing in __________.

nonpositive real numbers

$(-\infty, 0]$, $[0, \infty)$

real numbers

$g(x)$

y axis, even

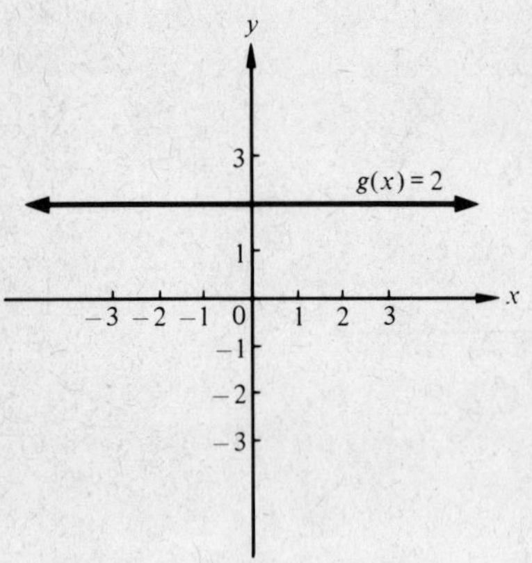

{2}

increasing, decreasing

$\{x \mid x \leqslant 9\}$

$f(x)$, $-f(x)$

y axis, origin

odd

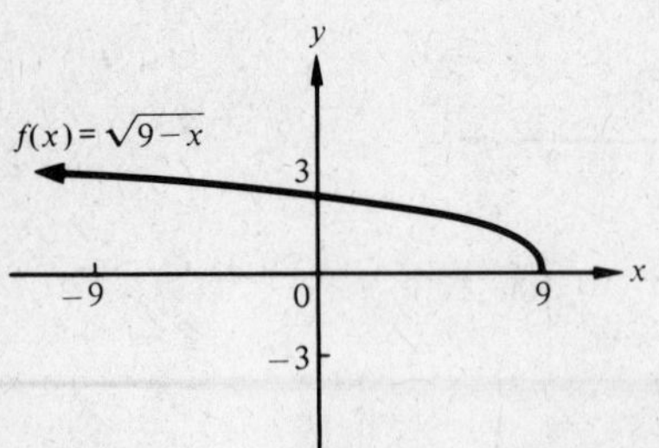

nonnegative real numbers

$-\infty, 9]$

122 $g(x) = 2$

The domain of g is the set of ________________. Since $g(-x) = 2 =$ ______, the graph is symmetric with respect to the ________ and g is an ________ function. The graph is

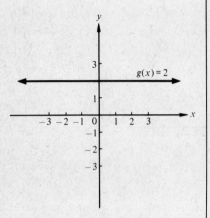

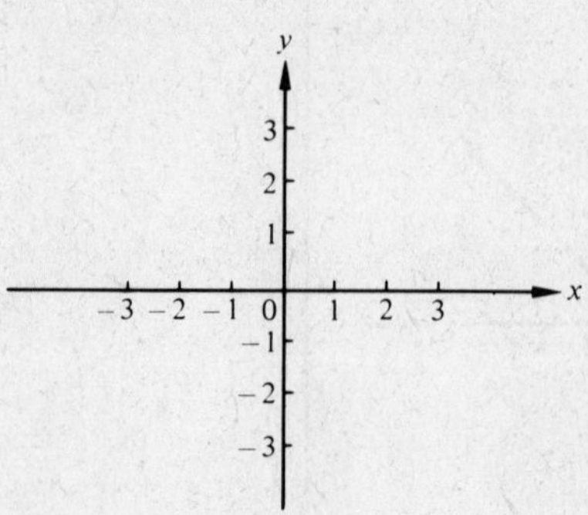

The graph shows that the range of g is the set ____ and g is neither ________________ nor ________________.

123 $f(x) = \sqrt{9 - x}$

The domain of f is the set ____________. Since $f(-x) = \sqrt{9 - (-x)} = \sqrt{9 + x} \neq$ ____ or ______, f is not symmetric with respect to the ________ or the ________ and f is neither even nor ____. The graph is

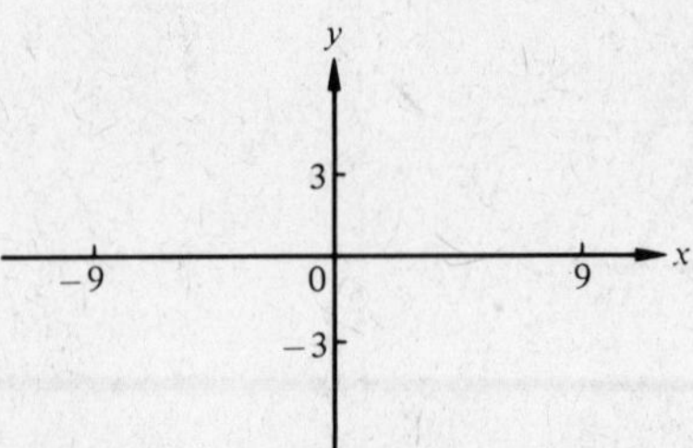

The graph shows that the range of f is the set of ________________________________, that is, $\{ y \mid y \geqslant 0\}$, and f is decreasing in ________.

124 $G(x) = \sqrt{9 - x^2}$

$\{x \mid -3 \leqslant x \leqslant 3\}$

The domain of the function G is ________________.

$G(-x) = \sqrt{9 - (-x)^2} = G(x)$, so that G is symmetric with respect to the ________ and G is an ________ function. The graph is

y axis, even

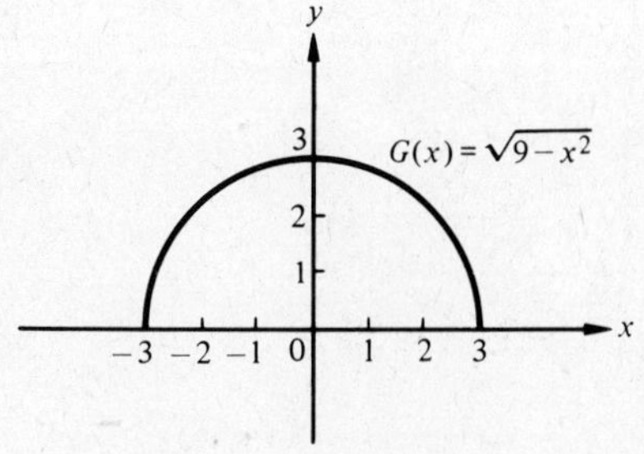

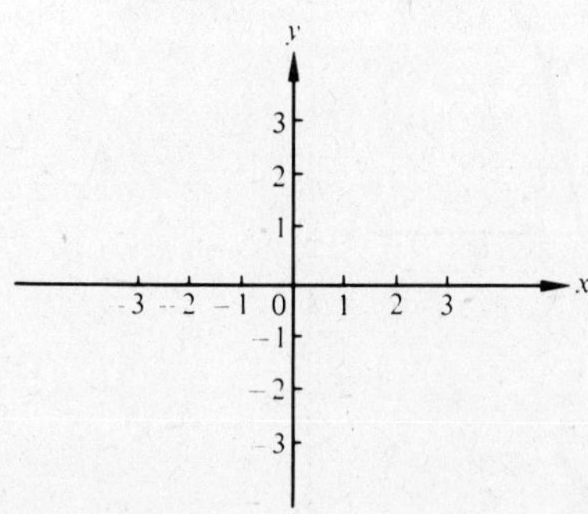

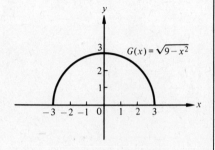

$\{y \mid 0 \leqslant y \leqslant 3\}$

$[-3, 0]$, $[0, 3]$

The graph shows that the range of G is ________________ and G is increasing in ________ and decreasing in ________.

125 $h(x) = [\![2x]\!]$

real numbers

$h(x)$, $-h(x)$

even

odd

The domain of h is the set of ________________. Since $h(-x) = [\![-2x]\!] \neq$ ____ or ____, h is not symmetric with respect to the y axis or the origin and h is neither an ______ nor an ______ function. The graph is

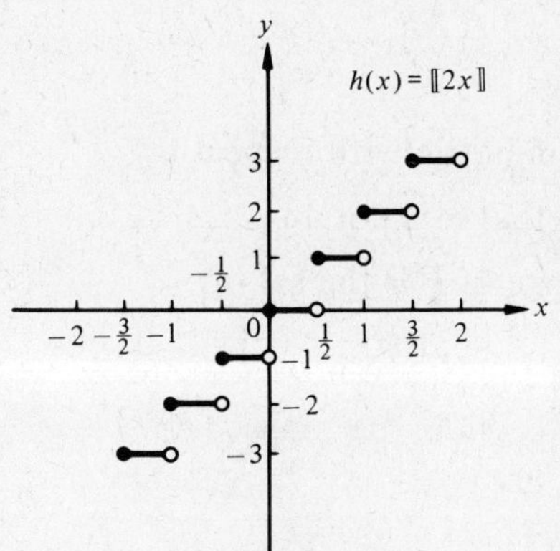

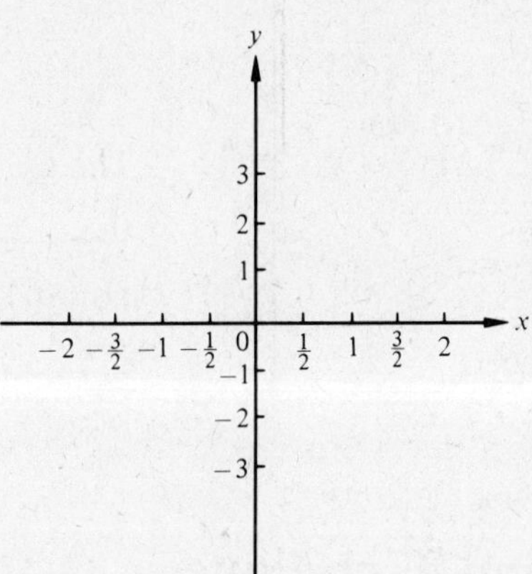

integers, decreasing

The graph shows that the range of h consists of the set of all ____________ and h is neither increasing nor ____________.

126 $g(x) = \begin{cases} 3x - 1 & \text{if } x \geqslant 2 \\ 2x^2 - 3 & \text{if } x < 2 \end{cases}$

The domain of the function g is the ______________.

set of real numbers

The graph is

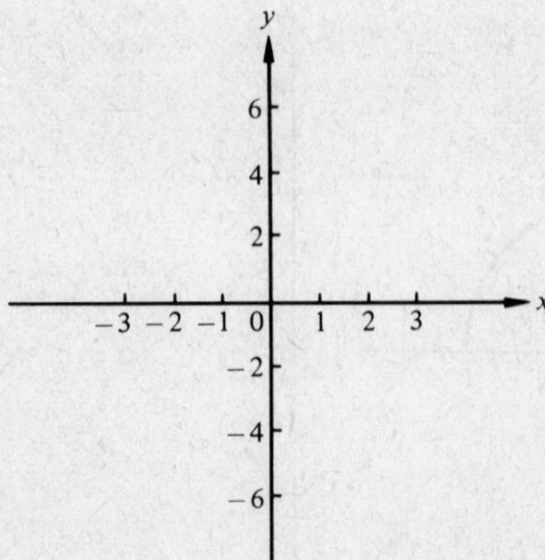

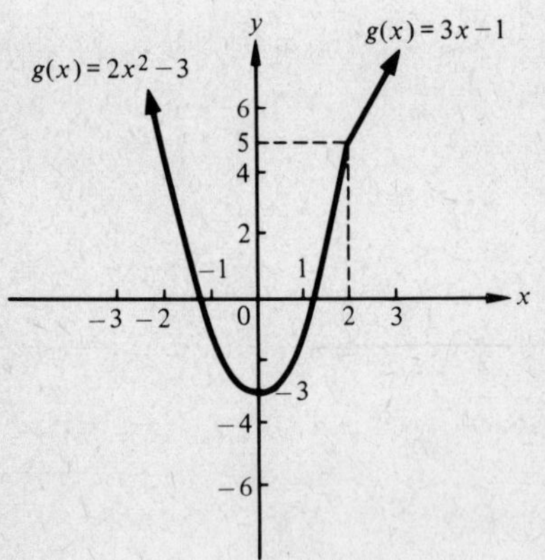

The graph shows that g is not symmetric with respect to either the ________ or the ________, and the range of g is ________. Also, g is increasing in ________ and decreasing in ________.

y axis, origin, $\{y \mid y \geqslant -3\}$

$[0, \infty)$, $(-\infty, 0]$

127 $F(x) = |x| + x$

The domain of F is the set of ______________. The graph is

real numbers

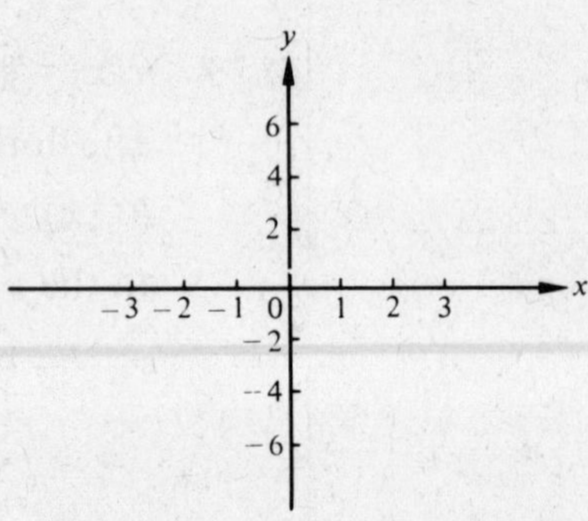

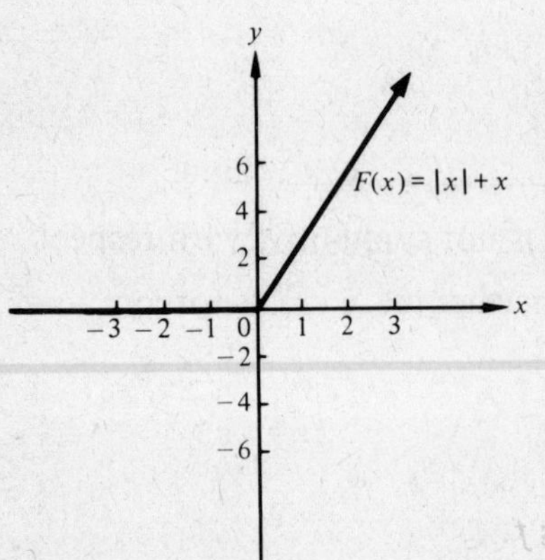

The graph indicates that F is not symmetric with respect to the ________ or the ________, so that F is not an ________ nor an ________ function. The range of F is the set of ______________, that is, $\{y \mid y \geqslant 0\}$. Also, F is increasing in ________.

y axis, origin, even

odd

nonnegative real numbers

$[0, \infty)$

4.4 Graphing Techniques

In Problems 128–132, let $y = f(x)$ and $c > 0$.

128 The graph of $y = f(x) + c$ is the graph of $y = f(x)$ shifted vertically ________ upward.

c units

c units

129 The graph of $y = f(x) - c$ is the graph of $y = f(x)$ shifted vertically ________ downward.

horizontally, right

130 The graph of $y = f(x - c)$ is the graph of $y = f(x)$ shifted ________ c units to the ________.

horizontally, left

131 The graph of $y = f(x + c)$ is the graph of $y = f(x)$ shifted ________ c units to the ______.

stretching, vertically

132 The graph of $y = cf(x)$ is obtained from the graph of $y = f(x)$ by ________ $y = f(x)$ by a factor of c ________.

x axis

133 The graph of $y = -f(x)$ is obtained by reflecting the graph of $y = f(x)$ across the ________.

In Problems 134–138, use the given known graph to sketch the graph of each function.

1 unit

134 $F(x) = |x| - 1$ if the known graph is $f(x) = |x|$

The graph of F is obtained by shifting the graph of f vertically ________ downward. The graph is

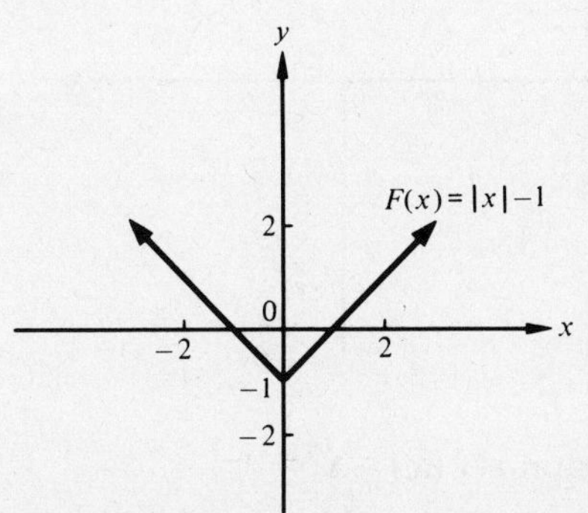

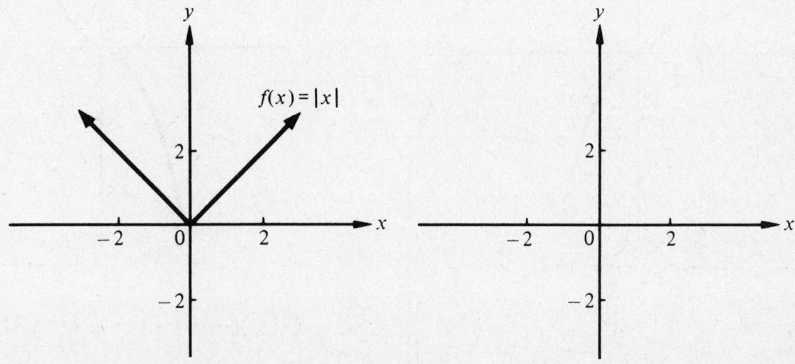

2 units

135 $F(x) = |x + 2|$ if the known graph is $f(x) = |x|$

The graph of F is obtained by shifting the graph of f horizontally ________ to the left. The graph is

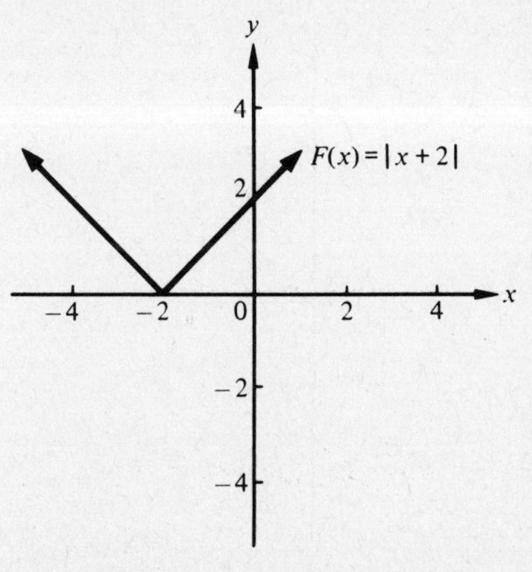

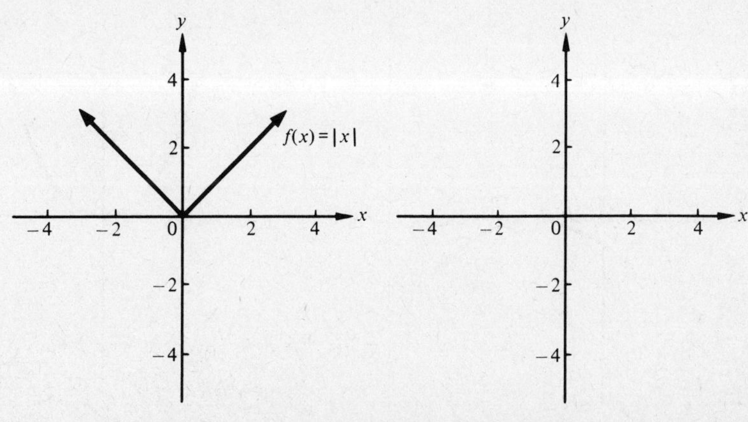

136 $G(x) = \frac{1}{2}x^2$ if the known graph is $g(x) = x^2$

$\frac{1}{2}$

The graph of G is obtained by shrinking the graph of g by a factor of ____ vertically. The graph is

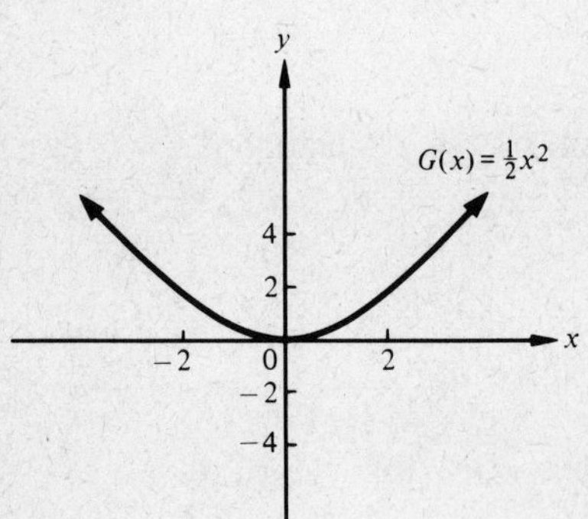

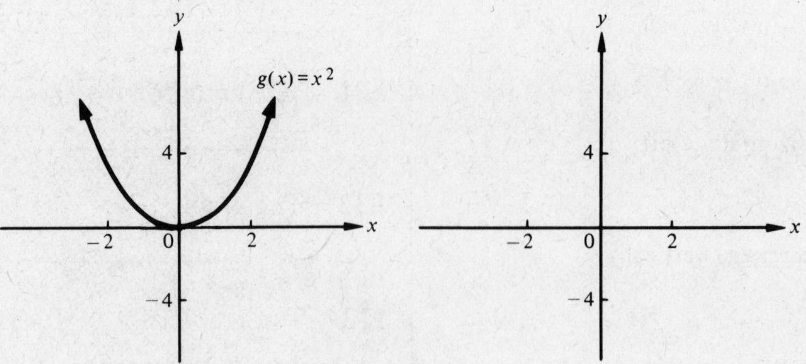

137 $H(x) = -x^3$ if the known graph is $h(x) = x^3$

the x axis

The graph of H is obtained by reflecting the graph of h across ________________. The graph is

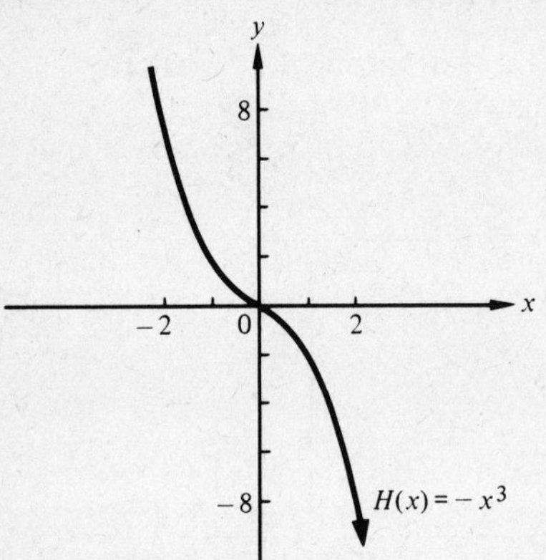

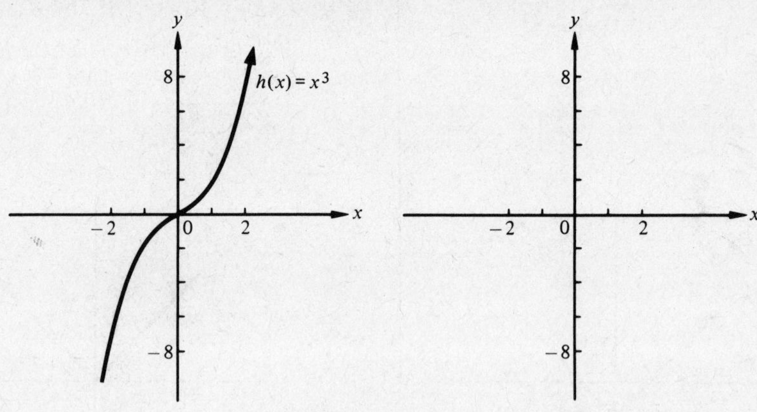

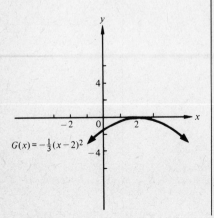

138 $G(x) = -\frac{1}{3}(x - 2)^2$ if the known graph is $g(x) = x^2$

$\frac{1}{3}$

x axis

2 units

The graph of $y = \frac{1}{3}x^2$ is obtained from $g(x) = x^2$ by shrinking the graph of g by a factor of ____ vertically. The graph of $y = -\frac{1}{3}x^2$ is then obtained by reflection across the ________. The graph of $G(x) = -\frac{1}{3}(x - 2)^2$ is obtained by shifting the graph of $y = -\frac{1}{3}x^2$ horizontally ________ to the right. The graph is

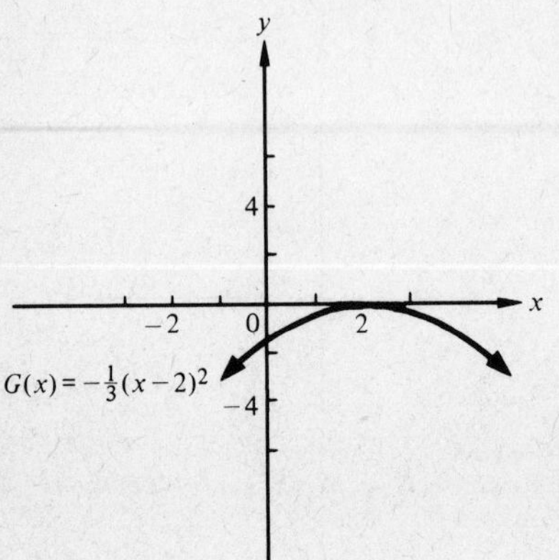

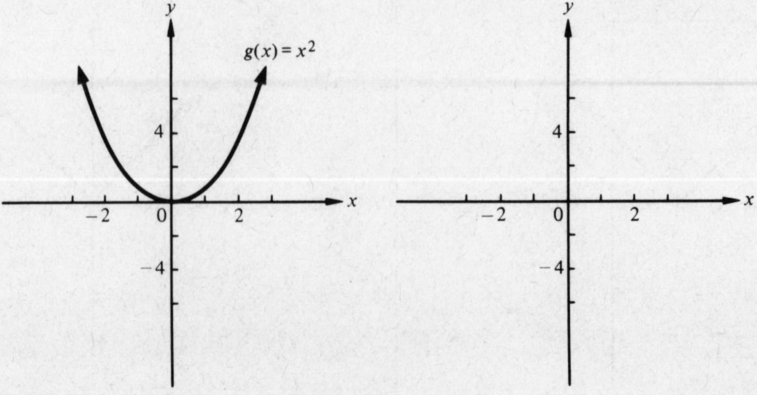

5 Algebra of Functions and Composition of Functions

5.1 Algebra of Functions

139 Let f and g be two functions with domains that intersect. We define the sum, the difference, the product, and the quotient of f and g, respectively, as follows:

$(f+g)(x) =$ ______

$(f-g)(x) =$ ______

$(f \cdot g)(x) =$ ______

$\left(\frac{f}{g}\right)(x) = \frac{(\_\_\_)}{g(x)}$, if $g(x) \neq 0$

Answers: $f(x) + g(x)$; $f(x) - g(x)$; $f(x) \cdot g(x)$; $f(x)$

In Problems 140–145, use the given pair of functions to find $f+g, f-g, f \cdot g$, and f/g.

140 $f(x) = \sqrt{4 - x^2}$ and $g(x) = \frac{2}{x}$

$(f+g)(x) = f(x) + g(x) =$ ______

$(f-g)(x) = f(x) - g(x) =$ ______

$(f \cdot g)(x) = f(x) \cdot g(x) =$ ______

$\left(\frac{f}{g}\right)(x) = \frac{f(x)}{g(x)} =$ ______

Answers: $\sqrt{4-x^2} + \frac{2}{x}$; $\sqrt{4-x^2} - \frac{2}{x}$; $\frac{2}{x}\sqrt{4-x^2}$; $\frac{x\sqrt{4-x^2}}{2}$

141 $f(x) = 2x$ and $g(x) = 3x^2$

$(f+g)(x) = f(x) + g(x) =$ ______

$(f-g)(x) = f(x) - g(x) =$ ______

$(f \cdot g)(x) = f(x) \cdot g(x) =$ ______

$\left(\frac{f}{g}\right)(x) = \frac{f(x)}{g(x)} =$ ______

Answers: $2x + 3x^2$; $2x - 3x^2$; $6x^3$; $\frac{2}{3x}$

142 $f(x) = x^2 - 4$ and $g(x) = 2x + 7$

$(f+g)(x) = f(x) + g(x) =$ ______

$(f-g)(x) = f(x) - g(x) =$ ______

$(f \cdot g)(x) = f(x) \cdot g(x) =$ ______

$\left(\frac{f}{g}\right)(x) = \frac{f(x)}{g(x)} =$ ______

Answers: $x^2 + 2x + 3$; $x^2 - 2x - 11$; $2x^3 + 7x^2 - 8x - 28$; $\frac{x^2 - 4}{2x + 7}$

143 $f(x) = x^3$ and $g(x) = -3x^2 + 1$

$(f+g)(x) = f(x) + g(x) =$ ______

$(f-g)(x) = f(x) - g(x) =$ ______

$(f \cdot g)(x) = f(x) \cdot g(x) =$ ______

$\left(\frac{f}{g}\right)(x) = \frac{f(x)}{g(x)} =$ ______

Answers: $x^3 - 3x^2 + 1$; $x^3 + 3x^2 - 1$; $-3x^5 + x^3$; $\frac{x^3}{-3x^2 + 1}$

Answers	Frames
	144 $f(x) = x^3$ and $g(x) = 5x^2$
$x^3 + 5x^2$	$(f+g)(x) = f(x) + g(x) =$ ________
$x^3 - 5x^2$	$(f-g)(x) = f(x) - g(x) =$ ________
$5x^5$	$(f \cdot g)(x) = f(x) \cdot g(x) =$ ____
$\frac{x}{5}$	$\left(\frac{f}{g}\right)(x) = \frac{f(x)}{g(x)} =$ ____
	145 $f(x) = x^2 + 1$ and $g(x) = x$
$x^2 + 1 + x$	$(f+g)(x) = f(x) + g(x) =$ ________
$x^2 + 1 - x$	$(f-g)(x) = f(x) - g(x) =$ ________
$x(x^2 + 1)$, $x^3 + x$	$(f \cdot g)(x) = f(x) \cdot g(x) =$ ________ = ________
$\frac{x^2+1}{x}$	$\left(\frac{f}{g}\right)(x) = \frac{f(x)}{g(x)} =$ ______
	146 The function f consists of the ordered pairs (2, 2), (3, 3), (4, 2), and (5, 1). The function g consists of the ordered pairs (2, 4), (3, 9), (4, 6), and (5, 7). The domain of f is the set consisting of the numbers
2, 3, 4, and 5	____________.
	The domain of g is the set consisting of the numbers
2, 3, 4, and 5	____________.
	Thus,
(3, 12), (4, 8), (5, 8)	$f + g$ consists of the ordered pairs (2, 6), ____________;
(4, −4), (5, −6)	$f - g$ consists of the ordered pairs (2, −2), (3, −6), ____________;
(2, 8), (3, 27)	$f \cdot g$ consists of the ordered pairs ____________, (4, 12), (5, 7);
$(4, \frac{1}{3}), (5, \frac{1}{7})$	and $\frac{f}{g}$ consists of the ordered pairs $(2, \frac{1}{2}), (3, \frac{1}{3})$, ____________.

5.2 Composition of Functions

Answers	Frames
composite	**147** Let f and g be two functions. The function $f \circ g$, defined by $(f \circ g)(x) = f[g(x)]$, is called a ____________ function. The domain of $f \circ g$ is the subset of the domain of g containing those values for which $f \circ g$ is defined.

In Problems 148–152, let $f(x) = 5x + 2$, $g(x) = x^2$, and $h(x) = \frac{1}{5}(x - 2)$. Evaluate the given compositions.

Answers	Frames
	148 $(f \circ g)(3)$
9	$(f \circ g)(3) = f[g(3)] = f($____$)$
9, 47	$= 5($____$) + 2 =$ ____
	149 $g[f(3)]$
17	$g[f(3)] = g($____$)$
17, 289	$= ($____$)^2 =$ ____

	150 $f[g(x)]$
x^2	$f[g(x)] = f(\underline{\hspace{1cm}})$
x^2, $5x^2 + 2$	$= 5(\underline{\hspace{1cm}}) + 2 = \underline{\hspace{2cm}}$
	151 $(f \circ h)(x)$ and $(h \circ f)(x)$
$\frac{1}{5}(x - 2)$	$(f \circ h)(x) = f[h(x)] = f[\underline{\hspace{3cm}}]$
$\frac{1}{5}(x - 2)$, x	$= 5[\underline{\hspace{3cm}}] + 2 = \underline{\hspace{1cm}}$
$5x + 2$	$(h \circ f)(x) = h[f(x)] = h(\underline{\hspace{2cm}})$
$5x + 2$, x	$= \frac{1}{5}[(\underline{\hspace{2cm}}) - 2] = \underline{\hspace{1cm}}$
	152 $(f \circ f)(x)$
$f(x)$, $5x + 2$	$(f \circ f)(x) = f[\underline{\hspace{1.5cm}}] = f(\underline{\hspace{2cm}})$
$5x + 2$, $25x + 12$	$= 5(\underline{\hspace{2cm}}) + 2 = \underline{\hspace{2.5cm}}$

In Problems 153–157, find $f[g(x)]$ and $g[f(x)]$ for the given pair of functions.

	153 $f(x) = 2$ and $g(x) = 2x^2 - 5$
$2x^2 - 5$, 2	$f[g(x)] = f(\underline{\hspace{2cm}}) = \underline{\hspace{1cm}}$
2, 3	$g[f(x)] = g(\underline{\hspace{1cm}}) = 2(2)^2 - 5 = \underline{\hspace{1cm}}$
	Therefore, $f[g(x)] \neq g[f(x)]$.
	154 $f(x) = x^2$ and $g(x) = 2x + 3$
$2x + 3$, $(2x + 3)^2$	$f[g(x)] = f(\underline{\hspace{2cm}}) = \underline{\hspace{3cm}}$
x^2, $2x^2 + 3$	$g[f(x)] = g(\underline{\hspace{1cm}}) = \underline{\hspace{2.5cm}}$
	155 $f(x) = 2x^2 + 6$ and $g(x) = 7x + 2$
$7x + 2$	$f[g(x)] = f(\underline{\hspace{2cm}})$
$(7x + 2)^2$, $98x^2 + 56x + 14$	$= 2[\underline{\hspace{3cm}}] + 6 = \underline{\hspace{5cm}}$
$2x^2 + 6$	$g[f(x)] = g(\underline{\hspace{2.5cm}})$
$2x^2 + 6$, $14x^2 + 44$	$= 7(\underline{\hspace{2.5cm}}) + 2 = \underline{\hspace{3cm}}$
	156 $f(x) = x + 5$ and $g(x) = x - 5$
$x - 5$, x	$f[g(x)] = f(\underline{\hspace{1.5cm}}) = \underline{\hspace{1cm}}$
$x + 5$, x	$g[f(x)] = g(\underline{\hspace{1.5cm}}) = \underline{\hspace{1cm}}$
	157 $f(x) = \sqrt{x}$ and $g(x) = 1 - x^2$
$1 - x^2$, $\sqrt{1 - x^2}$	$f[g(x)] = f(\underline{\hspace{2cm}}) = \underline{\hspace{2cm}}$
$\sqrt{x}$, $1 - x$	$g[f(x)] = g(\underline{\hspace{1cm}}) = \underline{\hspace{1.5cm}}$
	158 Determine k so that $(f \circ g)(x) = (g \circ f)(x)$ when $f(x) = 3x - 7$ and $g(x) = 2x + k$.
$2x + k$, $6x + 3k - 7$	$(f \circ g)(x) = f[g(x)] = f(\underline{\hspace{2cm}}) = \underline{\hspace{3cm}}$
$3x - 7$, $6x - 14 + k$	$(g \circ f)(x) = g[f(x)] = g(\underline{\hspace{2cm}}) = \underline{\hspace{3cm}}$
$6x - 14 + k$	Since $(f \circ g)(x) = (g \circ f)(x)$, we have $6x + 3k - 7 = \underline{\hspace{3cm}}$.
$-\frac{7}{2}$	Solving for k, we obtain $k = \underline{\hspace{1cm}}$.

In Problems 159–161, find two functions f and g that will produce the composite function h such that $h(x) = f[g(x)]$.

Answers	Frames
	159 $h(x) = (3x + 7)^4$
u^4, $3x + 7$	If we let $f(u) =$ ____ and $u = g(x) =$ ________, then
$f[g(x)]$, $(3x + 7)^4$	$h(x) =$ __________ $= f(3x + 7) =$ ____________.
	160 $h(x) = \left(\dfrac{x+3}{x-3}\right)^{-6}$
u^{-6}, $\dfrac{x+3}{x-3}$	If we let $f(u) =$ ____ and $u = g(x) =$ ________, then
$f[g(x)]$, $\left(\dfrac{x+3}{x-3}\right)^{-6}$	$h(x) =$ __________ $= f\left(\dfrac{x+3}{x-3}\right) =$ ____________.
	161 $h(x) = (x^2 + x^{-2})^3$
u^3, $x^2 + x^{-2}$	If we let $f(u) =$ ____ and $u = g(x) =$ ____________, then
$f[g(x)]$, $x^2 + x^{-2}$	$h(x) =$ __________ $= f($____________$) = (x^2 + x^{-2})^3$.
	162 Let $f(x) = 5x + 3$ and $g(x) = -2x + 7$. Solve $f[g(x)] = 8$ for x.
$-2x + 7$, 8	$f[g(x)] = f[$__________$] = 5(-2x + 7) + 3 =$ ____, so that
$-10x + 38$, -30, 3	____________ $= 8$ or $-10x =$ ____. Therefore, $x =$ ____.
$5f(x)$	**163** If f is the identity function, then $f(5x) = 5x =$ ________ and
$9f(w)$	$f(5x + 9w) = 5f(x) +$ ________.
	164 If f is the absolute value function, does $f(a + b) = f(a) + f(b)$?
$\lvert a + b\rvert$	$f(a + b) =$ __________
$\lvert b\rvert$	$f(a) + f(b) = \lvert a\rvert +$ ____
$f(a) + f(b)$	Since $\lvert a + b\rvert \leqslant \lvert a\rvert + \lvert b\rvert$, $f(a + b) \neq$ ______________.
	165 If f is the greatest integer function, does $f(4a) = 4f(a)$?
	$f(x) = [\![x]\!]$
$4[\![a]\!]$	$f(4a) = [\![4a]\!] \neq$ ______ $= 4f(a)$
	For example, if $a = \frac{1}{3}$,
1	$f(4a) = [\![\frac{4}{3}]\!] =$ ____, while
0, 0	$4f(a) = 4[\![\frac{1}{3}]\!] = 4($____$) =$ ____.
	166 If f is the absolute value function, does $f(ab) = f(a)\,f(b)$?
	$f(x) = \lvert x\rvert$
$\lvert ab\rvert$	$f(ab) =$ ______
$\lvert a\rvert \cdot \lvert b\rvert$	$f(a)\,f(b) =$ ____________
$f(a)\,f(b)$	Since $\lvert ab\rvert = \lvert a\rvert \cdot \lvert b\rvert$, $f(ab) =$ ____________.
	167 Let $f(x) = 3x + 2$. Does $f(a + b) = f(a) + f(b)$?
$a + b$, $3a + 3b + 2$	$f(a + b) = 3($______$) + 2 =$ ______________
$3b + 2$, $3a + 3b + 4$	$f(a) + f(b) = 3a + 2 +$ __________ $=$ ________________
$\neq$	Therefore, $f(a + b)$ ______ $f(a) + f(b)$.
	168 Suppose that a cylindrical vessel has a circular base of radius 2 inches, and further suppose that the height of its contents after t seconds is expressed by the function $h = 3t + 5$. Construct a function that expresses the volume of the contents of the vessel as a function of time.
2, $4\pi h$	$V = \pi r^2 h = \pi\,($_____$)^2 h =$ ____
$4\pi(3t + 5)$	But $h = 3t + 5$, so that $V =$ ______________.

6 Inverse Functions

169 Two functions f and g are said to be inverses of each other if the following conditions hold:

(i) The range of g is contained in the ______ of f. — domain

(ii) For every number x in the domain of g, $(f \circ g)(x) =$ ____. — x

(iii) The range of f is contained in the ______ of g. — domain

(iv) For every number x in the domain of f, $(g \circ f)(x) =$ ____. — x

170 Suppose that f is an invertible function. Then we define the inverse of the function f, written f^{-1}, to be the function whose graph is the reflection of the graph of f across the straight line ______. — $y = x$

In Problems 171–176, show that the function f is the inverse of the function g by showing that $(f \circ g)(x) = (g \circ f)(x) = x$.

171 $f(x) = 7x$ and $g(x) = \dfrac{x}{7}$

$(f \circ g)(x) = f[g(x)] = f\left(\dfrac{x}{7}\right) = 7(\_\_\_) =$ ____ — $\dfrac{x}{7}$, x

$(g \circ f)(x) = g[f(x)] = g(7x) = \frac{1}{7}(\_\_\_) =$ ____ — $7x$, x

Thus, $(f \circ g)(x) = (g \circ f)(x) =$ ____. — x

172 $f(x) = 5x + 1$ and $g(x) = \dfrac{x-1}{5}$

$(f \circ g)(x) = f[g(x)] = f(\_\_\_\_) =$ ____ — $\dfrac{x-1}{5}$, x

$(g \circ f)(x) = g[f(x)] = g(\_\_\_\_) =$ ____ — $5x + 1$, x

Therefore, $(f \circ g)(x) = (g \circ f)(x) =$ ____. — x

173 $f(x) = 2x + 2$ and $g(x) = \dfrac{x}{2} - 1$

$(f \circ g)(x) = f[g(x)] = f(\_\_\_\_) =$ ____ — $\dfrac{x}{2} - 1$, x

$(g \circ f)(x) = g[f(x)] = g(\_\_\_\_) =$ ____ — $2x + 2$, x

Therefore, $(f \circ g)(x) = (g \circ f)(x) =$ ____. — x

174 $f(x) = 7x + 2$ and $g(x) = \dfrac{x-2}{7}$

$(f \circ g)(x) = f[g(x)] = f(\_\_\_\_) =$ ____ — $\dfrac{x-2}{7}$, x

$(g \circ f)(x) = g[f(x)] = g(\_\_\_\_) =$ ____ — $7x + 2$, x

Therefore, f and g are ________ of each other: — inverses

$f^{-1} =$ ____ and $g^{-1} =$ ____ — g, f

175 $f(x) = -\frac{1}{2}x + 8$ and $g(x) = -2x + 16$

$(f \circ g)(x) = f[g(x)] = f($____________$) =$ ____

$(g \circ f)(x) = g[f(x)] = g($____________$) =$ ____

Therefore, f and g are ____________ of each other:

$f^{-1} =$ ____ and $g^{-1} =$ ____

$-2x + 16,\ x$

$-\frac{1}{2}x + 8,\ x$

inverses

$g,\ f$

176 $f(x) = 3x - 4$ and $g(x) = \dfrac{x+4}{3}$

$(f \circ g)(x) = f[g(x)] = f\left(\dfrac{\quad}{\quad}\right) =$ ____

$(g \circ f)(x) = g[f(x)] = g($________$) =$ ____

Therefore, f and g are ____________ of each other.

$\dfrac{x+4}{3},\ x$

$3x - 4,\ x$

inverses

6.1 Existence of the Inverse Function

177 In order for a function $y = f(x)$ to be invertible, it must be true that for each y in the range of f there is one and only one ____ in the domain of f; that is, each possible horizontal line intersects the graph of $y = f(x)$ no more than ______.

x

once

In Problems 178–180, use the graph of the given function to decide whether or not that function is invertible.

178 $f(x) = -2x^2$

The graph of f is

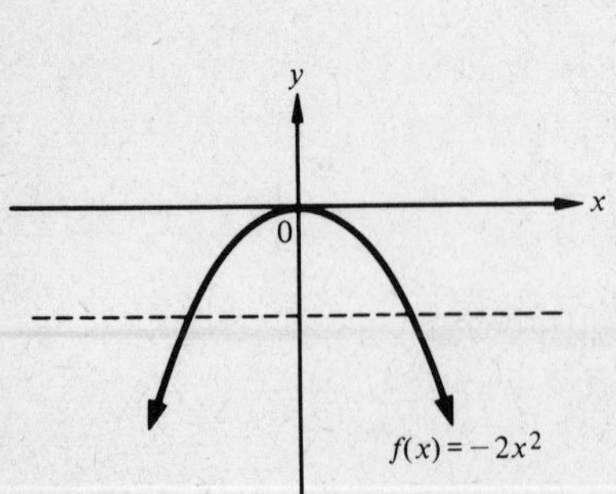

The graph shows that any horizontal line below the x axis intersects the graph ______. Therefore, f is ________________.

twice, not invertible

179 $g(x) = \frac{1}{2}x^3$

The graph of g is

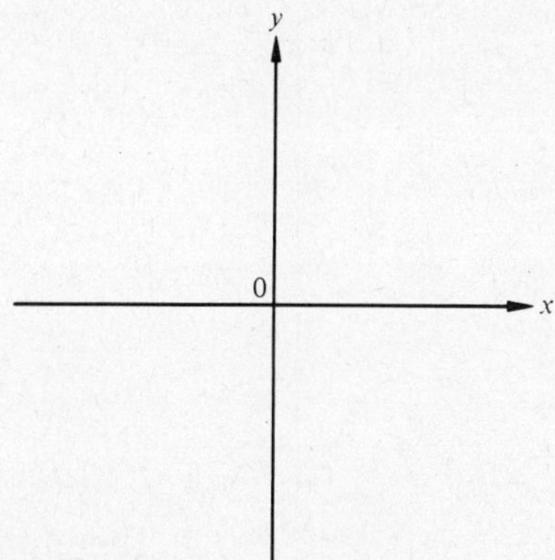

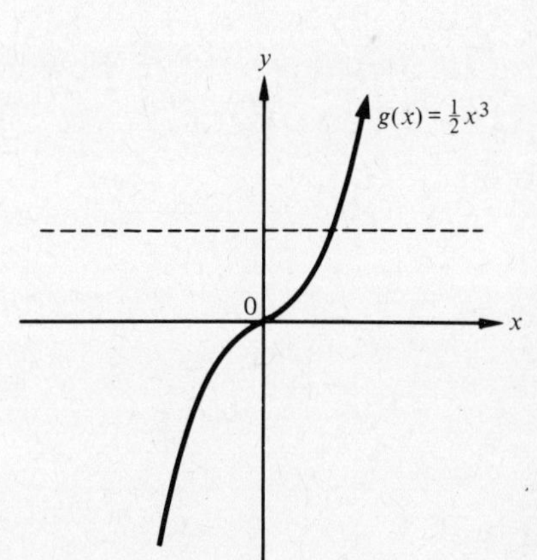

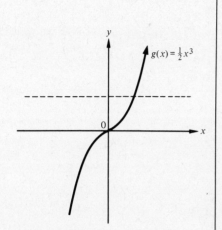

The graph of g shows that any horizontal line intersects the graph ____. Therefore, g is ____________.

once, invertible

180 $h(x) = 5x - 2$

The graph of h is

y

0 x

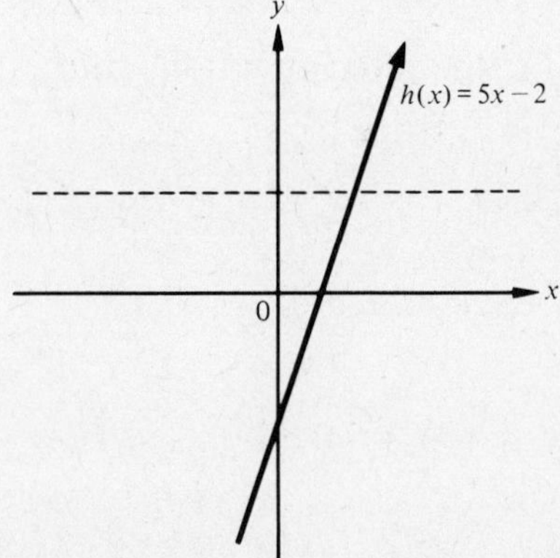

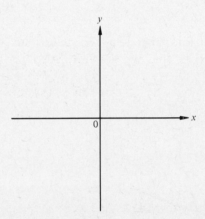

The graph of h shows that any horizontal line intersects the graph ____. Therefore, h is ____________.

once, invertible

6.2 Construction of the Inverse Function and Its Graph

181 If $y = f(x)$ is defined by an equation and if f is invertible, the equation that defines the inverse function can be constructed by interchanging the roles of x and y; in other words, we express x in terms of y to obtain f^{-1}.

$f[f^{-1}(x)] = f^{-1}[f(x)] =$ ____

x

182 If f^{-1} exists, and $y = f(x)$, then $x =$ ________.

$f^{-1}(y)$

In Problems 183–184, find f^{-1} and verify that $f[f^{-1}(x)] = f^{-1}[f(x)] = x$.

Answers	Frames
	183 $f(x) = 3x - 4$
$y + 4$, $\frac{y+4}{3}$	Let $y = f(x) = 3x - 4$. Then $3x =$ ______ or $x =$ ______. Now
$\frac{y+4}{3}$	$y = f(x)$ if and only if $x = f^{-1}(y)$, so that $f^{-1}(y) =$ ______ or
$\frac{x+4}{3}$	$f^{-1}(x) =$ ______.
$\frac{x+4}{3}$, x	$f[f^{-1}(x)] = f\left(\text{\_\_\_\_\_\_}\right) =$ ____
$3x - 4$, x	$f^{-1}[f(x)] = f^{-1}($______$) =$ ____
x	Therefore, $(f \circ f^{-1})(x) = (f^{-1} \circ f)(x) =$ ____.
	184 $f(x) = x^3$
$\sqrt[3]{y}$	If $y = f(x) = x^3$, then $x =$ ____. Now $y = f(x)$ if and only if
$\sqrt[3]{y}$, $\sqrt[3]{x}$	$x = f^{-1}(y)$, so that $f^{-1}(y) =$ ____ or $f^{-1}(x) =$ ____.
x	$f[f^{-1}(x)] = f(\sqrt[3]{x}) =$ ____
x	$f^{-1}[f(x)] = f^{-1}(x^3) =$ ____
$f^{-1} \circ f$	Therefore, $(f \circ f^{-1})(x) = ($______$)(x) = x$.

In Problems 185–188, determine whether or not f^{-1} exists for the given function. If f^{-1} exists, find and graph it. Also, verify that $f[f^{-1}(x)] = f^{-1}[f(x)] = x$.

185 $f(x) = 3x - 2$

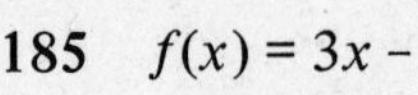

The graph is

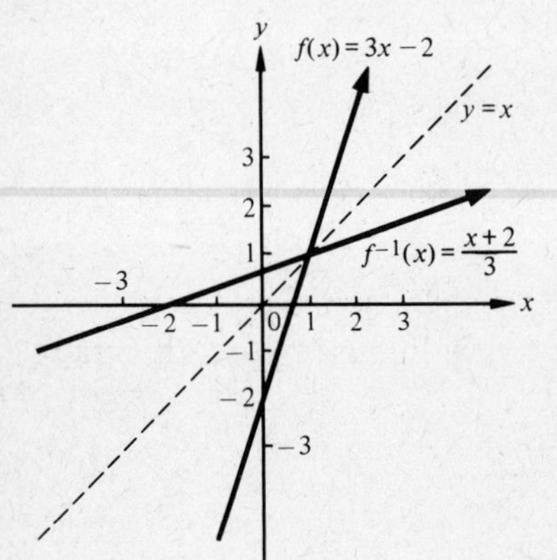

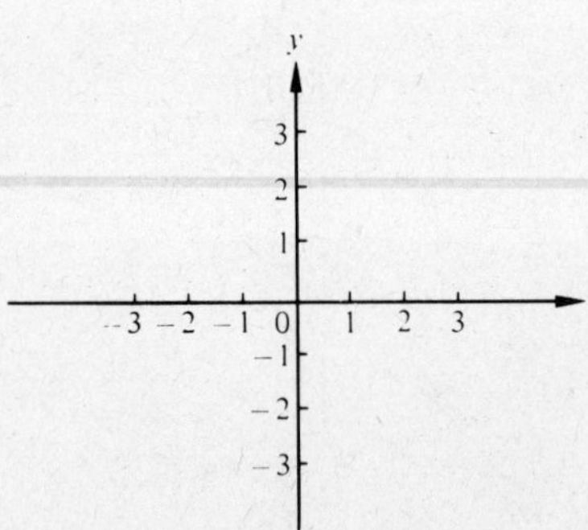

Answers	Frames
f^{-1} exists	The graph indicates that f is invertible, therefore ______.
$\frac{y+2}{3}$	Let $y = 3x - 2$, then $x =$ ______.
$\frac{y+2}{3}$, $\frac{x+2}{3}$	Therefore, $f^{-1}(y) =$ ______ or $f^{-1}(x) =$ ______.
x	$f[f^{-1}(x)] = f\left(\frac{x+2}{3}\right) =$ ____
x	$f^{-1}[f(x)] = f^{-1}(3x - 2) =$ ____
x	The graph of f^{-1} can be obtained by reflecting the graph of f across the line $y =$ ____.

186 $f(x) = x^2 - 4$, for $x \in [0, \infty)$

The graph is

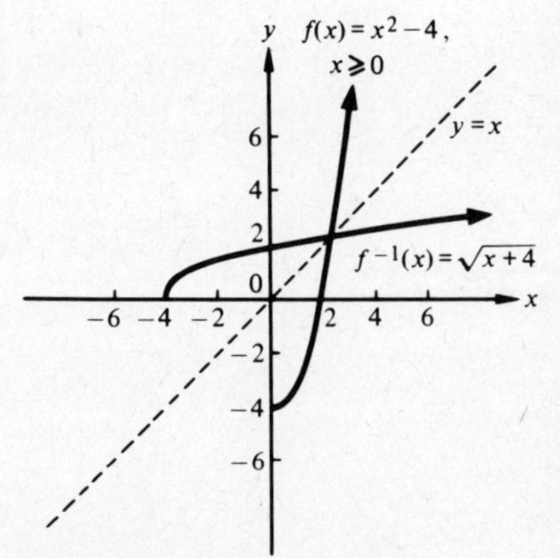

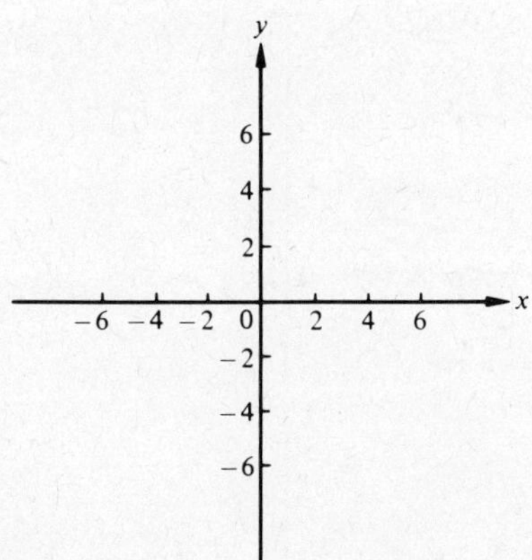

increasing

The graph of f indicates that f is an ______________ function, hence f^{-1} exists.

$y + 4$, $\sqrt{y+4}$

Let $y = x^2 - 4$, then $x^2 =$ ________ or $x =$ ________.

$\sqrt{x+4}$

Therefore, $f^{-1}(y) = \sqrt{y+4}$ or $f^{-1}(x) =$ ________.

x

$f[f^{-1}(x)] = f(\sqrt{x+4}) = (\sqrt{x+4})^2 - 4 =$ ____

x

$f^{-1}[f(x)] = f^{-1}(x^2 - 4) = \sqrt{x^2 - 4 + 4} =$ ____

$y = x$

The graph of f^{-1} is a reflection of the graph of f across ________.

187 $f(x) = x^2 - 1$, for $x \in (-\infty, -1]$

The graph is

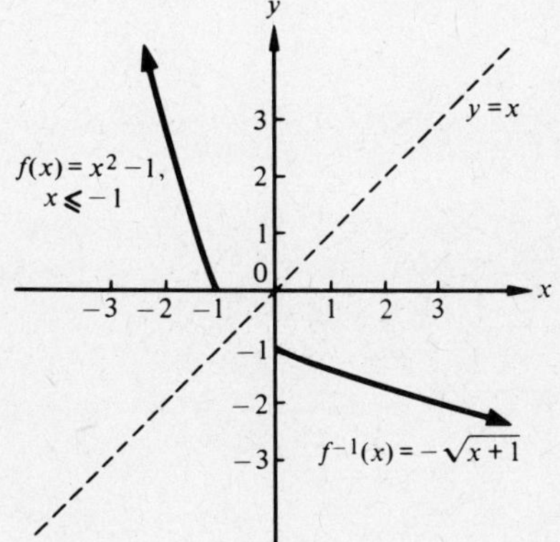

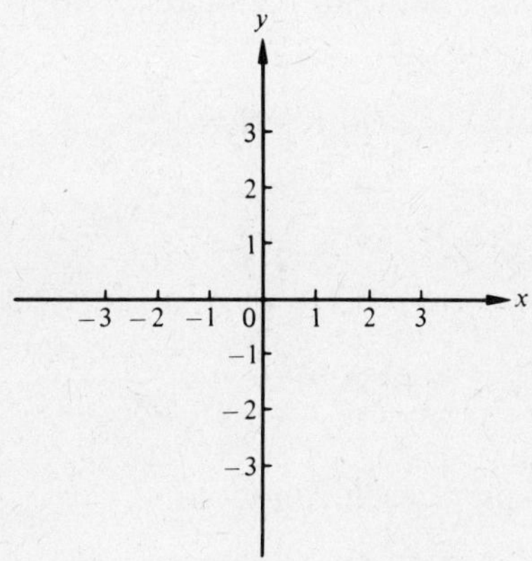

decreasing

The graph of f indicates that f is a ______________ function, hence f^{-1} exists.

$y + 1$, $-\sqrt{y+1}$

Let $y = x^2 - 1$, then $x^2 =$ ________ or $x =$ ________.

$-\sqrt{x+1}$

Therefore, $f^{-1}(y) = -\sqrt{y+1}$ or $f^{-1}(x) =$ ________.

x

$f[f^{-1}(x)] = f(-\sqrt{x+1}) = (-\sqrt{x+1})^2 - 1 =$ ____

$f^{-1}[f(x)] = f^{-1}(x^2 - 1) = -\sqrt{x^2 - 1 + 1} = -\sqrt{x^2}$

$-x$, x

$= -($____$) =$ ____ $(\sqrt{x^2} = -x \text{ if } x < 0)$

reflection

The graph of f^{-1} is a ______________ of the graph of f across $y = x$.

188 $f(x) = x^3 + 1$

The graph is

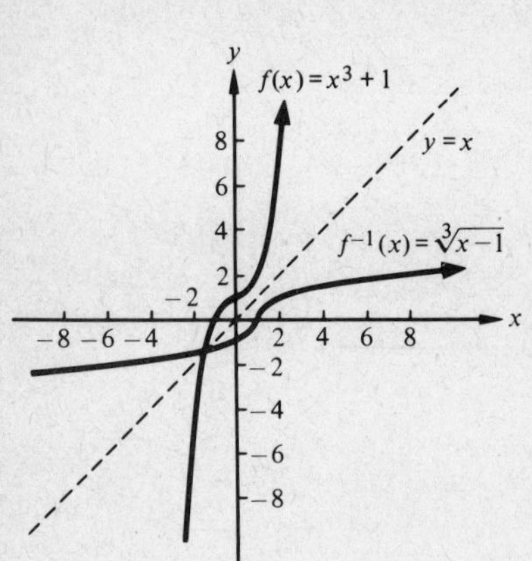

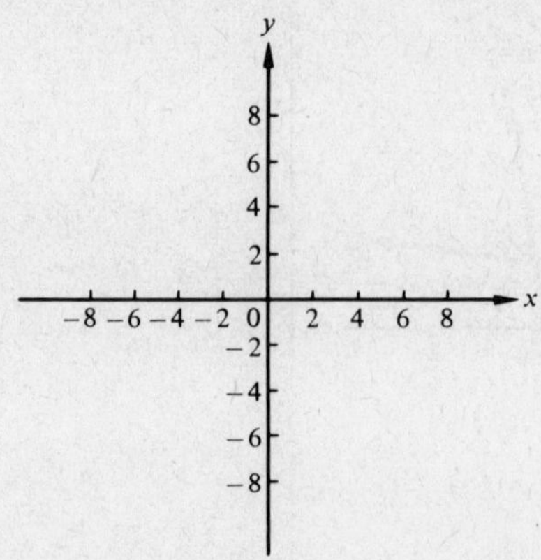

increasing	The graph of f indicates that f is an ____________ function, hence f^{-1} exists.
$y - 1$, $\sqrt[3]{y - 1}$	Let $y = x^3 + 1$, then $x^3 =$ ______ or $x =$ _______.
$\sqrt[3]{x - 1}$	Therefore, $f^{-1}(y) = \sqrt[3]{y - 1}$ or $f^{-1}(x) =$ _______.
x	$f[f^{-1}(x)] = f(\sqrt[3]{x - 1}) =$ ____
x	$f^{-1}[f(x)] = f^{-1}(x^3 + 1) =$ ____
$y = x$	The graph of f^{-1} is a reflection of the graph of f across ______.

Chapter Test

1 Find the distance between each pair of points.

(a) $(3, -1)$ and $(7, 2)$ (b) $(2, 7)$ and $(2, -5)$

2 Sketch the graph of each relation.

(a) $\{(x, y) \mid |x| \geq 1 \text{ and } y \leq 0\}$ (b) $|x| + |y| = 2$

3 Let $f(x) = 2x + 5$. Compute

(a) $f(0)$ (b) $f(2)$ (c) $f(-3)$ (d) $f(\frac{1}{2})$ (e) $f\left(\frac{1}{a}\right)$

(f) $\dfrac{f(x + h) - f(x)}{h}$

4 Find the domain of each function.

(a) $f(x) = \sqrt{5 - 2x}$ (b) $g(x) = \dfrac{x}{x + 2}$ (c) $h(x) = \dfrac{1}{\sqrt{1 - x}}$

5 For each of the following functions, determine if

(i) f is even or odd
(ii) f is increasing or decreasing
(iii) f is symmetric with respect to the y axis or the origin

(a) $f(x) = 3x^4$ (b) $f(x) = 5x^3$ (c) $f(x) = \sqrt{9 - x^2}$

6 Sketch the graph of each function in Problem 5 and use the graph to find the range of each function.

7 Let f be the function defined by $f(x) = |3x|$.

(a) Find the domain and the range of f. (b) Find $f(-2)$, $f(0)$, and $f(2)$.

(c) Is $f(7b) = 7f(b)$? (d) Sketch the graph of f.

8 Let $f(x) = \sqrt{x+1}$ and $g(x) = \sqrt{x-4}$. Find:
(a) $(f+g)(x)$ (b) $(f-g)(x)$ (c) $(f \cdot g)(x)$ (d) $\left(\frac{f}{g}\right)(x)$

9 Let $f(x) = \sqrt{x}$ and $g(x) = x^2 - 1$. Find:
(a) $f \circ f$ (b) $g \circ g$ (c) $f[g(x)]$ (d) $g[f(x)]$

10 Suppose that $f(x) = -4x^3$.
(a) Find the domain and the range of f.
(b) Is f an even or odd function?
(c) Is f an increasing or decreasing function?
(d) Is f symmetric with respect to the y axis or the origin?
(e) Does f have an inverse? If so, find it.
(f) Sketch the graph of f^{-1}.

Answers

1 (a) 5 (b) 12

2 (a)

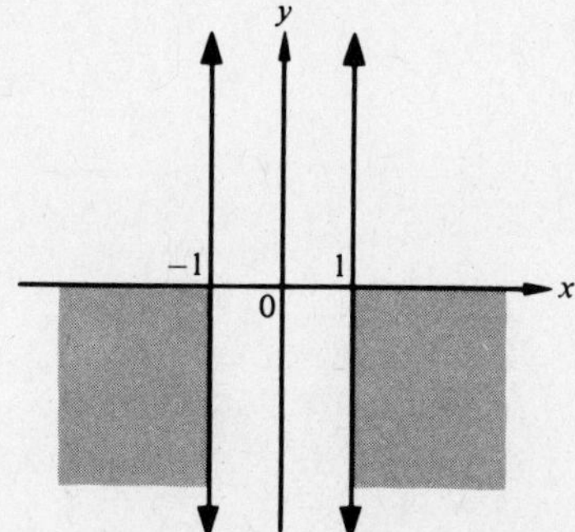

(b)

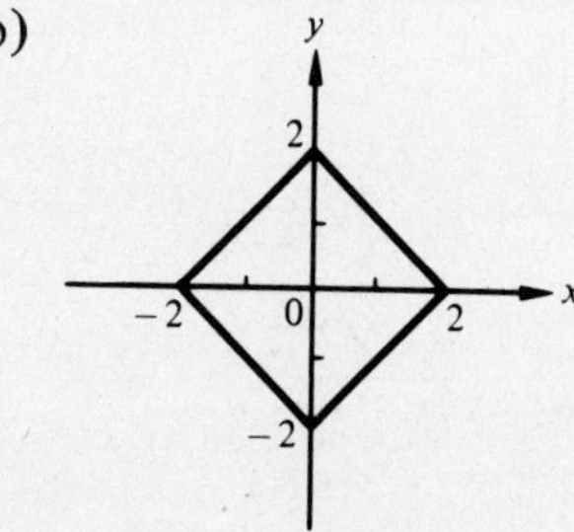

3 (a) 5 (b) 9 (c) −1 (d) 6 (e) $\frac{2}{a} + 5$ (f) 2

4 (a) $\{x \mid x \leqslant \frac{5}{2}\}$ (b) $\{x \mid x \neq -2\}$ (c) $\{x \mid x < 1\}$

5 (a) f is even; f is increasing in $[0, \infty)$ and decreasing in $(-\infty, 0]$; f is symmetric with respect to y axis.
(b) f is odd; f is increasing; f is symmetric with respect to the origin.
(c) f is even; f is decreasing in $[0, 3]$ and increasing in $[-3, 0]$; f is symmetric with respect to the y axis.

6 (a) range = $\{y \mid y \geqslant 0\}$ (b) range = $\{y \mid y \text{ is a real number}\}$

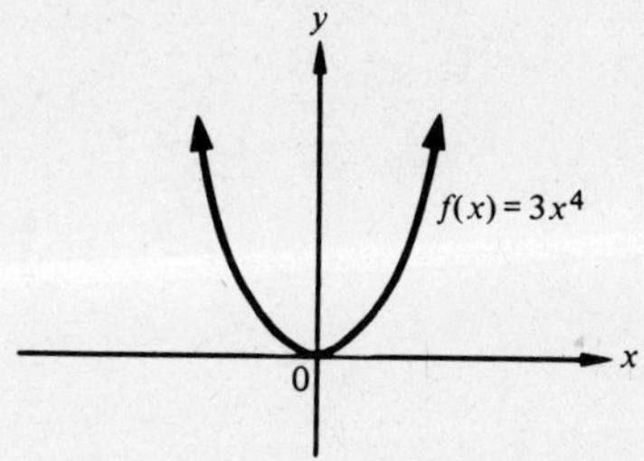

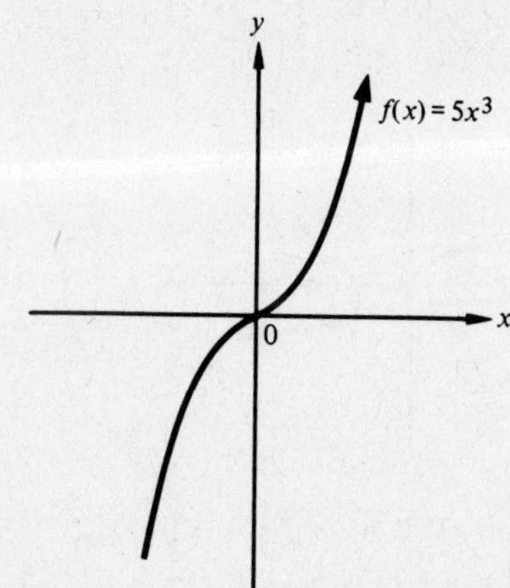

(c) range = $\{y \mid 0 \leqslant y \leqslant 3\}$

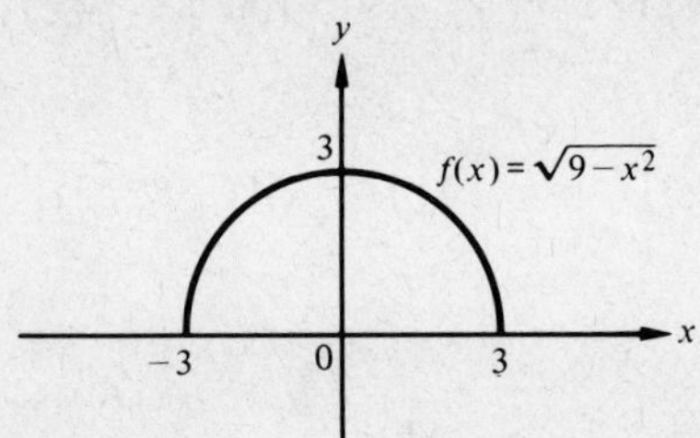

7 (a) The domain of $f = R$; the range of $f = \{y \mid y$ is a nonnegative real number$\}$.
(b) $f(-2) = 6, f(0) = 0$, and $f(2) = 6$.
(c) Yes
(d)

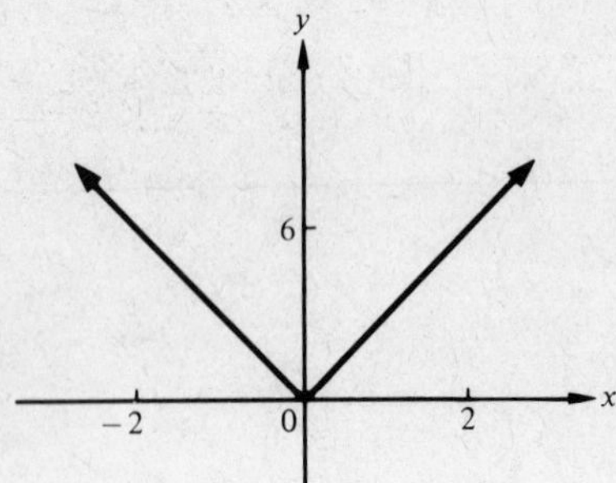

8 (a) $(f + g)(x) = \sqrt{x + 1} + \sqrt{x - 4}$
(b) $(f - g)(x) = \sqrt{x + 1} - \sqrt{x - 4}$
(c) $(f \cdot g)(x) = \sqrt{x^2 - 3x - 4}$
(d) $\left(\frac{f}{g}\right)(x) = \sqrt{\frac{x + 1}{x - 4}}$

9 (a) $(f \circ f)(x) = \sqrt[4]{x}$
(b) $(g \circ g)(x) = x^4 - 2x^2$
(c) $f[g(x)] = \sqrt{x^2 - 1}$
(d) $g[f(x)] = x - 1$

10 (a) The domain of $f = \{x \mid x \in R\}$; the range of $f = \{y \mid y \in R\}$.
(b) f is odd.
(c) f is decreasing.
(d) f is symmetric with respect to origin.
(e) f^{-1} exists, $f^{-1}(x) = -\sqrt[3]{\frac{x}{4}}$.
(f)

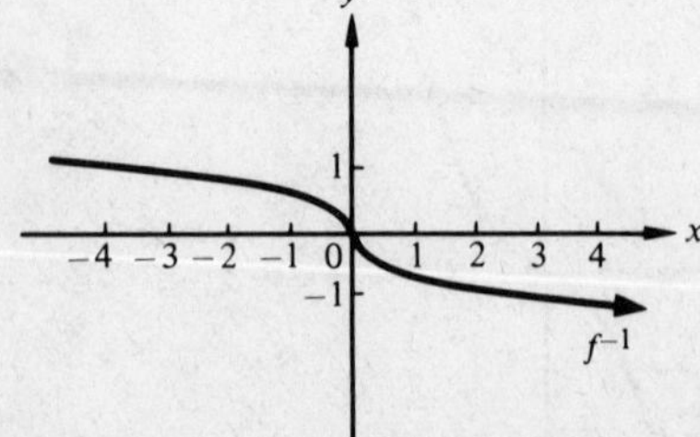

Chapter 3 POLYNOMIAL AND RATIONAL FUNCTIONS

In this chapter we investigate two particular types of functions—polynomial functions and rational functions. The objectives of this chapter are that the student will be able to:

1 Sketch the graphs of linear functions.

2 Sketch the graphs of quadratic functions and solve quadratic equations.

3 Find the solution sets of quadratic inequalities.

4 Sketch the graphs of polynomial functions of degree greater than 2 and divide polynomials by using synthetic division.

5 Sketch the graphs of rational functions.

1 Linear Functions

linear

1 A function written in the form $f(x) = mx + b$, where m and b are constant real numbers, is called a ________ function.

straight line

2 The graph of a linear function $f(x) = mx + b$ is a ____________.

linear, -3, 7

3 $f(x) = -3x + 7$ is a ______ function, with $m =$ _____ and $b =$ _____.

constant

4 A function of the form $f(x) = b$, where b is a constant real number, is called a ____________ function.

In Problems 5-9, find the domain and the range of each given linear function and sketch its graph.

real numbers

real numbers

straight

0, 3

$\frac{3}{2}$, 0

5 $f(x) = -2x + 3$

The domain of f is the set of ________________. The range of f is the set of ________________. Since the graph of a linear function is a ____________ line, two points are enough to determine the graph.

If $x = 0$, then $f(0) = -2(\_\_\_\_) + 3 = \_\_\_\_$

If $x = \frac{3}{2}$, then $f(\frac{3}{2}) = -2(\_\_\_\_) + 3 = \_\_\_\_$

The graph is

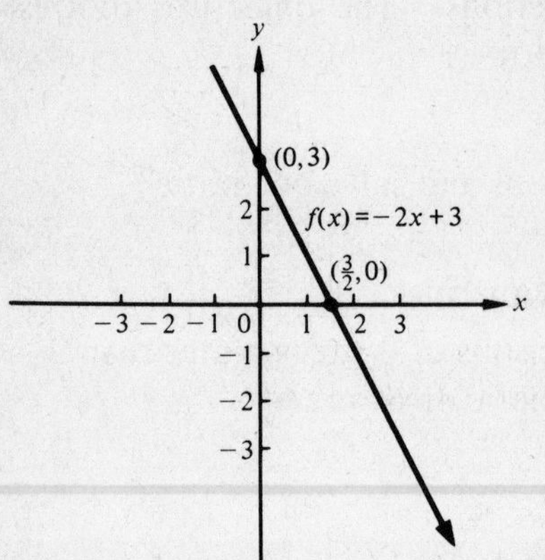

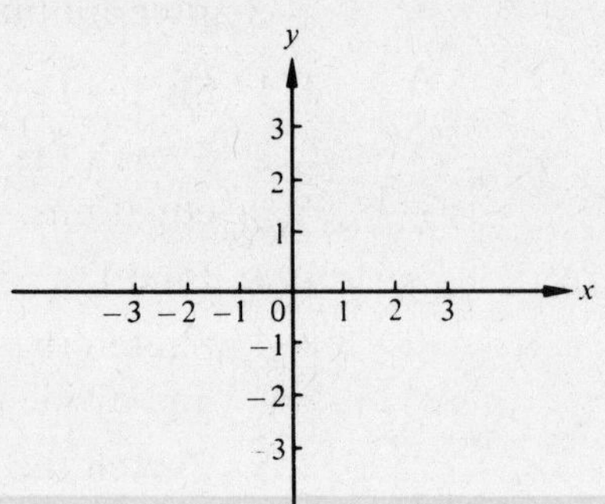

real numbers

real numbers

0, -2

2, 0

6 $f(x) = x - 2$

The domain of f is the set of ________________. The range of f is the set of ________________.

If $x = 0$, then $f(0) = \_\_\_\_ - 2 = \_\_\_\_$

If $x = 2$, then $f(2) = \_\_\_\_ - 2 = \_\_\_\_$

The graph is

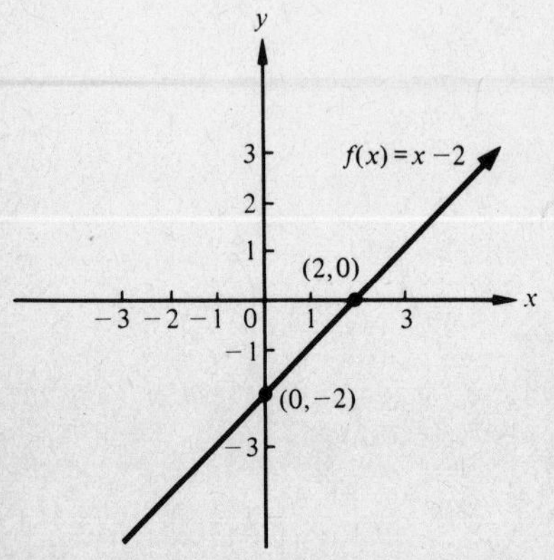

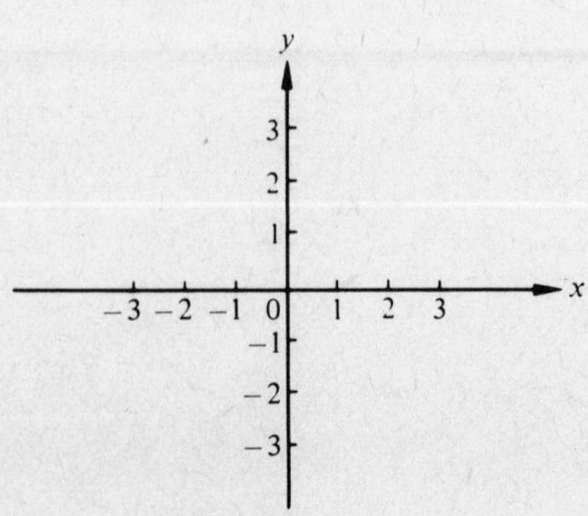

7 f is a linear function whose graph contains $(3, -2)$ and with $m = -3$ and $y = f(x)$.

Since the point $(3, -2)$ lies on the line, it must satisfy the equation ______________. Since $m =$ ____, we have $f(x) = -3x + b$. Using the point $(3, -2)$, the equation becomes $-2 = -3(3) + b$, so $b =$ ____. That is, $f(x) =$ __________.

The domain of f is the set ____. The range of f is the set ____.

The graph is

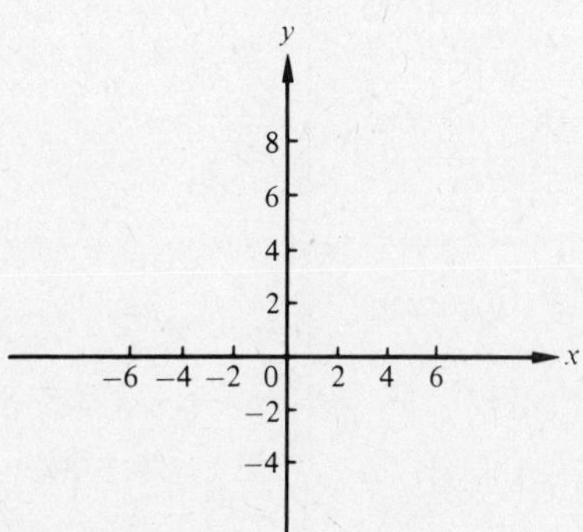

$f(x) = mx + b$, -3

7, $-3x + 7$

R, R

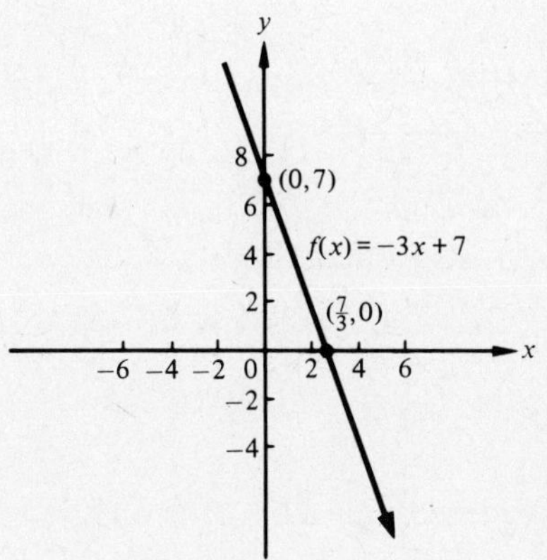

8 $f(x) = -1$

The domain of f is the set ____. The range of f is ______.

The graph is

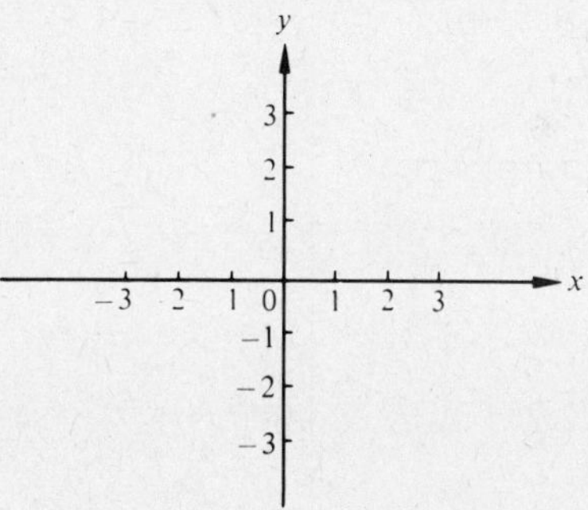

R, $\{-1\}$

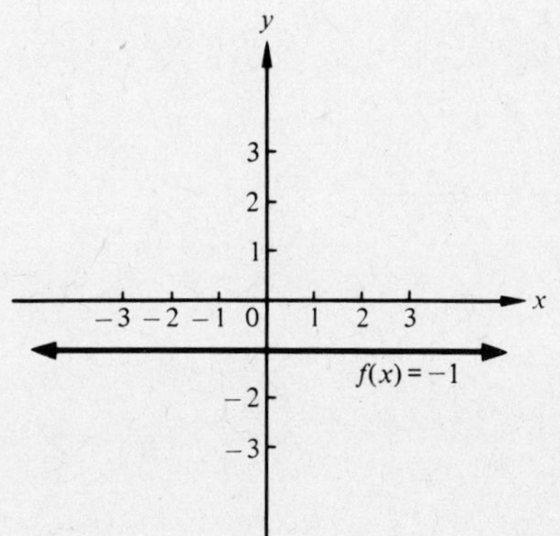

9 $f(x) = 2$

The domain of f is the set ____. The range of f is ____. The graph is

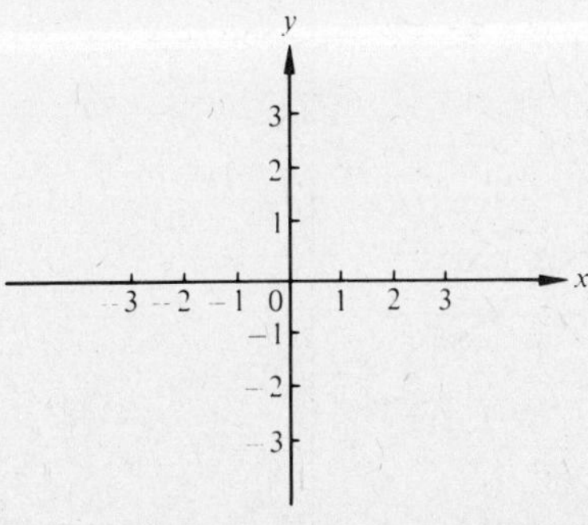

R, $\{2\}$

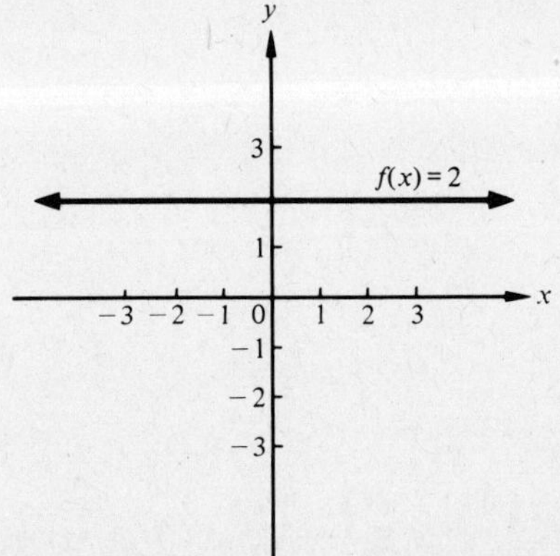

In Problems 10–13, graph the given linear equation.

0

graph, 2

line

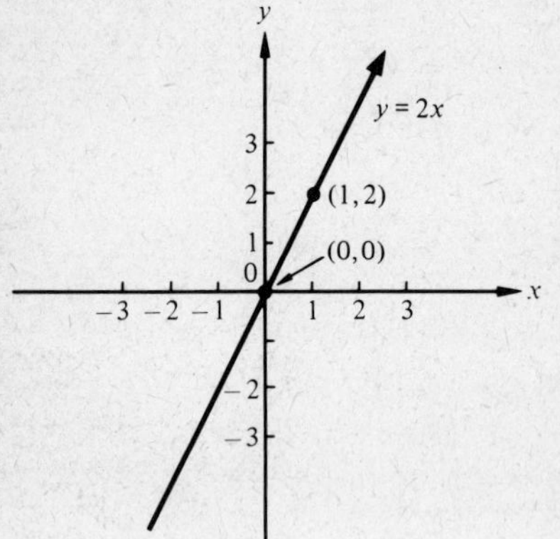

10 $y = 2x$

If $x = 0$, then $y =$ ____. Thus, the ordered pair (0, 0) is on the ________. If $x = 1$, then $y =$ ____, so that the ordered pair (1, 2) is on the graph of the _____. The graph is

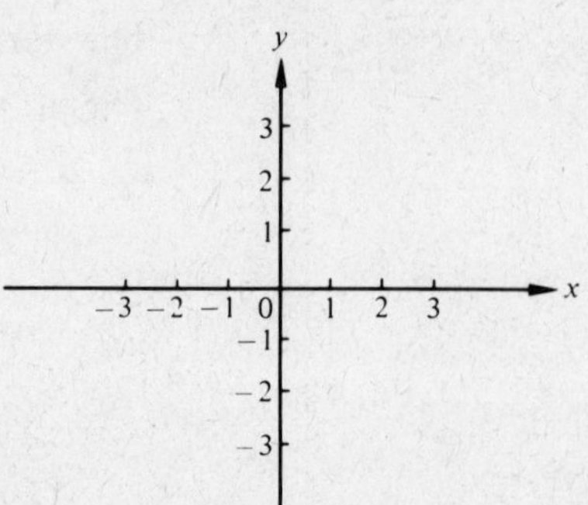

-2

line, 0

graph

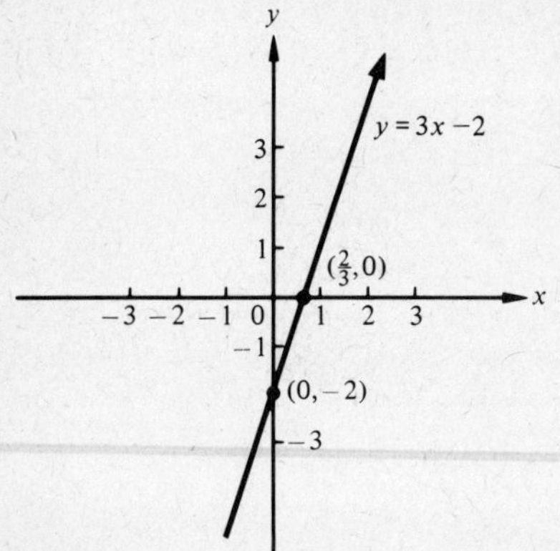

11. $y = 3x - 2$

If $x = 0$, then $y =$ ____, so that the ordered pair (0, −2) is on the graph of the _____. If $x = \frac{2}{3}$, then $y =$ ____, so that the ordered pair ($\frac{2}{3}$, 0) is on the ________ of the line. The graph is

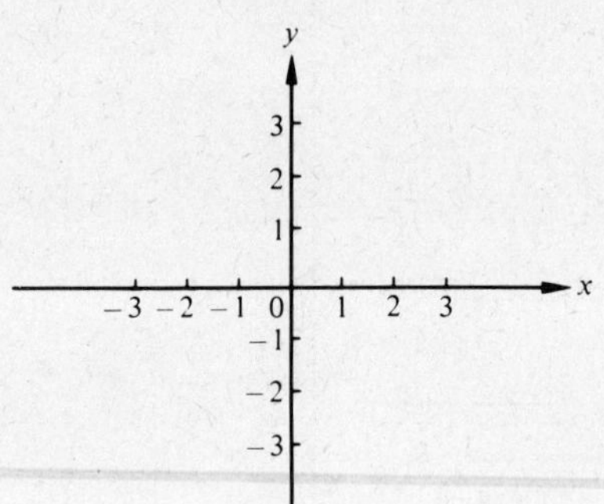

$-\frac{2}{3}x + 2$

2

line, 0

line

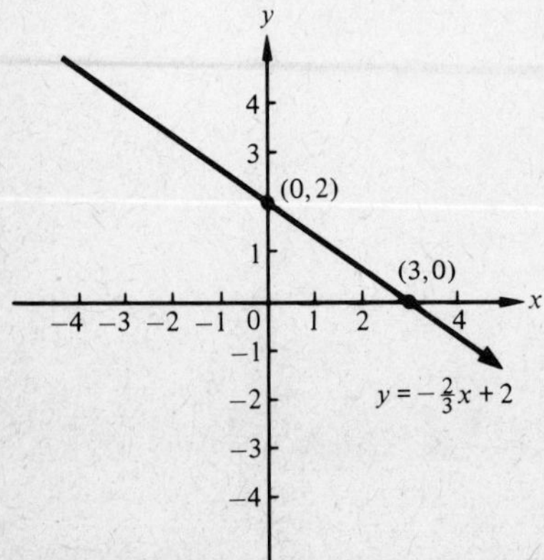

12 $2x + 3y = 6$

This equation can be written as $y =$ _________. If $x = 0$, then $y =$ ____, so that the ordered pair (0, 2) is on the graph of the _____. If $x = 3$, then $y =$ ____, so that the ordered pair (3, 0) is on the graph of the _____. The graph is

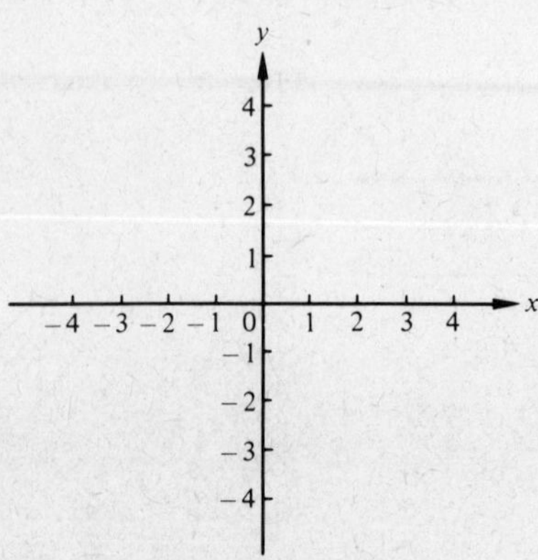

-2, -5

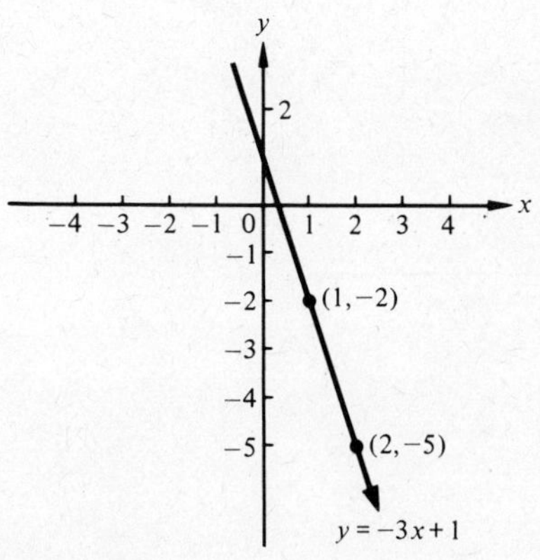

13 $y = -3x + 1$

If $x = 1$, then $y =$ ____, and if $x = 2$, then $y =$ ____. The graph is

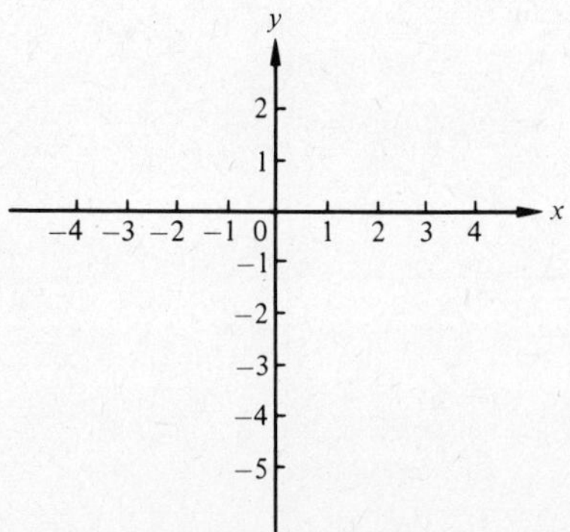

x intercept

y intercept

14 The x coordinate of the point where the graph of a linear equation crosses the x axis is called the ____________ of the graph, and the y coordinate of the point where the graph of a linear equation crosses the y axis is called the ____________ of the graph.

In Problems 15–18, find the x intercept and the y intercept of the graph of each equation and sketch the graph.

6, 6

-2, -2

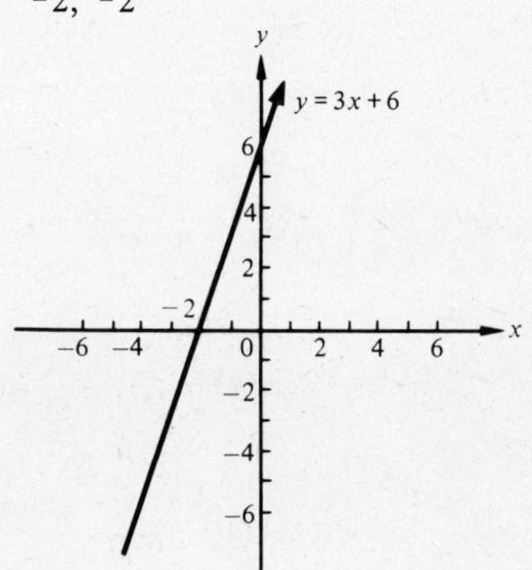

15 $y = 3x + 6$

If $x = 0$, then $y =$ ____, so the y intercept is ____. If $y = 0$, then $x =$ ____, so the x intercept is ____. The graph is

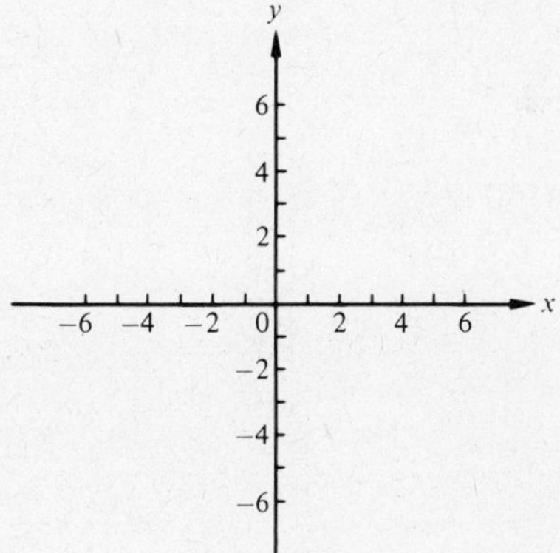

-3, -3

4, 4

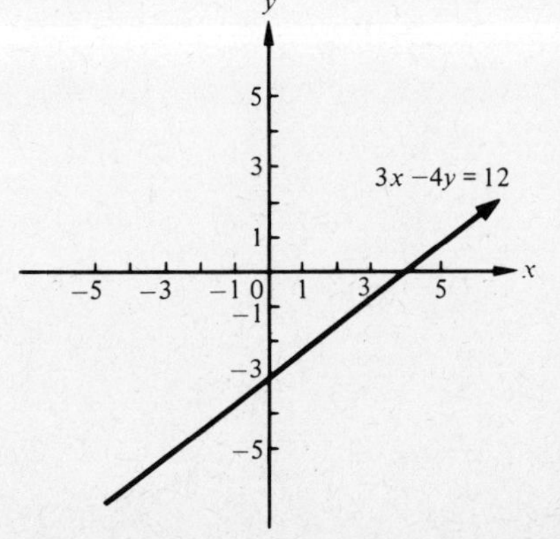

16 $3x - 4y = 12$

If $x = 0$, then $y =$ ____, so the y intercept is ____. If $y = 0$, then $x =$ ____, so the x intercept is ____. The graph is

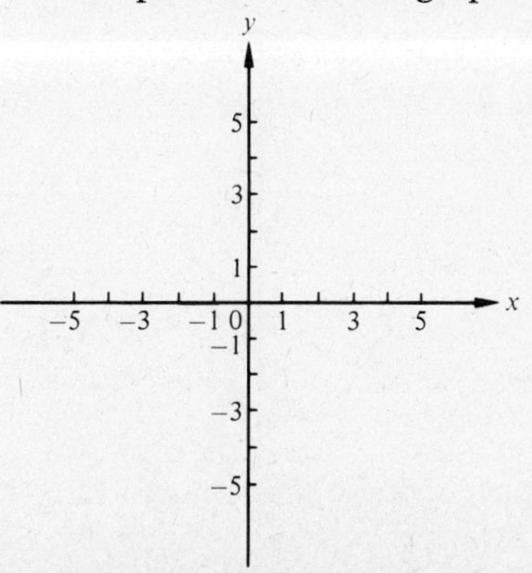

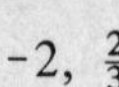

$-2, \frac{2}{3}$

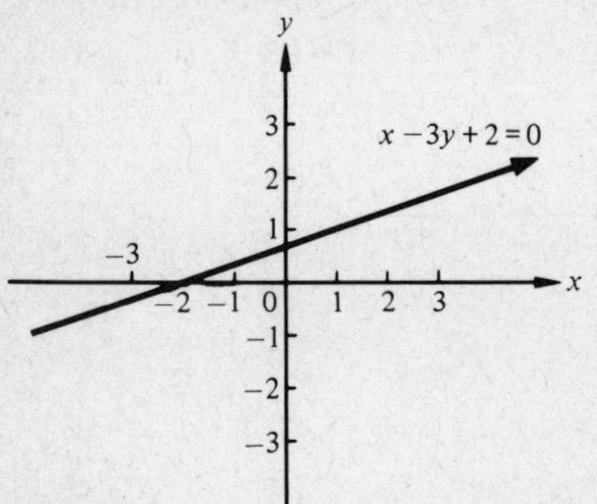

17 $x - 3y + 2 = 0$

The x intercept is ____, and the y intercept is ____. The graph is

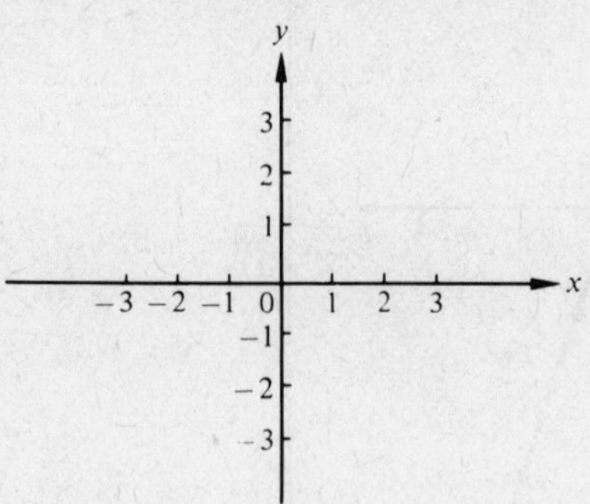

$-\frac{3}{2}, 3$

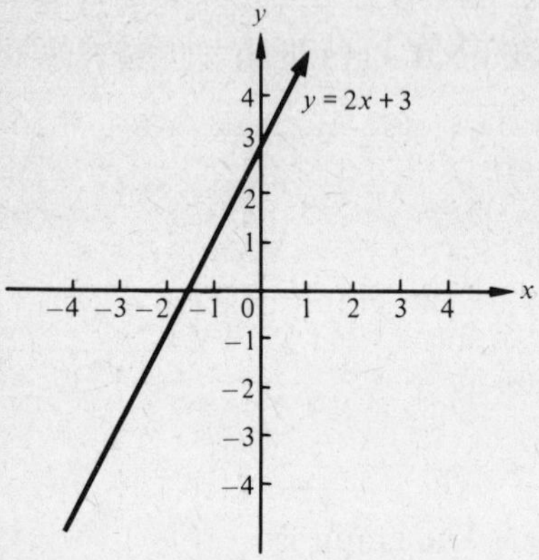

18 $y = 2x + 3$

The x intercept is ____, and the y intercept is ____. The graph is

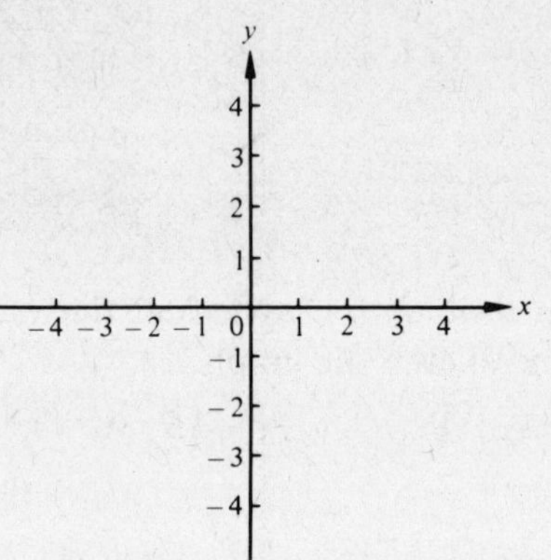

1.1 Slope of a Line

$\frac{y_1 - y_2}{x_1 - x_2}$ or $\frac{y_2 - y_1}{x_2 - x_1}$

$x_1 \neq x_2$

$y_1 - y_2, x_2 - x_1$

19 The slope of the line segment containing the points $P_1 = (x_1, y_1)$ and $P_2 = (x_2, y_2)$ is given by the formula $m =$ __________, provided that ________.

20 $\frac{y_1 - y_2}{x_1 - x_2} = \frac{-(\_\_\_\_\_\_)}{-(x_1 - x_2)} = \frac{y_2 - y_1}{\_\_\_\_\_\_} = m$

In Problems 21–24, find the slope of the line containing each pair of points.

3, 2

1, -8

-8 - 2, -10, 5

21 (3, 2) and (1, -8)

$x_1 =$ ____ $y_1 =$ ____

$x_2 =$ ____ $y_2 =$ ____

$m = \frac{\_\_\_\_}{1 - 3} = \frac{\_\_\_\_}{-2} =$ ____

2, 5

-4, 3

$3 - 5, \frac{-2}{-6}, \frac{1}{3}$

22 (2, 5) and (-4, 3)

$x_1 =$ ____ $y_1 =$ ____

$x_2 =$ ____ $y_2 =$ ____

$m = \frac{\_\_\_\_}{-4 - 2} =$ ____ $=$ ____

23 (−5, 2) and (3, 7)

$x_1 =$ ____ $y_1 =$ ____

$x_2 =$ ____ $y_2 =$ ____

$m =$ ____

−5, 2

3, 7

$\frac{5}{8}$

24 (−3, 6) and (9, 11)

$x_1 =$ ____ $y_1 =$ ____

$x_2 =$ ____ $y_2 =$ ____

$m =$ ____

−3, 6

9, 11

$\frac{5}{12}$

25 A geometric interpretation of the slope of the line containing the points P_1 and P_2 is the ratio of the vertical distance from P_1 to P_2 (called the rise) to the horizontal distance from P_1 to P_2 (called the run), so that the slope $m =$ ____.

$\frac{\text{rise}}{\text{run}}$

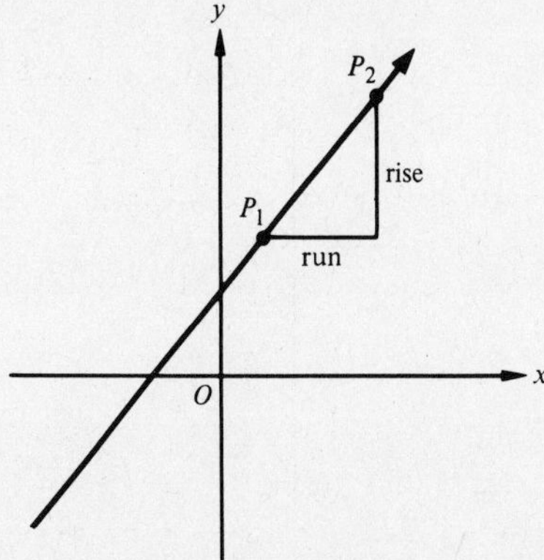

26 If the rise is positive and the run is positive, the slope is ____________, and the function defined by the equation is ________________.

positive

increasing

27 If the rise is negative and the run is positive, the slope is ____________, and the function defined by the equation is ________________.

negative

decreasing

28 If $x_1 = x_2$, then $x_2 - x_1 =$ ____ for each pair of points on the line. Therefore, $m = \frac{y_2 - y_1}{x_2 - x_1}$ has no meaning, and we say that the slope is ____________.

0

undefined

29 If $y_1 = y_2$ and $x_1 \neq x_2$, then $y_2 - y_1 =$ ____ for each pair of points on the line. Therefore, $m = \frac{y_2 - y_1}{x_2 - x_1} = \frac{\_\_\_\_}{x_2 - x_1} =$ ____.

0

0, 0

30 The slope of the line determined by the function $f(x) = mx + b$ is ____.

m

In Problems 31–32, find the slope of the given line, then decide if the function defined by the equation is increasing or decreasing.

31 $y = 3x - 7$

The slope of the line $y = 3x - 7$ is ____, so the function is ____________.

3

increasing

32 $y = -4x + 2$

The slope of the line $y = -4x + 2$ is ____, so the function is ____________.

-4

decreasing

1.2 Forms of Equations of Lines

33 The point-slope form for the equation of the line L containing the point $P_1 = (x_1, y_1)$ and having slope m is ____________.

$y - y_1 = m(x - x_1)$

In Problems 34–40, find a point-slope form of the equation of the line with the given slope and containing the given point. Sketch the graph of the line.

34 $m = -\frac{1}{2}$ and $(x_1, y_1) = (1, 3)$

In point-slope form, the equation is

$y -$ ____ $= -\frac{1}{2}(x -$ ____$)$

The graph is

3, 1

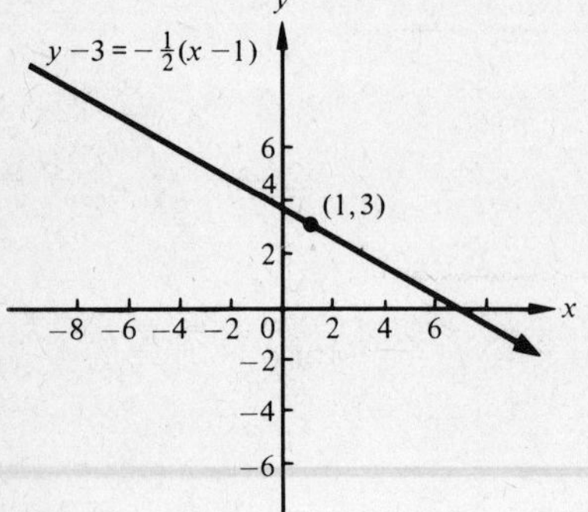

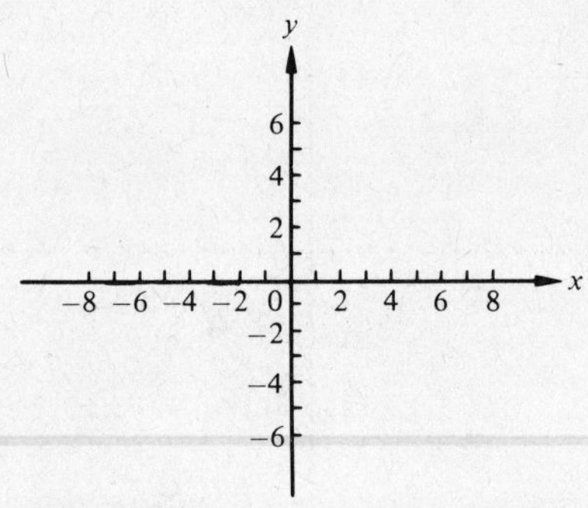

35 $m = -\frac{3}{4}$ and $(x_1, y_1) = (-3, 2)$

The point-slope form of the equation is

$y - 2 = -\frac{3}{4}[x - ($____$)]$ or $y - 2 = -\frac{3}{4}($____$)$

The graph is

-3, $x + 3$

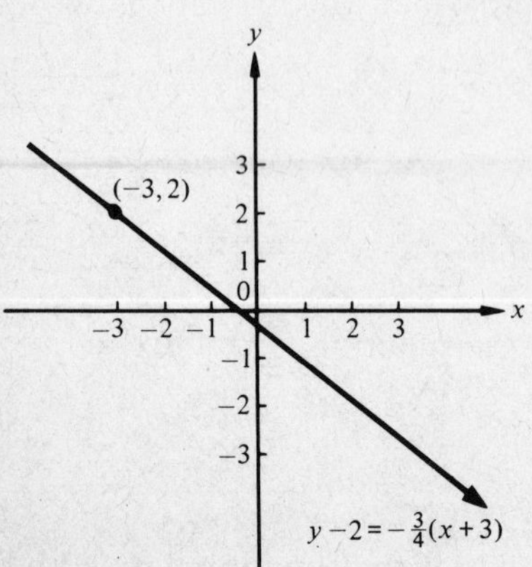

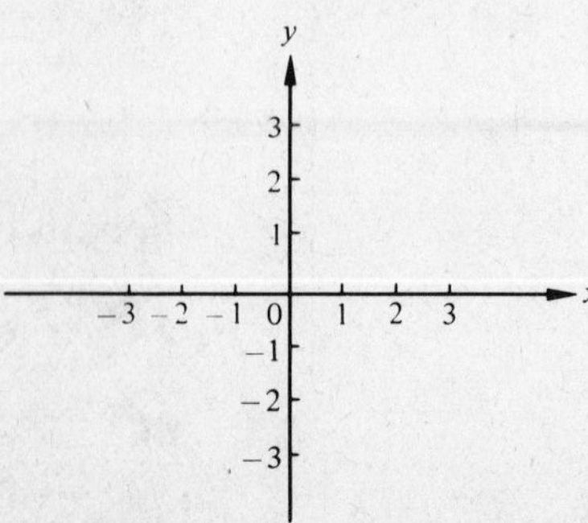

36 $m = \frac{2}{3}$ and $(x_1, y_1) = (-5, 2)$

The point-slope form of the equation is

$y - 2 = \frac{2}{3}(\_\_\_\_\_)$

The graph is

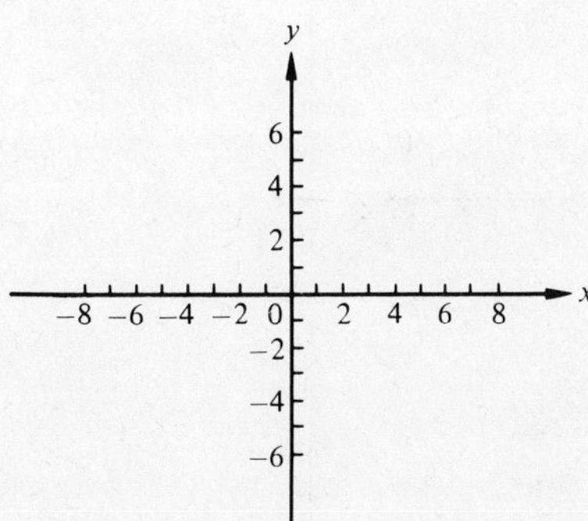

$x + 5$

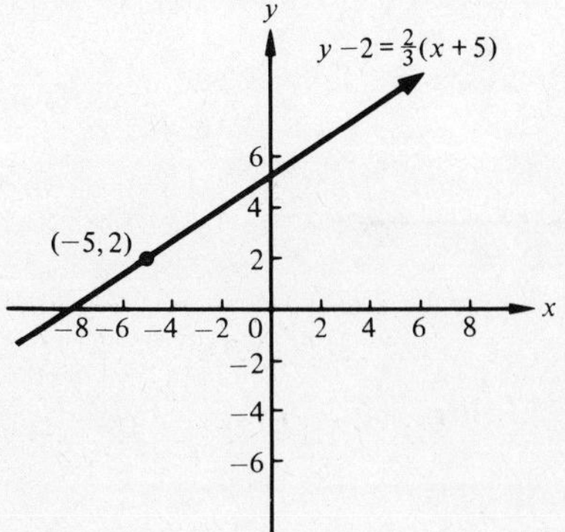

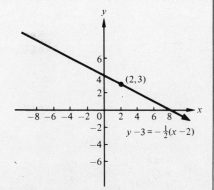

37 $m = -\frac{1}{2}$ and $(x_1, y_1) = (2, 3)$

The point-slope form of the equation is

$y - 3 = \_\_\_\_\_(x - 2)$

The graph is

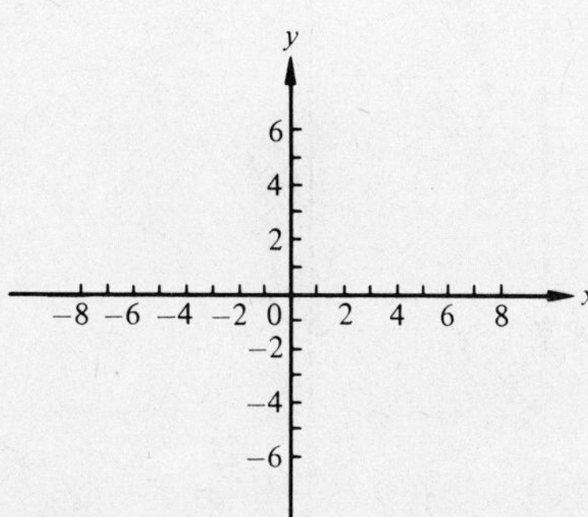

$-\frac{1}{2}$

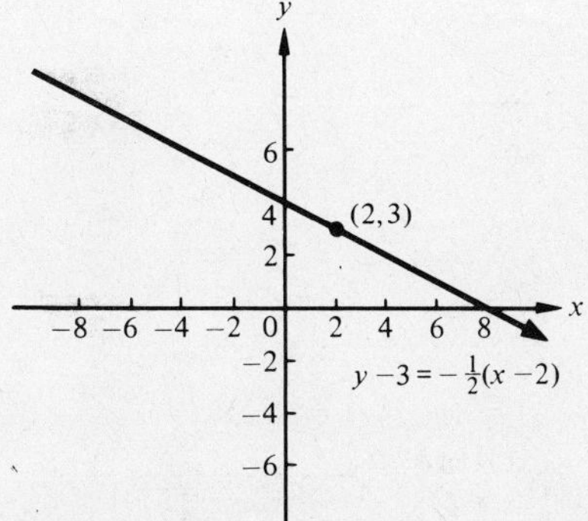

38 $m = 0$ and $(x_1, y_1) = (2, 5)$

The point-slope form of the equation is

$y - 5 = 0(x - \_\_\_\_\_)$ or $y = \_\_\_\_\_$

The graph is

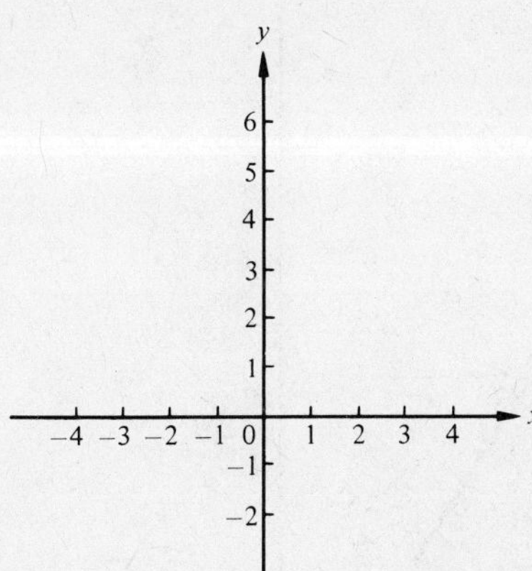

2, 5

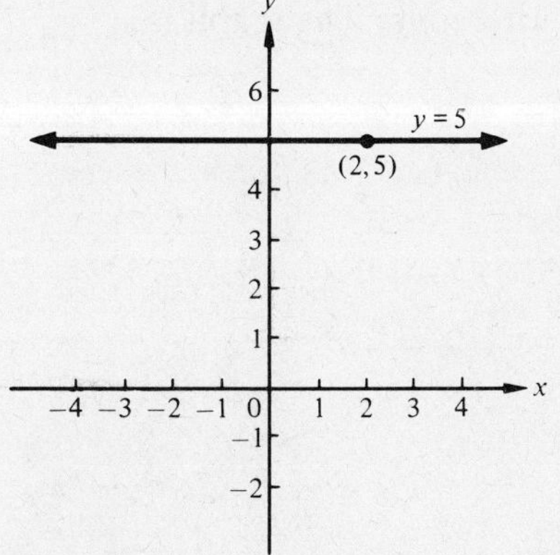

39 $m = 0$ and $(x_1, y_1) = (3, -2)$

The point-slope form of the equation is

$x - 3$, -2

$y - (-2) = 0(\_\_\_\_)$ or $y = \_\_\_\_$

The graph is

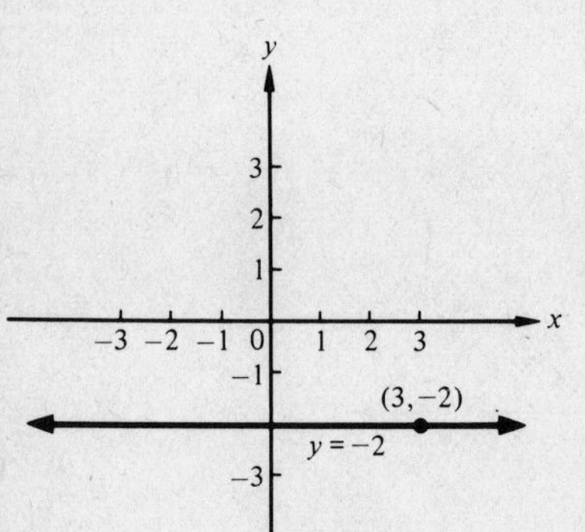

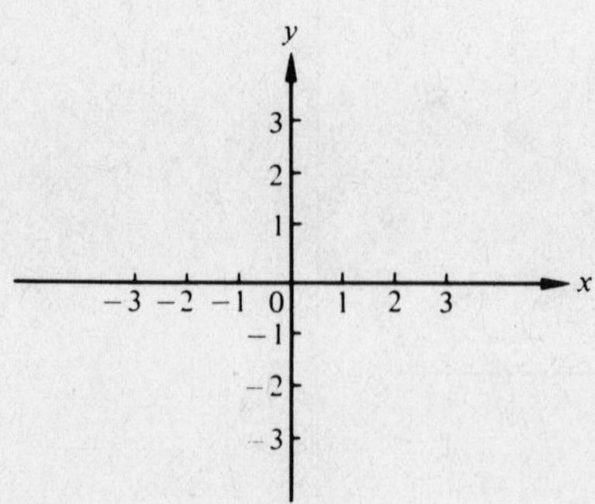

40 m undefined and $(x_1, y_1) = (2, 1)$

The point-slope form of the equation does not apply in this case. However, m undefined means that the line is parallel to the y axis;

2

that is, its equation is $x = \_\_\_\_$. The graph is

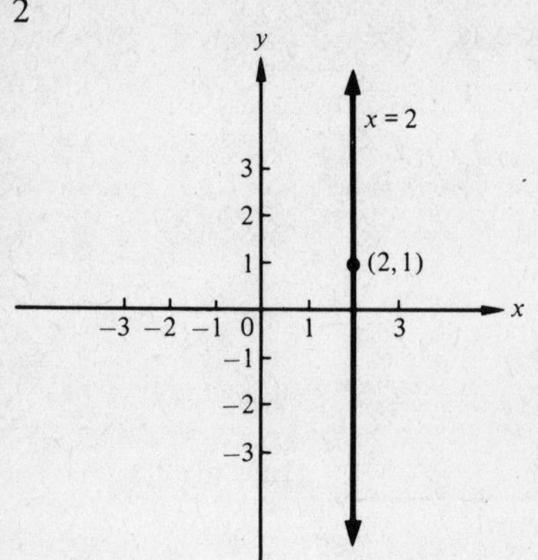

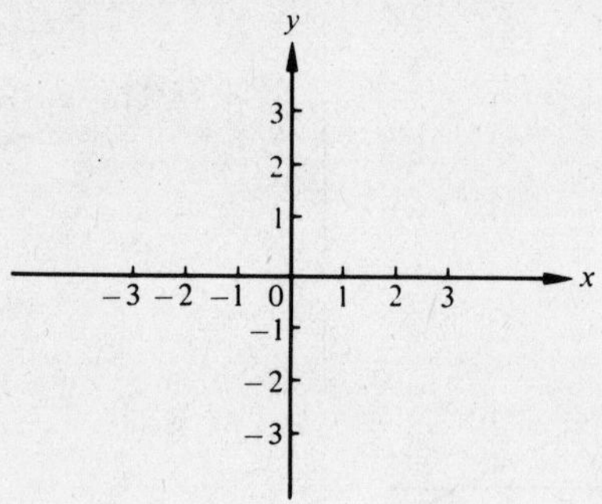

$y = mx + b$

m, b

41 **The slope-intercept form of a linear equation is ________, where ____ is the slope and ____ is the y intercept.**

In Problems 42–44, write the given equation as an equivalent equation in slope-intercept form, and determine the slope and the y intercept. Sketch the graph of the line.

42 $5x + 3y = 6$

$-\frac{5}{3}x + 2$,

$-\frac{5}{3}$, 2

Solving for y explicitly, we obtain $y = \_\_\_\_\_\_\_\_$, where ____ is the slope and ____ is the y intercept. The graph is

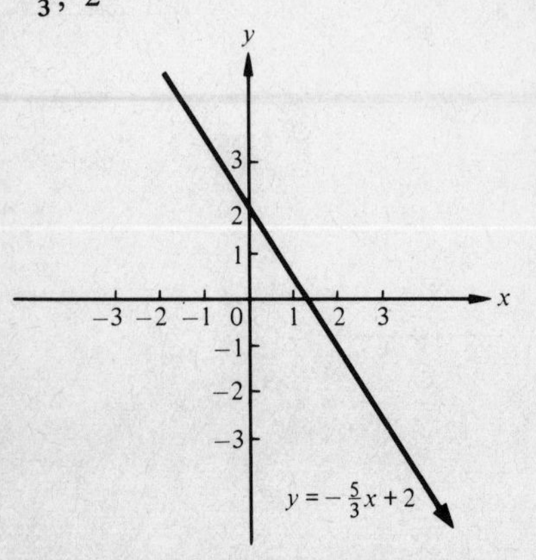

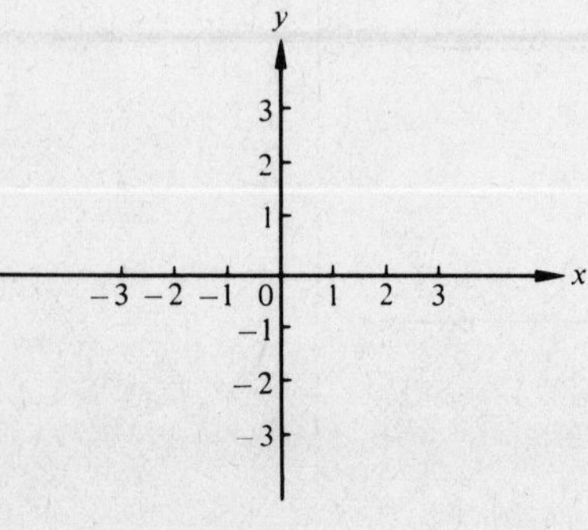

$\frac{2}{3}x - \frac{5}{3}$, $\frac{2}{3}$
$-\frac{5}{3}$

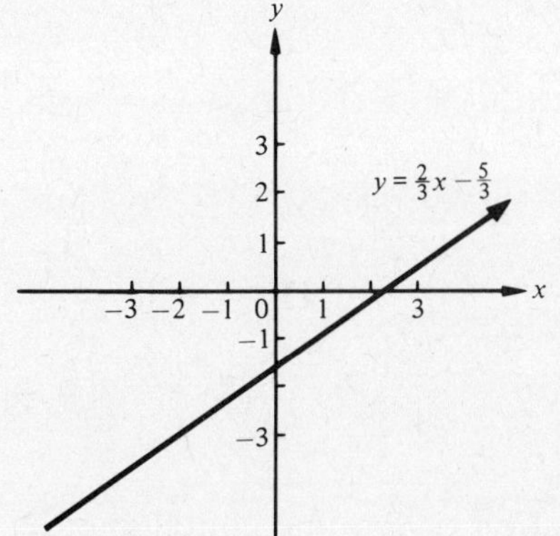

43 $2x - 3y - 5 = 0$

Solving for y explicitly, we obtain $y =$ __________, where _____ is the slope and _____ is the y intercept. The graph is

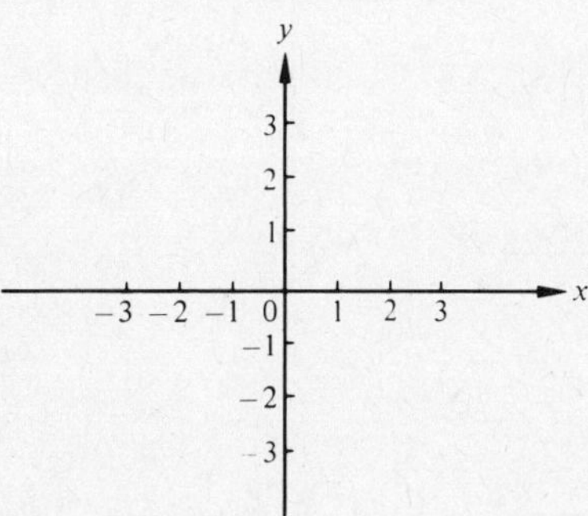

$3x + 4$, 3
4

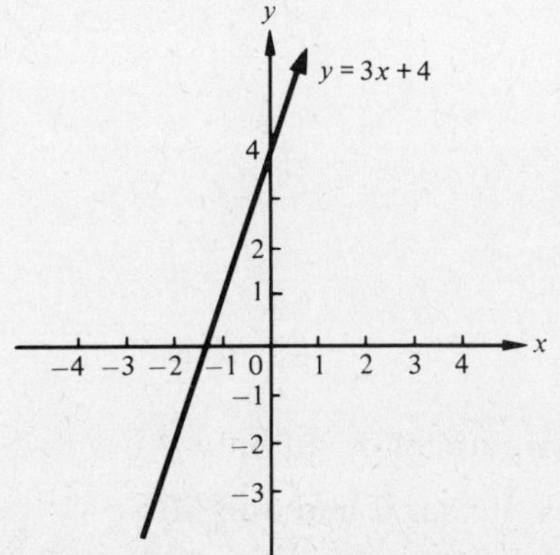

44 $3x - y + 4 = 0$

Solving for y explicitly, we have $y =$ __________, where _____ is the slope and _____ is the y intercept. The graph is

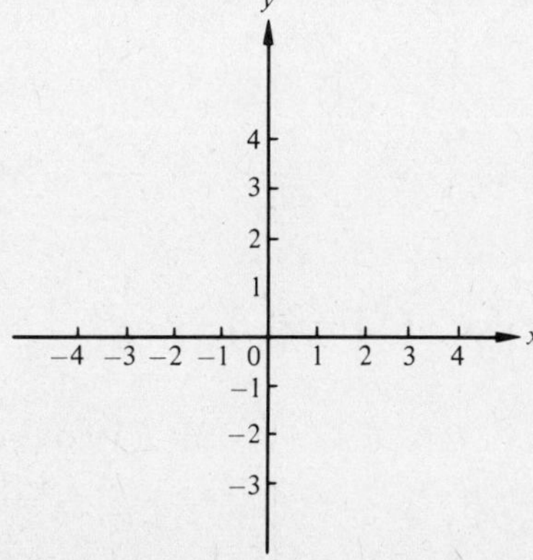

$\dfrac{y_2 - y_1}{x_2 - x_1}$

45 The equation of the line containing the two points $P_1 = (x_1, y_1)$ and $P_2 = (x_2, y_2)$, providing that $x_1 \neq x_2$, is

$y - y_1 =$ __________ $(x - x_1)$

In Problems 46–47, find a point-slope equation of the line that contains the two given points.

$\dfrac{6+2}{5+3}$

$x + 3$

46 $P_1 = (-3, -2)$ and $P_2 = (5, 6)$

The equation of the line is

$y + 2 =$ __________ $(x + 3)$

or

$y + 2 =$ __________

47 $P_1 = (2, -5)$ and $P_2 = (-3, -1)$

The equation of the line is

$\frac{-1+5}{-3-2}$

$y + 5 = \underline{\qquad\qquad} (x - 2)$

or

$-\frac{4}{5}$

$y + 5 = \underline{\quad} (x - 2)$

48 The equation of a line whose intercepts are the points $(a, 0)$ and $(0, b)$, with $a \neq 0$ and $b \neq 0$, is __________. This is called the ________________ of the line.

$\frac{x}{a} + \frac{y}{b} = 1$

intercept form

In Problems 49–50, use the intercept form to find the equation of the line containing the given points.

49 $(3, 0)$ and $(0, 6)$

The intercept form of the line is

$\frac{x}{3} + \frac{y}{\underline{\quad}} = 1$

6

50 $(-2, 0)$ and $(0, 5)$

The intercept form of the line is

$\frac{x}{\underline{\quad}} + \frac{y}{5} = 1$

-2

1.3 Geometry of Two Lines

51 Two lines with slopes m_1 and m_2 are parallel if and only if ________; the two lines are perpendicular if and only if ________ or, equivalently, $m_1 =$ ______.

$m_1 = m_2$

$m_1 m_2 = -1$, $-\frac{1}{m_2}$

52 The slope of line L_1, containing the points $P_1 = (-5, 4)$ and $P_2 = (7, -11)$, is ____, and the slope of line L_2, containing the points $P_3 = (12, 25)$ and $P_4 = (0, 40)$, is ____. Therefore, line L_1 is ________ to line L_2.

$-\frac{5}{4}$

$-\frac{5}{4}$

parallel

53 The slope of line L_1, containing the points $P_1 = (3, 3)$ and $P_2 = (5, 6)$, is ____, and the slope of line L_2, containing the points $P_3 = (-1, 1)$ and $P_4 = (1, 4)$, is ____. Therefore, line L_1 is ________ to line L_2.

$\frac{3}{2}$

$\frac{3}{2}$

parallel

54 The slope of line L_1, containing the points $P_1 = (3, 3)$ and $P_2 = (5, 6)$, is ____, and the slope of line L_2, containing $P_3 = (1, 4)$ and $P_4 = (-2, 6)$, is ____. Since $(\frac{3}{2})(-\frac{2}{3}) =$ ____, line L_1 is __________ to line L_2.

$\frac{3}{2}$

$-\frac{2}{3}$, -1

perpendicular

$-\frac{3}{2}$

$\frac{2}{3}$, -1

perpendicular

55 The slope of line L_1, containing the points $P_1 = (0, 7)$ and $P_2 = (8, -5)$, is ____, and the slope of line L_2, containing the points $P_3 = (5, 5)$ and $P_4 = (2, 3)$, is ____. Since $(-\frac{3}{2})(\frac{2}{3}) =$ ____, line L_1 is ____________ to line L_2.

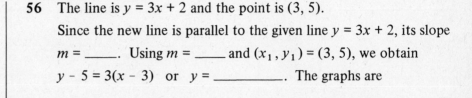

In Problems 56–60, find an equation of the line that is parallel to the given line and that contains the given point. Graph both lines.

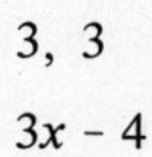

3, 3

$3x - 4$

56 The line is $y = 3x + 2$ and the point is $(3, 5)$.
Since the new line is parallel to the given line $y = 3x + 2$, its slope $m =$ ____. Using $m =$ ____ and $(x_1, y_1) = (3, 5)$, we obtain $y - 5 = 3(x - 3)$ or $y =$ ________. The graphs are

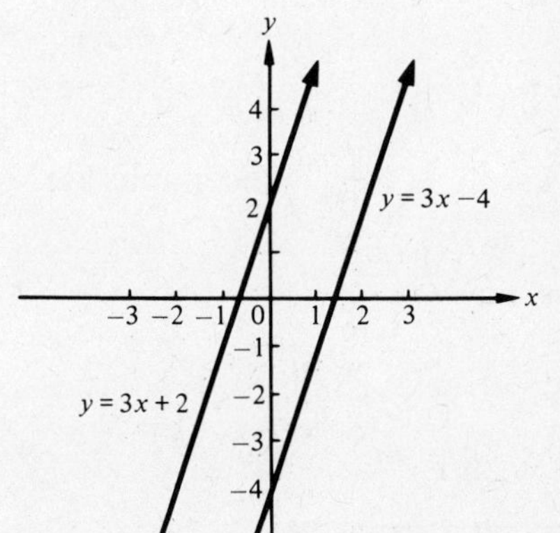

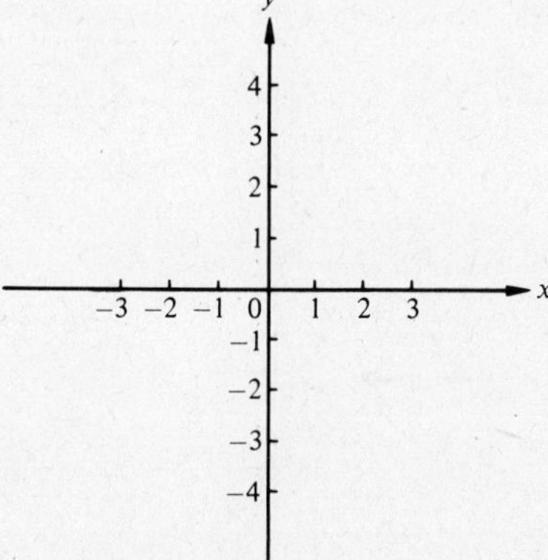

$-2x + 3$

-2, $(-2, 1)$

$-2x - 3$

57 The line is $-2x - y + 3 = 0$ and the point is $(-2, 1)$.
Writing the given equation in slope-intercept form, we obtain $y =$ ________. Using the point-slope form, $y - y_1 = m(x - x_1)$, with $m =$ ____ and $(x_1, y_1) =$ ________, we obtain $y - 1 = -2(x + 2)$ or $y =$ ________. The graphs are

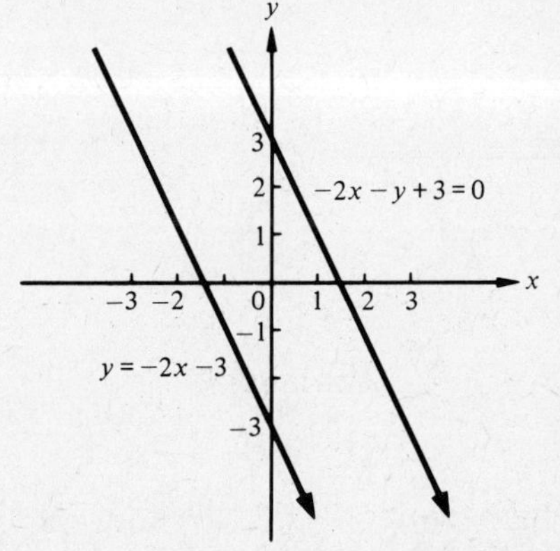

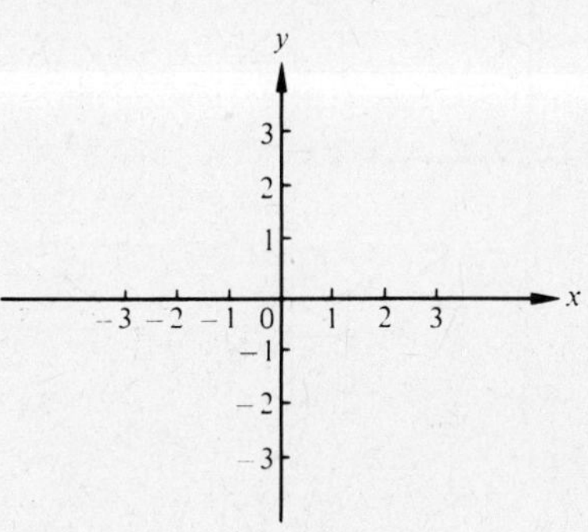

$y + 5 = -1(x - 2)$

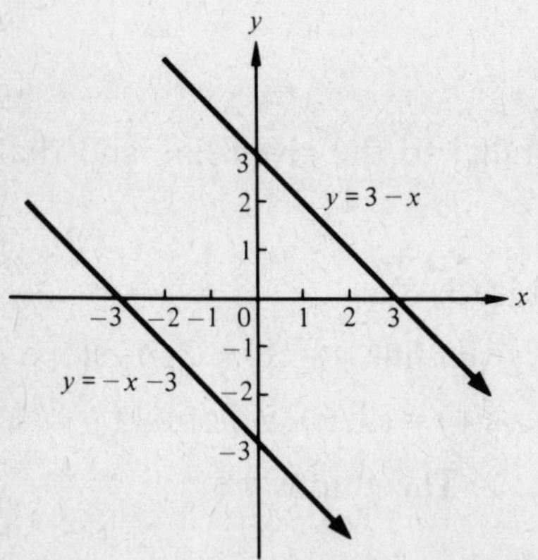

58 The line is $y = 3 - x$ and the point is $(2, -5)$.

Using the point-slope form, $y - y_1 = m(x - x_1)$, with $m = -1$ and $(x_1, y_1) = (2, -5)$, we obtain ____________________ or $y = -x - 3$. The graphs are

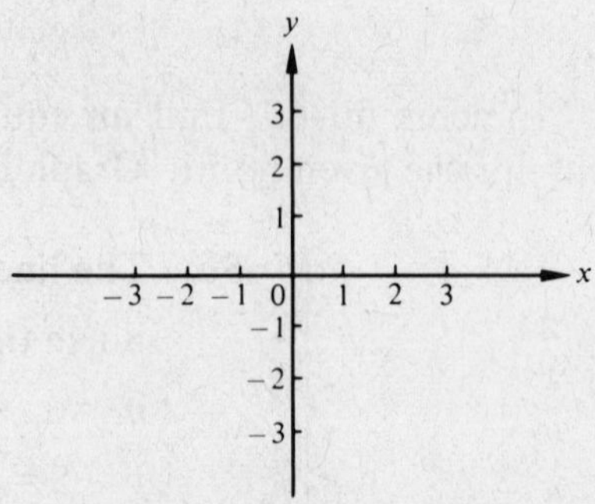

0

$y = -2$

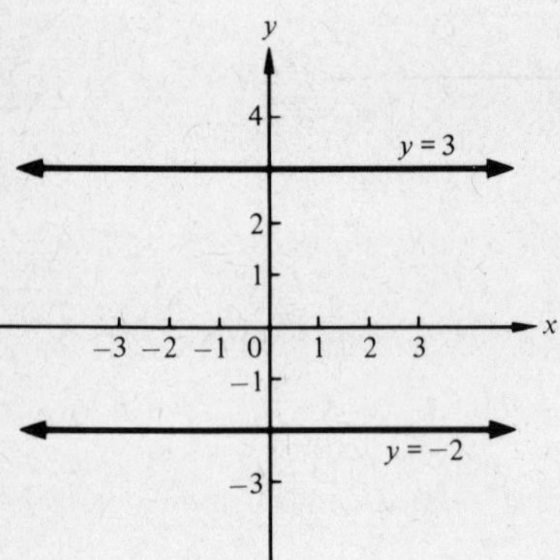

59 The line is $y = 3$ and the point is $(2, -2)$.

Since the line is parallel to the line $y = 3$, its slope $m =$ ____.

Hence, an equation of the new line is ________. The graphs are

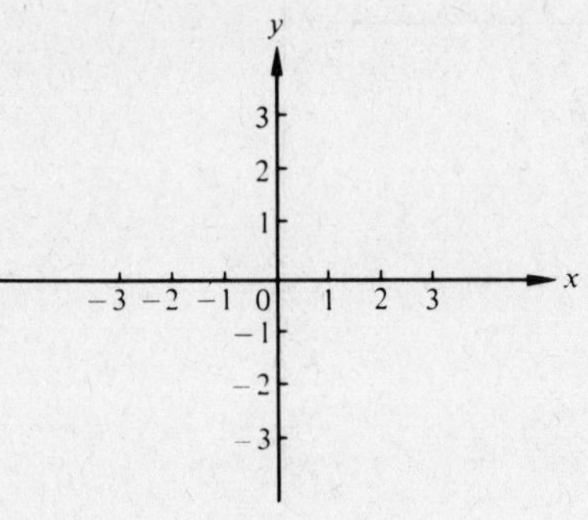

undefined

$x = -1$

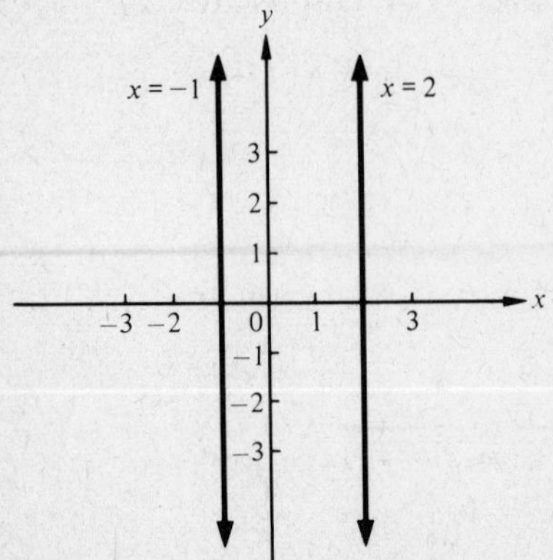

60 The line is $x = 2$ and the point is $(-1, 3)$.

Since the line is parallel to the line $x = 2$, its slope is __________.

Hence, an equation of the new line is ________. The graphs are

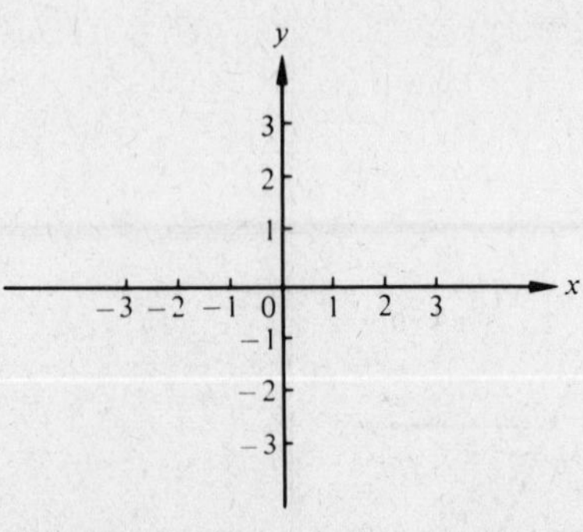

In Problems 61-63, find an equation of the line that is perpendicular to the given line and that contains the given point. Graph both lines.

$-\frac{1}{2}$

$-\frac{1}{2}$, $(-1, 1)$

$-\frac{1}{2}x + \frac{1}{2}$

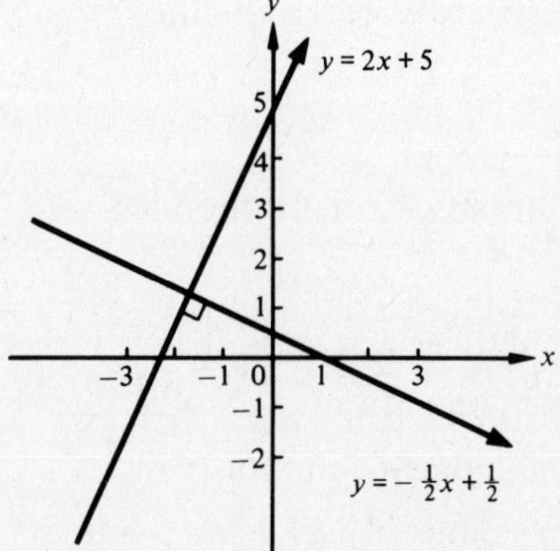

61 The line is $y = 2x + 5$ and the point is $(-1, 1)$.
Since the new line is perpendicular to the line $y = 2x + 5$, its slope $m =$ ____. Using the point-slope form, $y - y_1 = m(x - x_1)$, with $m =$ ____ and $(x_1, y_1) =$ ________, we obtain $y - 1 = -\frac{1}{2}(x + 1)$ or $y =$ __________. The graphs are

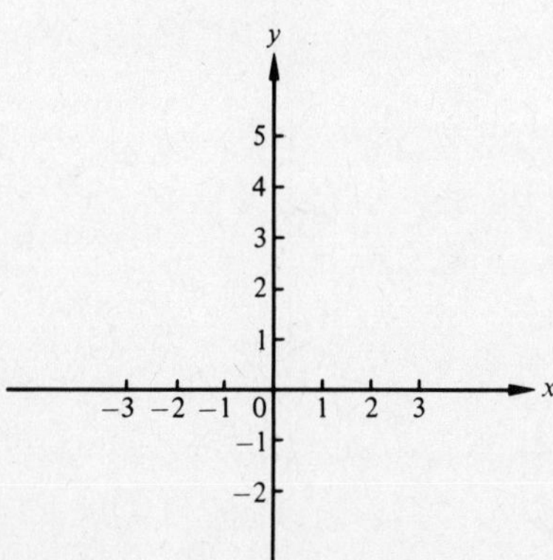

$-\frac{1}{3}$

$-\frac{1}{3}$, $(3, 1)$

$-\frac{1}{3}x + 2$

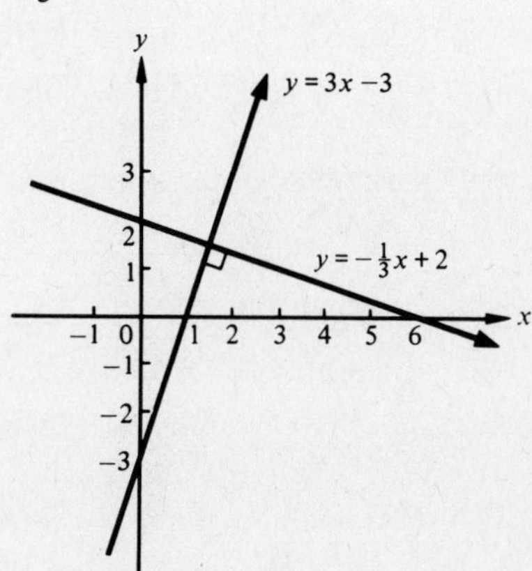

62 The line is $y = 3x - 3$ and the point is $(3, 1)$.
The slope of the new line is $m =$ ____, since it is perpendicular to the line $y = 3x - 3$. Using the point-slope form, $y - y_1 = m(x - x_1)$, with $m =$ ____ and $(x_1, y_1) =$ ________, we obtain $y - 1 = -\frac{1}{3}(x - 3)$ or $y =$ __________. The graphs are

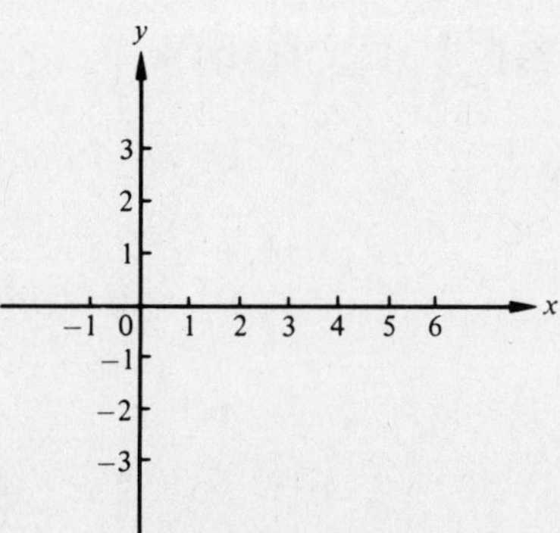

$\frac{1}{2}$

$x - 1$

$\frac{1}{2}x - \frac{7}{2}$

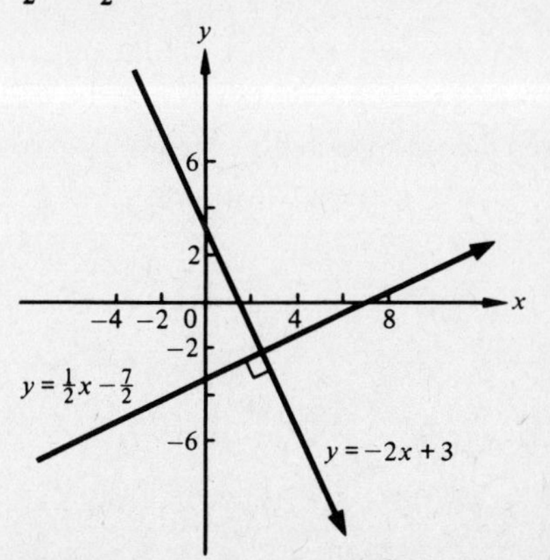

63 The line is $y = -2x + 3$ and the point is $(1, -3)$.
The slope of the new line is $m =$ ____. Using the point-slope form, $y - y_1 = m(x - x_1)$, we obtain $y + 3 = \frac{1}{2}($__________$)$ or $y =$ __________. The graph is

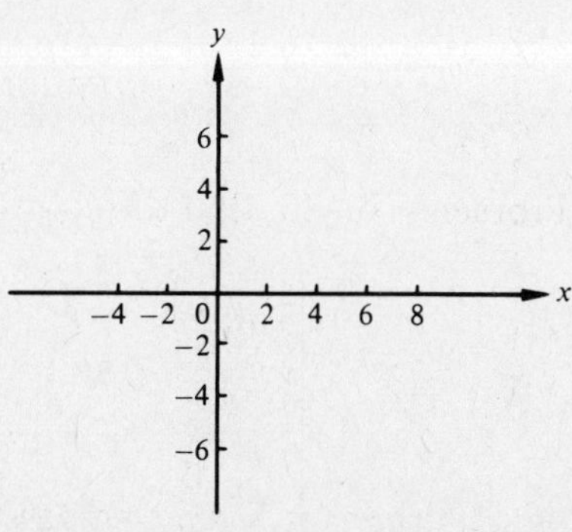

64 A car dealer sells 50 automobiles per week. He had 450 cars in stock at the beginning of a particular month, and the factory normally delivers 20 new cars to him each week. Express the dealer's stockpile as a linear function. When will he run out of his stock of automobiles?

Let N be the stockpile function of t, where t represents the number of weeks. Then

$50t$, $450 - 30t$

$N(t) = 450 + 20t -$ _____ $=$ __________

To find out when he will run out of his stock of automobiles, we must find when $N(t) = 0$.

450, 15

$450 - 30t = 0$, so that $30t =$ _____, and $t =$ _____.

15 weeks

Thus, his stockpile of new cars will be gone after __________.

65 The total cost of an item is given by the linear function $f(x) = 4x + 158$, where x is the cost per unit of packaging the item. Find the total cost per unit of an item if the packaging costs \$12.

$f(x) = 4x + 158$, so that

12

$f(12) = 4($_____$) + 158$

206

$=$ _____

\$206

Therefore, _____ is the total cost of each item.

2 Quadratic Functions

quadratic function

66 A function f defined by the equation $f(x) = ax^2 + bx + c$, where a, b, and c are real numbers and $a \neq 0$, is called a __________.

$f(x) = 0$

67 In graphing any quadratic function, we shall locate the x intercepts of f; that is, we shall identify those values of x for which __________.

2.1 Quadratic Equations

quadratic equation

68 An equation that is equivalent to an equation of the form $ax^2 + bx + c = 0$, where a, b, and c are real numbers and $a \neq 0$, is called a __________ in x.

Factor Method

0, 0

69 $ab = 0$ if and only if $a =$ _____ or $b =$ _____.

In Problems 70–80, find the solution set of each equation.

70 $2x(x - 2) = 0$

0

$2x = 0$ or $x - 2 =$ _____

2

$x = 0$ $\quad x =$ _____

$\{0, 2\}$

The solution set is _____.

$x + 3$
-3
$\{-3, 2\}$

71 $(x - 2)(x + 3) = 0$
$x - 2 = 0$ or ________ $= 0$
$x = 2$ $\quad$ $x =$ ____
The solution set is __________.

0
2
$-\frac{1}{3}$
$\{-\frac{1}{3}, \frac{2}{5}\}$

72 $(3x + 1)(5x - 2) = 0$
$3x + 1 = 0$ or $5x - 2 =$ ____
$3x = -1$ $\quad$ $5x =$ ____
$x =$ ____ $\quad$ $x = \frac{2}{5}$
The solution set is ___________.

$7x + 4$
$7x + 4$
$\frac{4}{7}, -\frac{4}{7}$
$\{-\frac{4}{7}, \frac{4}{7}\}$

73 $49x^2 - 16 = 0$
$(7x - 4)($_________$) = 0$
$7x - 4 = 0$ or _________ $= 0$
$x =$ ____ $\quad$ $x =$ ____
The solution set is ___________.

$x + 9,\ x - 8$
$x - 8$
$-9,\ 8$
$\{-9, 8\}$

74 $x^2 + x - 72 = 0$
$($_______$)($_______$) = 0$
$x + 9 = 0$ or _______ $= 0$
$x =$ ____ $\quad$ $x =$ ____
The solution set is __________.

$5x + 7$
$5x + 7$
$-\frac{7}{5}, -\frac{7}{5}$
$\{-\frac{7}{5}\}$

75 $25x^2 + 70x + 49 = 0$
$($_________$)^2 = 0$
$5x + 7 = 0$ or _________ $= 0$
$x =$ ____ $\quad$ $x =$ ____
The solution set is _____.

$x - 7$
$x - 7$
$-3,\ 7$
$\{-3, 7\}$

76 $x^2 - 4x - 21 = 0$
$(x + 3)($________$) = 0$
$x + 3 = 0$ or _______ $= 0$
$x =$ ____ $\quad$ $x =$ ____
The solution set is __________.

x
$x + 1$
$0,\ 0$
$0,\ -1$
$\{-1, 0\}$

77 $(x - 9)(x + 9) + x + 81 = 0$
$x^2 +$ ____ $= 0$
$x($_______$) = 0$
$x =$ ____ or $x + 1 =$ ____
$x =$ ____ $\quad$ $x =$ ____
The solution set is __________.

78 $\dfrac{1}{x} = \dfrac{x-3}{4}$

Answers	
$4x$	$4x\left(\dfrac{1}{x}\right) = \_\_\_\_\left(\dfrac{x-3}{4}\right)$
$3x$	$4 = x^2 - \_\_\_\_$
4	$x^2 - 3x - \_\_\_\_ = 0$
$x+1$	$(x-4)(\_\_\_\_) = 0$
$x+1$	$x - 4 = 0$ or $\_\_\_\_ = 0$
$4, -1$	$x = \_\_\_\_$ $x = \_\_\_\_$
$\{-1, 4\}$	The solution set is ____.

79 $(2x-1)(x+7) = 2x - 7$

Answers	
$13x$	$2x^2 + \_\_\_\_ - 7 = 2x - 7$
$11x$	$2x^2 + \_\_\_\_ = 0$
$2x + 11$	$x(\_\_\_\_) = 0$
$2x + 11$	$x = 0$ or $\_\_\_\_ = 0$
$0, -\frac{11}{2}$	$x = \_\_\_\_$ $x = \_\_\_\_$
$\{-\frac{11}{2}, 0\}$	The solution set is ____.

80 $\dfrac{x^2 - 4}{4} = x - 1$

Answers	
$4x - 4$	$x^2 - 4 = \_\_\_\_$
$4x$	$x^2 - \_\_\_\_ = 0$
$x - 4$	$x(\_\_\_\_) = 0$
$x - 4$	$x = 0$ or $\_\_\_\_ = 0$
$0, 4$	$x = \_\_\_\_$ $x = \_\_\_\_$
$\{0, 4\}$	The solution set is ____.

Completing-the-Square Method

81 If an expression is a perfect square trinomial, then it can be expressed as the square of a ____.

binomial

82 In order to complete the square when the two terms $x^2 \pm 2dx$ are given, add the term ____ to it. That is, add the square of one-half the coefficient of the ____.

d^2

x term

In Problems 83–85, complete the square of the binomial and express the result as the square of a binomial.

83 $x^2 + 3x$

Add $(\_\_\_\_)^2$ to the expression $x^2 + 3x$. The resulting perfect square is ____.

$x^2 + 3x + \frac{9}{4} = (\_\_\_\_)^2$

Answers: $\frac{3}{2}$; $x^2 + 3x + \frac{9}{4}$; $x + \frac{3}{2}$

84 $x^2 + 4x$

Add (____)^2 to the expression $x^2 + 4x$. The resulting perfect square is $x^2 + 4x + 4 =$ (______)^2.

2

$x + 2$

85 $x^2 - \frac{4}{5}x$

Add (____)^2 to the expression $x^2 - \frac{4}{5}x$. The resulting perfect square is $x^2 - \frac{4}{5}x + \frac{4}{25} =$ (______)^2.

$-\frac{2}{5}$

$x - \frac{2}{5}$

In Problems 86-90, use the method of completing the square to find the solution set of each equation.

86 $x^2 - 5x + 1 = 0$

$x^2 - 5x =$ ____

Completing the square,

$x^2 - 5x +$ ____ $= -1 +$ ____

$(x - \frac{5}{2})^2 =$ ____

$x - \dfrac{5}{2} =$ ______ or $x - \dfrac{5}{2} = \dfrac{\sqrt{21}}{2}$

$x =$ ______ $x =$ ______

The solution set is ____________.

-1

$\frac{25}{4}, \frac{25}{4}$

$\frac{21}{4}$

$-\dfrac{\sqrt{21}}{2}$

$\dfrac{5}{2} - \dfrac{\sqrt{21}}{2}, \dfrac{5}{2} + \dfrac{\sqrt{21}}{2}$

$\left\{\dfrac{5 - \sqrt{21}}{2}, \dfrac{5 + \sqrt{21}}{2}\right\}$

87 $x^2 + 4x + 3 = 0$

$x^2 + 4x =$ ____

Completing the square,

$x^2 + 4x +$ ____ $= -3 +$ ____

$(x + 2)^2 =$ ____

$x + 2 =$ ____ or $x + 2 =$ ____

$x =$ ____ $x =$ ____

The solution set is ________.

-3

$4, 4$

1

$-1, 1$

$-3, -1$

$\{-3, -1\}$

88 $x^2 - 9x + 3 = 0$

$x^2 - 9x =$ ____

Completing the square,

$x^2 - 9x +$ ____ $= -3 +$ ____

$(x - \frac{9}{2})^2 =$ ____

$x - \dfrac{9}{2} =$ ______ or $x - \dfrac{9}{2} =$ ____

$x =$ ______ $x =$ ______

The solution set is ____________.

-3

$\frac{81}{4}, \frac{81}{4}$

$\frac{69}{4}$

$-\dfrac{\sqrt{69}}{2}, \dfrac{\sqrt{69}}{2}$

$\dfrac{9}{2} - \dfrac{\sqrt{69}}{2}, \dfrac{9}{2} + \dfrac{\sqrt{69}}{2}$

$\left\{\dfrac{9 - \sqrt{69}}{2}, \dfrac{9 + \sqrt{69}}{2}\right\}$

89 $2x^2 - 5x + 1 = 0$

Divide both sides of the equation by ____ in order to make the coefficient of the x^2 term 1.

$x^2 - \frac{5}{2}x + \_\_\_\_ = 0$

$x^2 - \frac{5}{2}x = \_\_\_\_$

$x^2 - \frac{5}{2}x + \_\_\_\_ = -\frac{1}{2} + \_\_\_\_$

$(\_\_\_\_)^2 = \frac{17}{16}$

$x - \frac{5}{4} = \_\_\_\_$ or $x - \frac{5}{4} = \_\_\_\_$

$x = \_\_\_\_$ $x = \_\_\_\_$

The solution set is ____________.

Answers: 2; $\frac{1}{2}$; $-\frac{1}{2}$; $\frac{25}{16}, \frac{25}{16}$; $x - \frac{5}{4}$; $-\frac{\sqrt{17}}{4}, \frac{\sqrt{17}}{4}$; $\frac{5}{4} - \frac{\sqrt{17}}{4}, \frac{5}{4} + \frac{\sqrt{17}}{4}$; $\left\{\frac{5-\sqrt{17}}{4}, \frac{5+\sqrt{17}}{4}\right\}$

90 $5x^2 - 6x - 4 = 0$

Divide both sides of the equation by ____ in order to make the coefficient of the x^2 term 1.

$x^2 - \frac{6}{5}x + (\_\_\_\_) = 0$

$x^2 - \frac{6}{5}x = \_\_\_\_$

$x^2 - \frac{6}{5}x + \_\_\_\_ = \frac{4}{5} + \_\_\_\_$

$(\_\_\_\_)^2 = \frac{29}{25}$

$x - \frac{3}{5} = \_\_\_\_$ or $x - \frac{3}{5} = \_\_\_\_$

$x = \_\_\_\_$ $x = \_\_\_\_$

The solution set is ____________.

Answers: 5; $-\frac{4}{5}$; $\frac{4}{5}$; $\frac{9}{25}, \frac{9}{25}$; $x - \frac{3}{5}$; $-\frac{\sqrt{29}}{5}, \frac{\sqrt{29}}{5}$; $\frac{3-\sqrt{29}}{5}, \frac{3+\sqrt{29}}{5}$; $\left\{\frac{3-\sqrt{29}}{5}, \frac{3+\sqrt{29}}{5}\right\}$

Quadratic Formula

91 The solutions of the quadratic equation $ax^2 + bx + c = 0$, where a, b, and $c \in R$ and $a \neq 0$, are expressed by the formula

$x = \_\_\_\_\_\_\_\_$

Answer: $\frac{-b \pm \sqrt{b^2 - 4ac}}{2a}$

In Problems 92–96, find the solution set of each equation by using the quadratic formula.

92 $x^2 + 6x + 5 = 0$

$a = \_\_\_\_$ $b = \_\_\_\_$ $c = \_\_\_\_$

$x = \frac{-\_\_\_\_ \pm \sqrt{6^2 - 4(1)(5)}}{2(1)}$

$= \frac{-6 \pm \sqrt{\_\_\_\_}}{2} = \frac{-6 \pm \_\_\_\_}{2}$

Answers: 1, 6, 5; 6; 16, 4

$$x = \frac{-6 - \_\_\_\_}{2} \quad \text{or} \quad x = \frac{-6+4}{2}$$

$x =$ ____ $\qquad x =$ ____

The solution set is ____________.

4

$-5, -1$

$\{-5, -1\}$

93 $x^2 - 2x = 15$

The standard form of the equation is ________________ = 0.

$a =$ ____ $\quad b =$ ____ $\quad c =$ ____

$$x = \frac{-(-2) \pm \sqrt{(\_\_\_\_)^2 - 4(1)(-15)}}{2(1)}$$

$$= \frac{2 \pm \sqrt{4 + \_\_\_\_}}{2} = \frac{2 \pm \_\_\_\_}{2}$$

$$x = \frac{2 - \_\_\_\_}{2} \quad \text{or} \quad x = \frac{2 + \_\_\_\_}{2}$$

$x =$ ____ $\qquad x =$ ____

The solution set is ___________.

$x^2 - 2x - 15$

$1, -2, -15$

-2

$60, 8$

$8, 8$

$-3, 5$

$\{-3, 5\}$

94 $4x^2 - 8x = 13$

The standard form of the equation is ________________ = 0.

$a =$ ____ $\quad b =$ ____ $\quad c =$ ____

$$x = \frac{-(\_\_\_\_) \pm \sqrt{(\_\_\_\_)^2 - 4(\_\_\_\_)(\_\_\_\_)}}{2(4)}$$

$$= \frac{8 \pm \sqrt{\_\_\_\_}}{2(4)} = \frac{8 \pm \_\_\_\_}{8}$$

$$x = \frac{8 - \_\_\_\_}{8} \quad \text{or} \quad x = \frac{8 + \_\_\_\_}{8}$$

$x =$ ________ $\qquad x =$ ________

The solution set is ______________________.

$4x^2 - 8x - 13$

$4, -8, -13$

$-8, -8, 4, -13$

$272, 4\sqrt{17}$

$4\sqrt{17}, 4\sqrt{17}$

$\frac{2 - \sqrt{17}}{2}, \frac{2 + \sqrt{17}}{2}$

$\left\{\frac{2 - \sqrt{17}}{2}, \frac{2 + \sqrt{17}}{2}\right\}$

95 $5x^2 + 13x = 6$

The standard form of the equation is ________________ = 0.

$a =$ ____ $\quad b =$ ____ $\quad c =$ ____

$$x = \frac{-\_\_\_\_ \pm \sqrt{(\_\_\_\_)^2 - 4(\_\_\_\_)(\_\_\_\_)}}{2(5)}$$

$$= \frac{-13 \pm \sqrt{\_\_\_\_}}{10} = \frac{-13 \pm \_\_\_\_}{10}$$

$x =$ ____ or $x =$ ____

The solution set is __________.

$5x^2 + 13x - 6$

$5, 13, -6$

$13, 13, 5, -6$

$289, 17$

$-3, \frac{2}{5}$

$\{-3, \frac{2}{5}\}$

96 $\dfrac{x^2 - 3}{3} + \dfrac{x}{4} = 2$

$12\left(\dfrac{x^2 - 3}{3} + \dfrac{x}{4}\right) = (\_\_\_\_)(2)$

$4(x^2 - 3) + \_\_\_\_ x = 24$

$4x^2 - 12 + 3x = \_\_\_\_$

The standard form of the equation is ____________________ = 0.

$a = \_\_\_\_ \quad b = \_\_\_\_ \quad c = \_\_\_\_$

$x = \dfrac{-\_\_\_\_ \pm \sqrt{(\_\_\_\_)^2 - 4(\_\_\_\_)(\_\_\_\_)}}{2(4)}$

$= \dfrac{-3 \pm 3\sqrt{\_\_\_\_}}{8}$

$x = \_\_\_\_\_\_\_\_$ or $x = \_\_\_\_\_\_\_\_$

The solution set is ______________________.

Answers: 12; 3; 24; $4x^2 + 3x - 36$; 4, 3, −36; 3, 3, 4, −36; 65; $\dfrac{-3 - 3\sqrt{65}}{8}$, $\dfrac{-3 + 3\sqrt{65}}{8}$; $\left\{\dfrac{-3 - 3\sqrt{65}}{8}, \dfrac{-3 + 3\sqrt{65}}{8}\right\}$

97 If $ax^2 + bx + c = 0$, with $a \neq 0$, we define the discriminant of the quadratic equation to be ________.

Answer: $b^2 - 4ac$

98 If $b^2 - 4ac > 0$, there are two unequal ______ roots.

Answer: real

99 If $b^2 - 4ac = 0$, the two roots are ______ and equal.

Answer: real

100 If $b^2 - 4ac < 0$, there are two unequal ________ roots.

Answer: nonreal

In Problems 101–106, find the discriminant of each quadratic equation and determine the type of roots of the equation by using the discriminant.

101 $3x^2 - 4x + 1 = 0$

$a = \_\_\_\_ \quad b = \_\_\_\_ \quad c = \_\_\_\_$

$b^2 - 4ac = (\_\_\_\_)^2 - 4(\_\_\_\_)(\_\_\_\_) = \_\_\_\_$

Since $4 > 0$, the roots are ____________________.

Answers: 3, −4, 1; −4, 3, 1, 4; real and unequal

102 $2x^2 - 7x - 4 = 0$

$a = \_\_\_\_ \quad b = \_\_\_\_ \quad c = \_\_\_\_$

$b^2 - 4ac = (\_\_\_\_)^2 - 4(\_\_\_\_)(\_\_\_\_) = \_\_\_\_$

Since $81 > 0$, the roots are ____________________.

Answers: 2, −7, −4; −7, 2, −4, 81; real and unequal

103 $25x^2 + 20x + 4 = 0$

$a = \_\_\_\_ \quad b = \_\_\_\_ \quad c = \_\_\_\_$

$b^2 - 4ac = (\_\_\_\_)^2 - 4(\_\_\_\_)(\_\_\_\_) = \_\_\_\_$

Therefore, the roots are ____________________.

Answers: 25, 20, 4; 20, 25, 4, 0; real and equal

104 $x^2 + 14x + 49 = 0$

$a =$ ____ $b =$ ____ $c =$ ____

$b^2 - 4ac =$ ____ $- 4($____$)($____$) =$ ____

Therefore, the roots are ____________.

1, 14, 49

196, 1, 49, 0

real and equal

105 $2x^2 + 5 = 6x$

____________ $= 0$

$a =$ ____ $b =$ ____ $c =$ ____

$b^2 - 4ac = ($____$)^2 - 4($____$)($____$) =$ ____

Therefore, the roots are ________ and unequal.

$2x^2 - 6x + 5$

2, −6, 5

−6, 2, 5, −4

nonreal

106 $5x^2 + 2x + 2 = 0$

$a =$ ____ $b =$ ____ $c =$ ____

$b^2 - 4ac = ($____$)^2 - 4($____$)($____$) =$ ____

Therefore, the roots are ____________.

5, 2, 2

2, 5, 2, −36

unequal and nonreal

In Problems 107–111, find the solution set of each equation.

107 $x^4 - 13x^2 + 36 = 0$

Let $u =$ ____; then $u^2 =$ ____. Substitute u into the equation to obtain ____________ $= 0$. Then

$($________$)($________$) = 0$

$u =$ ____ or $u =$ ____

Since $u = x^2$ and $u = 4$ or $u = 9$, we have

$x^2 =$ ____ or $x^2 =$ ____

$x =$ ____ $x =$ ____

or or

$x =$ ____ $x =$ ____

The solution set is ____________.

x^2, x^4

$u^2 - 13u + 36$

$u - 4$, $u - 9$

4, 9

4, 9

−2, −3

2, 3

$\{-2, 2, -3, 3\}$

108 $x^{-2} + x^{-1} - 6 = 0$

Let $u =$ ____; then $u^2 =$ ____. Substitute u into the equation to obtain ____________ $= 0$. Then

$($________$)($________$) = 0$

$u =$ ____ or $u =$ ____

Since $u = x^{-1}$ and $u = 2$ or $u = -3$, we have

$x^{-1} = 2$ or $x^{-1} =$ ____

$\frac{1}{x} =$ ____ $\frac{1}{x} =$ ____

$x =$ ____ $x =$ ____

The solution set is ________.

x^{-1}, x^{-2}

$u^2 + u - 6$

$u - 2$, $u + 3$

2, −3

−3

2, −3

$\frac{1}{2}$, $-\frac{1}{3}$

$\{-\frac{1}{3}, \frac{1}{2}\}$

Answers	Frames
	109 $x^{-4} - 17x^{-2} + 16 = 0$
x^{-2}, x^{-4}	Let $u =$ ____; then $u^2 =$ ____. Substitute u into the equation to
u	obtain $u^2 - 17($____$) + 16 = 0$. Then
$u - 1, u - 16$	(____)(____) $= 0$
1, 16	$u =$ ____ or $u =$ ____
	Since $u = x^{-2}$ and $u = 1$ or $u = 16$, we have
1, 16	$x^{-2} =$ ____ or $x^{-2} =$ ____
1, 16	$\frac{1}{x^2} =$ ____ $\quad \frac{1}{x^2} =$ ____
$1, \frac{1}{16}$	$x^2 =$ ____ $\quad x^2 =$ ____
$1, \frac{1}{4}$	$x =$ ____ $\quad x =$ ____
	or $\quad$ or
$-1, -\frac{1}{4}$	$x =$ ____ $\quad x =$ ____
$\{1, -1, \frac{1}{4}, -\frac{1}{4}\}$	The solution set is ____.
	110 $\frac{x^2}{x+1} + \frac{2(x+1)}{x^2} = 3$
$\frac{1}{u}$	Let $u = \frac{x^2}{x+1}$, then $\frac{x+1}{x^2} =$ ____, so that the resulting equation is
$u + \frac{2}{u} - 3, 3u$	____ $= 0$ or $u^2 -$ ____ $+ 2 = 0$. Then
$u - 1, u - 2$	(____)(____) $= 0$
1, 2	$u =$ ____ or $u =$ ____
	Since $u = \frac{x^2}{x+1}$ and $u = 1$ or $u = 2$, we have
1, 2	$\frac{x^2}{x+1} =$ ____ or $\frac{x^2}{x+1} =$ ____
$x^2 - x - 1, x^2 - 2x - 2$	____ $= 0$ $\quad$ ____ $= 0$
$1 \pm \sqrt{3}$	$x = \frac{1 \pm \sqrt{5}}{2}$ $\quad x =$ ____
	The solution set is
$\left\{\frac{1-\sqrt{5}}{2}, \frac{1+\sqrt{5}}{2}, 1 - \sqrt{3}, 1 + \sqrt{3}\right\}$	____.
	111 $\left(3x - \frac{2}{x}\right)^2 + 6\left(3x - \frac{2}{x}\right) + 5 = 0$
$3x - \frac{2}{x}, \left(3x - \frac{2}{x}\right)^2$	Let $u =$ ____; then $u^2 =$ ____. Substituting u in
$u^2 + 6u + 5$	the equation, we have ____ $= 0$, so
$u + 1$	$(u + 5)($____$) = 0$
$-1, -5$	$u =$ ____ or $u =$ ____
	Since $u = 3x - \frac{2}{x}$ and $u = -1$ or $u = -5$, we have
$-1, -5$	$3x - \frac{2}{x} =$ ____ or $3x - \frac{2}{x} =$ ____

	The quadratic equations formed are
$3x^2 + x - 2,\ 3x^2 + 5x - 2$	________ = 0 or ________ = 0
$\frac{-1 \pm 5}{6}, \frac{-5 \pm 7}{6}$	$x =$ ______ $x =$ ______
$\{-1, \frac{2}{3}, -2, \frac{1}{3}\}$	The solution set is ________.

2.2 Properties of Quadratic Functions

In Problems 112–115, find the domain of f and the x intercept(s) and the y intercept.

	112 $f(x) = 80x - 16x^2$
R	The domain of f is the set ____.
0, 5, 0	The x intercepts are ____ and ____. The y intercept is ____.
	113 $f(x) = x^2 - 3x + 2$
R	The domain of f is the set ____.
1, 2, 2	The x intercepts are ____ and ____. The y intercept is ____.
	114 $f(x) = x^2 - 4x + 4$
R	The domain of f is the set ____.
2, 4	The x intercept is ____. The y intercept is ____.
	115 $f(x) = 600 + 10x - x^2$
R	The domain of f is the set ____.
-20, 30, 600	The x intercepts are ____ and ____. The y intercept is ____.
	116 The graph of any quadratic function of the form
parabola	$f(x) = ax^2 + bx + c$, where $a \neq 0$, is a ________.
upward	**117** The graph of $f(x) = ax^2 + bx + c$ opens ________ when $a > 0$ and
downward	we have a minimum point; the graph opens ________ when
maximum	$a < 0$ and we have a ________ point.
	118 The minimum point or the maximum point on the graph of
extreme	$f(x) = ax^2 + bx + c$, where $a \neq 0$, is called the ________ point or
vertex	________ of the parabola, and it is expressed by $\left(-\frac{b}{2a}, f\left(-\frac{b}{2a}\right)\right)$.
	119 Find the extreme point of $y = x^2 - 3x + 2$.
	By completing the square in the right-hand member, we obtain
$\frac{9}{4}$	$y = (x^2 - 3x + \_\_\_\_) - \frac{1}{4} = (x - \frac{3}{2})^2 - \frac{1}{4}$
	From this equation we can see that the extreme point occurs
0	when the term $(x - \frac{3}{2})^2 =$ ____, that is, when
0, $\frac{3}{2}$	$x - \frac{3}{2} =$ ____ or $x =$ ____
$(\frac{3}{2}, -\frac{1}{4})$	Thus the extreme point is ________.
	Using $\left(-\frac{b}{2a}, f\left(-\frac{b}{2a}\right)\right)$, with $a = 1$ and $b = -3$, the extreme point is
$f(\frac{3}{2}),\ (\frac{3}{2}, -\frac{1}{4})$	$(\frac{3}{2}, \_\_\_\_) =$ ________.

In Problems 120–124, find the domain, the x and y intercepts, and the extreme point of f. Find the range and sketch the graph of the function.

120 $f(x) = 6 - x - x^2$

The domain is the set ____. The x intercepts are ____ and ____. The y intercept is ____.

Since $x = -\dfrac{b}{2a} =$ ____, the coordinates of the extreme point are $\left(-\dfrac{b}{2a}, f\left(-\dfrac{b}{2a}\right)\right) =$ ________

The range of f is the set $\{y \mid$ ________$\}$. The graph is a parabola that opens ____________. The graph is

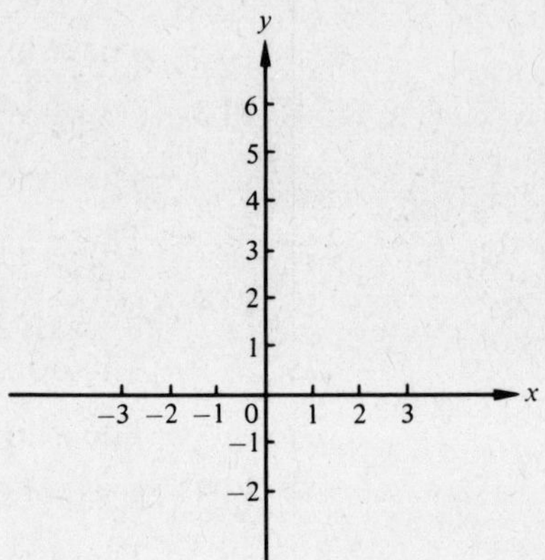

Answers:

R, -3, 2

6

$-\dfrac{1}{2}$

$\left(-\dfrac{1}{2}, \dfrac{25}{4}\right)$

$y \leqslant \frac{25}{4}$

downward

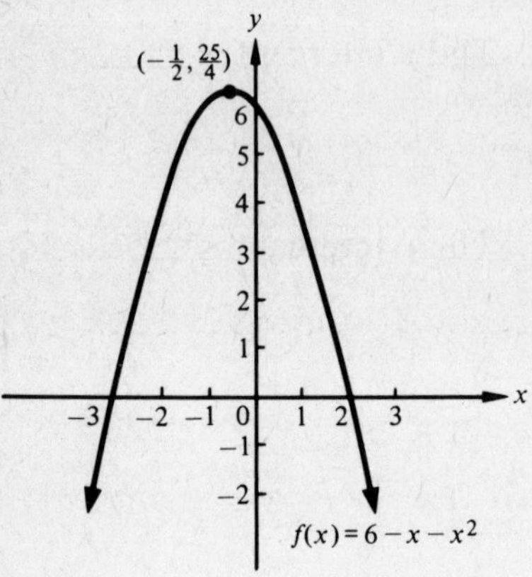

121 $f(x) = x^2 + 2x - 3$

The domain is the set ____. The x intercepts are ____ and ____. The y intercept is ____. Since $a = 1$, $b = 2$, and $c =$ ____, the coordinates of the extreme point are $\left(-\dfrac{b}{2a}, f\left(-\dfrac{b}{2a}\right)\right) =$ ________.

The range of f is the set $\{y \mid$ ________$\}$. The graph of f is a parabola that opens ________. The graph is

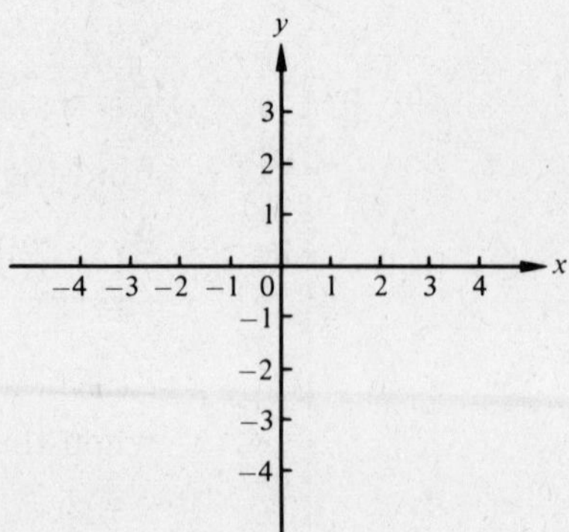

Answers:

R, -3, 1

-3, -3

$(-1, -4)$

$y \geqslant -4$

upward

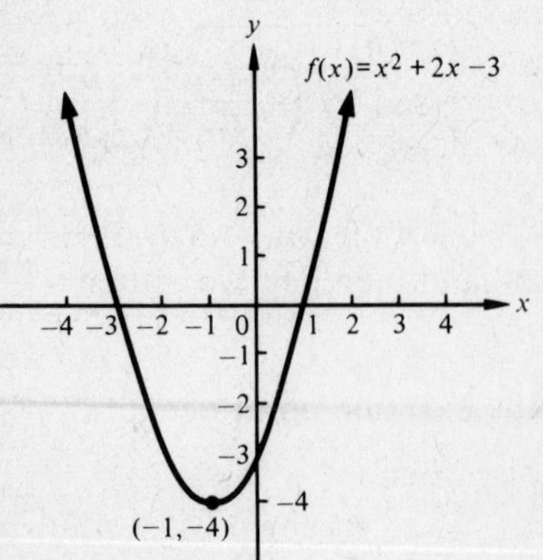

122 $f(x) = 2 + x - x^2$

The domain is the set ____. The x intercepts are ____ and ____. The y intercept is ____. The extreme point is determined by

Answers:

R, -1, 2

2

1

$(\frac{1}{2}, \frac{9}{4})$, $y \leqslant \frac{9}{4}$

downward

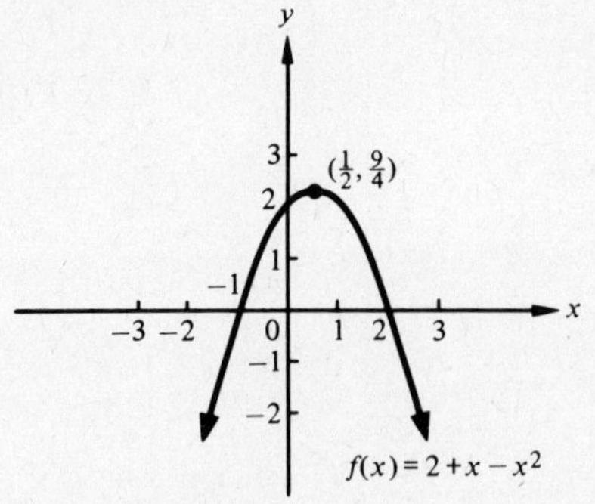

$\left(-\frac{b}{2a}, f\left(-\frac{b}{2a}\right)\right)$, with $a = -1$ and $b =$ ____. Thus the extreme point is ________. The range of f is the set $\{y|$________$\}$. The graph of f is a parabola that opens ____________. The graph is

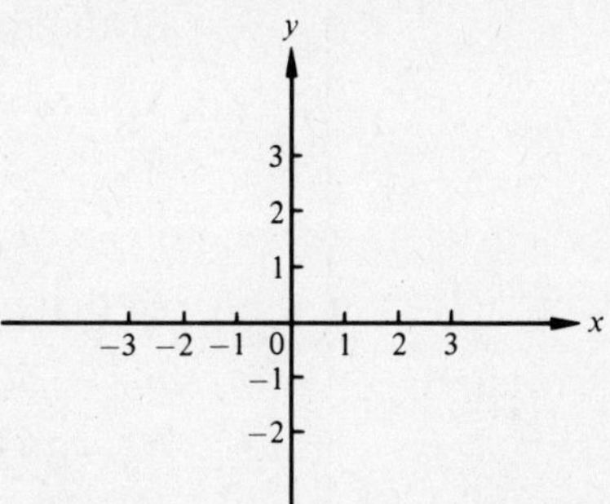

R, 1

1

$(1, 0)$, $y \geqslant 0$

upward

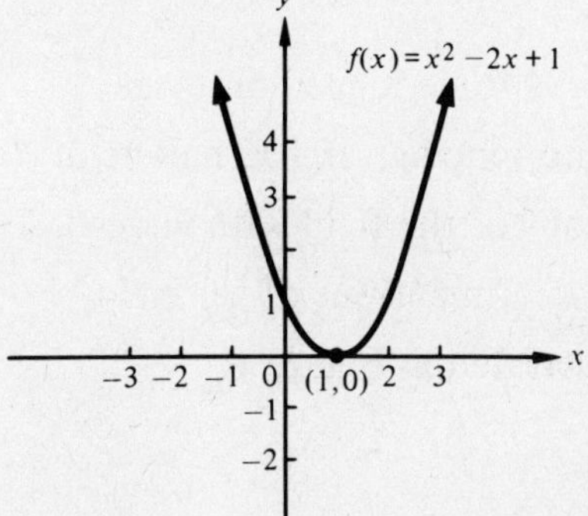

123 $f(x) = x^2 - 2x + 1$

The domain of f is the set ____. The x intercept is ____. The y intercept is ____. The extreme point of f is determined by $\left(-\frac{b}{2a}, f\left(-\frac{b}{2a}\right)\right) =$ ________. The range of f is the set $\{y|$________$\}$. The graph of f is a parabola that opens ________. The graph is

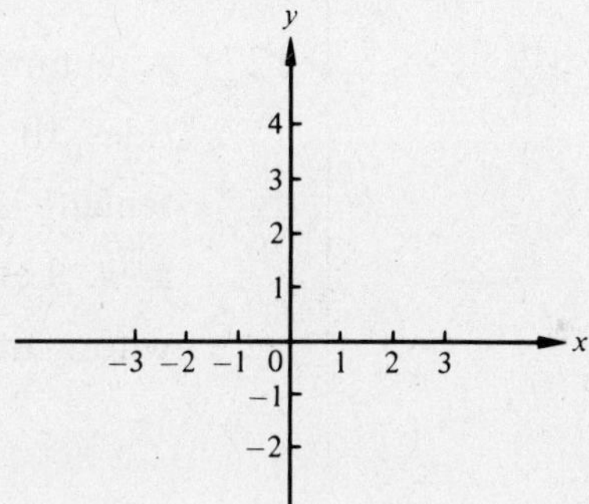

R

nonreal, 4

$(0, 4)$

$y \geqslant 4$, upward

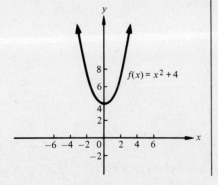

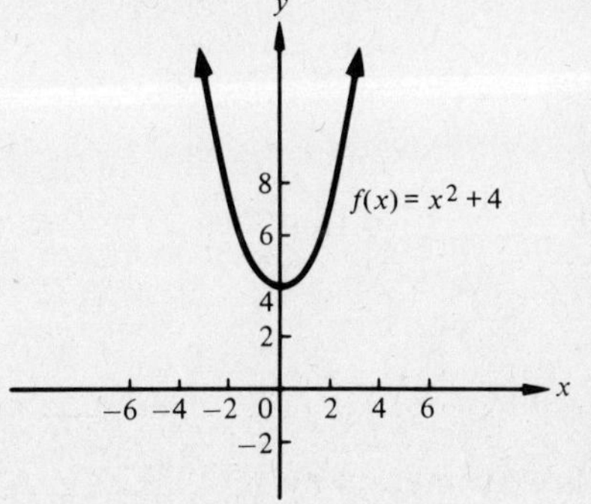

124 $f(x) = x^2 + 4$

The domain of f is the set ____. There are no x intercepts, since $x^2 + 4 = 0$ will give __________ solutions. The y intercept is ____. The extreme point is $\left(-\frac{b}{2a}, f\left(-\frac{b}{2a}\right)\right) =$ ________. The range of f is the set $\{y|$______$\}$. The graph is a parabola that opens ______. The graph is

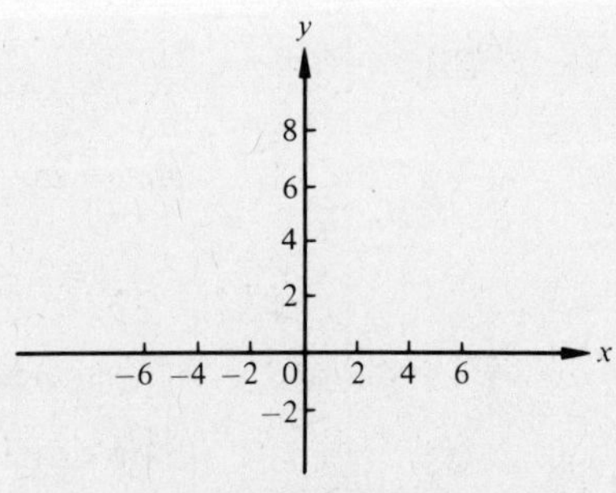

125 A rocket is fired upward from an airplane flying at 4800 feet such that, t seconds after launching, its height above the ground is given by the quadratic function $h = -15t^2 + 150t + 4800$, where its initial velocity is 150 feet per second. Find the maximum height of the rocket's flight.

To find the maximum of h, we have

5 $\quad t = -\frac{b}{2a} = -\frac{150}{-30} =$ ____

so that

5 $\quad h = -15(5)^2 + 150(\_\_\_\_) + 4800$

5175 $\quad = -375 + 750 + 4800 =$ _____

5175 feet $\quad$ The maximum height reached is ____________.

126 When will the rocket in Problem 125 reach the ground?

0 $\quad$ Set $h =$ ____; then $-15t^2 + 150t + 4800 = 0$

320, 23.6 $\quad$ or $t^2 - 10t -$ ____ $= 0$, so that $t = \frac{10 \pm \sqrt{100 + 1280}}{2} =$ _____

23.6 seconds $\quad$ Therefore, the rocket will reach the ground ____________ after reaching its maximum height.

127 A rectangular field adjacent to a river is to be fenced on three sides, the side on the river requiring no fencing. If 100 meters of fencing is available, find the dimensions of the field with largest area. Let x meters and y meters be the dimensions of the field whose area is A square meters, as illustrated in Figure 1.

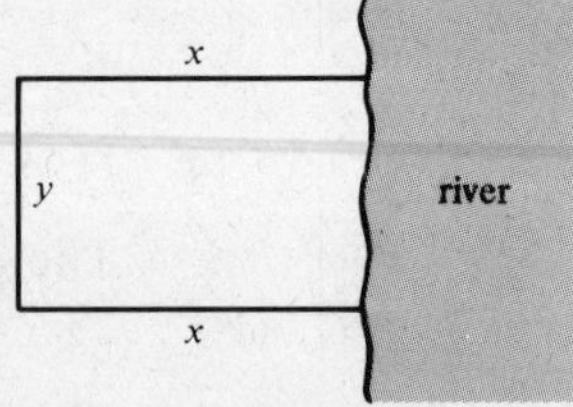

Figure 1

Then

$2x + y,\ 100 - 2x$ $\quad$ ____________ $= 100$ or $y =$ ____________

xy $\quad$ Since $A =$ ____, then

$100 - 2x,\ 100x - 2x^2$ $\quad$ $A = x(\_\_\_\_\_\_\_\_\_\_\_\_) =$ ____________

To maximize the area, we have

100 $\quad$ $A = -2x^2 + 100x$, with $a = -2$ and $b =$ ____, so that

25 $\quad$ $-\frac{b}{2a} = -\frac{100}{-4} =$ ____

25, 50 $\quad$ Thus, $y = 100 - 2(\_\_\_\_) =$ ____.

50 meters $\quad$ The dimensions of the largest field are 50 meters by ____________.

3 Quadratic Inequalities

128 Inequalities of the form $ax^2 + bx + c > 0$, $ax^2 + bx + c \geqslant 0$, $ax^2 + bx + x < 0$, or $ax^2 + bx + x \leqslant 0$, where $a \neq 0$, are called ______________.

quadratic inequalities

129 To solve quadratic inequalities like those in Problem 128, consider the graph of $f(x) = ax^2 + bx + c$ and find all values of x for which $f(x) > 0$, ________, ________, or $f(x) \leqslant 0$.

$f(x) \geqslant 0$, $f(x) < 0$

In Problems 130–132, use the graph of the quadratic function f to find the solution set of the corresponding inequality.

130 $x^2 - 4x + 3 < 0$

The graph of $f(x) = x^2 - 4x + 3$ is

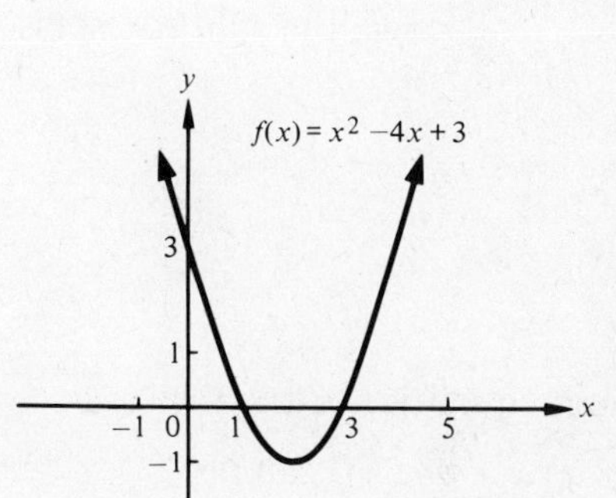

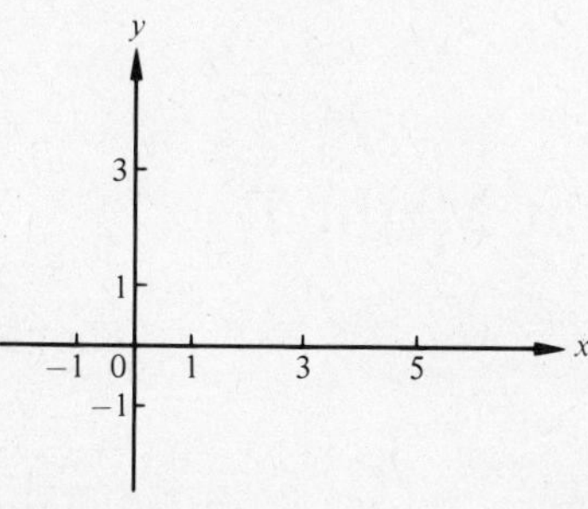

From the graph, notice that the corresponding inequality is satisfied by all values of x of the graph of f that have ordinates below the x axis, that is, for all x such that __________. The x intercepts of the graph of $f(x) = x^2 - 4x + 3$ are ____ and ____, so that the solution set is = $\{x|$______________$\}$.

$f(x) < 0$

1, 3

$1 < x < 3$

131 $-4 + 5x - x^2 > 0$

The graph of $f(x) = -4 + 5x - x^2$ is

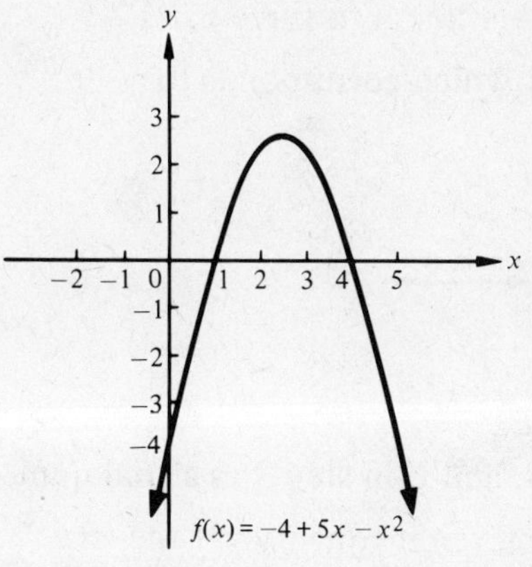

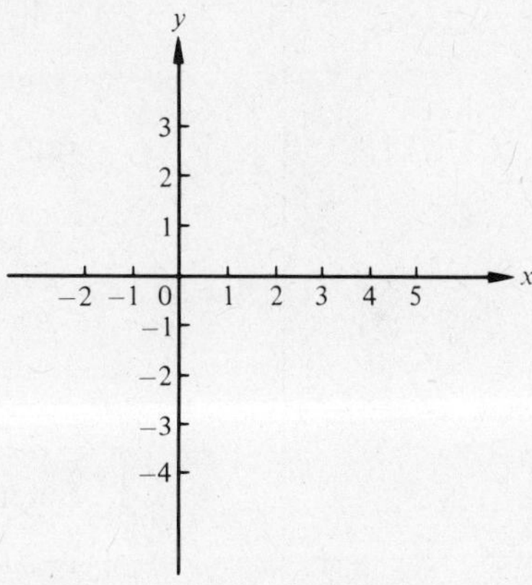

From the graph, we see that the inequality is satisfied by all values of x of the graph of f that have ordinates above the x axis, that is, by all x such that __________. Hence, the solution set is $\{x|$______________$\}$.

$f(x) > 0$

$1 < x < 4$

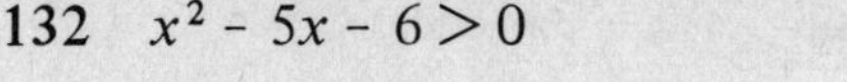

132 $x^2 - 5x - 6 > 0$

The graph of $f(x) = x^2 - 5x - 6$ is

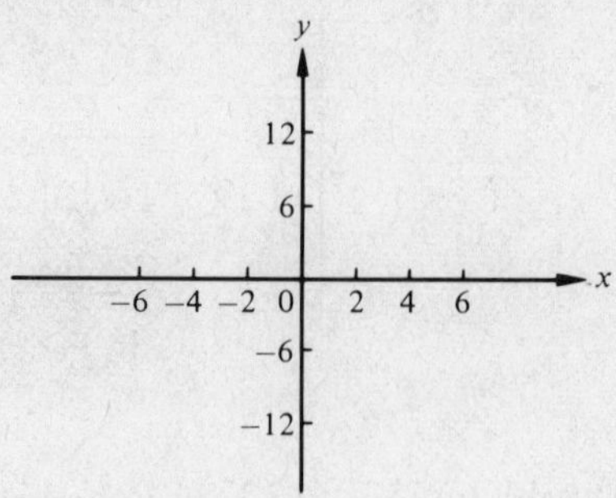

From the graph, we see that the inequality is satisfied by all values of x of the graph of f where $f(x) = x^2 - 5x - 6$ lies above the x axis, that is, by all x such that ________. Hence, the solution set is $\{x|$____________________$\}$.

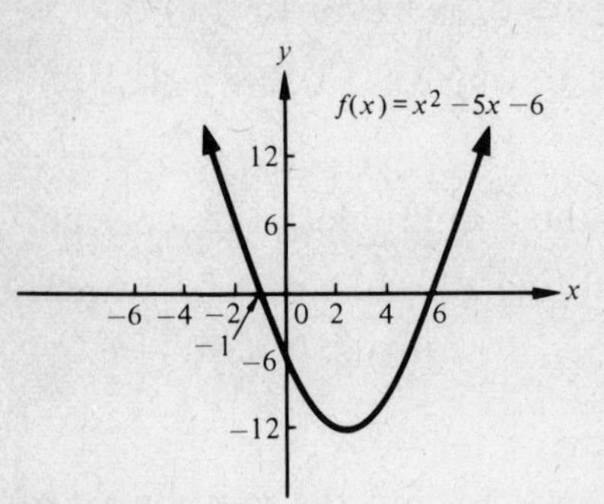

$f(x) > 0$

$x > 6$ or $x < -1$

3.1 Cut-Point Method

The techniques for solving quadratic inequalities of the form $x^2 - (p + q)x + pq < 0$ or $x^2 - (p + q)x + pq > 0$ can be simplified by adopting the following steps:

1 Change the given inequality into an equation by replacing the sign of inequality with an equal sign. In either case, the resulting equation is $x^2 - (p + q)x + pq = 0$.

2 Find the two real solutions of the quadratic equation formed in step 1. These solutions are $x = p$ and $x = q$.

3 Locate the two solutions $x = p$ and $x = q$ of the quadratic equation $x^2 - (p + q)x + pq = 0$ on the number line. These solutions will partition the number line into three parts, denoted by A, B, and C, each of which corresponds to a set of real numbers (Figure 1).

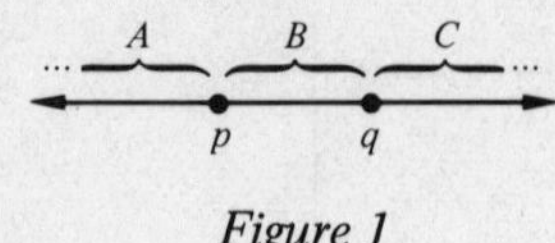

Figure 1

4 Select a number from parts A, B, and C in step 3 as a trial number and substitute it for x in the given inequality. If the trial number satisfies the given inequality, then all other numbers in its set also satisfy the inequality.

In Problems 133–140, use the 4 steps outlined above to find the solution set of the inequality.

	133 $x(x-3) \leqslant 0$
$x(x-3)$	*Step 1.* Set ________ = 0.
	Step 2. Solve the equation in step 1 so that
0, 0	$x =$ ____ or $x - 3 =$ ____
0, 3	$x =$ ____ $x =$ ____
	Step 3. The numbers 0 and 3 partition the number line into
three	______ parts, denoted by A, B, and C.
	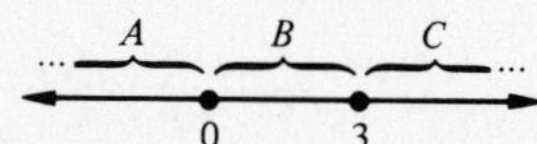
	Step 4. Select numbers from each part and check them as follows:
$-1(-1-3)$ or 4	A: If $x = -1$, then ________ $\nleqslant 0$.
$1(1-3)$ or -2	B: If $x = 1$, then ________ $\leqslant 0$.
$4(4-3)$ or 4	C: If $x = 4$, then ________ $\nleqslant 0$.
B	Thus, the solution set includes numbers in part ____, and $x = 0$
$\{x \mid 0 \leqslant x \leqslant 3\}$	and $x = 3$. This is written as ________. The graph is
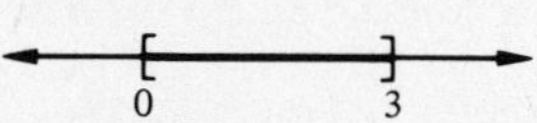	
	134 $x^2 - 5x + 4 > 0$
$(x-1)(x-4)$	*Step 1.* Set $x^2 - 5x + 4 =$ ________ = 0.
	Step 2. Solve the equation in step 1 so that
0, 0	$x - 1 =$ ____ or $x - 4 =$ ____
1, 4	$x =$ ____ $x =$ ____
	Step 3. The numbers 1 and 4 partition the number line into
three	______ parts, denoted by A, B, and C.
	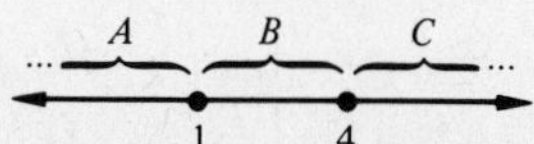
	Step 4. Select numbers from each part and check them as follows:
$(0-1)(0-4)$ or 4	A: If $x = 0$, then ________ > 0.
$(2-1)(2-4)$ or -2	B: If $x = 2$, then ________ $\ngtr 0$.
$(5-1)(5-4)$ or 4	C: If $x = 5$, then ________ > 0.
A, C	The solution set includes numbers in parts ____ and ____. This is
$\{x \mid x < 1 \text{ or } x > 4\}$	written as ________. The graph is
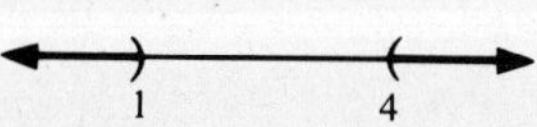	
	135 $x^2 + x - 12 \geqslant 0$
$(x+4)(x-3)$	*Step 1.* Set $x^2 + x - 12 =$ ________ = 0.
	Step 2. Solve the equation in step 1 so that
0, 0	$x + 4 =$ ____ or $x - 3 =$ ____
-4, 3	$x =$ ____ $x =$ ____

three

Step 3. The numbers -4 and 3 partition the number line into ________ parts, denoted by A, B, and C.

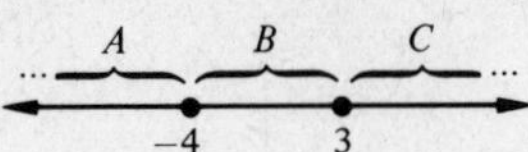

$(-5+4)(-5-3)$ or 8

$(0+4)(0-3)$ or -12

$(4+4)(4-3)$ or 8

A, C

$\{x \mid x \leqslant -4 \text{ or } x \geqslant 3\}$

Step 4. Select numbers from each part and check them as follows:

A: If $x = -5$, then ________________ $\geqslant 0$.

B: If $x = 0$, then ________________ $\ngeqslant 0$.

C: If $x = 4$, then ________________ $\geqslant 0$.

The solution set includes numbers in parts ____ and ____, and $x = -4$ and $x = 3$. This is written as ________________. The graph is

136 $10x^2 + 11x < 6$

$(5x-2)(2x+3)$

Step 1. Set $10x^2 + 11x - 6 =$ ________________ $= 0$.

0, 0

$\frac{2}{5}$, $-\frac{3}{2}$

Step 2. Solve the equation in step 1 so that

$5x - 2 =$ ____ or $2x + 3 =$ ____

$x =$ ____ $\qquad x =$ ____

three

Step 3. The numbers $\frac{2}{5}$ and $-\frac{3}{2}$ partition the number line into ________ parts, denoted by A, B, and C.

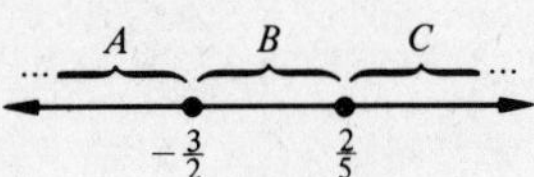

$(-10-2)(-4+3)$ or 12

$(0-2)(0+3)$ or -6

$(5-2)(2+3)$ or 15

B

$\{x \mid -\frac{3}{2} < x < \frac{2}{5}\}$

Step 4. Select numbers from each part and check them as follows:

A: If $x = -2$, then ________________ $\nless 0$.

B: If $x = 0$, then ________________ < 0.

C: If $x = 1$, then ________________ $\nless 0$.

The solution set includes numbers in part ____. This is written as ________________. The graph is

137 $2x^2 - x - 1 \geqslant 0$

$(2x+1)(x-1)$

Step 1. Set $2x^2 - x - 1 =$ ________________ $= 0$.

0, 0

$-\frac{1}{2}$, 1

Step 2. Solve the equation in step 1 so that

$2x + 1 =$ ____ or $x - 1 =$ ____

$x =$ ____ $\qquad x =$ ____

three

Step 3. The numbers $-\frac{1}{2}$ and 1 partition the number line into ________ parts, denoted by A, B, and C.

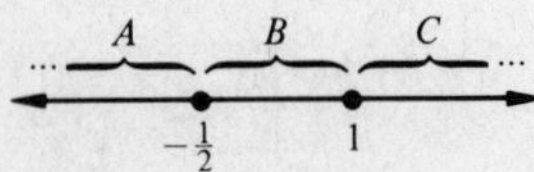

$(-2+1)(-1-1)$ or 2

$(0+1)(0-1)$ or -1

$(4+1)(2-1)$ or 5

C

$\{x \mid x \leqslant -\frac{1}{2} \text{ or } x \geqslant 1\}$

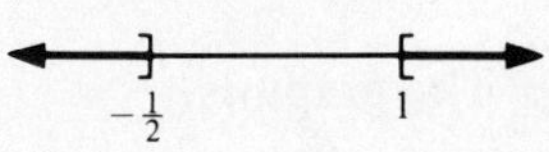

Step 4. Select numbers from each part and check them as follows:

A: If $x = -1$, then ____________________ $\geqslant 0$.

B: If $x = 0$, then ____________________ $\not\geqslant 0$.

C: If $x = 2$, then ____________________ $\geqslant 0$.

The solution set includes numbers in parts A and _____, and $x = -\frac{1}{2}$ or $x = 1$. This is written as ____________________. The graph is

0, 0

1, 3

three

0 - 1

2 - 3

4 - 1, 4 - 3

B

$\{x \mid 1 < x < 3\}$

138 $(x-1)(x-3) < 0$

Step 1. Set $(x-1)(x-3) = 0$.

Step 2. Solve the equation in step 1 so that

$x - 1 =$ _____ or $x - 3 =$ _____

$x =$ _____ $\qquad x =$ _____

Step 3. The numbers 1 and 3 from step 2 partition the number line into _______ parts, denoted by A, B, and C.

A B C
1 3

Step 4. Select numbers from each part and check them as follows:

A: If $x = 0$, then (_______)$(0-3) \not< 0$.

B: If $x = 2$, then $(2-1)$(_______)< 0.

C: If $x = 4$, then (_______)(_______)$\not< 0$.

Thus, the solution set includes numbers in part _____. This is written as ____________________. The graph is

$(x+3)^2$

positive

$(x+3)^2$

$(x+3)(3x-2)$

$(x+3)(3x-2)$

$-3, \frac{2}{3}$

three

139 $\dfrac{3x-2}{x+3} < 0$

Multiply both sides of the inequality by __________. Since $(x+3)^2$ is always ____________, we have

$(x+3)^2 \left(\dfrac{3x-2}{x+3}\right) <$ __________ $\cdot\, 0$

Then ____________________ < 0.

Step 1. Set ____________________ $= 0$.

Step 2. Solve for x so that

$x =$ _____ or $x =$ _____

Step 3. The numbers -3 and $\frac{2}{3}$ partition the number line into _______ parts, which we denote by A, B, and C.

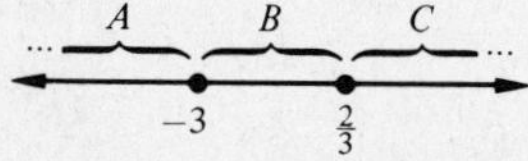

$-12 - 2$

$0 - 2$

$3 - 2$

$\{x | -3 < x < \frac{2}{3}\}$

Step 4. Select numbers in each part and check them as follows:

A: If $x = -4$, then $\frac{\_\_\_\_\_\_}{-4 + 3} \nless 0$.

B: If $x = 0$, then $\frac{\_\_\_\_\_\_}{0 + 3} < 0$.

C: If $x = 1$, then $\frac{\_\_\_\_\_\_}{1 + 3} \nless 0$.

The solution set is ______________________. The graph is

$(x - 2)^2$

$x - 2,\ 0$

$(x - 2)(5x + 1)$

$(x - 2)(5x + 1)$

$2,\ -\frac{1}{5}$

three

$-5 + 1$

$0 + 1$

$15 + 1$

$\{x | x \leqslant -\frac{1}{5} \text{ or } x > 2\}$

140 $\frac{5x + 1}{x - 2} \geqslant 0$

Multiply both sides of the inequality by __________. Since $(x - 2)^2 > 0$, we have

$(\_\_\_\_\_\_)^2 \left(\frac{5x + 1}{x - 2}\right) \geqslant (x - 2)^2(\_\_\_\_\_\_)$

So ____________________ $\geqslant 0$.

Step 1. Set ____________________ $= 0$.

Step 2. Solve for x, so that

$x =$ _____ or $x =$ _____

Step 3. The numbers $-\frac{1}{5}$ and 2 partition the number line into _______ parts, which we denote by A, B, and C.

Step 4. Select numbers in each part and check them as follows:

A: If $x = -1$, then $\frac{\_\_\_\_\_\_}{-1 - 2} \geqslant 0$.

B: If $x = 0$, then $\frac{\_\_\_\_\_\_}{0 - 2} \ngtr 0$.

C: If $x = 3$, then $\frac{\_\_\_\_\_\_}{3 - 2} \geqslant 0$.

The solution set is ______________________. The graph is

4 Polynomial Functions of Degree Greater Than 2

polynomial

zero

polynomial

4

141 Functions that can be expressed in the form

$f(x) = a_n x^n + a_{n-1} x^{n-1} + \cdots + a_1 x + a_0$

where n is a positive integer and $a_n \neq 0$,

are called ______________ functions of degree n in x.

$f(x) = 0$ is called the _____ polynomial.

142 The function $f(x) = 3x^4 - 2x^3 + x - 1$ is a ______________ function of degree ____.

polynomial, 3

143 $f(x) = 3x - 7x^3$ is a __________ function of degree ____.

polynomial, 0

144 $f(x) = 7$ is a __________ function of degree ____.

not

145 $f(x) = 2x + \dfrac{1}{x^2} = 2x + x^{-2}$ is ____ a polynomial function in x because it contains a negative exponent.

zero, root

146 Let $f(x) = a_n x^n + a_{n-1} x^{n-1} + \cdots + a_1 x + a_0$, for $a_n \neq 0$. If $f(c) = 0$, then c is called a ____ of f or a ____ of $f(x) = 0$.

0

147 Let $f(x) = x^3 - 6x^2 + x - 6$; then 6 is a zero of f, since $f(6) =$ ____.

0, zero

148 If $f(x) = 2x^{73} - x^{37} - 1$, then $f(1) =$ ____, and 1 is a ____ of f.

0, 0

zeros

149 If $f(x) = x^3 - x^2$, then $f(0) =$ ____ and $f(1) =$ ____. Therefore, 0 and 1 are _____ of f.

0, 1, 3, and −2

150 The zeros of $f(x) = (x - 1)(x + 2)(x - 3)x$ are __________.

4.1 Division of Polynomials—Synthetic Division

$2x + 1$

$x + 5,\ 2x + 1$

151 Since $2x^2 + 11x + 5 = (x + 5)($_______$)$, then $2x^2 + 11x + 5$ is divisible by both _______ and _______.

$D(x) \cdot Q(x) + R(x)$

152 If $f(x)$ and $D(x)$ are polynomials such that the degree of $f(x)$ is greater than or equal to the degree of $D(x)$, with $D(x) \neq 0$, then there exist unique polynomials $Q(x)$ and $R(x)$ such that $f(x) =$ _______________, where the degree of $R(x)$ is less than or equal to the degree of $D(x)$ or $R(x) = 0$. This property is called the division algorithm property.

In Problems 153–157, divide each of the given polynomials by the given term. That is, find $Q(x)$ and $R(x)$ as defined in Problem 152 above.

153 $2x^2 + 9x + 9$ by $x + 3$

$$\begin{array}{r} 2x + 3 \\ x + 3\,\overline{)\,2x^2 + 9x + 9} \\ 2x^2 + \_\_\_ \\ \hline 3x + 9 \\ 3x + 9 \\ \hline 0 \end{array}$$

$6x$

$2x + 3,\ 0$

Therefore, $Q(x) =$ _______ and $R(x) =$ ____.

154 $2x^3 - 9x^2 + 19x + 15$ by $2x - 3$

$$\begin{array}{r} x^2 - 3x + 5 \\ 2x - 3\,\overline{)\,2x^3 - 9x^2 + 19x + 15} \\ 2x^3 - \_\_\_ \\ \hline -6x^2 + 19x \\ -6x^2 + \_\_\_ \\ \hline 10x + 15 \\ 10x - \_\_\_ \\ \hline 30 \end{array}$$

$3x^2$

$9x$

15

$x^2 - 3x + 5,\ 30$

Therefore, $Q(x) =$ __________ and $R(x) =$ ____.

155 $3x^3 - 4x^2 - 36x + 16$ by $x - 4$

$$\begin{array}{r l}
 & 3x^2 + 8x - 4 \\
x - 4 & \overline{)3x^3 - 4x^2 - 36x + 16} \\
 & 3x^3 - \underline{\qquad} \\
 & \quad 8x^2 - 36x \\
 & \quad 8x^2 - \underline{\qquad} \\
 & \qquad -4x + 16 \\
 & \qquad \underline{-4x + 16} \\
 & \qquad\qquad 0
\end{array}$$

Therefore, $Q(x) =$ ________________ and $R(x) =$ ____.

$12x^2$

$32x$

$3x^2 + 8x - 4,\ 0$

156 $21x^4 - 42x^3 - 45x^2 + 102x - 24$ by $3x - 6$

$$\begin{array}{r l}
 & 7x^3 \qquad - \underline{\qquad} + \ 4 \\
3x - 6 & \overline{)21x^4 - 42x^3 - 45x^2 + 102x - 24} \\
 & \underline{21x^4 - 42x^3} \\
 & \qquad\quad -45x^2 + 102x \\
 & \qquad\quad \underline{-45x^2 + \qquad} \\
 & \qquad\qquad\qquad 12x - 24 \\
 & \qquad\qquad\qquad \underline{12x - 24} \\
 & \qquad\qquad\qquad\qquad 0
\end{array}$$

Therefore, $Q(x) =$ ________________ and $R(x) =$ ____.

$15x$

$90x$

$7x^3 - 15x + 4,\ 0$

157 $8x^3 + 6x^2 - 29x + 15$ by $2x + 5$

$$\begin{array}{r l}
 & \underline{\qquad} - 7x + 3 \\
2x + 5 & \overline{)8x^3 + 6x^2 - 29x + 15} \\
 & \underline{8x^3 + 20x^2} \\
 & \quad \underline{\qquad} - 29x \\
 & \quad \underline{-14x^2 - 35x} \\
 & \qquad\qquad 6x + 15 \\
 & \qquad\qquad \underline{6x + 15} \\
 & \qquad\qquad\quad \underline{\quad}
\end{array}$$

Therefore, $Q(x) =$ ________________ and $R(x) =$ ____.

$4x^2$

$-14x^2$

0

$4x^2 - 7x + 3,\ 0$

158 The process of long division can be shortened by using synthetic division when dividing a polynomial $P(x)$ by a polynomial $D(x)$, where $D(x)$ is of the form ____.

$x - r$

159 To use synthetic division, you must arrange the dividend in descending powers of the variable, representing any missing power by the use of the coefficient ____.

zero

In Problems 160–163, divide each of the given polynomials by using synthetic division.

160 $3x^3 - 2x^2 + 4x - 2$ by $x - 2$

$$\begin{array}{r|rrrr}
2 & 3 & -2 & 4 & -2 \\
 & & 6 & 8 & 24 \\
\hline
 & 3 & 4 & 12 & \underline{\quad}
\end{array}$$

$Q(x) =$ ________________ $R =$ ____

22

$3x^2 + 4x + 12,\ 22$

161 $2x^4 - x^2 + 3x - 5$ by $x + 3$

$$\begin{array}{r|rrrrr} -3 & 2 & 0 & -1 & 3 & -5 \\ & & -6 & 18 & -51 & 144 \\ \hline & 2 & -6 & 17 & -48 & \underline{\quad} \end{array}$$

139

$Q(x) =$ ____________ $R =$ ____

$2x^3 - 6x^2 + 17x - 48$, 139

162 $14x^3 - 55x^2 - 34x - 9$ by $x + 2$

$$\begin{array}{r|rrrr} -2 & 14 & -55 & -34 & -9 \\ & & -28 & 166 & -264 \\ \hline & 14 & -83 & 132 & \underline{\quad} \end{array}$$

-273

$Q(x) =$ ____________ $R =$ ____

$14x^2 - 83x + 132$, -273

163 $x^5 + 20x^2 + 8$ by $x - 3$

$$\begin{array}{r|rrrrrr} 3 & 1 & 0 & 0 & 20 & 0 & 8 \\ & & 3 & 9 & 27 & 141 & 423 \\ \hline & 1 & 3 & 9 & 47 & 141 & \underline{\quad} \end{array}$$

431

$Q(x) =$ ____________ $R =$ ____

$x^4 + 3x^3 + 9x^2 + 47x + 141$, 431

164 The *remainder theorem* states that if a polynomial function f of degree $n > 0$ is divided by $x - c$, where c is any real number, then $f(c) =$ ____________.

the remainder R

165 If $f(x) = x^3 - 7x^2 + 3x - 2$ is divided by $x - 2$, then the quotient $Q(x) =$ ________ and the remainder $R =$ ____ $= f($____$)$.

$x^2 - 5x - 7$, -16, 2

166 If $f(x) = x^5 - 4x^3 - 6x^2 - 9$ is divided by $x + 3$, then the quotient $Q(x) =$ ____________

$x^4 - 3x^3 + 5x^2 - 21x + 63$

and the remainder $R =$ ____ $= f($____$)$.

-198, -3

In Problems 167–170, use synthetic division to determine the value of each of the given expressions at the indicated value of x.

167 $6x^4 + 35x^3 + 39x^2 - 27x - 28$ at $x = 1$

$$\begin{array}{r|rrrrr} 1 & 6 & 35 & 39 & -27 & -28 \\ & & 6 & 41 & 80 & 53 \\ \hline & 6 & 41 & 80 & 53 & \underline{\quad} \end{array}$$

25

168 $x^4 + 3x^3 - 5x^2 + 9$ at $x = -3$

$$\begin{array}{r|rrrrr} -3 & 1 & 3 & -5 & 0 & 9 \\ & & -3 & 0 & 15 & -45 \\ \hline & 1 & 0 & -5 & 15 & \underline{\quad} \end{array}$$

-36

169 $3x^3 - 4x^2 + 7$ at $x = 2$

$$\begin{array}{r|rrrr} 2 & 3 & -4 & 0 & 7 \\ & & 6 & 4 & 8 \\ \hline & 3 & 2 & 4 & \underline{\quad} \end{array}$$

15

170 $3x^5 - 5x^3 + 1$ at $x = 2$

$$\begin{array}{r|rrrrrr} 2 & 3 & 0 & -5 & 0 & 0 & 1 \\ & & 6 & 12 & 14 & 28 & 56 \\ \hline & 3 & 6 & 7 & 14 & 28 & \underline{\quad} \end{array}$$

57

factor — **171** $f(c) = 0$ if and only if $x - c$ is a ________ of the polynomial $f(x)$. This is called the *factor theorem.*

0 — **172** $x - 6$ is a factor of $f(x) = x^3 - 6x^2 + x - 6$, since $f(6) =$ ____.

0 — **173** $x + 2$ is a factor of $f(x) = x^4 - 2x^2 + 3x - 2$, since $f(-2) =$ ____.

4 — **174** If $x - 3$ is a factor of $f(x) = 3x^3 - 4x^2 - kx - 33$, then $k =$ ____.

4.2 Rational Zeros

175 If $f(x) = a_n x^n + a_{n-1} x^{n-1} + \cdots + a_1 x + a_0 = 0$, if the coefficients are integers, and if p/q is a rational zero in lowest
a_n — terms of $f(x) = 0$, then p is a divisor of a_0 and q is a divisor of ____.
This theorem is called the *rational zero theorem.*

In Problems 176–179, write the possible rational roots in each case.

176 $f(x) = 3x^3 - 8x^2 + 3x + 2$
0 — Let p/q be a rational root of $f(x) = 0$, that is, $f(p/q) =$ ____. Then
2, 3 — p is a divisor of ____ and q is a divisor of ____. The possibilities are
1, 2; 1, 3; 1, 2, $\frac{1}{3}$, $\frac{2}{3}$ — $p = \pm\{$____$\}$ and $q = \pm\{$____$\}$ so that $p/q = \pm\{$____________$\}$.

177 $f(x) = 3x^4 - 8x^3 - 28x^2 + 64x - 15$
0 — Let p/q be a rational root of $f(x) = 0$, that is, $f(p/q) =$ ____.
-15, 3 — Then p is a divisor of ____ and q is a divisor of ____. The possi-
1, 3, 5, 15; 1, 3 — bilities are $p = \pm\{$____________$\}$ and $q = \pm\{$____$\}$ so that
1, 3, 5, 15, $\frac{1}{3}$, $\frac{5}{3}$ — $p/q = \pm\{$____________$\}$

178 $f(x) = x^4 + 2x^3 + 2x^2 - 4x - 8$
0 — Let p/q be a rational root of $f(x) = 0$, that is, $f(p/q) =$ ____.
-8, 1 — Then p is a divisor of ____ and q is a divisor of ____. The possi-
1, 2, 4, 8 — bilities are $p = \pm\{$____________$\}$ and $q = \pm\{1\}$, so that
1, 2, 4, 8 — $p/q = \pm\{$____________$\}$

179 $f(x) = 2x^3 + 3x^2 + 2x + 3$
0 — Let p/q be a rational root of $f(x) = 0$, that is, $f(p/q) =$ ____.
3, 2 — Then p is a divisor of ____ and q is a divisor of ____. The possi-
1, 3; 1, 2 — bilities are $p = \pm\{$____$\}$ and $q = \pm\{$____$\}$, so that
1, 3, $\frac{1}{2}$, $\frac{3}{2}$ — $p/q = \pm\{$____________$\}$

In Problems 180–183, use the rational root theorem to find all rational zeros of the given polynomial function.

180 $f(x) = x^4 - 10x^2 + 9$

Assume that p/q is a rational zero of f. By the rational zero

9, 1 — theorem, p is a divisor of ____ and q is a divisor of ____. The

1, 3, 9; 1 — possibilities are $p = \pm\{$______$\}$ and $q = \pm\{$____$\}$, so that

1, 3, 9 — $p/q = \pm\{$______$\}$. Testing these possible zeros by synthetic division, we find

```
 1 | 1   0  -10    0    9
   |     1    1   -9   -9
   ----------------------
     1   1   -9   -9 |  0        f(1) = ____
```
0

```
-1 | 1   0  -10    0    9
   |    -1    1    9   -9
   ----------------------
     1  -1   -9    9 |  0        f(-1) = ____
```
0

```
 3 | 1   0  -10    0    9
   |     3    9   -3   -9
   ----------------------
     1   3   -1   -3 |  0        f(3) = ____
```
0

```
-3 | 1   0  -10    0    9
   |    -3    9    3   -9
   ----------------------
     1  -3   -1    3 |  0        f(-3) = ____
```
0

```
 9 | 1   0  -10    0     9
   |     9   81  639  5751
   -----------------------
     1   9   71  639 | 5760      f(9) = ____
```
5760

```
-9 | 1   0  -10     0     9
   |    -9   81  -639  5751
   ------------------------
     1  -9   71  -639 | 5760     f(-9) = ____
```
5760

-1, 1, -3, 3 — Hence, the rational zeros of f are ____, ____, ____, and ____.

181 $f(x) = x^3 - 2x^2 - x + 2$

Assume that p/q is a rational zero of f. By the rational zero

2, 1 — theorem, p is a divisor of ____ and q is a divisor of ____. The

1, 2; 1 — possibilities are $p = \pm\{$____$\}$ and $q = \pm\{$____$\}$, so that

1, 2 — $p/q = \pm\{$____$\}$. Using synthetic division, we find

```
 1 | 1  -2  -1   2
   |     1  -1  -2
   ---------------
     1  -1  -2 | 0      f(1) = ____
```
0

```
-1 | 1  -2  -1   2
   |    -1   3  -2
   ---------------
     1  -3   2 | 0      f(-1) = ____
```
0

```
 2 | 1  -2  -1   2
   |     2   0  -2
   ---------------
     1   0  -1 | 0      f(2) = ____
```
0

```
-2 | 1  -2  -1    2
   |    -2   8  -14
   ----------------
     1  -4   7 | -12    f(-2) = ____
```
-12

-1, 1, 2 — Hence, the rational zeros of f are ____, ____, and ____.

182 $f(x) = x^3 + 6x^2 + x + 6$

Assume that p/q is a rational zero of f. Then p is a divisor

6; 1; 1, 2, 3, 6 — of ____ and q is a divisor of ____, so that $p = \pm\{$________$\}$

1; 1, 2, 3, 6 — and $q = \pm\{$____$\}$ and $p/q = \pm\{$________$\}$.

Testing these possible zeros by synthetic division, we find that -6 is a zero of f, since

$$\begin{array}{r|rrrr} -6 & 1 & 6 & 1 & 6 \\ & & -6 & 0 & -6 \\ \hline & 1 & 0 & 1 & 0 \end{array}$$

0 — $f(-6) =$ ____

Testing the other possible zeros, we find that -6 is the only zero of f.

183 $f(x) = 3x^4 - 11x^3 + 10x - 4$

Assume that p/q is a rational zero of f. Then p is a divisor

-4; 3; 1, 2, 4 — of ____ and q is a divisor of ____, so that $p = \pm\{$________$\}$ and

1, 3; 4, 2, 1, $\frac{4}{3}$, $\frac{2}{3}$, $\frac{1}{3}$ — $q = \pm\{$____$\}$ and $p/q = \pm\{$________$\}$.

Testing these possibilities, we find that -1 and $\frac{2}{3}$ are zeros of f, since

$$\begin{array}{r|rrrrr} -1 & 3 & -11 & 0 & 10 & -4 \\ & & -3 & 14 & -14 & 4 \\ \hline & 3 & -14 & 14 & -4 & 0 \end{array}$$

0 — $f(-1) =$ ____

$$\begin{array}{r|rrrr} \frac{2}{3} & 3 & -14 & 14 & -4 \\ & & 2 & -8 & 4 \\ \hline & 3 & -12 & 6 & 0 \end{array}$$

0 — $f(\frac{2}{3}) =$ ____

Testing the other possible zeros, we find that -1 and $\frac{2}{3}$ are the only rational zeros of f.

4.3 Graphs of Polynomial Functions of Degree Greater Than 2

184 In order to graph a polynomial function f, we shall attempt to find the zeros of f by using the fact that if $a < b$ and $f(a) < 0 < f(b)$;

$f(c) = 0$ — then there is a number c such that $a < c < b$ and ________.

In Problems 185–188, sketch the graph of each polynomial function and find the solution set of the corresponding inequality.

185 $f(x) = x^3 - 2x^2 - x + 2$ and $x^3 - 2x^2 - x + 2 < 0$

-1, 1 — The graph of f intercepts the x axis only at $x =$ ____, $x =$ ____,

2 — and $x =$ ____. A few additional points are needed to complete the graph, which is

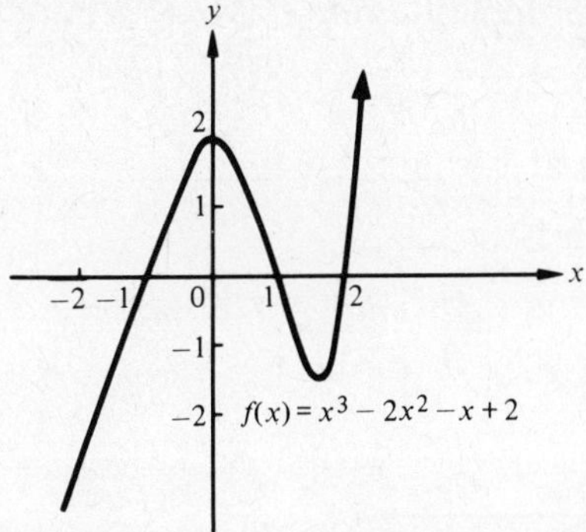

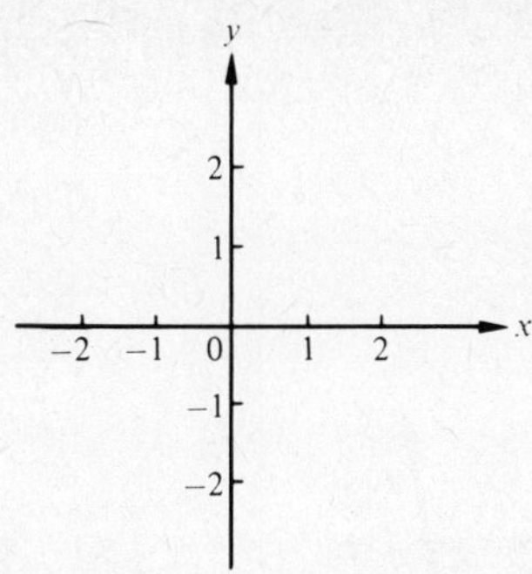

Notice that the portion of the graph below the x axis suggests that these values of x satisfy the inequality, so that the solution set is $\{x \mid$ ____________________ $\}$.

$x < -1$ or $1 < x < 2$

186 $f(x) = 2x^3 + x^2 - 5x + 2$ and $2x^3 + x^2 - 5x + 2 > 0$

The graph of f intercepts the x axis at $x =$ ____, $x =$ ____, and $x =$ ____. A few additional points are needed to complete the graph, which is

$-2, \frac{1}{2}$

1

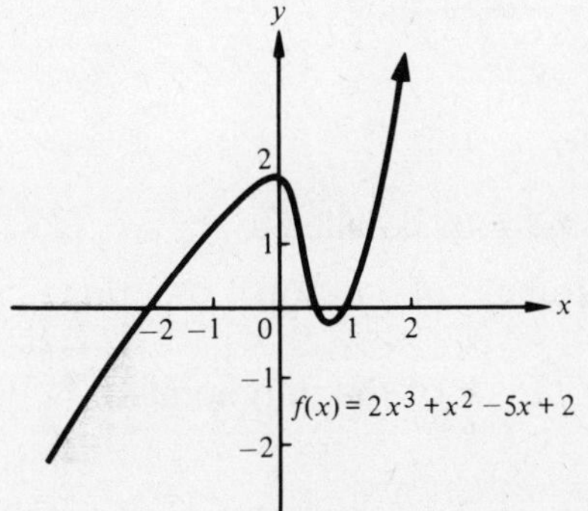

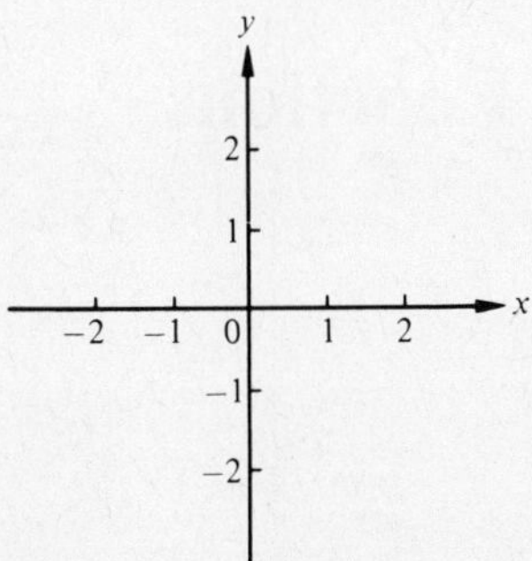

Notice that the portion of the graph above the x axis suggests that these values of x satisfy the inequality, so that the solution set is $\{x \mid$ ____________________ $\}$.

$-2 < x < \frac{1}{2}$ or $x > 1$

187 $f(x) = (x + 2)(x - 1)(x - 3)$ and $(x + 2)(x - 1)(x - 3) > 0$

The graph of f intercepts the x axis at $x =$ ____, $x =$ ____, and $x =$ ____. The graph is

$-2, 1$

3

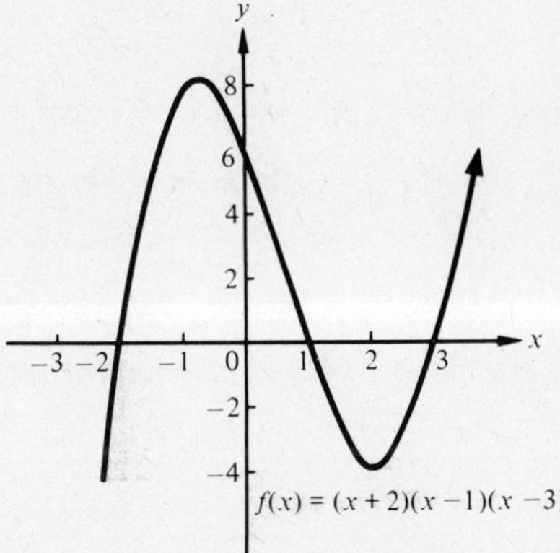

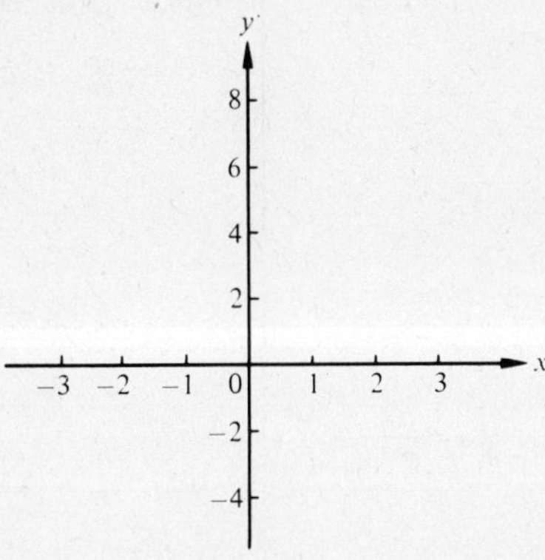

The portion of the graph above the x axis suggests that these values of x satisfy the inequality, so that the solution set is $\{x \mid$ ____________________ $\}$.

$-2 < x < 1$ or $x > 3$

188 $f(x) = (x + 4)(x + 1)(x - 1)(x - 3)$ and $(x + 4)(x + 1)(x - 1)(x - 3) < 0$

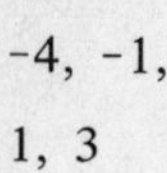

-4, -1, 1, 3

The graph of f intercepts the x axis at x = ____, x = ____, x = ____, and x = ____. The graph is

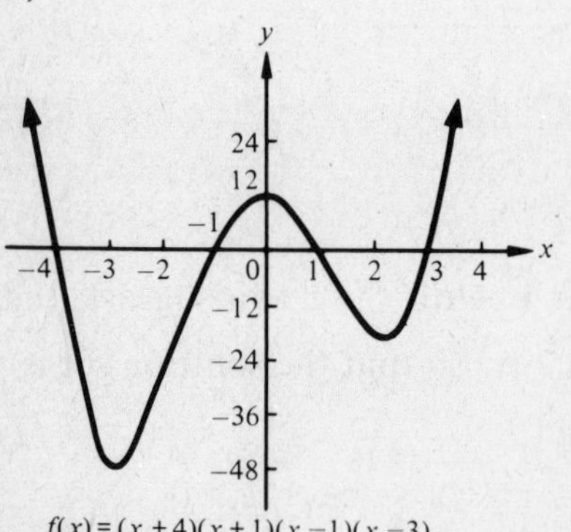

$f(x) = (x + 4)(x + 1)(x - 1)(x - 3)$

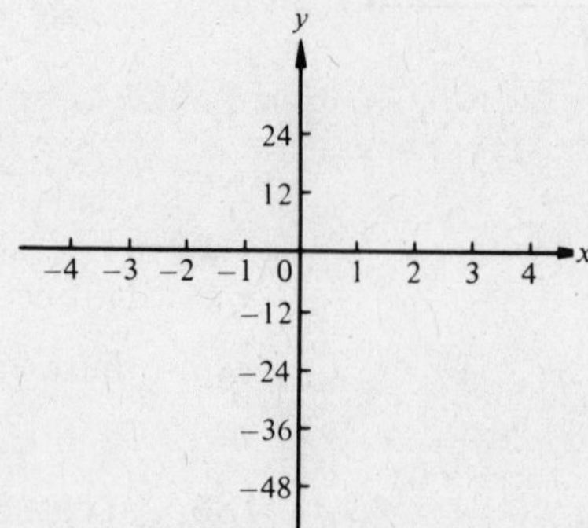

The portion of the graph below the x axis suggests that these values of x satisfy the inequality, so that the solution set is $\{x \mid$ ____________________ $\}$.

$-4 < x < -1$ or $1 < x < 3$

5 Rational Functions

189 The quotient of two polynomial functions is called a ____________. The domain of a rational function $R(x) = \dfrac{f(x)}{g(x)}$ does not contain the ______ of the polynomial function g, since division by zero is not defined.

rational function

zeros

190 If a is a real number, such that $f(a) = 0$ and $g(a) = 0$, then

$$R(x) = \frac{f(x)}{g(x)} = \frac{f_1(x)(x - a)}{g_1(x)(x - a)} = \frac{f_1(x)}{g_1(x)}, \quad \text{for } x \neq \_\_\_\_$$

a

In Problems 191–193, simplify the given rational function, state for what value the function is undefined, and graph the function.

191 $R(x) = \dfrac{x^2 - 4}{x + 2}$

$= \dfrac{(x - 2)(x + 2)}{x + 2}$

$=$ ____, for $x \neq$ ____

$x - 2$, -2

The graph is

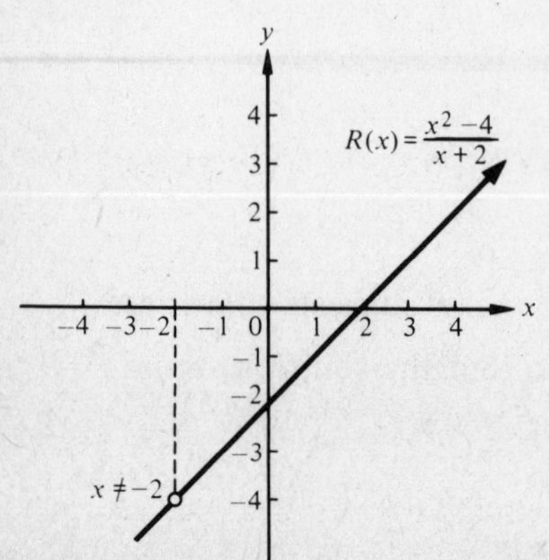

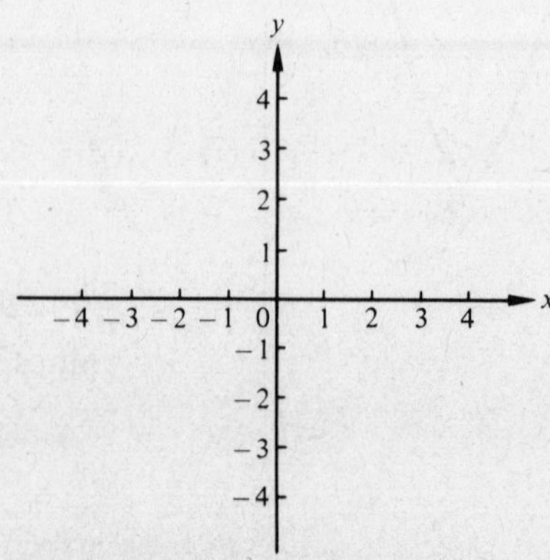

192 $R(x) = \dfrac{x^2 + 5x + 6}{x + 3}$

$= \dfrac{(x + 2)(x + 3)}{x + 3}$

$=$ ______, for $x \neq$ _____

$x + 2, -3$

The graph is

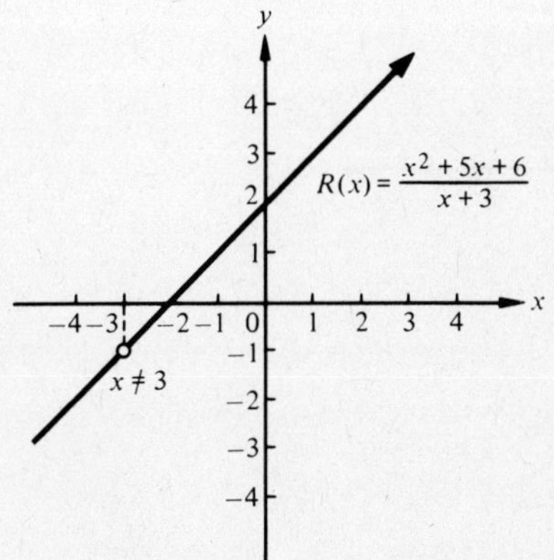

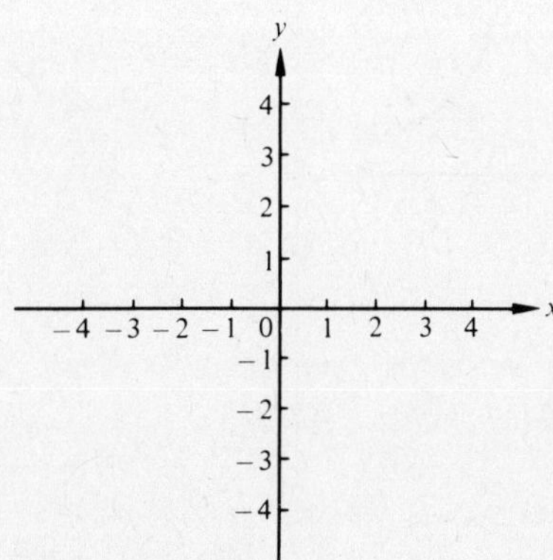

193 $R(x) = \dfrac{x^3 - 27}{x - 3}$

$= \dfrac{(x - 3)(x^2 + 3x + 9)}{x - 3}$

$=$ ________________, for $x \neq$ _____

$x^2 + 3x + 9, 3$

The graph is

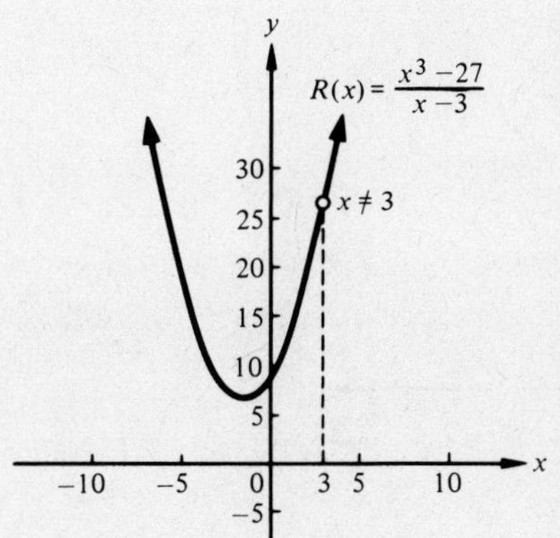

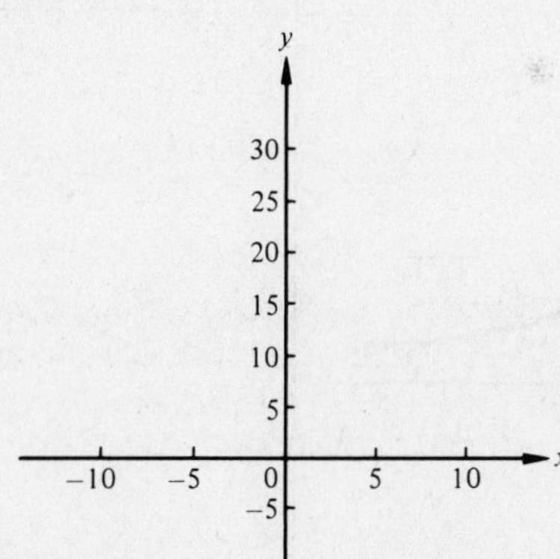

194 A vertical asymptote occurs for the graph of $R(x) = f(x)/g(x)$ at $x = a$ when $f(a) \neq 0$ and $g(a) =$ _____.

0

In Problems 195–202, find the domain, x intercept(s), y intercept, vertical asymptotes, and sketch the graph of the given rational function.

195 $f(x) = \dfrac{3}{x - 2}$

The domain of f is ________________. Letting $f(x) = 0$, we have

$\{x \mid x \neq 2\}$

$\dfrac{3}{x - 2} = 0$, which has no solutions, so that there is no ______________.

x intercept

Letting $x = 0$, we have $f(0) = \dfrac{3}{0-2} = -\dfrac{3}{2}$, so that the y intercept is ____. The vertical asymptote occurs when $x - 2 = 0$, so that the vertical asymptote is ______. The graph is

$-\frac{3}{2}$

$x = 2$

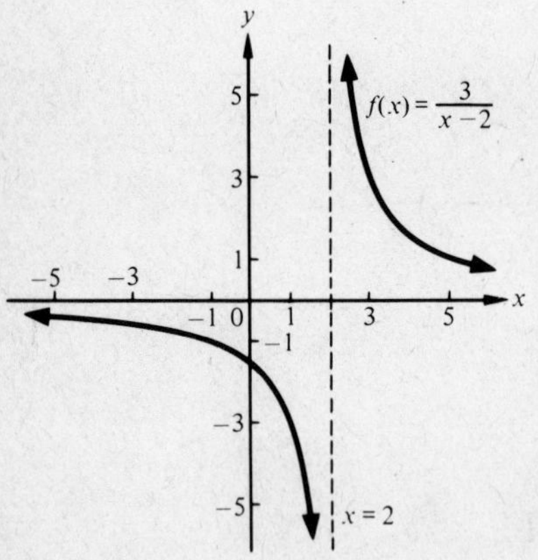

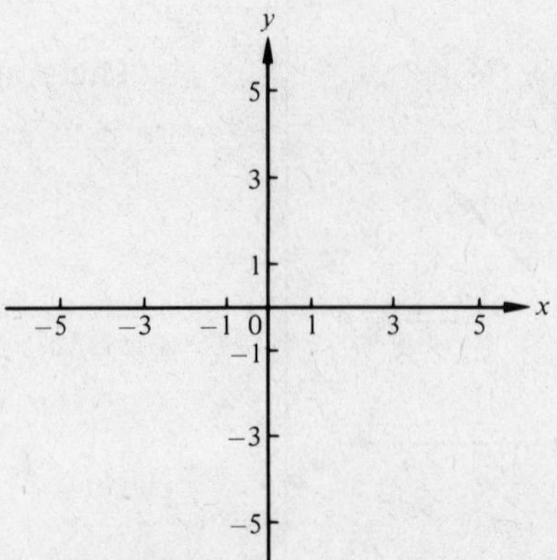

196 $f(x) = \dfrac{6}{x+3}$

The domain of f is ____________. There are no ____ intercepts, since $\dfrac{6}{x+3} = 0$ has no solutions. Letting $x = 0$, we have $f(0) = \dfrac{6}{0+3} = 2$, so that the y intercept is ____. The vertical asymptote occurs when ______ = 0, so that the vertical asymptote is _______. The graph is

$\{x \mid x \neq -3\}$, x

2

$x + 3$, $x = -3$

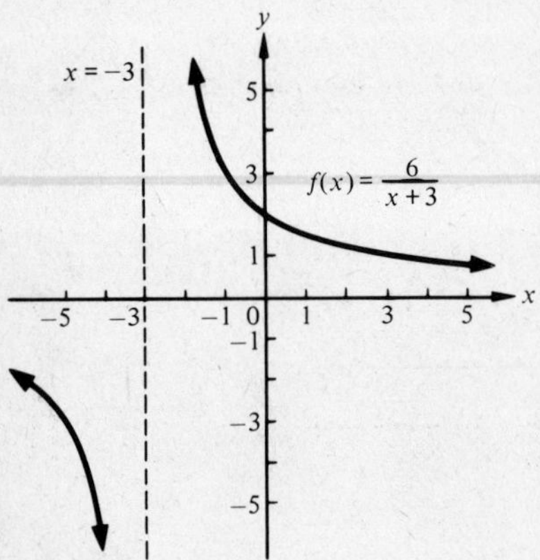

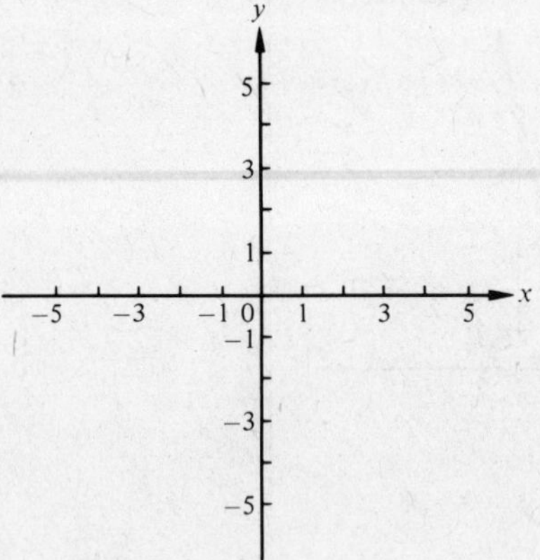

197 $f(x) = \dfrac{x}{x-2}$

The domain of f is ____________. For $f(x) = 0$ or $\dfrac{x}{x-2} = 0$, we have $x = 0$, which is the ____ intercept. Letting $x = 0$, we have $f(0) = \dfrac{0}{0-2} = 0$, so that the ____ intercept is 0. The vertical asymptote occurs when _____ = 0, so that _____ is the vertical

$\{x \mid x \neq 2\}$

x

y

$x - 2$, $x = 2$

asymptote. $f(x) = \dfrac{x}{x - 2}$ can also be expressed as $f(x) = 1 + \dfrac{2}{x - 2}$,

1 so that as $|x|$ becomes larger, $f(x)$ approaches ____ asymptotically. The graph is

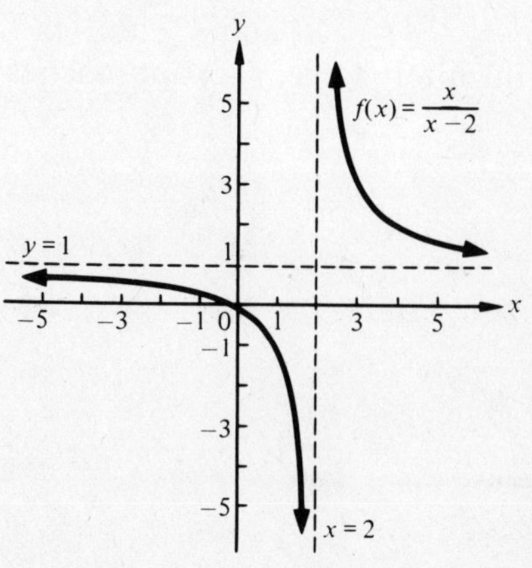

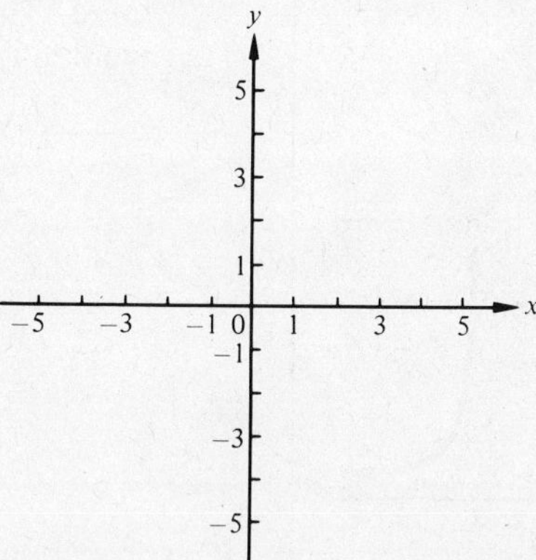

198 $f(x) = \dfrac{3x + 1}{x}$

$\{x \mid x \neq 0\}$ The domain of f is ____________. For $\dfrac{3x + 1}{x} = 0$, we have

$-\frac{1}{3}$ $3x + 1 = 0$ or $x =$ ____, which is the x intercept. $f(0)$ does not

y exist, so there is no ____ intercept. The vertical asymptote is

0 $x =$ ____. Dividing $3x + 1$ by x, we can express

$3 + \dfrac{1}{x}$ $f(x) = \dfrac{3x + 1}{x}$ as $f(x) =$ _____

3 As $|x|$ becomes larger, $f(x)$ approaches ____ asymptotically. The graph is

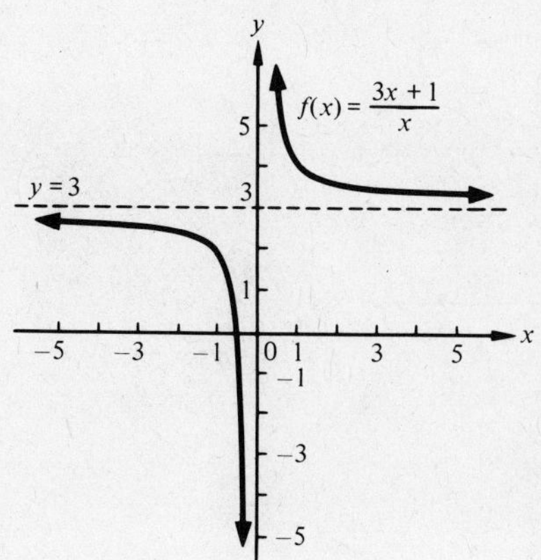

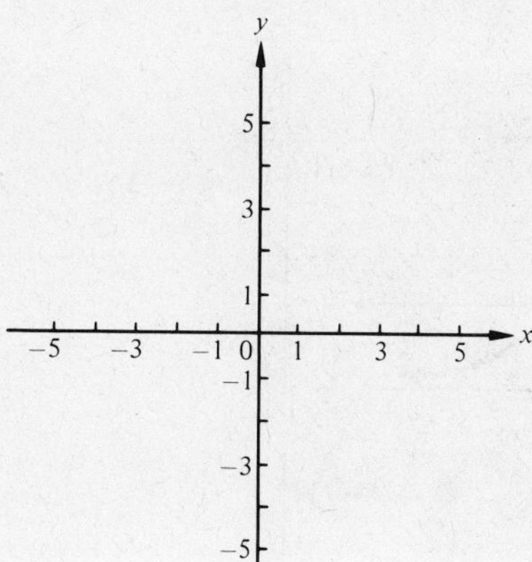

199 $f(x) = \dfrac{x + 1}{x^2 - 9}$

$\{x \mid x \neq \pm 3\}$ The domain of f is ____________. The x intercept is found by

-1 letting $\dfrac{x + 1}{x^2 - 9} = 0$, so that $x =$ ____. The y intercept is $f(0) =$

$-\dfrac{1}{9}$ $\dfrac{0 + 1}{0^2 - 9} =$ ____. The vertical asymptotes are those values of x for

0, ±3

which $x^2 - 9 =$ ____, or $x =$ ____. Rewriting

$$f(x) = \frac{x+1}{x^2-9} \quad \text{as} \quad f(x) = \frac{\frac{1}{x} + \frac{1}{x^2}}{1 - \frac{9}{x^2}}$$

0

we can observe that as $|x|$ becomes infinitely large, $f(x)$ approaches ____ asymptotically. The graph is

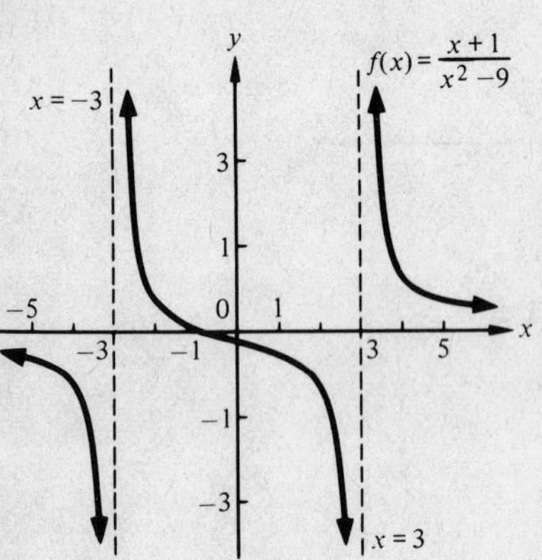

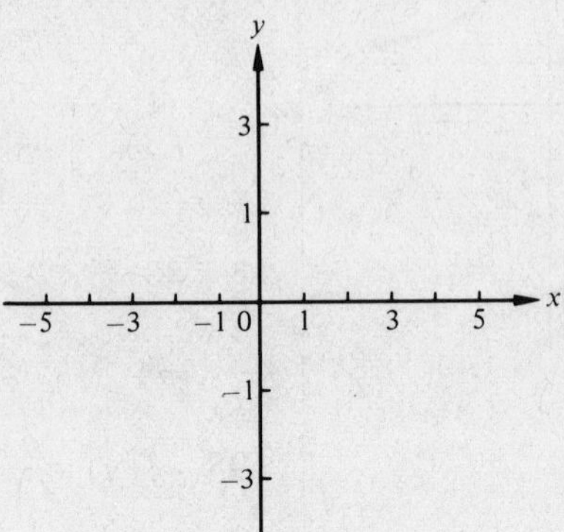

200 $f(x) = \dfrac{-x^2}{x^2+1}$

R

The domain of f is ____. The x intercept is found by letting

0

$\dfrac{-x^2}{x^2+1} = 0$, so that $x =$ ____. There is no vertical asymptote because $x^2 + 1$ has no real solution. The y intercept is

0

$f(0) = \dfrac{-0^2}{0^2+1} =$ ____. The graph is

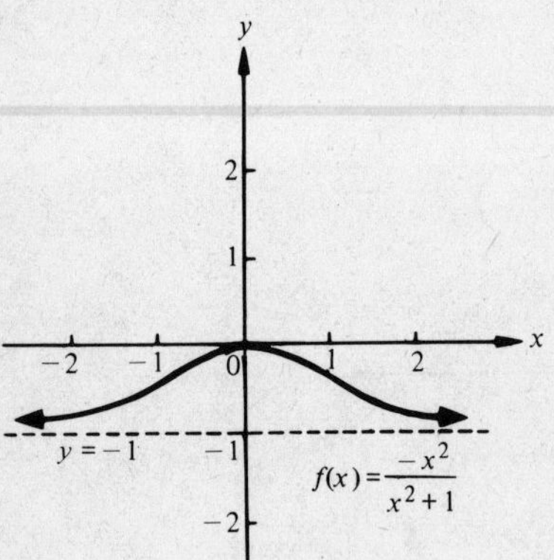

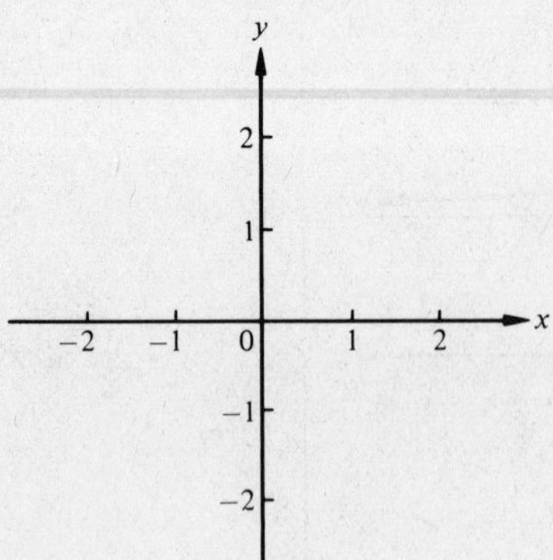

201 $f(x) = \dfrac{x^2}{x^2+1}$

$\{x \mid x \in R\}$

The domain of f is ________. The x intercept is found by

0

letting $\dfrac{x^2}{x^2+1} = 0$, so that $x =$ ____. The y intercept is determined

0

by $f(0) = \dfrac{0^2}{0^2+1} =$ ____. There are no vertical asymptotes, since

$x^2 + 1$

________ = 0 has no real solution. Rewriting

$$f(x) = \frac{x^2}{x^2 + 1} \quad \text{as} \quad f(x) = \frac{1}{1 + \dfrac{1}{x^2}}$$

1

we observe that $f(x)$ approaches ____ as $|x|$ becomes infinitely large. The graph is

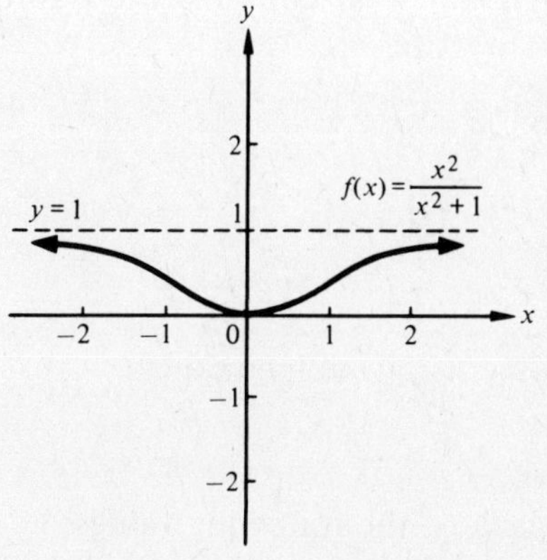

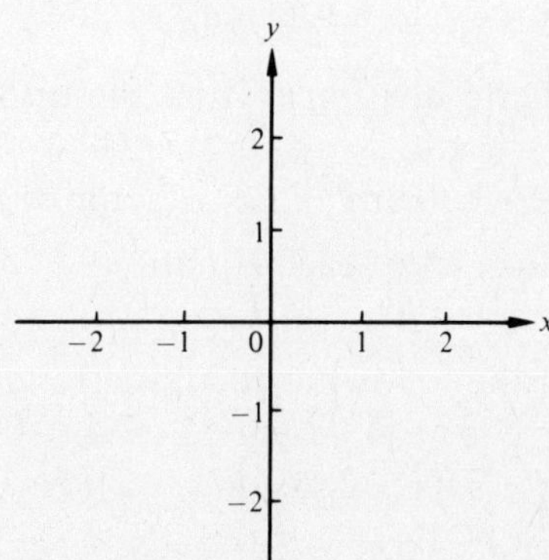

202 $f(x) = \dfrac{x^2 - 9}{x^2 - 3x + 2}$

1 or 2

The domain of f is $\{x \mid x^2 - 3x + 2 \neq 0\}$ or $\{x \mid x \neq$ ________$\}$.

The x intercepts are found by letting $\dfrac{x^2 - 9}{x^2 - 3x + 2} = 0$, so that

-3, 3

$x =$ ____ or $x =$ ____. The y intercept is determined by $f(0) =$

$-\dfrac{9}{2}$

$\dfrac{0^2 - 9}{0^2 - 3(0) + 2} =$ ____. The vertical asymptotes are determined

1, 2

by $x^2 - 3x + 2 = 0$, or $x =$ ____ and $x =$ ____. Rewriting

$$f(x) = \frac{x^2 - 9}{x^2 - 3x + 2} \quad \text{as} \quad f(x) = \frac{1 - \dfrac{9}{x^2}}{1 - \dfrac{3}{x} + \dfrac{2}{x^2}}$$

1

we observe that $f(x)$ approaches ____ as $|x|$ becomes infinitely large. The graph is

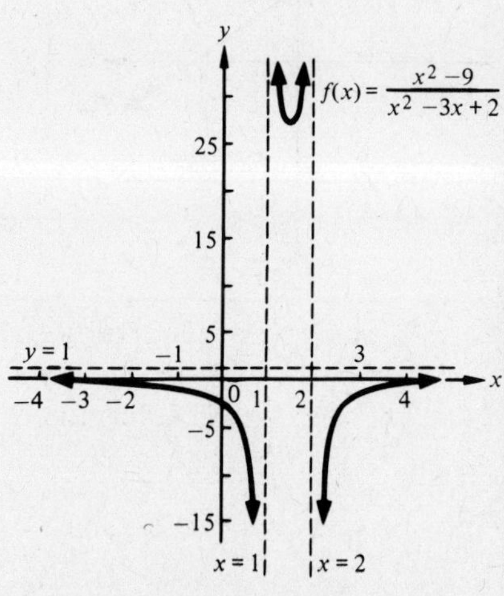

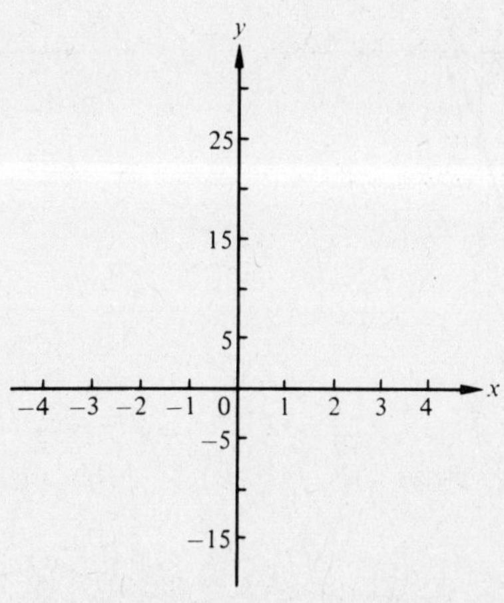

Chapter Test

1 Find an equation of the line and sketch the line in each case.
(a) slope $m = 3$ and containing the point (4, 5)
(b) containing the points (3, -2) and (5, 6)
(c) containing the points (0, 2) and (3, 0)

2 Find an equation of the line that is parallel to each line in Problem 1 and that contains the point (-2, 1).

3 (a) Use synthetic division to find the quotient $Q(x)$ and the remainder R if $f(x) = x^3 + 6x^2 + x - 4$ and $D(x) = x + 2$.
(b) Determine whether $x + 1$ is a factor of $f(x)$.

4 Find the solution set of each equation.
(a) $5x^2 - 13x - 6 = 0$ (b) $x + 3\sqrt{x} - 10 = 0$

5 Graph each of the following quadratic functions and use the graph to solve the given inequality:
(a) $f(x) = 6x^2 - 5x - 4$; $6x^2 - 5x - 4 \leq 0$
(b) $f(x) = (7x - 3)(1 - 2x)$; $(7x - 3)(1 - 2x) > 0$

6 Let $f(x) = x^5 + x^4 - 3x^3 - 5x^2 + x + 7$. Use synthetic division to find each of the following values:
(a) $f(2)$ (b) $f(3)$ (c) $f(-1)$ (d) $f(-2)$

7 Find all rational zeros of the function $f(x) = 2x^4 + 5x^3 - 9x^2 - 15x + 9$.

8 Let $f(x) = x^2 + x - 2$. Find the following:
(a) domain (b) range (c) x intercepts (d) y intercept (e) extreme point

9 Sketch the graph of each of the given rational functions. Also, find the intercepts and asymptotes.
(a) $f(x) = \dfrac{x^2 - 16}{x - 4}$ (b) $f(x) = \dfrac{2x - 6}{x + 1}$

Answers

1 (a) $y = 3x - 7$ (b) $y = 4x - 14$ (c) $y = -\frac{2}{3}x + 2$

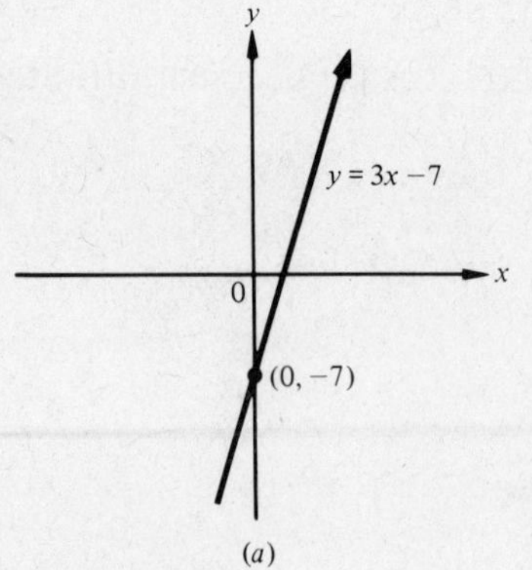

(a)

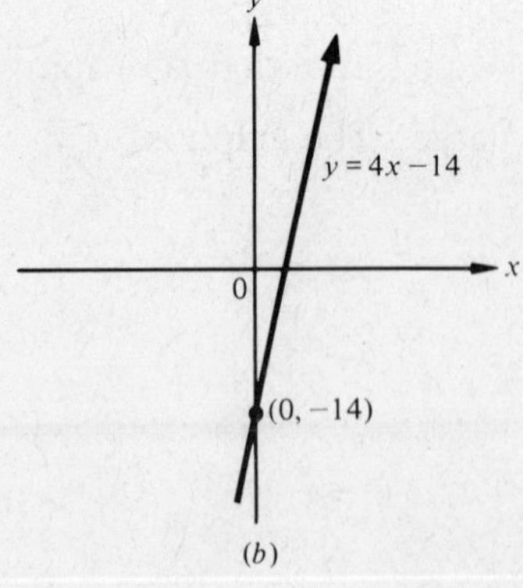

(b)

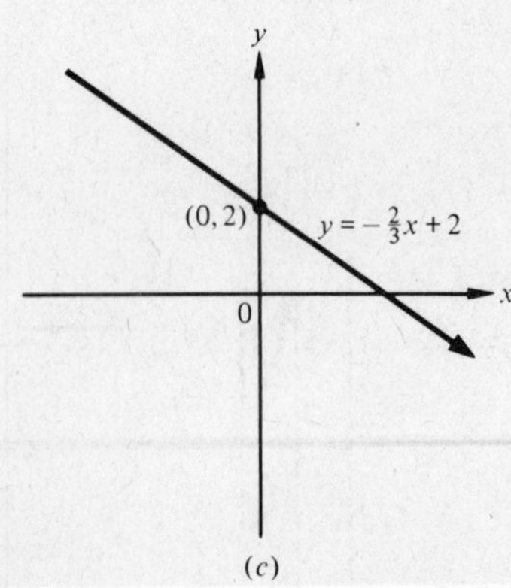

(c)

2 (a) $y = 3x + 7$ (b) $y = 4x + 9$ (c) $y = -\frac{2}{3}x - \frac{1}{3}$

3 (a) $Q(x) = x^2 + 4x - 7, R = 10$ (b) yes

4 (a) $\{-\frac{2}{5}, 3\}$ (b) $\{4\}$

5 (a) $\{x \mid -\frac{1}{2} \leqslant x \leqslant \frac{4}{3}\}$ (b) $\{x \mid \frac{3}{7} < x < \frac{1}{2}\}$

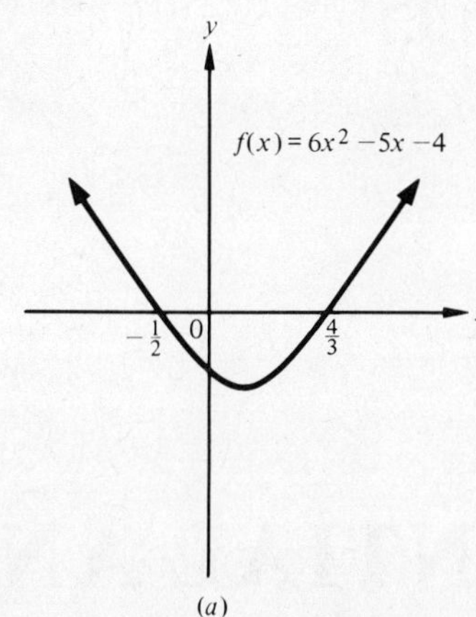

(a)

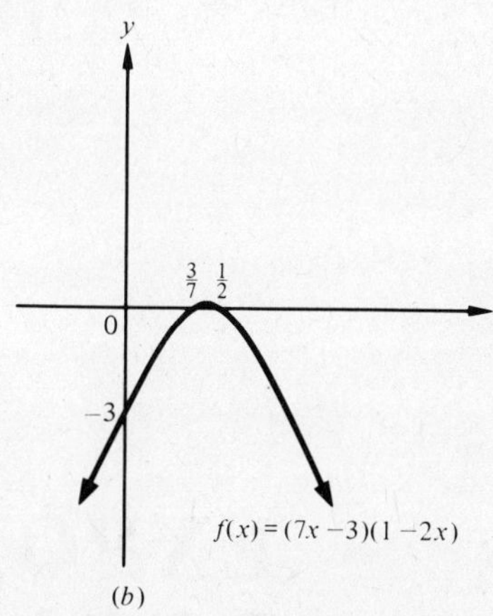

(b)

6 (a) $f(2) = 13$ (b) $f(3) = 208$ (c) $f(-1) = 4$ (d) $f(-2) = -7$

7 $\frac{1}{2}, -3$

8 (a) domain = $\{x \mid x \in R\}$ (b) range = $\{y \mid y \geqslant -\frac{9}{4}\}$ (c) $-2, 1$ (d) -2
(e) $(-\frac{1}{2}, -\frac{9}{4})$

9 (a) The x intercept is -4, and the y intercept is 4. There are no vertical asymptotes or horizontal asymptotes.
(b) The x intercept is 3; the y intercept is -6. The vertical asymptote is $x = -1$ and the horizontal asymptote is $y = 2$.

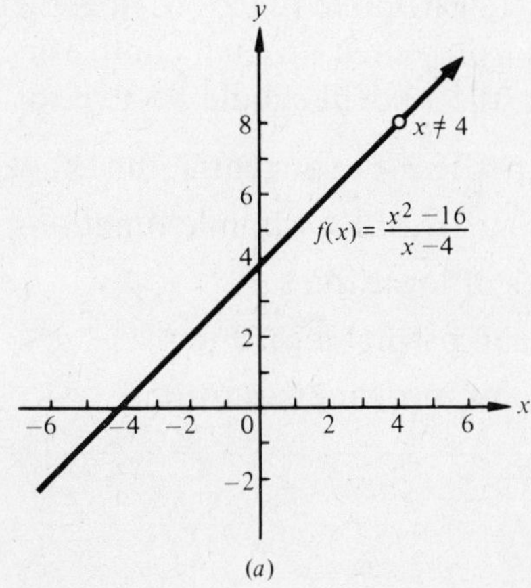

(a)

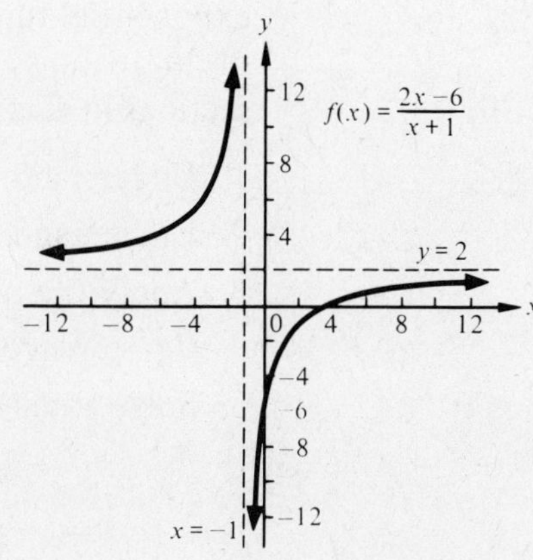

(b)

Chapter 4 EXPONENTIAL AND LOGARITHMIC FUNCTIONS

In this chapter, we make use of the algebra of exponents to introduce exponential functions. Logarithmic functions, together with the algebra of logarithms, are also presented. After completing the appropriate sections in this chapter, the student should be able to:

1 Understand the properties of exponential functions.
2 Understand the properties of logarithmic functions.
3 Apply the properties of logarithms.
4 Compute common and natural logarithms.
5 Solve applied problems involving logarithms.

1 Exponential Functions

exponential
R, $y > 0$

1 If b is a positive number, then the function $f(x) = b^x$ is called an ____________ function with base b. The domain of f is the set ____ and the range is the set $\{y \mid$ ________$\}$.

1.1 Exponential Function Properties

In Problems 2–5, find $f(0)$, $f(-1)$, and $f(1)$.

^{0}, 1
$^{-1}$, $\frac{1}{2}$
^{1}, 2

2 $f(x) = 2^x$

$f(0) = 2^{—} =$ ___
$f(-1) = 2^{—} =$ ___
$f(1) = 2^{—} =$ ___

^{0}, 1
$^{-1}$, $\frac{1}{5}$
^{1}, 5

3 $f(x) = 5^x$

$f(0) = 5^{—} =$ ___
$f(-1) = 5^{—} =$ ___
$f(1) = 5^{—} =$ ___

^{0}, 1
$^{-1}$, 3
^{1}, $\frac{1}{3}$

4 $f(x) = (\frac{1}{3})^x$

$f(0) = (\frac{1}{3})^{—} =$ ___
$f(-1) = (\frac{1}{3})^{—} =$ ___
$f(1) = (\frac{1}{3})^{—} =$ ___

^{0}, -1
$^{-1}$, $-\frac{1}{7}$
^{1}, -7

5 $f(x) = -7^x$

$f(0) = -7^{—} =$ ___
$f(-1) = -7^{—} =$ ___
$f(1) = -7^{—} =$ ___

$y = b^x$

6 The graph of an exponential function $f(x) = b^x$, where $b > 0$, is obtained by graphing ordered pairs that satisfy the equation ________.

increasing
decreasing

7 Consider the graph of an exponential function $f(x) = b^x$, for $b > 0$ and $b \neq 1$. If $b > 1$, then f is an ____________ function. If $b < 1$, then $f(x)$ is a ____________ function.

In Problems 8–11, determine the domain of f. Sketch the graph of f. Use the graph to determine if f is increasing or decreasing and to determine the range.

8 $f(x) = 2^x$

The domain of f is the set ____. The graph can be drawn by considering specific values of x, say, $-2, -1, 0, 1$, and 2, to determine $f(-2)$, $f(-1)$, $f(0)$, $f(1)$, and $f(2)$. Thus, $f(-2) =$ ____, $f(-1) =$ ____, $f(0) =$ ____, $f(1) =$ ____, and $f(2) =$ ____. Use these points to graph f. The graph is

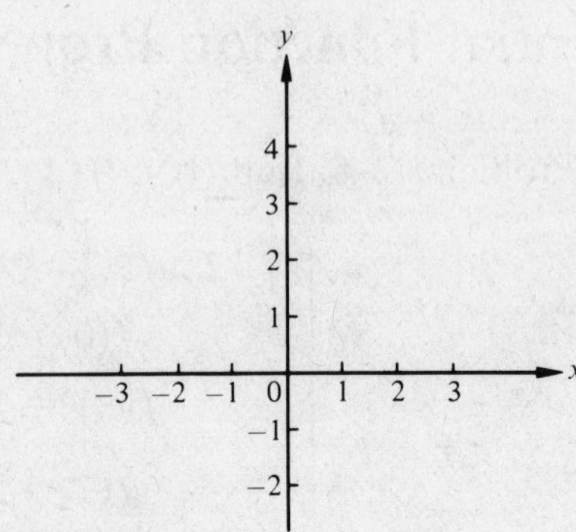

Notice that the graph of $f(x) = 2^x$ goes up to the right as x gets larger. Therefore, f is ________ function. The graph reveals that the range of f is $\{y|$____$\}$.

R

$\frac{1}{4}$

$\frac{1}{2}$, 1, 2, 4

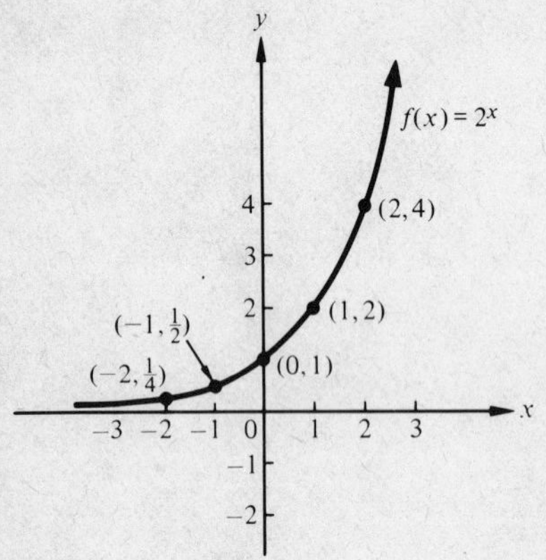

an increasing

$y > 0$

9 $f(x) = 2^{-x}$

The domain of f is the set ____. We can locate specific points of the graph by considering specific values of x, say, $-2, -1, 0, 1$, and 2, to determine $f(-2)$, $f(-1)$, $f(0)$, $f(1)$, and $f(2)$. That is, $f(-2) =$ ____, $f(-1) =$ ____, $f(0) =$ ____, $f(1) =$ ____, and $f(2) =$ ____. Use these points to graph f. The graph is

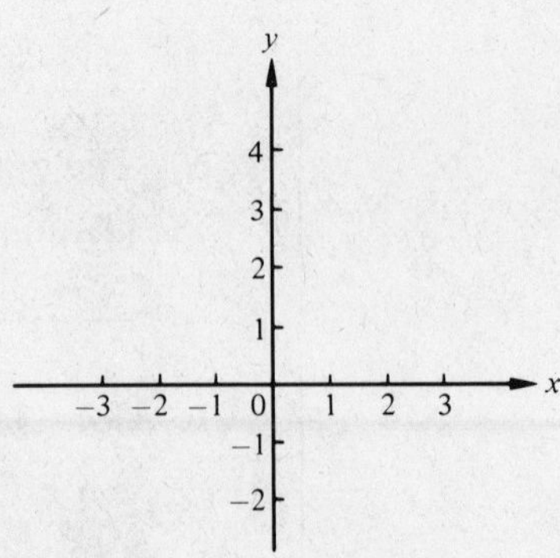

Notice that the graph of $f(x) = 2^{-x}$ goes down to the right as x gets larger. Therefore, f is ________ function. The range of f is $\{y|$____$\}$.

R

4, 2, 1, $\frac{1}{2}$

$\frac{1}{4}$

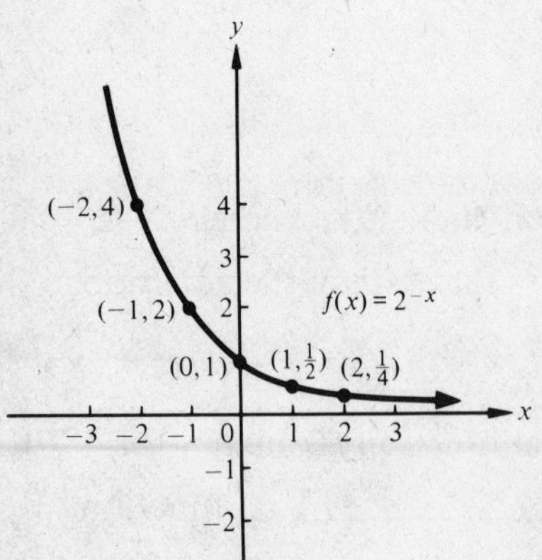

a decreasing

$y > 0$

10 $f(x) = -3^x$

The domain of f is the set _____. Consider specific values of x, say, $-2, -1, 0, 1$, and 2, in order to determine $f(-2), f(-1), f(0), f(1)$, and $f(2)$. We obtain $f(-2) =$ _____, $f(-1) =$ _____, $f(0) =$ _____, $f(1) =$ _____, and $f(2) =$ _____. Use these points to graph f. The graph is

R

$-\frac{1}{9}, -\frac{1}{3}, -1$

$-3, -9$

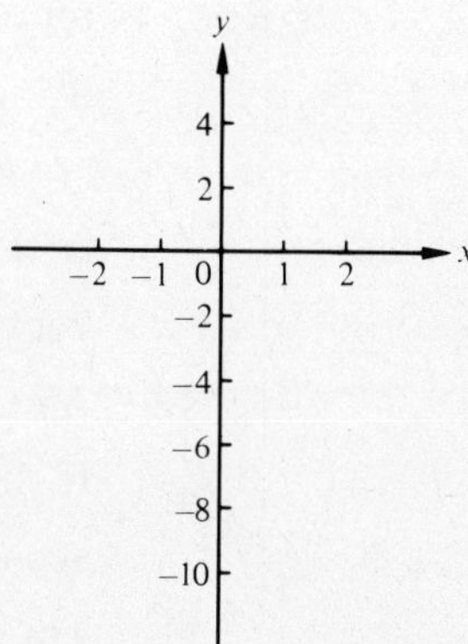

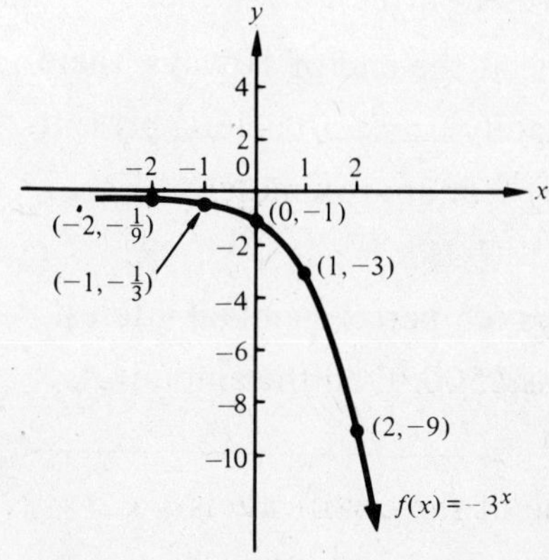

Notice from the graph of f that the function is __________.
The range of f is $\{y|$______$\}$.

decreasing

$y < 0$

11 $f(x) = -3^{-x}$

The domain of f is the set _____. Consider specific values of x, say, $-2, -1, 0, 1$, and 2, to determine $f(-2) =$ _____, $f(-1) =$ _____, $f(0) =$ _____, $f(1) =$ _____, and $f(2) =$ _____. Use these points to graph f. The graph is

R

$-9, -3$

$-1, -\frac{1}{3}, -\frac{1}{9}$

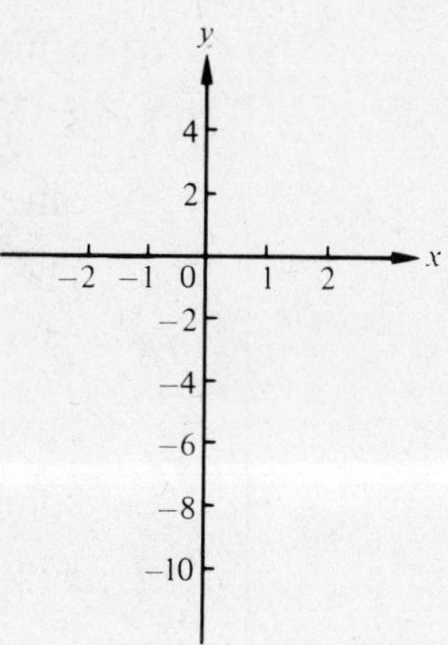

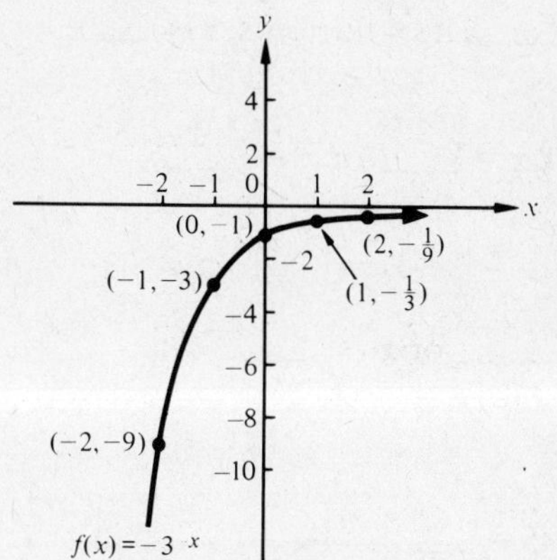

Notice from the graph that the function is __________.
The range of f is $\{y|$______$\}$.

increasing

$y < 0$

1.2 Exponential Functions—Applications

	12 Suppose that the number of bacteria in a certain laboratory colony grows at the rate of 5 percent per day. If there are 1000 bacteria
1.05	present initially, then after 1 day, there are $1000 \cdot$ ______ bacteria
$(1.05)^2$	present. After 2 days there are $1000 \cdot$ ______ present, and so
$(1.05)^n$	on, until after n days there are $1000 \cdot$ ______ present. If B represents the number of bacteria present after n days, then
$1000 \cdot (1.05)^n$	$B =$ ______. Thus, at the end of 10 days, there
1629	are $B = 1000 \cdot (1.05)^{10}$ or approximately ______ bacteria present.
4322	After 30 days there are $1000 \cdot (1.05)^{30}$ or approximately ______ bacteria present.
	13 Assume that a savings certificate pays 6.5 percent annual interest. If the initial value of the certificate is \$500, then the amount A
$500 \cdot (1.065)^n$	accumulated in n years is given by $A =$ ______: Thus, at the end of 5 years, the value of the certificate is
^{5}, 1.3701, 685.05	$500 \cdot (1.065)^{\_} = 500 \cdot$ ______ = ______ dollars. At the end of 10 years the accumulated value of the certificate is
$(1.065)^{10}$, 1.8771, 938.55	$500 \cdot$ ______ $= 500 \cdot$ ______ = ______ dollars.

1.3 Special Exponential Equations

	14 Equations in which the variable appears in the exponent and the
exponential	base is a constant are called ______ equations.
y	**15** If $a^x = a^y$, then $x =$ ______, for $a > 0$ and $a \neq 1$.

In Problems 16–21, find the solution set of each exponential equation.

	16 $3^x = 27$
3^3, 3, {3}	Since $3^x = 27 =$ ______, then $x =$ ______. The solution set is ______.
	17 $3^{-(3/2)x} = 27$
3^3, 3, -2	Since $3^{-(3/2)x} = 27 =$ ______, then $-\frac{3}{2}x =$ ______ or $x =$ ______.
{-2}	The solution set is ______.
	18 $2^{3-x} = 16$
^{4}, 4, -1	Since $2^{3-x} = 16 = 2^{\_}$, then $3 - x =$ ______ or $x =$ ______. The
{-1}	solution set is ______.
	19 $5^{4x-5} = 625$
^{4}, 4, 9	Since $5^{4x-5} = 625 = 5^{\_}$, then $4x - 5 =$ ______ or $4x =$ ______, so
$\frac{9}{4}$, $\{\frac{9}{4}\}$	$x =$ ______. The solution set is ______.
	20 $4^{3x-8} = 2^{3x+9}$
$3x + 9$	Since $4^{3x-8} = 2^{6x-16} = 2^{3x+9}$, then $6x - 16 =$ ______, so
25, $\frac{25}{3}$, $\{\frac{25}{3}\}$	$3x =$ ______ or $x =$ ______. The solution set is ______.

Answer	Problem
	21 $3^{5x-3} = 243$
^{5}, 5, 8, $\frac{8}{5}$	$3^{5x-3} = 3^{-}$, so $5x - 3 =$ ____ or $5x =$ ____ or $x =$ ____. The
$\{\frac{8}{5}\}$	solution set is _____.

2 Logarithmic Functions

Answer	Problem
	22 If $b > 0$ and $b \neq 1$, then $y = \log_b x$, which is read as "y equals the
logarithm, x	__________ of x to the base b," is equivalent to $b^y =$ ____.
base	b is called the _____ of the logarithm.

In Problems 23–35, write each exponential statement as an equivalent logarithmic statement.

Answer	Problem
4	23 $2^4 = 16$ is equivalent to $\log_2 16 =$ ____.
100	24 $10^2 = 100$ is equivalent to $\log_{10}$ ____ $= 2$.
5	25 $5^{-2} = \frac{1}{25}$ is equivalent to $\log_{\_} \frac{1}{25} = -2$.
3	26 $(\frac{1}{3})^3 = \frac{1}{27}$ is equivalent to $\log_{1/3} \frac{1}{27} =$ ____.
-4	27 $10^{-4} = 0.0001$ is equivalent to $\log_{10} 0.0001 =$ ____.
2, $\frac{1}{5}$	28 $32^{1/5} = 2$ is equivalent to $\log_{32}$ ____ $=$ ____.
$\log_{4/9} \frac{27}{8} = -\frac{3}{2}$	29 $(\frac{4}{9})^{-3/2} = \frac{27}{8}$ is equivalent to ________________.
$\log_{1/8} 4 = -\frac{2}{3}$	30 $(\frac{1}{8})^{-2/3} = 4$ is equivalent to ________________.
$\log_7 \frac{1}{49} = -2$	31 $7^{-2} = \frac{1}{49}$ is equivalent to ________________.
$\log_{4/25} \frac{125}{8} = -\frac{3}{2}$	32 $(\frac{4}{25})^{-3/2} = \frac{125}{8}$ is equivalent to ________________.
$\log_{10} 1 = 0$	33 $10^0 = 1$ is equivalent to ________________.
$\log_a b = c$	34 $a^c = b$ is equivalent to ________________.
$\log_x z = 2y$	35 $x^{2y} = z$ is equivalent to ________________.

In Problems 36–49, write each logarithmic statement as an equivalent exponential statement.

Answer	Problem
2	36 $\log_6 36 = 2$ is equivalent to $6^{-} = 36$.
2	37 $\log_2 32 = 5$ is equivalent to (____)$^5 = 32$.
9	38 $\log_{27} 9 = \frac{2}{3}$ is equivalent to $27^{2/3} =$ ____.
4	39 $\log_8 4 = \frac{2}{3}$ is equivalent to $8^{2/3} =$ ____.
216	40 $\log_6 216 = 3$ is equivalent to $6^3 =$ ____.
256	41 $\log_4 256 = 4$ is equivalent to $4^4 =$ ____.
81	42 $\log_{27} 81 = \frac{4}{3}$ is equivalent to $27^{4/3} =$ ____.
$\frac{1}{49}$	43 $\log_7 \frac{1}{49} = -2$ is equivalent to $7^{-2} =$ ____.
$\frac{1}{243}$	44 $\log_3 \frac{1}{243} = -5$ is equivalent to $3^{-5} =$ ____.
0.001	45 $\log_{10} 0.001 = -3$ is equivalent to $10^{-3} =$ _____.
$100^{-3/2} = 0.001$	46 $\log_{100} 0.001 = -\frac{3}{2}$ is equivalent to ________________.
$2^{-3} = \frac{1}{8}$	47 $\log_2 \frac{1}{8} = -3$ is equivalent to __________.
$a^{1/2} = \sqrt{a}$	48 $\log_a \sqrt{a} = \frac{1}{2}$ is equivalent to ____________.
$y^2 = y^2$	49 $\log_y y^2 = 2$ is equivalent to __________

In Problems 50–57, find the value of the logarithm.

Answers	Problems
	50 $\log_2 16$
16, ^{4}, 4	Let $\log_2 16 = x$, so $2^x =$ ____ $= 2^{-}$. Then $x =$ ____.
4	Therefore, $\log_2 16 =$ ____.
	51 $\log_{10} 100$
100, 2	Let $\log_{10} 100 = x$, so $10^x =$ ____ $= 10^2$. Then $x =$ ____.
2	Therefore, $\log_{10} 100 =$ ____.
	52 $\log_8 \frac{1}{64}$
$\frac{1}{64}$, 8^{-2}, -2	Let $\log_8 \frac{1}{64} = x$, so $8^x =$ ____ $=$ ____. Then $x =$ ____.
-2	Therefore, $\log_8 \frac{1}{64} =$ ____.
	53 $\log_{10} 1$
0, 1	The value of $\log_{10} 1 =$ ____, since $10^0 =$ ____.
	54 $\log_{0.5} 0.25$
2, ^{2}	The value of $\log_{0.5} 0.25 =$ ____, since $0.5^{-} = 0.25$.
	55 $\log_5 \sqrt{5}$
$\frac{1}{2}$, $\sqrt{5}$	The value of $\log_5 \sqrt{5} =$ ____, since $5^{1/2} =$ ____.
	56 $\log_9 27$
$\frac{3}{2}$, 27	The value of $\log_9 27 =$ ____, since $9^{3/2} =$ ____.
	57 $\log_a \sqrt[3]{a}$
$\frac{1}{3}$, $\sqrt[3]{a}$	The value of $\log_a \sqrt[3]{a} =$ ____, since $a^{1/3} =$ ____.

In Problems 58–65, find the value of the variable in each equation.

Answers	Problems
	58 $\log_b 16 = \frac{2}{3}$
$b^{2/3} = 16$	In exponential form, this equation is equivalent to ____________.
64	Then write $(b^{2/3})^{3/2} = 16^{3/2}$ and obtain $b =$ ____.
	59 $\log_b \frac{1}{64} = -2$
$b^{-2} = \frac{1}{64}$	In exponential form, this equation is equivalent to ____________
8	or $\frac{1}{b^2} = \frac{1}{64}$, so $b =$ ____.
	60 $\log_b 243 = 5$
$b^5 = 243$	In exponential form, this equation is equivalent to ____________
3	or $b^5 = 3^5$, so $b =$ ____.
	61 $\log_b \frac{1}{16} = \frac{4}{5}$
$b^{4/5} = \frac{1}{16}$	In exponential form, this equation is equivalent to ____________.
$\frac{1}{32}$	Then write $(b^{4/5})^{5/4} = (\frac{1}{16})^{5/4}$ and obtain $b =$ ____.
	62 $\log_b \frac{1}{9} = -\frac{2}{3}$
$b^{-2/3} = \frac{1}{9}$	In exponential form, this equation is equivalent to ____________.
27	Then write $(b^{-2/3})^{-3/2} = (\frac{1}{9})^{-3/2}$ and obtain $b =$ ____.
	63 $\log_2 N = 3$
2^3, 8	$\log_2 N = 3$ is equivalent to $N =$ ____, so $N =$ ____.

64 $\log_3 N = -4$

3^{-4}, $\frac{1}{81}$

$\log_3 N = -4$ is equivalent to $N =$ ____, so $N =$ ____.

65 $\log_{1/5} N = 3$

$(\frac{1}{5})^3$, $\frac{1}{125}$

$\log_{1/5} N = 3$ is equivalent to $N =$ ____, so $N =$ ____.

In Problems 66–67, evaluate the given expression by finding each logarithm and then performing the indicated operations.

66 $\dfrac{\log_{1/5} 25 + \log_4 64}{\log_6 36}$

-2, 3, 2

$\log_{1/5} 25 =$ ____ $\log_4 64 =$ ____ $\log_6 36 =$ ____

3, $\frac{1}{2}$

$\dfrac{\log_{1/5} 25 + \log_4 64}{\log_6 36} = \dfrac{-2 + \_\_\_\_}{2} =$ ____

67 $\dfrac{\log_8 4 + \log_{27} 81}{\log_7 \frac{1}{49} + \log_4 256}$

$\frac{2}{3}$, $\frac{4}{3}$

$\log_8 4 =$ ____ $\log_{27} 81 =$ ____

-2, 4

$\log_7 \frac{1}{49} =$ ____ $\log_4 256 =$ ____

$\frac{2}{3}$, 1

$\dfrac{\log_8 4 + \log_{27} 81}{\log_7 \frac{1}{49} + \log_4 256} = \dfrac{\_\_\_\_ + \frac{4}{3}}{-2 + 4} =$ ____

$b^{f(x)}$

$\log_b x$

68 The logarithmic function f with base b is defined by the exponential equation $x =$ _____, where $b > 0$ and $b \neq 1$, or, solving explicitly for $f(x)$, we get $f(x) =$ _______, where $x > 0$, $b > 0$, and $b \neq 1$.

$\{x | x > 0\}$

real numbers

69 The domain of $f(x) = \log_b x$ is the set ________, and the range is the set of ________________.

In Problems 70–71, find the indicated values.

70 $f(\frac{1}{9}), f(\frac{1}{3}), f(1), f(3)$, and $f(9)$, if $f(x) = \log_3 x$

$\frac{1}{9}$, -2 — $f(\frac{1}{9}) = \log_3$ ____ = ____

$\frac{1}{3}$, -1 — $f(\frac{1}{3}) = \log_3$ ____ = ____

1, 0 — $f(1) = \log_3$ ____ = ____

3, 1 — $f(3) = \log_3$ ____ = ____

9, 2 — $f(9) = \log_3$ ____ = ____

71 $f(\frac{1}{4}), f(\frac{1}{2}), f(1), f(2)$, and $f(4)$, if $f(x) = \log_2 x$

$\frac{1}{4}$, -2 — $f(\frac{1}{4}) = \log_2$ ____ = ____

$\frac{1}{2}$, -1 — $f(\frac{1}{2}) = \log_2$ ____ = ____

1, 0 — $f(1) = \log_2$ ____ = ____

2, 1 — $f(2) = \log_2$ ____ = ____

4, 2 — $f(4) = \log_2$ ____ = ____

2.1 Logarithmic Function Properties

positive

72 It is only possible to compute the logarithm of __________ numbers in the real number system.

In Problems 73–75, find the domain of each function.

$x > 0$

73 The domain of the function $y = \log_7 x$ is $\{x \mid$ ________ $\}$.

$x > -\frac{2}{3}$

74 The domain of the function $y = \log_{10}(3x + 2)$ is the set of all x such that $3x + 2 > 0$, namely, $\{x \mid$ ________ $\}$.

> 0, $-3 < x < 3$

75 The domain of $y = \log_2(9 - x^2)$ is the set of all x such that $9 - x^2$ ________, that is, $\{x \mid$ ________ $\}$.

2.2 Graphing Logarithmic Functions

In Problems 76–78, find the domain of f and sketch its graph. Use the graph to determine whether f is increasing or decreasing and to find the range of f.

$x > 0$

-2

$-1, 0, 1, 2$

76 $f(x) = \log_3 x$

The domain of f is the set $\{x \mid$ ________ $\}$. Consider specific values of x, say, $\frac{1}{9}, \frac{1}{3}$, 1, 3, and 9, to determine $f(\frac{1}{9}) =$ ____, $f(\frac{1}{3}) =$ ____, $f(1) =$ ____, $f(3) =$ ____, and $f(9) =$ ____. Use these points to graph f. The graph is

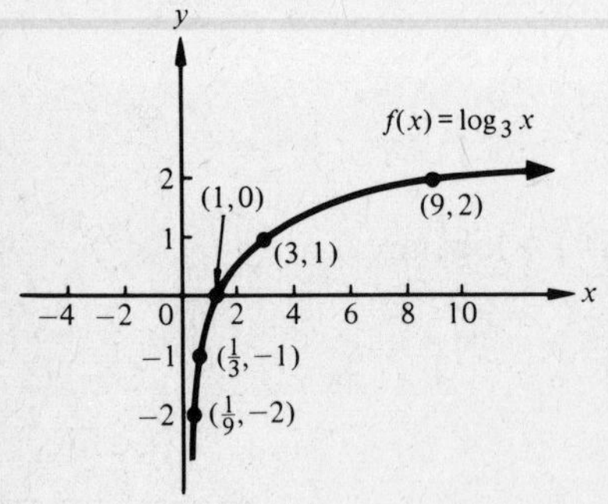

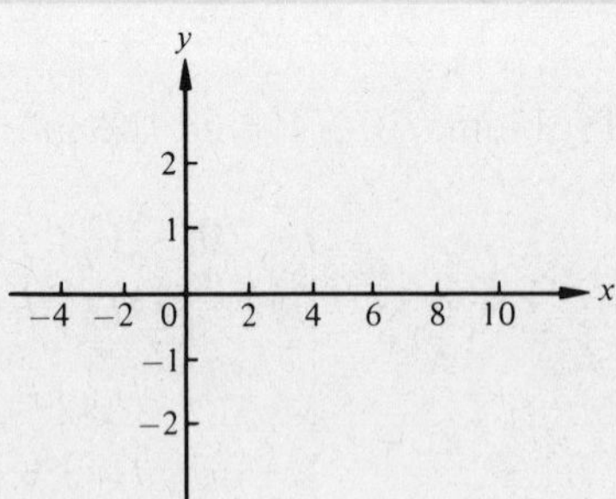

increasing

R

Notice from the graph of f that the function is __________. The range of f is the set ____.

$x > 0$

2

$1, 0, -1$

77 $f(x) = \log_{1/3} x$

The domain of f is the set $\{x \mid$ ________ $\}$. Consider specific values of x, say, $\frac{1}{9}, \frac{1}{3}$, 1, and 3, to determine $f(\frac{1}{9}) =$ ____, $f(\frac{1}{3}) =$ ____, $f(1) =$ ____, and $f(3) =$ ____. Use these points to

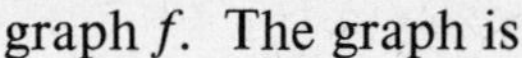

graph f. The graph is

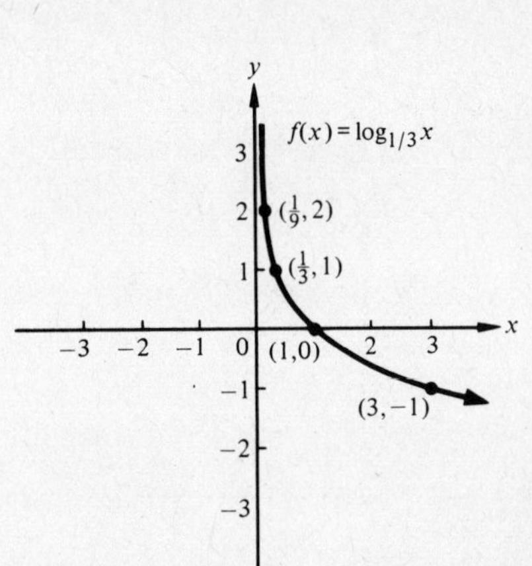

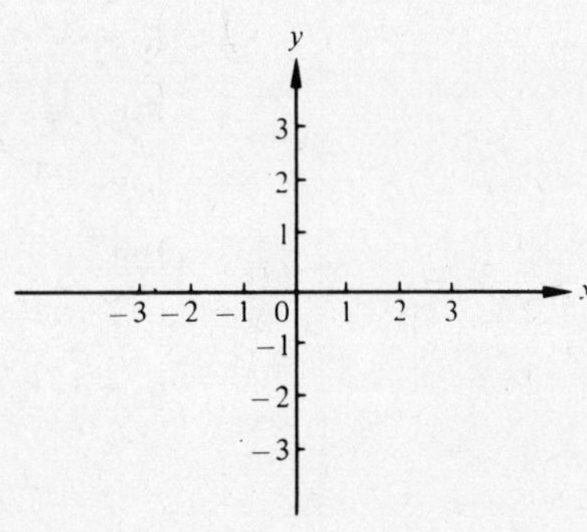

decreasing, R

The function is ______________. The range of f is ____.

78 $f(x) = -\log_3 x$

$x > 0$

The domain of f is the set $\{x \mid$ ________$\}$. Consider specific values of x, say, $\frac{1}{9}$, $\frac{1}{3}$, 1, and 3, to determine $f(\frac{1}{9}) =$ ____,

2

$f(\frac{1}{3}) =$ ____, $f(1) =$ ____, and $f(3) =$ ____. Use these points to graph f. The graph is

1, 0, −1

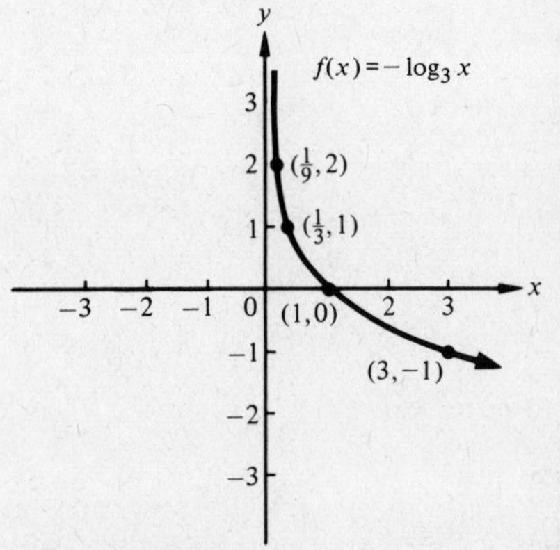

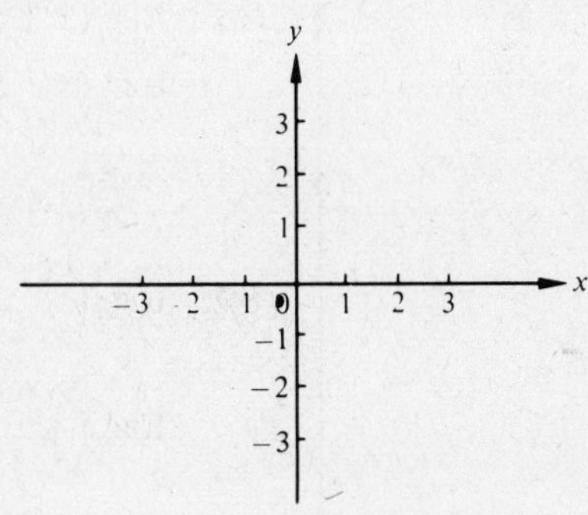

decreasing, R

The function is ______________. The range of f is ____.

3 Properties of Logarithms

In Problems 79–81, state the properties of logarithms that are valid for M and N (positive real numbers) where $b > 0$, $b \neq 1$, and r is any real number.

$\log_b M + \log_b N$

79 $\log_b MN =$ ______________________

$\log_b M - \log_b N$

80 $\log_b \dfrac{M}{N} =$ ______________________

$r \log_b N$

81 $\log_b N^r =$ ______________

In Problems 82–89, consider all the variables to be positive real numbers. Use the properties of logarithms to express each expression as a sum, difference, or multiple of logarithms.

Answers	Problems
	82 $\log_5 (7)(11)$
$\log_5 11$	$\log_5 (7)(11) = \log_5 7 + $ ________
	83 $\log_2 \frac{13}{7}$
$\log_2 7$	$\log_2 \frac{13}{7} = \log_2 13 - $ ________
	84 $\log_3 7^{21}$
$\log_3 7$	$\log_3 7^{21} = 21$ ________
	85 $\log_4(x^3y^4)$
$\log_4 y^4$	$\log_4(x^3y^4) = \log_4 x^3 + $ ________
4	$= 3 \log_4 x + $ ____ $\log_4 y$
	86 $\log_a \sqrt[6]{xy}$
1/6, $\frac{1}{6}$	$\log_a \sqrt[6]{xy} = \log_a (xy)^{\text{—}} = ($____$) \log_a (xy)$
$\log_a y$	$= \frac{1}{6} (\log_a x + $ ________$)$
	87 $\log_3 \sqrt[3]{\frac{x}{y^2}}$
$\frac{1}{3}$	$\log_3 \sqrt[3]{\frac{x}{y^2}} = \log_3 \left(\frac{x}{y^2}\right)^{1/3} = ($____$) \log_3 \left(\frac{x}{y^2}\right)$
$\log_3 y^2$	$= \frac{1}{3} \log_3 x - \frac{1}{3}$ ________
$\frac{2}{3}$	$= \frac{1}{3} \log_3 x - $ ____ $\log_3 y$
	88 $\log_a(x^m \sqrt[n]{y})$
$\log_a \sqrt[n]{y}$	$\log_a(x^m \sqrt[n]{y}) = \log_a x^m + $ ________
$\frac{1}{n}$	$= m \log_a x + \frac{\text{\_\_\_\_}}{\text{\_\_\_\_}} \log_a y$
	89 $\log_a\left(\frac{xy^n}{\sqrt[n]{z}}\right)$
$\log_a \sqrt[n]{z}$	$\log_a\left(\frac{xy^n}{\sqrt[n]{z}}\right) = \log_a(xy^n) - $ ________
$\log_a y^n$, $\frac{1}{n}$	$= \log_a x + $ ________ $- \frac{\text{\_\_\_\_}}{\text{\_\_\_\_}} \log_a z$
$\log_a y$	$= \log_a x + n$ ________ $- \frac{1}{n} \log_a z$

In Problems 90–101, let $\log_{10} 2 = 0.3010$, $\log_{10} 3 = 0.4771$, and $\log_{10} 7 = 0.8451$. Use the properties of logarithms to find each value.

Answers	Problems
	90 $\log_{10} 6$
	$\log_{10} 6 = \log_{10}(2 \cdot 3)$
$\log_{10} 3$	$= \log_{10} 2 + $ ________
0.4771	$= 0.3010 + $ ________
0.7781	$=$ ________
	91 $\log_{10} \frac{3}{2}$
$\log_{10} 2$	$\log_{10} \frac{3}{2} = \log_{10} 3 - $ ________
0.3010	$= 0.4771 - $ ________
0.1761	$=$ ________

92 $\log_{10} \frac{7}{2}$

$\log_{10} \frac{7}{2} = \log_{10} 7 -$ ________

$= 0.8451 -$ ________

$=$ ________

$\log_{10} 2$

0.3010

0.5441

93 $\log_{10} 3^5$

$\log_{10} 3^5 = 5 \log_{10} 3$

$= 5($________$)$

$=$ ________

0.4771

2.3855

94 $\log_{10} \frac{7}{6}$

$\log_{10} \frac{7}{6} = \log_{10} 7 -$ ________

$= 0.8451 -$ ________

$=$ ________

$\log_{10} 6$

0.7781

0.0670

95 $\log_{10} 8$

$\log_{10} 8 = \log_{10} 2^3 = 3$ ________

$= 3($________$)$

$=$ ________

$\log_{10} 2$

0.3010

0.9030

96 $\log_{10} 56$

$\log_{10} 56 = \log_{10} (7 \cdot 8)$

$= \log_{10} 7 +$ ________

$= 0.8451 +$ ________

$=$ ________

$\log_{10} 8$

0.9030

1.7481

97 $\log_{10} \sqrt[5]{6}$

$\log_{10} \sqrt[5]{6} = \frac{1}{5}$ ________

$= \frac{1}{5}$ (________)

$=$ ________

$\log_{10} 6$

0.7781

0.1556

98 $\log_{10} (0.06)$

$\log_{10} (0.06) = \log_{10}$ ____

$= \log_{10} 6 -$ ________

$= 0.7781 -$ ____

$=$ ________

$\frac{6}{100}$

$\log_{10} 100$

2

-1.2219

99 $\log_{10} \sqrt[3]{49}$

$\log_{10} \sqrt[3]{49} = \log_{10} \sqrt[3]{7^2}$

$=$ ____ $\log_{10} 7$

$= \frac{2}{3}$ (________)

$=$ ________

$\frac{2}{3}$

0.8451

0.5634

100 $\log_{10} 5$

$\log_{10} 5 = \log_{10}$ ____

$= \log_{10} 10 -$ ________

$=$ ____ $- 0.3010$

$=$ ________

$\frac{10}{2}$

$\log_{10} 2$

1

0.6990

Answers	Problems
	101 $\log_{10} 25$
5^2	$\log_{10} 25 = \log_{10}$ ____
$\log_{10} 5$	$= 2$ ________
0.6990	$= 2($________$)$
1.3980	$=$ ________

In Problems 102–109, use the properties of logarithms to write each expression as a single logarithm.

Answers	Problems
	102 $\log_{10} 2 + \log_{10} 25$
25, $\log_{10} 50$	$\log_{10} 2 + \log_{10} 25 = \log_{10} [(2)(\_\_\_\_)] =$ ________
	103 $\log_5 18 - \log_5 3$
$\frac{18}{3}$, $\log_5 6$	$\log_5 18 - \log_5 3 = \log_5$ ____ $=$ ________
	104 $3 \log_3 5 - 2 \log_3 7$
7^2	$3 \log_3 5 - 2 \log_3 7 = \log_3 5^3 - \log_3$ ____
$\frac{5^3}{7^2}$, $\log_3 \frac{125}{49}$	$= \log_3\left(\_\_\_\_\right) =$ ________
	105 $5 \log_a x + 3 \log_a y$
$\log_a y^3$, x^5y^3	$5 \log_a x + 3 \log_a y = \log_a x^5 +$ ________ $= \log_a$ ________
	106 $\log_a x^3 + \log_a\left(\frac{b}{\sqrt[3]{x}}\right)$
$\frac{x^3b}{\sqrt[3]{x}}$, $\log_a(x^{8/3}b)$	$\log_a x^3 + \log_a\left(\frac{b}{\sqrt[3]{x}}\right) = \log_a\left(\_\_\_\_\right) =$ ________
	107 $\log_a x^3 - \log_a \sqrt{x}$
$\sqrt{x}$, $x^{5/2}$, $\frac{5}{2} \log_a x$	$\log_a x^3 - \log_a \sqrt{x} = \log_a\left(\frac{x^3}{\_\_\_\_}\right) = \log_a$ ____ $=$ ________
	108 $\log_c\left(\frac{a}{\sqrt{x}}\right) - \log_c\left(\frac{\sqrt{x}}{a}\right)$
$\frac{\sqrt{x}}{a}$, $\frac{a^2}{x}$	$\log_c\left(\frac{a}{\sqrt{x}}\right) - \log_c\left(\frac{\sqrt{x}}{a}\right) = \log_c\left[\frac{\frac{a}{\sqrt{x}}}{\_\_\_\_}\right] = \log_c\left(\_\_\_\_\right)$
	109 $\log_a (x^2 - y^2) - \log_a (x - y)$
$x - y$	$\log_a (x^2 - y^2) - \log_a (x - y) = \log_a\left(\frac{x^2 - y^2}{\_\_\_\_}\right)$
$x - y$	$= \log_a\left[\frac{(x - y)(x + y)}{(\_\_\_\_)}\right]$
$x + y$	$= \log_a($________$)$

In Problems 110–115, let all the variables be constant positive real numbers. Answer true or false.

Answers	Problems
False	**110** $\log_a (M + N) = \log_a M + \log_a N$ ________
False	**111** $\log_a (M - N) = \log_a M - \log_a N$ ________

True	**112** $\log_a M^2N = 2\log_a M + \log_a N$ ________
False	**113** $\log_a M \cdot \log_a N = \log_a M + \log_a N$ ________
True	**114** $\log_a M^{p^2} = p^2 \log_a M$ ________
False	**115** $\dfrac{\log_a M}{\log_a N} = \log_a M - \log_a N$ ________
y	**116** If $\log_a x = \log_a y$, then $x =$ ____.

In Problems 117–128, find the solution set of each equation. Check the solutions in Problems 124–128.

	117 $\log_2 (x - 3) = 5$
2^5	$\log_2 (x - 3) = 5$ is equivalent to ____ $= x - 3$ or $32 = x - 3$, so
35, {35}	$x =$ ____. The solution set is _____.
	118 $\log_3 (2x - 1) = 2$
3^2	$\log_3 (2x - 1) = 2$ is equivalent to ____ $= 2x - 1$ or $2x - 1 = 9$,
10, 5, {5}	so $2x =$ ____ or $x =$ ____. The solution set is ____.
	119 $\log_{\sqrt{2}} x = -6$
$(\sqrt{2})^{-6}$, $\frac{1}{8}$	$\log_{\sqrt{2}} x = -6$ is equivalent to ________ $= x$, so $x =$ ____. The
$\{\frac{1}{8}\}$	solution set is ____.
	120 $\log_9 (7x - 12) = 1$
9, 21	$\log_9 (7x - 12) = 1$ is equivalent to ____ $= 7x - 12$, so $7x =$ ____
3, {3}	or $x =$ ____. The solution set is ____.
	121 $\log_3 (x^2 - 2x) = 1$
3	$\log_3 (x^2 - 2x) = 1$ is equivalent to ____ $= x^2 - 2x$, so
0, 0, 3	$x^2 - 2x - 3 =$ ____ or $(x - 3)(x + 1) =$ ____. Hence, $x =$ ____ or
-1, $\{-1, 3\}$	$x =$ ____. The solution set is ________.
	122 $\log_3 \left(\dfrac{x}{7}\right) = 2$
3^2, 9, 63	$\log_3 \left(\dfrac{x}{7}\right) = 2$ is equivalent to $\dfrac{x}{7} =$ ____ or $\dfrac{x}{7} =$ ____, so $x =$ ____.
{63}	The solution set is ____.
	123 $\log_5 (x^2 + 5x + 1) = 2$
5^2	$\log_5 (x^2 + 5x + 1) = 2$ is equivalent to $x^2 + 5x + 1 =$ ____, so
0, $x - 3$	$x^2 + 5x - 24 =$ ____ or $(x + 8)($________$) = 0$
-8, 3	$x =$ ____ $x =$ ____
$\{-8, 3\}$	The solution set is ________.
	124 $\log_2 (x - 1) - \log_2 3 = 3$
	Writing the left side of the equation as a single logarithmic expres-
3	sion, we obtain $\log_2 \left(\dfrac{x - 1}{3}\right) =$ ____. In exponential form, this
$\dfrac{x - 1}{3} = 2^3$, 25	equation is ____________. Solving for x, we have $x =$ ____.

		Check:
		$\log_2 (25 - 1) - \log_2 3 \stackrel{?}{=} 3$
24		$\log_2$ ____ $- \log_2 3 \stackrel{?}{=} 3$
8		$\log_2$ ____ $\stackrel{?}{=} 3$ (Yes)
{25}		Hence, the solution set is ____.
	125	$\log_3 (x + 1) + \log_3 (x + 3) = 1$
		Writing the left side of the equation as a single logarithmic expres-
$\log_3 (x + 1)(x + 3)$		sion, we obtain the equation ____________ = 1.
$(x + 1)(x + 3) = 3$		In exponential form, this equation is ____________.
0, -4		Solving for x, we have x = ____ or x = ____.
		Check:
		$x = -4$:
		$\log_3 (-4 + 1) + \log_3 (-4 + 3) \stackrel{?}{=} 1$
-3, -1		$\log_3$ (____) + $\log_3$ (____) $\stackrel{?}{=} 1$
		But $\log_3 (-3)$ and $\log_3 (-1)$ are not defined. Therefore, $x = -4$
is not		________ a solution to the equation.
		$x = 0$:
		$\log_3 (0 + 1) + \log_3 (0 + 3) \stackrel{?}{=} 1$
0		____ $+ \log_3 3 \stackrel{?}{=} 1$ (Yes)
{0}		Hence, the solution set is ____.
	126	$\log_{10} (x^2 - 121) - \log_{10} (x + 11) = 1$
		Writing the left side of the equation as a single logarithmic expres-
$\log_{10} \dfrac{x^2 - 121}{x + 11}$		sion, we obtain the equation ____________ = 1. This
$\log_{10} (x - 11) = 1$		equation can be written in a simpler form as ____________.
$x - 11 = 10$		In exponential form, this equation is ____________.
21		Solving for x, we have x = ____.
		Check:
		$x = 21$:
		$\log_{10} (441 - 121) - \log_{10} (21 + 11) \stackrel{?}{=} 1$
320		$\log_{10}$ ____ $- \log_{10} 32 \stackrel{?}{=} 1$
32		$\log_{10} \dfrac{320}{\_\_} \stackrel{?}{=} 1$
10		$\log_{10}$ ____ $\stackrel{?}{=} 1$ (Yes)
{21}		Thus, the solution set is ____.
	127	$\log_3 (x + 4) - \log_3 (x - 1) = 2$
		Writing the left side of the equation as a single logarithmic expres-
$\log_3 \left(\dfrac{x + 4}{x - 1}\right)$		sion, we obtain ________ = 2. In exponential form, this equation
$\dfrac{x + 4}{x - 1} = 3^2$, $\frac{13}{8}$		is ____________. Solving for x, we have x = ____.

Check:

$x = \frac{13}{8}$:

$\log_3 (\frac{13}{8} + 4) - \log_3$ ____ $\stackrel{?}{=} 2$

$\log_3 \frac{\frac{45}{8}}{\frac{5}{8}} = \log_3 \frac{45}{5} =$ ________ $\stackrel{?}{=}$ ____ (Yes)

Thus, the solution set is ____.

$\frac{5}{8}$

$\log_3 9$, 2

$\{\frac{13}{8}\}$

128 $\log_2 2 + \log_2 (x + 2) - \log_2 (3x - 5) = 3$

The left side of the equation can be written as the single logarithmic expression ________ = 3. In exponential form, this equation is __________. Solving for x, we have $x =$ ____.

Check:

$x = 2$:

$\log_2 2 + \log_2 4 - \log_2$ ____ $\stackrel{?}{=} 3$

$1 +$ ____ $- 0 \stackrel{?}{=}$ ____ (Yes)

Thus, the solution set is ____.

$\log_2\left[\frac{2(x+2)}{3x-5}\right]$

$\frac{2x+4}{3x-5} = 2^3$, 2

1

2, 3

$\{2\}$

4 Common and Natural Logarithms

129 Two logarithmic bases used most often for purposes of computation are base 10 and base e. Logarithms with base 10 are called ________ logarithms. Logarithms with base e are called __________ logarithms. The notation used for $\log_{10} x$ is ________, and the notation used for $\log_e x$ is ________.

common

natural

$\log x$, $\ln x$

4.1 Logarithms—Base 10

130 For any positive number x that can be represented in scientific notation as $x = s \times 10^n$, where $1 \leqslant s < 10$ and n is an integer, we have $\log x = \log s +$ ____. The latter form is called the ________ form of $\log x$, where $\log s$ is called the ________ of $\log x$ and n is called the __________ of $\log x$.

n, standard

mantissa

characteristic

131 The task of determining the value of $\log x$ involves determining ________, when $1 \leqslant s < 10$ and $x = s \times 10^n$. However, the approximate values of $\log s$ can be found from the __________ tables.

$\log s$

common log

In Problems 132–139, use Appendix Table I to find the value of each logarithm. Indicate the characteristic and the mantissa.

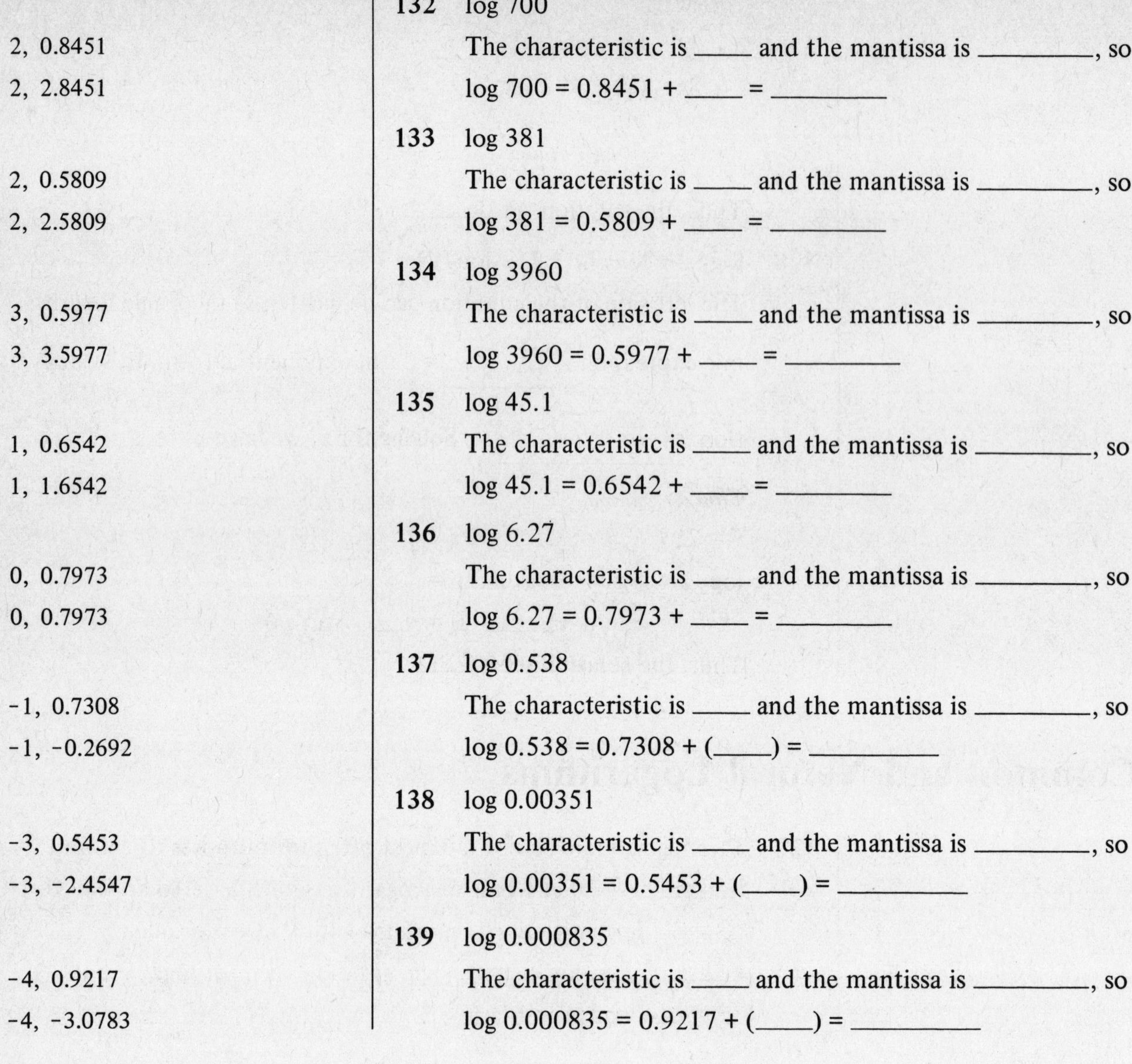

132 log 700

The characteristic is ____ and the mantissa is ________, so

log 700 = 0.8451 + ____ = ________

2, 0.8451

2, 2.8451

133 log 381

The characteristic is ____ and the mantissa is ________, so

log 381 = 0.5809 + ____ = ________

2, 0.5809

2, 2.5809

134 log 3960

The characteristic is ____ and the mantissa is ________, so

log 3960 = 0.5977 + ____ = ________

3, 0.5977

3, 3.5977

135 log 45.1

The characteristic is ____ and the mantissa is ________, so

log 45.1 = 0.6542 + ____ = ________

1, 0.6542

1, 1.6542

136 log 6.27

The characteristic is ____ and the mantissa is ________, so

log 6.27 = 0.7973 + ____ = ________

0, 0.7973

0, 0.7973

137 log 0.538

The characteristic is ____ and the mantissa is ________, so

log 0.538 = 0.7308 + (____) = ________

-1, 0.7308

-1, -0.2692

138 log 0.00351

The characteristic is ____ and the mantissa is ________, so

log 0.00351 = 0.5453 + (____) = ________

-3, 0.5453

-3, -2.4547

139 log 0.000835

The characteristic is ____ and the mantissa is ________, so

log 0.000835 = 0.9217 + (____) = ________

-4, 0.9217

-4, -3.0783

4.2 Antilogarithms—Base 10

140 If $\log x = r$, where r is a given number, then x is referred to as the antilog ____.

r

In Problems 141–147, find the value of the given antilogarithm by using Appendix Table I.

141 antilog 1.7308

Let x = antilog 1.7308, so that $\log x$ = ________.

$\log x = 0.7308 +$ ____

Since log 5.38 = ________,

$\log x = \log 5.38 + 1 = \log 5.38 +$ ________

= log (____________) = log ____

Therefore, x = antilog 1.7308 = _____.

1.7308

1

0.7308

log 10

5.38 × 10, 53.8

53.8

142 antilog 4.5977

Let x = antilog 4.5977, so that log x = ________.

log x = 0.5977 + ____

Since log 3.96 = ________,

log x = log 3.96 + ________

= log (3.96×10^4) = log ________

Therefore, x = antilog 4.5977 = ________.

Answers: 4.5977; 4; 0.5977; log 10^4; 39,600; 39,600

143 antilog 2.4969

Let x = antilog 2.4969, so that log x = ________.

log x = 0.4969 + ____

Since log 3.14 = ________,

log x = log 3.14 + ________ = log (3.14×10^2)

= log ____

Therefore, x = antilog 2.4969 = ____.

Answers: 2.4969; 2; 0.4969; log 10^2; 314; 314

144 antilog 3.8785

Let x = antilog 3.8785, so that log x = ________.

log x = 0.8785 + ____

Since log 7.56 = ________,

log x = log 7.56 + ________ = log (7.56×10^3)

= log ______

Therefore, x = antilog 3.8785 = _______.

Answers: 3.8785; 3; 0.8785; log 10^3; 7560; 7560

145 antilog [0.5682 + (-2)]

Let x = antilog [0.5682 + (-2)], so that

log x = 0.5682 + (____)

Since log 3.70 = ________,

log x = log 3.70 + log 10^{-2} = log (______________)

= log ________

Therefore, x = antilog [0.5682 + (-2)] = ________.

Answers: -2; 0.5682; 3.70×10^{-2}; 0.0370; 0.0370

146 antilog [0.5079 + (-4)]

Let x = antilog [0.5079 + (-4)], so that log x = ________ + (-4).

Since log 3.22 = ________,

log x = log 3.22 + __________

= log (3.22×10^{-4}) = ____________________

Therefore, x = antilog [0.5079 + (-4)] = ___________.

Answers: 0.5079; 0.5079; log 10^{-4}; log 0.000322; 0.000322

147 antilog (-2.1002)

Let x = antilog (-2.1002), so that log x = ___________.

Since the mantissa is always positive, we have (-2.1002 + 3) - 3 = ________ + (-3), so that log x = 0.8998 + (-3).

Since log 7.94 = ________,

log x = log 7.94 + ___________ = log (7.94×10^{-3})

= log __________

Therefore, x = antilog (-2.1002) = ___________.

Answers: -2.1002; 0.8998; 0.8998; log 10^{-3}; 0.00794; 0.00794

4.3 Linear Interpolation

In Problems 148–150, use Appendix Table I and linear interpolation to find the approximate value of the given logarithm.

148 log 1.234

0 The characteristic of log 1.234 is ____. Set up the following table:

	x	$\log x$
0.0934	1.240	______
	1.234	?
0.0899	1.230	______

Compute the differences indicated by the pairings to obtain:

$$\begin{array}{ccc} & x & \log x \\ 0.010\left\{\begin{array}{l} \\ 0.004\left\{\begin{array}{l}\\ \\ \end{array}\right. \end{array}\right. & \begin{array}{l}1.240\\1.234\\1.230\end{array} & \left.\begin{array}{l}0.0934\\ \left.\begin{array}{l}?\\0.0899\end{array}\right\} d\end{array}\right\} 0.0035 \end{array}$$

0.0014 Setting up the proportion $\frac{0.004}{0.010} = \frac{d}{0.0035}$, we get $d =$ ______.

0.0014 Then log 1.234 = 0.0899 + d = 0.0899 + ______.

0.0913 Therefore, log 1.234 = ______.

149 log 2.563

0 The characteristic of log 2.563 is ____. Set up the following table:

	x	$\log x$
	2.570	0.4099
	2.563	?
0.4082	2.560	______

Compute the difference indicated by the pairings to obtain:

$$\begin{array}{ccc} & x & \log x \\ 0.010\left\{\begin{array}{l} \\ 0.003\left\{\begin{array}{l}\\ \\ \end{array}\right. \end{array}\right. & \begin{array}{l}2.570\\2.563\\2.560\end{array} & \left.\begin{array}{l}0.4099\\ \left.\begin{array}{l}?\\0.4082\end{array}\right\} d\end{array}\right\} 0.0017 \end{array}$$

0.0005 From the proportion $\frac{0.003}{0.010} = \frac{d}{0.0017}$, we see that $d =$ ______.

0.0005 Then log 2.563 = 0.4082 + d = 0.4082 + ______.

0.4087 Therefore, log 2.563 = ______.

150 log 0.003528

Since log 0.003528 and log 3.528 differ only in their characteristic values, we first set up the necessary table to find log 3.528:

	x	$\log x$
0.5478	3.53	______
	3.528	?
0.5465	3.52	______

Compute the differences indicated by the pairings to obtain:

$$0.010\left\{\begin{array}{l} \\ 0.008\left\{\begin{array}{l} \\ \\ \end{array}\right. \end{array}\right.\begin{array}{cc} x & \log x \\ 3.53 & 0.5478 \\ 3.528 & ? \\ 3.52 & 0.5465 \end{array}\left.\begin{array}{l} \\ \left.\begin{array}{l} \\ \\ \end{array}\right\} d \end{array}\right\}0.0013$$

0.0010 From the proportion $\dfrac{0.008}{0.010} = \dfrac{d}{0.0013}$, we see that $d =$ ________.

0.5475 Then $\log 3.528 = 0.5465 + d =$ ________.

0.5475, -2.4525 Thus, $\log 0.003528 =$ ________ $+ (-3) =$ ________.

In Problems 151–153, use linear interpolation to find each antilogarithm.

151 antilog 0.2217

0.2217 Let $x = \text{antilog } 0.2217$, so that $\log x =$ ________.

From Appendix Table I, we find successive mantissa entries such

0.2201, 0.2227 that the given mantissa lies between ________ and ________.

Now, set up the following table:

y	antilog y
0.2227	1.67
0.2217	x
0.2201	____

1.66

Compute the differences indicated by the pairings below to obtain:

$$0.0026\left\{\begin{array}{l} \\ 0.0016\left\{\begin{array}{l} \\ \\ \end{array}\right. \end{array}\right.\begin{array}{cc} y & \text{antilog } y \\ 0.2227 & 1.67 \\ 0.2217 & x \\ 0.2201 & 1.66 \end{array}\left.\begin{array}{l} \\ \left.\begin{array}{l} \\ \\ \end{array}\right\} d \end{array}\right\}0.01$$

0.006 Setting up the proportion $\dfrac{d}{0.01} = \dfrac{0.0016}{0.0026}$, we get $d =$ ________.

Hence,

$x = 1.66 + d$

0.006 $= 1.66 +$ ________

1.666 $=$ ________

152 antilog 3.6129

First, we find antilog 0.6129; let $x = \text{antilog } 0.6129$, so that

0.6129 $\log x =$ ________

From Appendix Table I, we find successive mantissa entries such

0.6128, 0.6138 that the given mantissa lies between ________ and ________.

Now, set up the following table:

y	antilog y
0.6128	____
0.6129	log x
0.6138	____

4.10

4.11

Compute the differences indicated by the pairings below to obtain:

$$
\begin{array}{cccc}
 & & y & \text{antilog } y \\
0.0010\left\{\begin{array}{c} \\ \\ \\ \end{array}\right. & 0.0001\left\{\begin{array}{c} \\ \\ \end{array}\right. & \begin{array}{c} 0.6128 \\ 0.6129 \\ 0.6138 \end{array} & \begin{array}{c} 4.10 \\ x \\ 4.11 \end{array}\left.\begin{array}{c} \\ \\ \end{array}\right\} d \quad \left.\begin{array}{c} \\ \\ \\ \end{array}\right\} 0.01
\end{array}
$$

0.001 From the proportion $\dfrac{d}{0.01} = \dfrac{0.0001}{0.0010}$, we see that $d =$ ____. Hence

$x = 4.10 + d$

0.001 $= 4.10 +$ ______

4.101 $=$ ______

10^3, 4101 Thus, antilog $3.6129 = 4.101 \times$ ____ $=$ _____.

153 antilog (-2.4542)

-2.4542 Let $x =$ antilog (-2.4542), so that $\log x =$ _________. Since the mantissa must be positive, we write

$-2.4542 = (-2.4542 + 3) - 3$

0.5458 $=$ _________ $+ (-3)$

Thus, $\log x = 0.5458 + (-3)$

To find antilog 0.5458, we use linear interpolation, so that if

0.5458 $s =$ antilog 0.5458, then $\log s =$ ________. Set up the necessary table:

	y	antilog y
3.51	0.5453	____
	0.5458	s
3.52	0.5465	____

Compute the differences indicated by the pairings below to obtain:

$$
\begin{array}{cccc}
 & & y & \text{antilog } y \\
0.0012\left\{\begin{array}{c} \\ \\ \\ \end{array}\right. & 0.0005\left\{\begin{array}{c} \\ \\ \end{array}\right. & \begin{array}{c} 0.5453 \\ 0.5458 \\ 0.5465 \end{array} & \begin{array}{c} 3.51 \\ s \\ 3.52 \end{array}\left.\begin{array}{c} \\ \\ \end{array}\right\} d \quad \left.\begin{array}{c} \\ \\ \\ \end{array}\right\} 0.01
\end{array}
$$

0.004 From the proportion $\dfrac{d}{0.01} = \dfrac{0.0005}{0.0012}$, we see that $d =$ _______.

3.514 Hence, $s = 3.51 + 0.004 =$ ______. Therefore,

0.003514 $x =$ antilog $(-2.4542) =$ antilog $[0.5458 + (-3)] =$ __________

4.4 Base e Computations

In Problems 154–156, use Appendix Table II, and linear interpolation if necessary, to determine each value.

154 $\log_e 2.15 = \ln 2.15$

0.7655 $\ln 2.15 =$ ________

5

ln 5

1.6094

2.7080

155 $\ln 15 = \ln(3 \times \underline{\qquad})$

$= \ln 3 + \underline{\qquad}$

$= 1.0986 + \underline{\qquad}$

$= \underline{\qquad}$

2.1436

156 ln 8.537

First we set up the following table:

x	$\ln x$
8.54	2.1448
8.537	?
8.53	____

0.01, 0.0012

Compute the differences indicated by the pairings to obtain:

$$\underline{\qquad}\left\{\begin{array}{ll} & \begin{array}{cc} x & \ln x \\ 8.54 & 2.1448 \end{array} \\ 0.007\left\{\begin{array}{l} 8.537 \\ 8.53 \end{array}\right. & \left.\begin{array}{r} ? \\ 2.1436 \end{array}\right\} d \end{array}\right\}\underline{\qquad}$$

0.0008

2.1436, 2.1444

Setting up the proportion $\frac{d}{0.0012} = \frac{0.007}{0.01}$, we find that d is approximately equal to ______.

Then, $\ln 8.537 = \underline{\qquad} + d = \underline{\qquad}$.

$\log_a b$

157 $\log_b x = \dfrac{\log_a x}{\underline{\qquad}}$

$\log_{10} 2$, $\dfrac{1}{\log_{10} 2}$, 3.3223

158 $\log_2 10 = \dfrac{\log_{10} 10}{\underline{\qquad}} = \dfrac{\qquad}{\underline{\qquad}} = \underline{\qquad}$

$\log_b a$

159 $(\log_a b)(\underline{\qquad}) = 1$

$\log_{10} e$

160 $\ln x = \log_e x = \dfrac{\log_{10} x}{\underline{\qquad}}$

In Problems 161–163, use $e = 2.718$ and Appendix Table I to compute each of the given expressions to the nearest hundredth.

0.4343, 1.61

161 $\ln 5 = \log_e 5 = \dfrac{\log_{10} 5}{\log_{10} e} = \dfrac{0.6990}{\underline{\qquad}} = \underline{\qquad}$

2, 4.61

162 $\ln 100 = \log_e 100 = \dfrac{\log_{10} 100}{\log_{10} e} = \dfrac{\underline{\qquad}}{0.4343} = \underline{\qquad}$

1.6902, 3.89

163 $\ln 49 = \log_e 49 = \dfrac{\log_{10} 49}{\log_{10} e} = \dfrac{\underline{\qquad}}{0.4343} = \underline{\qquad}$

5 Applications of Logarithms

5.1 Computations Using Logarithms

In Problems 164–170, use the properties of logarithms and Appendix Table I to compute each expression.

164 53.7×0.83

Let $x = 53.7 \times 0.83$ and write $\log x =$ ________. — log (53.7 × 0.83)

Rewrite the right side of the equation as the sum of logarithms to obtain $\log x = \log 53.7 +$ ________, so that — log 0.83

$\log 53.7 =$ ________ and $\log 0.83 =$ ________. — 0.7300 + 1, 0.9191 − 1

Then, $\log x = 0.7300 + 1 + ($________$) =$ ________. — 0.9191 − 1, 1.6491

Thus, $x = \text{antilog } 1.6491 =$ ________. — 44.58

165 $0.0372 \times 4.81 \times 652$

Let $x = 0.0372 \times 4.81 \times 652$, so that

$\log x = \log 0.0372 + \log 4.81 +$ ________ — log 652

$= (0.5705 - 2) +$ ________ $+$ ________ — 0.6821, 2.8142

$=$ ________ — 2.0668

Thus, $x = \text{antilog } 2.0668 =$ ________. — 116.6

166 $\dfrac{0.817}{5.22}$

Let $x = \dfrac{0.817}{5.22}$, so that

$\log x = \log \dfrac{0.817}{5.22} = \log 0.817 -$ ________ — log 5.22

$= ($________$) - 0.7177$ — 0.9122 − 1

$=$ ________ — 0.1945 − 1

Thus, $x = \text{antilog } (0.1945 - 1) =$ ________. — 0.155

167 $\dfrac{3480 \times 1265}{0.00143}$

Let $x = \dfrac{3480 \times 1265}{0.00143}$, so that

$\log x =$ ________ — $\log \dfrac{3480 \times 1265}{0.00143}$

Rewrite the right side of the equation as a sum and difference of logarithms to obtain

$\log x = \log 3480 +$ ________ $- \log 0.00143$. Now, — log 1265

$\log 3480 =$ ________, $\log 1265 =$ ________, — 0.5416 + 3, 0.1021 + 3

and $\log 0.00143 =$ ________. — 0.1553 − 3

Then, $\log x = 0.5416 + 3 + ($________$) - 0.1553 + 3$ — 0.1021 + 3

$=$ ________ — 9.4884

Thus, $x = \text{antilog } 9.4884 =$ ________. — 3,079,000,000

168 $(2.14)^3$

Let $x = (2.14)^3$; write $\log x = \log (2.14)$—— $=$ ____ $\log 2.14$. — ^{3}, 3

Since $\log 2.14 =$ ________, then $\log x = 3(0.3304) =$ ________. — 0.3304, 0.9912

Hence, $x = \text{antilog } 0.9912 =$ ________. — 9.80

169 $\sqrt[5]{17}$

Let $x = \sqrt[5]{17}$ and write $\log x = \log (17)$____ = ____ log 17.

Since log 17 = ________, we can write

$\log x = \frac{1}{5}(1.2304) =$ ________

Thus, $x = \text{antilog } 0.2461 =$ ________.

1/5, $\frac{1}{5}$

0.2304 + 1

0.2461

1.76

170 $\sqrt[3]{\frac{(384)^2(723)}{291}}$

Let $x = \sqrt[3]{\frac{(384)^2(723)}{291}}$ and write

$\log x = \log \left[\frac{(384)^2(723)}{291}\right]^{\_\_\_\_} = \frac{1}{3} \log \left[\_\_\_\_\_\_\_\_\right]$

Rewrite the right side of the equation as a sum and difference of logarithms. We have

$\log x = \frac{1}{3}[\log (384)^2 + \log 723 -$ ________]

$= \frac{1}{3}($________ $+ \log 723 - \log 291)$

log 384 = ________, log 723 = ________, and

log 291 = ________. Thus,

$\log x = \frac{1}{3}[($________$) + (0.8591 + 2) - (0.4639 - 2)]$

$= \frac{1}{3}($________$)$

= ________

Thus, $x = \text{antilog } 1.8546 =$ ________.

1/3, $\frac{(384)^2(723)}{291}$

log 291

2 log 384

0.5843 + 2, 0.8591 + 2

0.4639 + 2

0.1686 + 5

5.5638

1.8546

71.55

5.2 Exponential Equations

In Problems 171–173, use logarithms and Appendix Table I to find the solution set of each equation.

171 $3^x = 11$

Write $\log 3^x =$ ________, so that

$x \log 3 =$ ________

$$x = \frac{\log 11}{\log 3} = \frac{\_\_\_\_}{0.4771} = \_\_\_\_$$

The solution set is ________.

log 11

log 11

1.0414, 2.18

{2.18}

172 $5^{-x} = 9$

Write $\log 5^{-x} =$ ________, so that

$-x \log 5 =$ ________

$$x = -\frac{\log 9}{\log 5} = -\frac{\_\_\_\_}{0.6990}$$

= ________

The solution set is ________.

log 9

log 9

0.9542

-1.37

{-1.37}

	173 $3^{2x-1} = 5$
log 3	Write $(2x - 1)$(________) = log 5. Now,
0.4771, 0.6990	log 3 = ________ and log 5 = ________, so that
	$2x - 1 = \dfrac{0.6990}{0.4771}$
2.47	$2x =$ ______
1.24	$x =$ ______
{1.24}	The solution set is ______.

In Problems 174–175, use logarithms and Appendix Table II to find the solution set of each equation.

	174 $e^x = 5$
ln 5	Since $e^x = 5$, we have $\ln e^x =$ ________, so that
1, 1.6094	$x \ln e = x \cdot$ ____ = ________
1.6094, {1.6094}	Thus, $x =$ ________, and the solution set is ________.
	175 $e^{-x/2} = 9.15$
ln 9.15	Since $e^{-x/2} = 9.15$, we have $\ln e^{-x/2} =$ ____________, so that
2.2138, -4.4276	$-\dfrac{x}{2} =$ ________ and $x =$ ________
{-4.4276}	The solution set is ________.

5.3 Applied Problems from Science and Business

In Problems 176–182, use logarithms to perform the required calculations in each situation.

	176 Find the volume of a circular cone of radius $r = 21.3$ inches and height $h = 79.6$ inches. The volume of the cone is given by
$\frac{1}{3}\pi r^2 h$	$V =$ ________, so $V = \frac{1}{3}\pi(21.3)^2(79.6)$. (Use $\pi = 3.14$.)
$\dfrac{(3.14)(21.3)^2(79.6)}{3}$	$\log V = \log \left[\rule{0pt}{1em}\quad\quad\quad\quad\quad\quad\right]$
log 3	$= \log 3.14 + \log (21.3)^2 + \log 79.6 -$ ________
log 3	$= \log 3.14 + 2 \log 21.3 + \log 79.6 -$ ________
0.4771	$= 0.4969 + 2(1.3284) + 1.9009 -$ ________
4.5775	= ________
37,800	Therefore, $V =$ ________ cubic inches.
	177 Newton's law of gravitation states that the gravitational force acting on two particles of mass m_1 and m_2, respectively, is given by $F = k\dfrac{m_1 m_2}{x^2}$, where k is a constant depending on the unit used and x is the distance between the two particles. Find the gravitational force acting on two particles of mass 67.3 grams and 89.5

grams if the distance between them is 83.6 centimeters and $k = 39.1$.

The force F is given by

$$F = \frac{39.1 \times 67.3 \times 89.5}{(83.6)^2}$$

$\frac{39.1 \times 67.3 \times 89.5}{(83.6)^2}$ — $\log F = \log \left[\underline{\qquad\qquad} \right]$

$\log (83.6)^2$ — $= \log 39.1 + \log 67.3 + \log 89.5 - \underline{\qquad}$

$2 \log 83.6$ — $= \log 39.1 + \log 67.3 + \log 89.5 - \underline{\qquad}$

1.9222 — $= 1.5922 + 1.8280 + 1.9518 - 2(\underline{\qquad})$

1.5276 — $= \underline{\qquad}$

33.70 — Therefore, $F = \underline{\qquad}$ dynes.

In Problems 178–179, consider the application of *compound interest*, which states that if P dollars represents the amount invested at an annual interest rate r (expressed as a decimal), the amount A accumulated in n years, when interest is compounded t times per year, is given by the formula

$$A = P\left(1 + \frac{r}{t}\right)^{nt}$$

178 Suppose that \$1000 is invested at a yearly interest rate of 6 percent compounded every 4 months. How much money is accumulated after 4 years?

1000, 0.06 — Here, $P = \underline{\qquad}$, $r = \underline{\qquad}$, $t = 3$, and $n = 4$, so

$1 + \frac{0.06}{3}$, $(1.02)^{12}$ — $A = 1000 \left(\underline{\qquad\qquad} \right)^{3 \cdot 4} = 1000 \left[\underline{\qquad\qquad} \right]$

Write

$\log A = \log[1000(1.02)^{12}]$

log 1.02 — $= \log 1000 + 12 \underline{\qquad}$

or

$\log A = 3 + 12 \log 1.02$

0.0086 — From Appendix Table I, $\log 1.02 = \underline{\qquad}$. Then

3.1032 — $\log A = 3 + 12(0.0086) = \underline{\qquad}$

1268 — $A = \text{antilog } 3.1032 = \underline{\qquad}$

1268, 4 — Therefore, \$$\underline{\qquad}$ is accumulated after $\underline{\qquad}$ years.

179 In how many years would \$1000 invested at an annual interest rate of 6 percent compounded every 4 months be doubled?

1000, 0.06 — Here, $P = \underline{\qquad}$, $r = \underline{\qquad}$, and $t = 3$, so

$1 + \frac{0.06}{3}$, $(1.02)^{3n}$ — $A = 1000 \left(\underline{\qquad\qquad} \right)^{3n} = 1000 \left[\underline{\qquad\qquad} \right]$

2 — If $A = 2000$, then $2000 = 1000(1.02)^{3n}$, or $\underline{\qquad} = (1.02)^{3n}$.

^{3n}, $3n$ — Write $\log 2 = \log (1.02)^{\underline{\quad}} = \underline{\qquad} \log 1.02$. Using Appendix

0.3010, 0.0086 — Table I, we find that $\log 2 = \underline{\qquad}$ and $\log 1.02 = \underline{\qquad}$.

Therefore,

0.0086 $$3n = \frac{\log 2}{\log 1.02} = \frac{0.3010}{\_\_\_\_\_\_}$$

11.7 Hence, $n =$ ______. Thus, 11.7 years later the original $1000

doubled would have ____________.

In Problems 180–181, consider the application of a *simple annuity*, which states that

> if the size of each payment is P dollars and the interest rate is r, the accumulated value S of all payments at the time of the nth payment is given by the formula
> $$S = P\left[\frac{(1+r)^n - 1}{r}\right]$$

180 At the end of each month a worker deposits $150 in a savings and loan association. If the annual interest rate is 6 percent converted monthly, find the amount in the account after 10 years.

120 The annuity consists of $n = (12)(10) =$ ____ payments.

0.005 The interest rate is $r = \frac{1}{12}(0.06) =$ _______.

150 The periodic payment $P =$ ____.

Then

30,000 $$S = 150\left[\frac{(1+0.005)^{120} - 1}{0.005}\right] = \_\_\_\_\_\_\_\_[(1.005)^{120} - 1]$$

30,000 $$= 30{,}000(1.005)^{120} - \_\_\_\_\_\_\_\_$$

To evaluate the expression $30{,}000(1.005)^{120}$, let

$x = 30{,}000(1.005)^{120}$, so

$\log x = \log[(30{,}000)(1.005)^{120}]$

$(1.005)^{120}$ $= \log 30{,}000 + \log[\_\_\_\_\_\_\_\_\_\_\_\_]$

log 1.005 $= \log 30{,}000 + 120(\_\_\_\_\_\_\_\_\_\_\_\_)$

0.0022 $= 4.4771 + 120(\_\_\_\_\_\_\_\_)$

4.7411 $= \_\_\_\_\_\_\_\_$

so

4.7411 $x = \text{antilog}\ \_\_\_\_\_\_\_\_$

55,090 $= \_\_\_\_\_\_\_\_$

25,090 Therefore, $S = 55{,}090 - 30{,}000 =$ ________. Thus the accumu-

25,090 lated value after 10 years is $________.

181 A family wishing to provide for the college education of a newborn child plans to deposit $30 at the end of every month into a fund that pays 8 percent annual interest converted monthly. Find the amount that would be on deposit after 18 years if this plan is carried out.

216 The annuity consists of $n = (12)(18) =$ ____ payments.

0.0067 The interest rate is $r = \frac{1}{12}(0.08) =$ ________.

Answers	
30	The periodic payment $P =$ ____.
	Then
	$S = 30\left[\dfrac{(1 + 0.0067)^{216} - 1}{0.0067}\right]$
4478	$= ($____$)[(1.0067)^{216} - 1]$
4478	$= (4478)(1.0067)^{216} -$ ____
	To evaluate the expression $(4478)(1.0067)^{216}$, we use a calculator to compute the required logarithmic values, since they cannot be found in our Appendix Table I.
	Let $x = (4478)(1.0067)^{216}$, so
	$\log x = \log [(4478)(1.0067)^{216}]$
$\log (1.0067)^{216}$	$= \log 4478 +$ ____
log 1.0067	$= \log 4478 + 216($____$)$
0.0029	$= 3.6511 + 216($____$)$
4.2775	$=$ ____
18,940	so $x = \text{antilog } 4.2775 =$ ____.
14,462	Therefore, $S = 18{,}940 - 4478 =$ ____. Thus the amount on
14,462	deposit after 18 years is approximately \$____.

In Problem 182, consider the application of the *present value A* of an annuity given by

$A = P\left[\dfrac{1 - (1 + r)^{-n}}{r}\right]$ where P, r, and n are defined as in Problem 180.

Answers	
	182 A family bought a home by making a down payment of \$10,000 and contracting for monthly payments of \$300 for 25 years. If the interest rate on the mortgage is 9 percent, what price did the family pay for the home? (Assume that the first mortgage payment occurs 1 month after the down payment has been made.)
300, 0.0075	$n = 25 \times 12 =$ ____ and $r = \frac{1}{12}(0.09) =$ ____
	Let A be the amount of mortgage on the home:
	$A = P\left[\dfrac{1 - (1 + r)^{-n}}{r}\right] = 300\left[\dfrac{1 - (1 + 0.0075)^{-300}}{0.0075}\right]$
	$= (40{,}000)[1 - (1.0075)^{-300}]$
$(40{,}000)(1.0075)^{-300}$	$= 40{,}000 -$ ____
	To evaluate $40{,}000(1.0075)^{-300}$, we use a calculator to compute the required logarithmic values, since they cannot be found in our Appendix Table I.
	Let $x = 40{,}000(1.0075)^{-300}$, so that
$(40{,}000)(1.0075)^{-300}$	$\log x = \log [$____$]$
$(1.0075)^{-300}$	$= \log 40{,}000 + \log$ ____
log 1.0075	$= \log 40{,}000 - 300($____$)$

0.0032	= 4.6021 - 300 (________)
3.6421	= ________
4386	x = antilog 3.6421 = ______
4386, 35,614	Therefore, A = 40,000 - ______ = ________
45,614	The price of the home was 35,614 + 10,000 = ________ dollars.

Chapter Test

1 Write each of the following statements in logarithmic form.

(a) $2^9 = 512$ (b) $3^4 = 81$ (c) $\frac{1}{10^3} = 0.001$

2 Write each of the following statements in exponential form.

(a) $\log_2 32 = 5$ (b) $\log_5 125 = 3$ (c) $\log_7 \frac{1}{49} = -2$

3 Find the solution set of each equation.

(a) $2^{x-1} = 8$ (b) $9(3^{1-x}) = 81$ (c) $\log_3 (2x - 1) = 4$ (d) $\log_5 (7x - 2) = 2$

4 Evaluate each expression.

(a) $\log_3 \frac{1}{9}$ (b) $\log_5 \frac{1}{125}$ (c) $\log_7 7\sqrt{7}$ (d) $\log_9 27\sqrt{3}$

5 Write each expression as a single logarithm.

(a) $\log_3 \frac{17}{2} + \log_3 \frac{5}{34}$ (b) $\log_a x^5 - 3 \log_a x^2$

(c) $\log_a (x^2 - 16) - \log_a (x - 4)$ (d) $3 \log_a x + 5 \log_a y^2$

6 Determine the value of x in each of the following equations.

(a) $\log_x 625 = 4$ (b) $\log_2 \frac{1}{8} = x$ (c) $\log_{32} x = \frac{1}{5}$ (d) $\log_b x = 0$

7 Find the solution set of each equation.

(a) $\log_3 (x + 1) + \log_3 (x + 3) = 1$ (b) $\log_4 (x + 3) - \log_4 x = 1$

(c) $\log_2 (x^2 - 9) - \log_2 (x + 3) = 2$

8 Evaluate each expression.

(a) log 3.21 (b) ln 12 (c) log 3.216

(d) antilog 2.1818 (e) antilog (−3.6162) (f) antilog 0.6925

9 In each of the following functions, find the domain, indicate whether f is increasing or decreasing. Sketch the graph. Use the graph to determine the range.

(a) $f(x) = 3^x$ (b) $f(x) = (\frac{1}{4})^x$

(c) $f(x) = \log_3 x$ (d) $f(x) = \log_{1/3} x$

10 Let $f(x) = \log_a x$, $f(2) = 0.35$, $f(3) = 0.55$, and $f(5) = 0.82$. Use the properties of logarithms to find:

(a) $f(\frac{2}{3})$ (b) $f(2^3)$ (c) $f(\sqrt{\frac{2}{3}})$ (d) $f(0.6)$

11 Use logarithms to compute the value of each expression.

(a) $(152)^3(0.000242)$ (b) $\sqrt{\frac{(0.000263)^3}{(0.0389)(9{,}420)}}$

12 If $500 is invested at an annual interest rate of 8 percent compounded quarterly, how much money is accumulated after 4 years?

Answers

1 (a) $\log_2 512 = 9$ (b) $\log_3 81 = 4$ (c) $\log_{10} 0.001 = -3$

2 (a) $2^5 = 32$ (b) $5^3 = 125$ (c) $7^{-2} = \frac{1}{49}$

3 (a) $\{4\}$ (b) $\{-1\}$ (c) $\{41\}$ (d) $\{\frac{27}{7}\}$

4 (a) -2 (b) -3 (c) $\frac{3}{2}$ (d) $\frac{7}{4}$

5 (a) $\log_3 \frac{5}{4}$ (b) $\log_a\left(\frac{1}{x}\right)$ (c) $\log_a (x+4)$ (d) $\log_a(x^3y^{10})$

6 (a) 5 (b) -3 (c) 2 (d) 1

7 (a) $\{0\}$ (b) $\{1\}$ (c) $\{7\}$

8 (a) 0.5065 (b) 2.4849 (c) 0.5073
(d) 152 (e) 0.000242 (f) 4.926

9 (a) Domain = R
Increasing
Range = $\{y \mid y > 0\}$

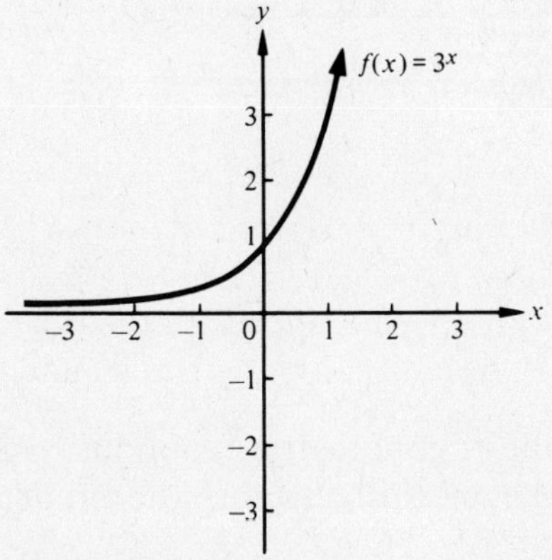

(b) Domain = R
Decreasing
Range = $\{y \mid y > 0\}$

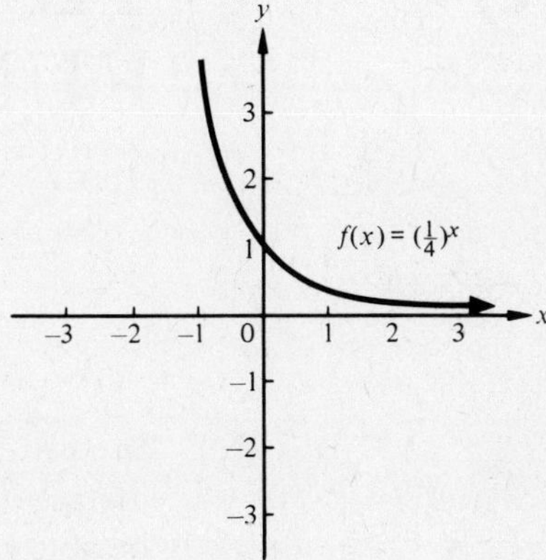

(c) Domain = $\{x \mid x > 0\}$
Increasing
Range = R

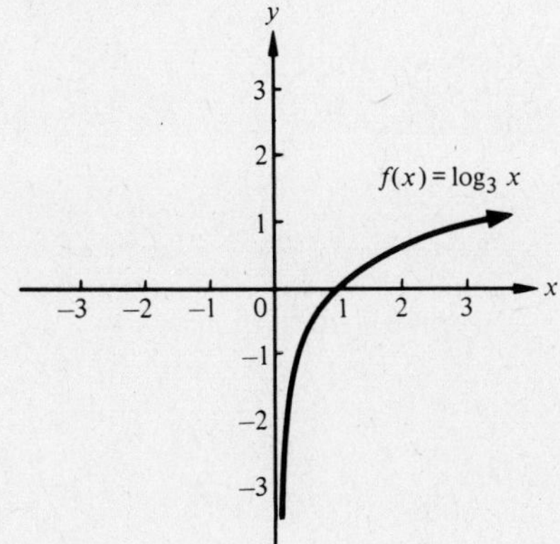

(d) Domain = $\{x \mid x > 0\}$
Decreasing
Range = R

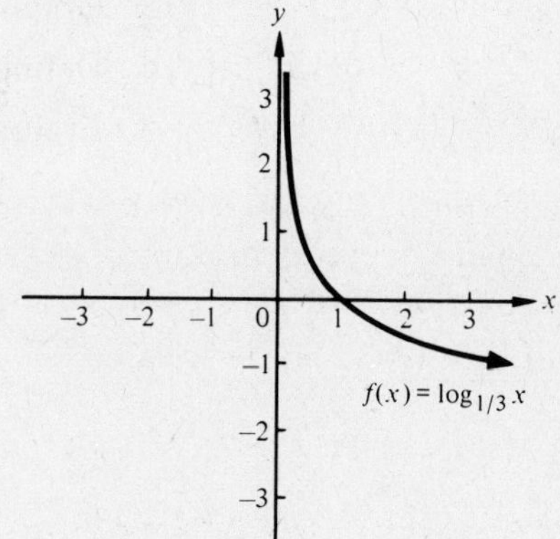

10 (a) -0.20 (b) 1.05 (c) -0.10 (d) -0.27

11 (a) 850 (b) 0.000000223

12 \$686.40

Chapter 5 TRIGONOMETRIC FUNCTIONS

In this chapter we introduce the trigonometric functions. After completing the appropriate sections in this chapter, the student should be able to:

1 Define the trigonometric (circular) functions on real numbers by using the wrapping function.

2 Determine properties of the trigonometric functions.

3 Graph trigonometric functions.

4 Define the trigonometric functions of angles.

5 Evaluate the trigonometric functions.

1 The Wrapping Function

1, (0, 0)
$x^2 + y^2 = 1$

$2\pi r$, 2π

1 The *unit circle* is the circle of radius ____ with center at ______ and with the equation ______________.

2 The circumference C of a circle of radius r is given by the formula $C =$ ____; hence, the unit circle has circumference ____.

In Problems 3–5, find the actual distance traveled along the circumference of the unit circle if one starts at the point (1, 0) and moves counterclockwise the specified distance.

π, π

$\frac{3}{10}$, $\frac{3\pi}{5}$, $\frac{3\pi}{5}$

2π, $\frac{9\pi}{2}$, $\frac{9\pi}{2}$

3 One-half of the way around the circumference
Since $\frac{1}{2}(2\pi) =$ ____, the actual distance traveled is ____ units.

4 Three-tenths of the way around the circumference
Since $\frac{\quad}{\quad}(2\pi) = \frac{\quad}{\quad}$, one travels $\frac{\quad}{\quad}$ units.

5 Nine-fourths of the way around the circumference
Since $\frac{9}{4}$(____) $= \frac{\quad}{\quad}$, one travels $\frac{\quad}{\quad}$ units.

unit circle
(1, 0), $|t|$
counterclockwise
$>$, clockwise, $<$

6 The *wrapping function P* associates each real number t with a point (x, y) on the unit circle as follows:
$P(t) = (x, y)$, where (x, y) is the point on the ______________ located by starting at point ________ and traversing ____ units along the circumference in the ______________________ sense, if t ____ 0, or in the ______________ sense, if t ____ 0.

In Problems 7–13, determine the quadrant in which $P(t)$ is located.

II

7 $P(3)$ is a point in quadrant ____, since $\frac{\pi}{2} < 3 < \pi$.

I

8 $P(1.5)$ is a point in quadrant ____, since $0 < 1.5 < \frac{\pi}{2}$.

III

9 $P(4)$ is a point in quadrant ____, since $\pi < 4 < \frac{3\pi}{2}$.

IV

10 $P(5)$ is a point in quadrant ____, since $\frac{3\pi}{2} < 5 < 2\pi$.

III

11 $P(-3)$ is a point in quadrant ____, since $-\pi < -3 < -\frac{\pi}{2}$.

II

12 $P\left(\frac{2\pi}{3}\right)$ is a point in quadrant ____, since $\frac{\pi}{2} < \frac{2\pi}{3} < \pi$.

I, $-\frac{5\pi}{3}$

13 $P\left(-\frac{5\pi}{3}\right)$ is a point in quadrant ____, since $-2\pi <$ ____ $< -\frac{3\pi}{2}$.

In Problems 14–15, find the coordinates of each of the points.

Answers	Problems
(-1, 0), (-1, 0)	**14** $P(\pi) =$ ______ and $P(-\pi) =$ ______
(0, -1), (0, 1)	**15** $P\left(\frac{3\pi}{2}\right) =$ ______ and $P\left(-\frac{3\pi}{2}\right) =$ ______

1.1 Trigonometric (Circular) Functions

Answers	Problems
	16 If $P(t) = (x, y)$, then the numbers that are defined on the real number t by the trigonometric (circular) functions are
y; $\frac{1}{y}$, y	$\sin t =$ ____ $\csc t = \frac{—}{—}$, ____ $\neq 0$
x; $\frac{1}{x}$, x	$\cos t =$ ____ $\sec t = \frac{—}{—}$, ____ $\neq 0$
$\frac{y}{x}$, x; $\frac{x}{y}$, y	$\tan t = \frac{—}{—}$, ____ $\neq 0$ $\cot t = \frac{—}{—}$, ____ $\neq 0$

In Problems 17–20, determine the six trigonometric function values at the given value of t.

Answers	Problems
	17 $t = 0$
(1, 0)	Since $P(0) =$ ______, we have
x, 1	$\cos 0 =$ ____ = ____
y, 0	$\sin 0 =$ ____ = ____
$\frac{y}{x}$, 0	$\tan 0 = \frac{—}{—} =$ ____
$\frac{x}{y}$, undefined	$\cot 0 = \frac{—}{—} =$ ______
$\frac{1}{x}$, 1	$\sec 0 = \frac{—}{—} =$ ____
$\frac{1}{y}$, undefined	$\csc 0 = \frac{—}{—} =$ ______
	18 $t = -\frac{\pi}{2}$
(0, -1)	Since $P\left(-\frac{\pi}{2}\right) =$ ______, we have
x, 0	$\cos\left(-\frac{\pi}{2}\right) =$ ____ = ____
y, -1	$\sin\left(-\frac{\pi}{2}\right) =$ ____ = ____
$\frac{y}{x}$, undefined	$\tan\left(-\frac{\pi}{2}\right) = \frac{—}{—} =$ ______
$\frac{x}{y}$, 0	$\cot\left(-\frac{\pi}{2}\right) = \frac{—}{—} =$ ____
$\frac{1}{x}$, undefined	$\sec\left(-\frac{\pi}{2}\right) = \frac{—}{—} =$ ______
$\frac{1}{y}$, -1	$\csc\left(-\frac{\pi}{2}\right) = \frac{—}{—} =$ ____

$(-1, 0)$

x, -1

y, 0

$\frac{y}{x}$, 0

$\frac{x}{y}$, undefined

$\frac{1}{x}$, -1

$\frac{1}{y}$, undefined

19 $t = -\pi$

Since $P(-\pi) =$ ________, we have

$\cos(-\pi) =$ ____ $=$ ____

$\sin(-\pi) =$ ____ $=$ ____

$\tan(-\pi) = \frac{\quad}{\_\_\_\_} =$ ____

$\cot(-\pi) = \frac{\quad}{\_\_\_\_} =$ ____________

$\sec(-\pi) = \frac{\quad}{\_\_\_\_} =$ ____

$\csc(-\pi) = \frac{\quad}{\_\_\_\_} =$ ____________

$\left(\frac{\sqrt{2}}{2}, \frac{\sqrt{2}}{2}\right)$

$\frac{\sqrt{2}}{2}$

$\frac{\sqrt{2}}{2}$

1

1

$\sqrt{2}$

$\sqrt{2}$

20 $t = \frac{\pi}{4}$

Since $P\left(\frac{\pi}{4}\right) =$ ____________, we have

$\cos \frac{\pi}{4} =$ ____

$\sin \frac{\pi}{4} =$ ____

$\tan \frac{\pi}{4} =$ ____

$\cot \frac{\pi}{4} =$ ____

$\sec \frac{\pi}{4} =$ ____

$\csc \frac{\pi}{4} =$ ____

1.2 Values of $P(t)$ in Quadrants II, III, and IV

$\left(\frac{\sqrt{3}}{2}, \frac{1}{2}\right)$

$\left(\frac{\sqrt{2}}{2}, \frac{\sqrt{2}}{2}\right), \left(\frac{1}{2}, \frac{\sqrt{3}}{2}\right)$

$\left(\frac{1}{2}, \frac{\sqrt{3}}{2}\right)$

$\left(-\frac{1}{2}, \frac{\sqrt{3}}{2}\right)$

$\frac{\sqrt{3}}{2}$

$-\frac{1}{2}$

$-\sqrt{3}$

21 Some special values of $P(t)$ are $P\left(\frac{\pi}{6}\right) =$ __________, $P\left(\frac{\pi}{4}\right) =$ __________, and $P\left(\frac{\pi}{3}\right) =$ __________.

22 Since $P\left(\frac{\pi}{3}\right) =$ __________, we can use the symmetry of the unit circle to get $P\left(\frac{2\pi}{3}\right) =$ ____________ so that

$\sin \frac{2\pi}{3} =$ ____

$\cos \frac{2\pi}{3} =$ ____

$\tan \frac{2\pi}{3} =$ ____

$-\frac{\sqrt{3}}{3}$

-2

$\frac{2\sqrt{3}}{3}$

$\left(\frac{\sqrt{3}}{2}, \frac{1}{2}\right)$

$\left(-\frac{\sqrt{3}}{2}, -\frac{1}{2}\right)$

$-\frac{1}{2}$

$-\frac{\sqrt{3}}{2}$

$\frac{\sqrt{3}}{3}$

$\sqrt{3}$

-2

$-\frac{2\sqrt{3}}{3}$

$\cot \frac{2\pi}{3} =$ ____

$\sec \frac{2\pi}{3} =$ ____

$\csc \frac{2\pi}{3} =$ ____

23 Since $P\left(\frac{\pi}{6}\right) =$ ________, we can use the symmetry of the unit circle to get $P\left(-\frac{5\pi}{6}\right) =$ ____________ so that

$\sin\left(-\frac{5\pi}{6}\right) =$ ____

$\cos\left(-\frac{5\pi}{6}\right) =$ ____

$\tan\left(-\frac{5\pi}{6}\right) =$ ____

$\cot\left(-\frac{5\pi}{6}\right) =$ ____

$\csc\left(-\frac{5\pi}{6}\right) =$ ____

$\sec\left(-\frac{5\pi}{6}\right) =$ ____

In Problems 24–29, use $P(t)$ to evaluate the given function.

24 $\cos\left(-\frac{2\pi}{3}\right)$

$\left(-\frac{1}{2}, -\frac{\sqrt{3}}{2}\right), -\frac{1}{2}$

Since $P\left(-\frac{2\pi}{3}\right) =$ __________, then $\cos\left(-\frac{2\pi}{3}\right) =$ ____.

25 $\sin \frac{5\pi}{4}$

$\left(-\frac{\sqrt{2}}{2}, -\frac{\sqrt{2}}{2}\right), -\frac{\sqrt{2}}{2}$

Since $P\left(\frac{5\pi}{4}\right) =$ ____________, then $\sin \frac{5\pi}{4} =$ ____.

26 $\cot\left(-\frac{7\pi}{6}\right)$

$\left(-\frac{\sqrt{3}}{2}, \frac{1}{2}\right), -\sqrt{3}$

Since $P\left(-\frac{7\pi}{6}\right) =$ __________, then $\cot\left(-\frac{7\pi}{6}\right) =$ ____.

27 $\tan \frac{4\pi}{3}$

$\left(-\frac{1}{2}, -\frac{\sqrt{3}}{2}\right), \sqrt{3}$

Since $P\left(\frac{4\pi}{3}\right) =$ __________, then $\tan \frac{4\pi}{3} =$ ____.

28 $\sec \frac{11\pi}{6}$

$\left(\frac{\sqrt{3}}{2}, -\frac{1}{2}\right), \frac{2\sqrt{3}}{3}$

Since $P\left(\frac{11\pi}{6}\right) =$ __________, then $\sec \frac{11\pi}{6} =$ ____.

29 $\csc\left(-\frac{7\pi}{4}\right)$

Since $P\left(-\frac{7\pi}{4}\right) =$ ________, then $\csc\left(-\frac{7\pi}{4}\right) =$ ____.

$\left(\frac{\sqrt{2}}{2}, \frac{\sqrt{2}}{2}\right)$, $\sqrt{2}$

In Problems 30–33, express the given function at t in terms of the sine and/or cosine of t.

30 $\tan t =$ ____, for $\cos t \neq 0$

$\frac{\sin t}{\cos t}$

31 $\cot t =$ ____, for $\sin t \neq 0$

$\frac{\cos t}{\sin t}$

32 $\sec t =$ ____, for $\cos t \neq 0$

$\frac{1}{\cos t}$

33 $\csc t =$ ____, for $\sin t \neq 0$

$\frac{1}{\sin t}$

34 $\cot t = \frac{1}{\_\_\_\_}$, for $\tan t \neq 0$

$\tan t$

In Problems 35–37, evaluate $\sin t$ and $\cos t$ under the given condition; then use the results in Problems 30–33 to evaluate the other four trigonometric functions at t.

35 $t = -\frac{\pi}{2}$

Since $P\left(-\frac{\pi}{2}\right) =$ ________, $\sin\left(-\frac{\pi}{2}\right) =$ ____, and

$(0, -1)$, -1

$\cos\left(-\frac{\pi}{2}\right) =$ ____, so that

0

$$\tan\left(-\frac{\pi}{2}\right) = \frac{\sin\left(-\frac{\pi}{2}\right)}{\_\_\_\_} = \frac{-1}{\_\_\_\_} = \_\_\_\_$$

$\cos\left(-\frac{\pi}{2}\right)$, 0, undefined

$$\cot\left(-\frac{\pi}{2}\right) = \frac{\cos\left(-\frac{\pi}{2}\right)}{\_\_\_\_} = \frac{0}{-1} = \_\_\_\_$$

$\sin\left(-\frac{\pi}{2}\right)$, 0

$$\sec\left(-\frac{\pi}{2}\right) = \frac{1}{\cos\left(-\frac{\pi}{2}\right)} = \frac{1}{\_\_\_\_} = \_\_\_\_$$

0, undefined

$$\csc\left(-\frac{\pi}{2}\right) = \frac{1}{\sin\left(-\frac{\pi}{2}\right)} = \frac{1}{\_\_\_\_} = \_\_\_\_$$

-1, -1

36 $P(t) = \left(-\frac{4}{5}, \frac{3}{5}\right)$

$x, -\frac{4}{5}$ — $\cos t = \_\_\_\_ = \_\_\_\_$

$y, \frac{3}{5}$ — $\sin t = \_\_\_\_ = \_\_\_\_$

$\sin t, -\frac{4}{5}, -\frac{3}{4}$ — $\tan t = \frac{\_\_\_\_}{\cos t} = \frac{\frac{3}{5}}{\_\_\_\_} = \_\_\_\_.$

$\sin t, \frac{3}{5}, -\frac{4}{3}$ — $\cot t = \frac{\cos t}{\_\_\_\_} = \frac{-\frac{4}{5}}{\_\_\_\_} = \_\_\_\_$

$\cos t, -\frac{4}{5}, -\frac{5}{4}$ — $\sec t = \frac{1}{\_\_\_\_} = \frac{1}{\_\_\_\_} = \_\_\_\_$

$\sin t, \frac{3}{5}, \frac{5}{3}$ — $\csc t = \frac{1}{\_\_\_\_} = \frac{1}{\_\_\_\_} = \_\_\_\_$

37 $P(t) = \left(\frac{2\sqrt{5}}{5}, \frac{\sqrt{5}}{5}\right)$

$x, \frac{2\sqrt{5}}{5}$ — $\cos t = \_\_\_\_ = \_\_\_\_$

$y, \frac{\sqrt{5}}{5}$ — $\sin t = \_\_\_\_ = \_\_\_\_$

$\frac{\sin t}{\cos t}, \frac{1}{2}$ — $\tan t = \_\_\_\_ = \_\_\_\_$

$\frac{\cos t}{\sin t}, 2$ — $\cot t = \_\_\_\_ = \_\_\_\_$

$\frac{1}{\cos t}, \frac{\sqrt{5}}{2}$ — $\sec t = \_\_\_\_ = \_\_\_\_$

$\frac{1}{\sin t}, \sqrt{5}$ — $\csc t = \_\_\_\_ = \_\_\_\_$

2 Properties of the Trigonometric Functions

2.1 Periodicity of the Trigonometric Functions

$x + a$

$f(x)$

fundamental

38 A function f is said to be a periodic function of period a, where a is a nonzero number, if, for all x in the domain of f, ______ is also in the domain of f, and $f(x + a) =$ ______. The smallest positive period is called the __________ period of f.

(1, 0)

39 $P(0) = P(2\pi) = P(4\pi) =$ ______.

2π
2π
$2\pi n$

40 The wrapping function P is a periodic function with period ____, so that for any real number t we have $P(t) = P(t + \_\_\_\_)$. Also, if n is any integer, then $P(t) = P(t + \_\_\_\_)$.

In Problems 41–44, use the periodicity to find a value for t, where $0 \leqslant t < 2\pi$, so that $P(t)$ coincides with the given point. (Use $\pi = 3.14$.)

Answers	Problem
6.28, 2.72	**41** $P(9) = P(2.72 + \_\_\_\_) = P(\_\_\_\_)$
0.44, 0.44	**42** $P(13) = P[\_\_\_\_ + 2(6.28)] = P(\_\_\_\_)$
6.28, 3.28	**43** $P(-3) = P(-3 + \_\_\_\_) = P(\_\_\_\_)$
6.28, 2.12	**44** $P(-23) = P[-23 + 4\,(\_\_\_\_)] = P(\_\_\_\_)$

In Problems 45–52, use periodicity to find the coordinates of each point.

$4\pi,\ 2\pi,\ \frac{\pi}{2},\ (0, 1)$

45 $P\left(\frac{9\pi}{2}\right) = P\left(\frac{\pi}{2} + \_\_\_\_\right) = P\left(\frac{\pi}{2} + 2 \cdot \_\_\_\_\right) = P\left(\_\_\_\_\right) = \_\_\_\_$

2π

$\frac{\pi}{3},\ \left(\frac{1}{2}, \frac{\sqrt{3}}{2}\right)$

46 $P\left(\frac{19\pi}{3}\right) = P\left(\frac{\pi}{3} + 6\pi\right) = P\left(\frac{\pi}{3} + 3 \cdot \_\_\_\_\right)$

$= P\left(\_\_\_\_\right) = \_\_\_\_$

2π

$\frac{\pi}{6},\ \left(\frac{\sqrt{3}}{2}, \frac{1}{2}\right)$

47 $P\left(-\frac{35\pi}{6}\right) = P\left(\frac{\pi}{6} - 6\pi\right) = P\left[\frac{\pi}{6} + (-3)\,(\_\_\_\_)\right]$

$= P\left(\_\_\_\_\right) = \_\_\_\_$

$2\pi,\ \frac{3\pi}{4}$

$\left(-\frac{1}{\sqrt{2}}, \frac{1}{\sqrt{2}}\right),\ \left(-\frac{\sqrt{2}}{2}, \frac{\sqrt{2}}{2}\right)$

48 $P\left(\frac{11\pi}{4}\right) = P\left(\frac{3\pi}{4} + \_\_\_\_\right) = P\left(\_\_\_\_\right)$

$= \_\_\_\_ = \_\_\_\_$

$-\frac{\pi}{4}$

$\left(\frac{\sqrt{2}}{2}, -\frac{\sqrt{2}}{2}\right)$

49 $P\left(-\frac{9\pi}{4}\right) = P\left(-\frac{\pi}{4} - 2\pi\right) = P\left(\_\_\_\_\right)$

$= \_\_\_\_$

$\frac{5\pi}{3} + 22\pi,\ \frac{5\pi}{3}$

$\left(\frac{1}{2}, -\frac{\sqrt{3}}{2}\right)$

50 $P\left(\frac{71\pi}{3}\right) = P\left(\_\_\_\_\right) = P\left(\_\_\_\_\right)$

$= \_\_\_\_$

$\frac{3\pi}{4},\ \left(-\frac{\sqrt{2}}{2}, \frac{\sqrt{2}}{2}\right)$

51 $P\left(\frac{27\pi}{4}\right) = P\left(\_\_\_\_\right) = \_\_\_\_$

$\frac{5\pi}{6},\ \left(-\frac{\sqrt{3}}{2}, \frac{1}{2}\right)$

52 $P\left(\frac{41\pi}{6}\right) = P\left(\_\_\_\_\right) = \_\_\_\_$

$t + 2\pi,\ t + 2\pi$

53 Since $P(t) = P(t + 2\pi)$, we have
$\sin t = \sin(\_\_\_\_)$ and $\cos t = \cos(\_\_\_\_)$
Thus, the sine and cosine are periodic functions of period 2π.

54 $\tan(t + 2\pi) = \dfrac{\sin(t + 2\pi)}{\_\_\_\_}$

$= \dfrac{\sin t}{\_\_\_\_}$

$= \_\_\_\_$

Hence, the tangent is a ________ function of period ____.

Answers: $\cos(t + 2\pi)$; $\cos t$; $\tan t$; periodic, 2π

55 $\cot(t + 2\pi) = \dfrac{\cos(t + 2\pi)}{\sin(t + 2\pi)} = \dfrac{\_\_\_\_}{\sin t} = \_\_\_\_$

Hence, the cotangent is a periodic function of period ____.

Answers: $\cos t$, $\cot t$; 2π

56 $\sec(t + 2\pi) = \dfrac{1}{\cos(t + 2\pi)} = \dfrac{1}{\_\_\_\_} = \_\_\_\_$

Hence, the secant is a periodic function of period ____.

Answers: $\cos t$, $\sec t$; 2π

57 $\csc(t + 2\pi) = \dfrac{1}{\_\_\_\_}$

$= \dfrac{1}{\_\_\_\_}$

$= \_\_\_\_$

Hence, the cosecant is a ________ function of period ____.

Answers: $\sin(t + 2\pi)$; $\sin t$; $\csc t$; periodic, 2π

In Problems 58–61, use periodicity to find the values of the trigonometric functions at the given value of t.

58 $t = \dfrac{9\pi}{2}$

Since the trigonometric functions are periodic functions of period ____, we have

$\sin\dfrac{9\pi}{2} = \sin\left(\dfrac{\pi}{2} + 4\pi\right) = \sin\dfrac{\pi}{2} = \_\_\_\_$

$\cos\dfrac{9\pi}{2} = \cos\left(\dfrac{\pi}{2} + 4\pi\right) = \_\_\_\_ = \_\_\_\_$

$\tan\dfrac{9\pi}{2} = \tan\left(\dfrac{\pi}{2} + 4\pi\right) = \_\_\_\_ = \dfrac{\sin\frac{\pi}{2}}{\_\_\_\_} = \_\_\_\_$

$\cot\dfrac{9\pi}{2} = \cot\left(\dfrac{\pi}{2} + 4\pi\right) = \_\_\_\_ = \dfrac{\_\_\_\_}{\sin\frac{\pi}{2}} = \_\_\_\_$

$\sec\dfrac{9\pi}{2} = \sec\left(\dfrac{\pi}{2} + 4\pi\right) = \_\_\_\_ = \dfrac{1}{\_\_\_\_} = \_\_\_\_$

$\csc\dfrac{9\pi}{2} = \csc\left(\dfrac{\pi}{2} + 4\pi\right) = \_\_\_\_ = \dfrac{1}{\_\_\_\_} = \_\_\_\_$

Answers: 2π; 1; $\cos\frac{\pi}{2}$, 0; $\tan\frac{\pi}{2}$, $\cos\frac{\pi}{2}$, undefined; $\cot\frac{\pi}{2}$, $\cos\frac{\pi}{2}$, 0; $\sec\frac{\pi}{2}$, $\cos\frac{\pi}{2}$, undefined; $\csc\frac{\pi}{2}$, $\sin\frac{\pi}{2}$, 1

59 $t = \frac{13\pi}{6}$

Since the trigonometric functions are periodic functions of period ____, we have

$\sin\frac{13\pi}{6} = \sin\left(\frac{\pi}{6} + 2\pi\right) = \sin\frac{\pi}{6} =$ ____

$\cos\frac{13\pi}{6} = \cos\left(\frac{\pi}{6} + 2\pi\right) =$ ______ $=$ ____

$\tan\frac{13\pi}{6} = \tan\left(\frac{\pi}{6} + 2\pi\right) =$ ______ $= \frac{\_\_\_\_}{\cos\frac{\pi}{6}} = \frac{\_\_\_\_}{\frac{\sqrt{3}}{2}} =$ ____

$\cot\frac{13\pi}{6} = \cot\left(\frac{\pi}{6} + 2\pi\right) =$ ______ $= \frac{\cos\frac{\pi}{6}}{\_\_\_\_} = \frac{\frac{\sqrt{3}}{2}}{\_\_\_\_} =$ ____

$\sec\frac{13\pi}{6} = \sec\left(\frac{\pi}{6} + 2\pi\right) =$ ______ $= \frac{1}{\_\_\_\_} = \frac{1}{\_\_\_\_} =$ ____

$\csc\frac{13\pi}{6} = \csc\left(\frac{\pi}{6} + 2\pi\right) =$ ______ $= \frac{1}{\_\_\_\_} = \frac{1}{\_\_\_\_} =$ ____

Answers: 2π; $\frac{1}{2}$; $\cos\frac{\pi}{6}, \frac{\sqrt{3}}{2}$; $\tan\frac{\pi}{6}, \sin\frac{\pi}{6}, \frac{1}{2}, \frac{\sqrt{3}}{3}$; $\cot\frac{\pi}{6}, \sin\frac{\pi}{6}, \frac{1}{2}, \sqrt{3}$; $\sec\frac{\pi}{6}, \cos\frac{\pi}{6}, \frac{\sqrt{3}}{2}, \frac{2\sqrt{3}}{3}$; $\csc\frac{\pi}{6}, \sin\frac{\pi}{6}, \frac{1}{2}, 2$

60 $t = \frac{17\pi}{4}$

Since the trigonometric functions are periodic functions of period ____, we have

$\sin\frac{17\pi}{4} = \sin\left(\frac{\pi}{4} + 4\pi\right) = \sin\frac{\pi}{4} =$ ____

$\cos\frac{17\pi}{4} = \cos\left(\frac{\pi}{4} + 4\pi\right) =$ ______ $=$ ____

$\tan\frac{17\pi}{4} = \tan\left(\frac{\pi}{4} + 4\pi\right) =$ ______ $= \frac{\sin\frac{\pi}{4}}{\_\_\_\_} = \frac{\frac{\sqrt{2}}{2}}{\_\_\_\_} =$ ____

$\cot\frac{17\pi}{4} = \cot\left(\frac{\pi}{4} + 4\pi\right) =$ ______ $= \frac{\cos\frac{\pi}{4}}{\_\_\_\_} = \frac{\frac{\sqrt{2}}{2}}{\_\_\_\_} =$ ____

$\sec\frac{17\pi}{4} = \sec\left(\frac{\pi}{4} + 4\pi\right) =$ ______ $= \frac{1}{\_\_\_\_} = \frac{1}{\_\_\_\_} =$ ____

$\csc\frac{17\pi}{4} = \csc\left(\frac{\pi}{4} + 4\pi\right) =$ ______ $= \frac{1}{\_\_\_\_} = \frac{1}{\_\_\_\_} =$ ____

Answers: 2π; $\frac{\sqrt{2}}{2}$; $\cos\frac{\pi}{4}, \frac{\sqrt{2}}{2}$; $\tan\frac{\pi}{4}, \cos\frac{\pi}{4}, \frac{\sqrt{2}}{2}, 1$; $\cot\frac{\pi}{4}, \sin\frac{\pi}{4}, \frac{\sqrt{2}}{2}, 1$; $\sec\frac{\pi}{4}, \cos\frac{\pi}{4}, \frac{\sqrt{2}}{2}, \sqrt{2}$; $\csc\frac{\pi}{4}, \sin\frac{\pi}{4}, \frac{\sqrt{2}}{2}, \sqrt{2}$

61 $t = -\frac{47\pi}{6}$

Since the trigonometric functions are periodic functions of period

2π | ____, we have

$\frac{1}{2}$ | $\sin\left(-\frac{47\pi}{6}\right) = \sin\left(-\frac{47\pi}{6} + 8\pi\right) = \sin\frac{\pi}{6} =$ ____

$\frac{\sqrt{3}}{2}$ | $\cos\left(-\frac{47\pi}{6}\right) = \cos\left(-\frac{47\pi}{6} + 8\pi\right) = \cos\frac{\pi}{6} =$ ____

$\frac{\sqrt{3}}{3}$ | $\tan\left(-\frac{47\pi}{6}\right) = \tan\left(-\frac{47\pi}{6} + 8\pi\right) = \tan\frac{\pi}{6} =$ ____

$\sqrt{3}$ | $\cot\left(-\frac{47\pi}{6}\right) = \cot\left(-\frac{47\pi}{6} + 8\pi\right) = \cot\frac{\pi}{6} =$ ____

$\frac{2\sqrt{3}}{3}$ | $\sec\left(-\frac{47\pi}{6}\right) = \sec\left(-\frac{47\pi}{6} + 8\pi\right) = \sec\frac{\pi}{6} =$ ____

2 | $\csc\left(-\frac{47\pi}{6}\right) = \csc\left(-\frac{47\pi}{6} + 8\pi\right) = \csc\frac{\pi}{6} =$ ____

In Problems 62–65, use the fact that the cosine and sine are periodic functions of period 2π to find the period of each function.

62 $f(x) = \cos\frac{x}{2}$

4π, $x + 4\pi$ | $f(x) = \cos\frac{x}{2} = \cos\left(\frac{x}{2} + 2\pi\right) = \cos\left[\frac{1}{2}(x + \_\_\_\_)\right] = f(\_\_\_\_\_\_\_\_)$,

4π | so f is periodic of period ____.

63 $f(x) = \cos 3x$

2π, $\frac{2\pi}{3}$, $x + \frac{2\pi}{3}$ | $f(x) = \cos 3x = \cos(3x + \_\_\_\_) = \cos\left[3\left(x + \_\_\_\_\right)\right] = f\left(\_\_\_\_\_\_\right)$,

$\frac{2\pi}{3}$ | so f is periodic of period ____.

64 $g(x) = \sin 7x$

$\frac{2\pi}{7}$, $x + \frac{2\pi}{7}$ | $g(x) = \sin 7x = \sin(7x + 2\pi) = \sin\left[7\left(x + \_\_\_\_\right)\right] = g\left(\_\_\_\_\_\_\right)$,

$\frac{2\pi}{7}$ | so g is periodic of period ____.

65 $g(x) = \sin\frac{3x}{4}$

$\frac{8\pi}{3}$, $x + \frac{8\pi}{3}$ | $g(x) = \sin\left(\frac{3x}{4} + 2\pi\right) = \sin\left[\frac{3}{4}\left(x + \_\_\_\_\right)\right] = g\left(\_\_\_\_\_\_\right)$,

$\frac{8\pi}{3}$ | so g is periodic of period ____.

2.2 Signs of the Values of the Trigonometric Functions

66 The signs of the values of the trigonometric functions are as follows:

	P(t) is in quadrant			
	I	II	III	IV
sin t	+	___	___	___
cos t	+	___	___	___
tan t	+	___	___	___
cot t	+	___	___	___
sec t	+	___	___	___
csc t	+	___	___	___

Answers (margin):
+, −, −
−, −, +
−, +, −
−, +, −
−, −, +
+, −, −

III — **67** If $\tan t > 0$ and $\cos t < 0$, then $P(t)$ is in quadrant ___.

IV — **68** If $\sec t > 0$ and $\sin t < 0$, then $P(t)$ is in quadrant ___.

I — **69** If $\sin t > 0$ and $\cot t > 0$, then $P(t)$ is in quadrant ___.

II — **70** If $\csc t > 0$ and $\cos t < 0$, then $P(t)$ is in quadrant ___.

2.3 Domains and Ranges of the Trigonometric Functions

R, R — **71** The domain of the cosine function is ___ and the domain of the sine function is ___.

$-1,\ 1$; $-1,\ 1$; $[-1, 1]$; $[-1, 1]$ — **72** Since the abscissa x on the unit circle satisfies ___ $\leqslant x \leqslant$ ___, and since $\cos t = x$, it follows that ___ $\leqslant \cos t \leqslant$ ___; that is, the range of the cosine function is ___. Similarly, the range of the sine function is ___.

$\cos t,\ \dfrac{\pi}{2} + n\pi$ for $n \in I$ — **73** The domain of both the tangent and the secant functions is $\{t \mid$ ___ $\neq 0\}$; that is, $\left\{t \mid t \neq \text{\_\_\_}\right\}$.

$\sin t,\ n\pi$ for $n \in I$ — **74** The domain of both the cotangent and the cosecant functions is $\{t \mid$ ___ $\neq 0\}$; that is, $\{t \mid t \neq$ ___ $\}$.

2.4 Relationships Between the Trigonometric Functions

$\sin^2 t$ — **75** $\cos^2 t +$ ___ $= 1$

$\tan^2 t$ — **76** ___ $+ 1 = \sec^2 t$

$\cot^2 t$ — **77** $1 +$ ___ $= \csc^2 t$

In Problems 78–80, use the given information to evaluate the other five trigonometric functions.

78 $\sin t = \frac{5}{13},\quad 0 < t < \frac{\pi}{2}$

$\frac{12}{13}$ — $\cos t = \sqrt{1 - \sin^2 t} = \sqrt{1 - (\frac{5}{13})^2} =$ ____

$\frac{5}{12}$ — $\tan t = \frac{\sin t}{\cos t} = \frac{\frac{5}{13}}{\frac{12}{13}} =$ ____

$\frac{12}{5}$ — $\cot t = \frac{\cos t}{\sin t} = \frac{\frac{12}{13}}{\frac{5}{13}} =$ ____

$\frac{13}{12}$ — $\sec t = \frac{1}{\cos t} = \frac{1}{\frac{12}{13}} =$ ____

$\frac{13}{5}$ — $\csc t = \frac{1}{\sin t} = \frac{1}{\frac{5}{13}} =$ ____

79 $\cos t = -\frac{4}{5},\quad \frac{\pi}{2} < t < \pi$

$-\frac{4}{5}, \frac{3}{5}$ — $\sin t = \sqrt{1 - \cos^2 t} = \sqrt{1 - (\_\_\_\_)^2} =$ ____

$-\frac{3}{4}$ — $\tan t = \frac{\sin t}{\cos t} = \frac{\frac{3}{5}}{-\frac{4}{5}} =$ ____

$-\frac{4}{3}$ — $\cot t = \frac{\cos t}{\sin t} = \frac{-\frac{4}{5}}{\frac{3}{5}} =$ ____

$-\frac{4}{5}, -\frac{5}{4}$ — $\sec t = \frac{1}{\cos t} = \frac{1}{\_\_\_\_} =$ ____

$\frac{3}{5}, \frac{5}{3}$ — $\csc t = \frac{1}{\sin t} = \frac{1}{\_\_\_\_} =$ ____

80 $\cot t = \frac{15}{8},\quad \pi < t < \frac{3\pi}{2}$

III, $<$ — $P(t)$ is in quadrant ____, where $\csc t$ ____ 0, so that

$\frac{15}{8}, -\frac{17}{8}$ — $\csc t = -\sqrt{1 + \cot^2 t} = -\sqrt{1 + (\_\_\_\_)^2} =$ ____

$-\frac{17}{8}, -\frac{8}{17}$ — $\sin t = \frac{1}{\csc t} = \frac{1}{\_\_\_\_} =$ ____

$\cot t$, $\frac{15}{8}, \frac{8}{15}$ — $\tan t = \frac{1}{\_\_\_\_} = \frac{1}{\_\_\_\_} =$ ____

$-\frac{8}{17}, -\frac{15}{17}$ — $\cos t = -\sqrt{1 - \sin^2 t} = -\sqrt{1 - (\_\_\_\_)^2} =$ ____

$-\frac{15}{17}, -\frac{17}{15}$ — $\sec t = \frac{1}{\cos t} = \frac{1}{\_\_\_\_} =$ ____

2.5 Even and Odd Trigonometric Functions

t, even — **81** $\cos(-t) = \cos$ ____. Hence, the cosine is an ____ function.

$-\sin t$, odd — **82** $\sin(-t) =$ ____. Hence, the sine is an ____ function.

$\cos(-t)$, $\cos t$, $-\tan t$

83 $\tan(-t) = \dfrac{\sin(-t)}{\_\_\_\_} = \dfrac{-\sin t}{\_\_\_\_} = \_\_\_\_.$

odd

Hence, the tangent is an ____ function.

$\sin(-t)$, $-\sin t$

84 $\cot(-t) = \dfrac{\cos(-t)}{\_\_\_\_} = \dfrac{\cos t}{\_\_\_\_} = -\cot t.$

odd

Hence, the cotangent is an ____ function.

$\cos(-t)$, $\cos t$, $\sec t$

85 $\sec(-t) = \dfrac{1}{\_\_\_\_} = \dfrac{1}{\_\_\_\_} = \_\_\_\_.$

even

Hence, the secant is an ____ function.

$\sin(-t)$, $\sin t$, $\csc t$

86 $\csc(-t) = \dfrac{1}{\_\_\_\_} = \dfrac{1}{-\_\_\_\_} = -\_\_\_\_.$

odd

Hence, the cosecant is an ____ function.

$-$, $-\dfrac{\sqrt{2}}{2}$

87 $\sin\left(-\dfrac{\pi}{4}\right) = \_\_\sin\dfrac{\pi}{4} = \_\_\_\_$

$\dfrac{\pi}{6}$, $\dfrac{\sqrt{3}}{2}$

88 $\cos\left(-\dfrac{\pi}{6}\right) = \cos \_\_\_\_ = \_\_\_\_$

$-$, $-\sqrt{3}$

89 $\tan\left(-\dfrac{\pi}{3}\right) = \_\_\tan\dfrac{\pi}{3} = \_\_\_\_$

3 Graphs of the Trigonometric Functions

3.1 Graphs of Sine and Cosine Functions

2π

2π

90 Since the sine and cosine functions are periodic of period ____, their graphs repeat over each interval of length ____.

$-1 \leqslant y \leqslant 1$

$y = -1$, $y = 1$

91 Since the range of the sine and cosine functions is the set $\{y | \_\_\_\_\_\_\}$, we know that the graphs are between the horizontal lines ______ and ______.

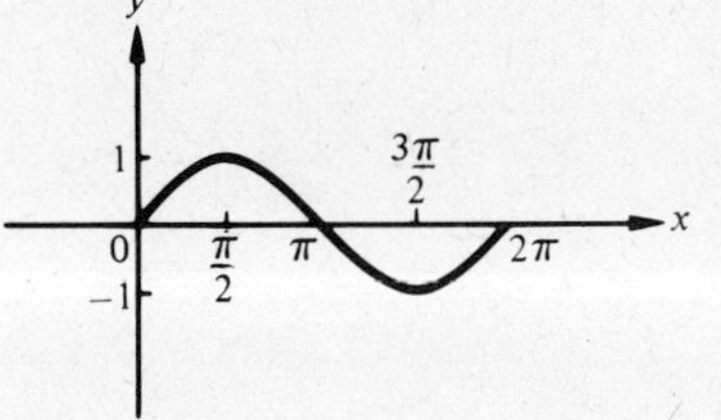

92 The graph of one cycle of $y = \sin x$ is

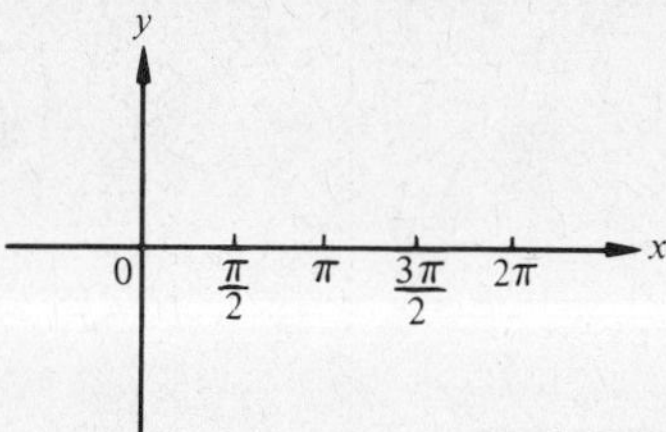

$\left[0, \dfrac{\pi}{2}\right]$ and $\left[\dfrac{3\pi}{2}, 2\pi\right]$

$\left[\dfrac{\pi}{2}, \dfrac{3\pi}{2}\right]$

From the graph of $y = \sin x$, we notice that the function is increasing in the intervals ______________ and decreasing in the interval ______.

Answer	Frame
odd origin	**93** Since $f(x) = \sin x$ is an ____ function, the graph of f is symmetric with respect to the ________.
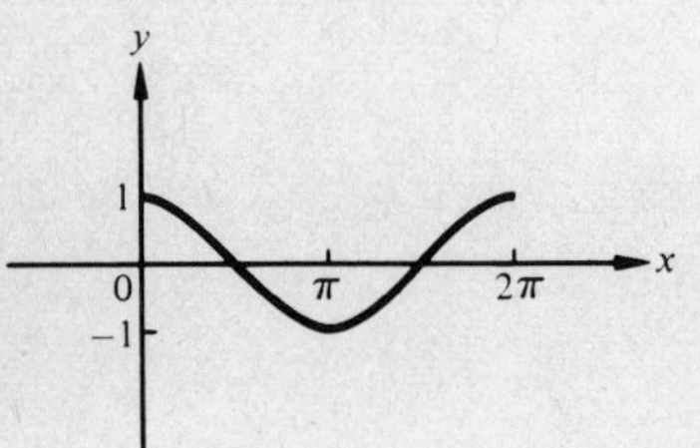	**94** The graph of one cycle of $g(x) = \cos x$ is 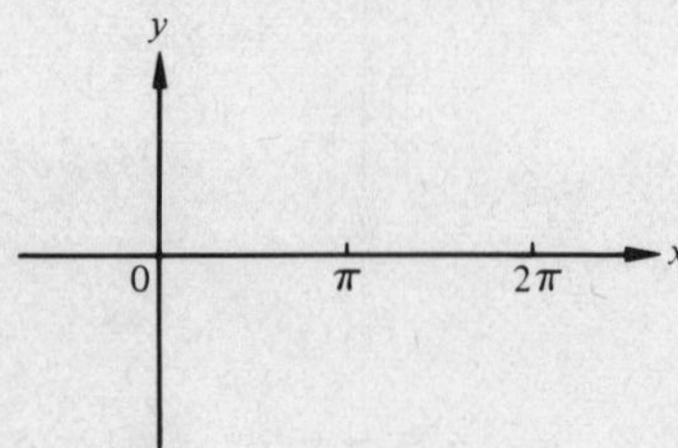
$[\pi, 2\pi]$, $[0, \pi]$	From the graph of g we notice that the function is increasing in the interval ________ and is decreasing in the interval ________.
even y axis	**95** Since $g(x) = \cos x$ is an ____ function, the graph of g is symmetric with respect to the ________.
amplitude	**96** The absolute value of half the difference of the maximum and the minimum values of the ordinates of a function is called the ________ of the function.
5	**97** The amplitude of $f(x) = 5 \sin x$ is ____.
2	**98** The amplitude of $f(x) = -2 \sin x$ is ____.
$\frac{3}{2}$	**99** The amplitude of $g(x) = \frac{3}{2} \cos x$ is ____.
7	**100** The amplitude of $g(x) = -7 \cos x$ is ____.
$-b$ phase shift	**101** The graph of $f(x) = \sin(x + b)$ can be obtained from the graph of $y = \sin x$ by shifting the graph of $y = \sin x$ ____ units; $-b$ is called the ________ of the function f.
$-b$ phase shift	**102** The graph of $g(x) = \cos(x + b)$ can be obtained from the graph of $y = \cos x$ by shifting the graph of $y = \cos x$ ____ units; $-b$ is called the ________ of the function g.

In Problems 103–112, graph one cycle of the indicated function and identify the following: (a) the amplitude (b) the period (c) the phase shift

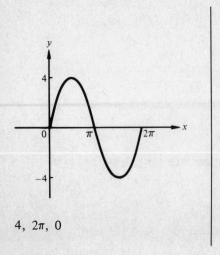

Answer	Frame
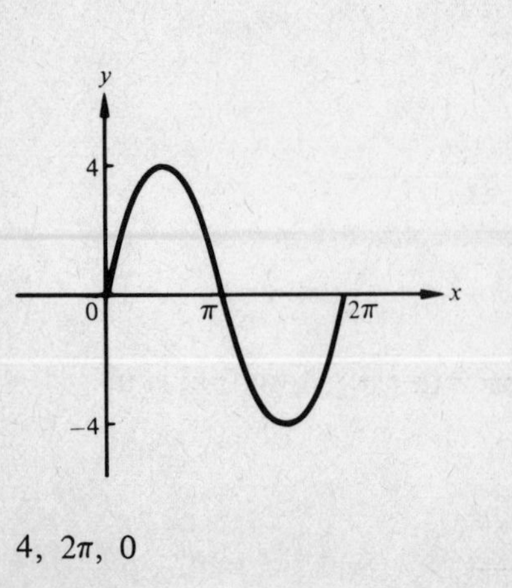 4, 2π, 0	**103** $y = 4 \sin x$ 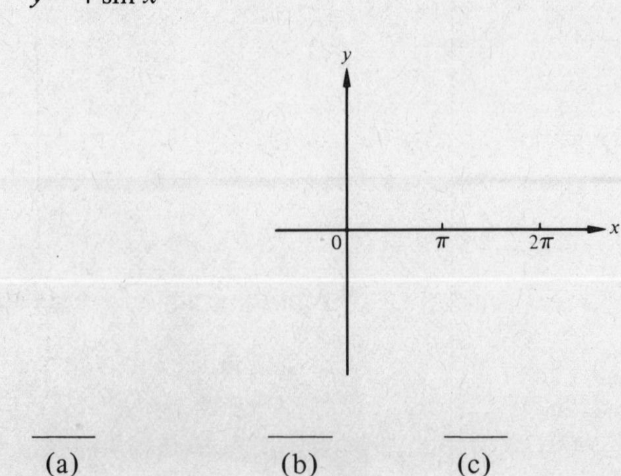 ____ (a) ____ (b) ____ (c)

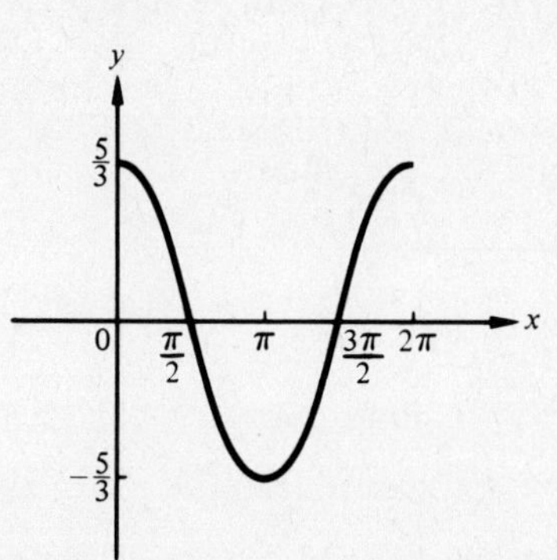

$\frac{5}{3}$, 2π, 0

104 $y = \frac{5}{3} \cos x$

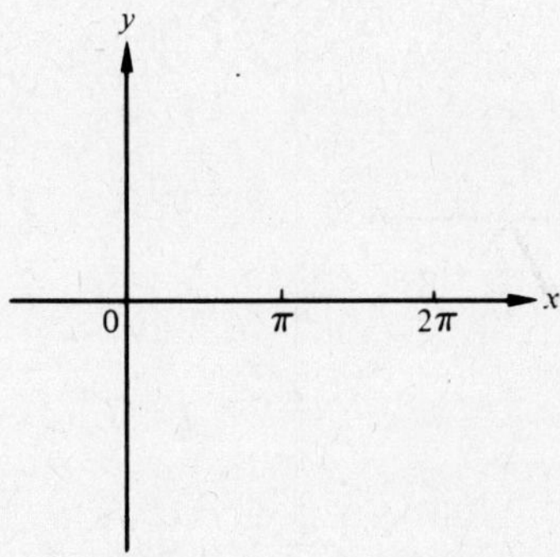

____ (a) ____ (b) ____ (c)

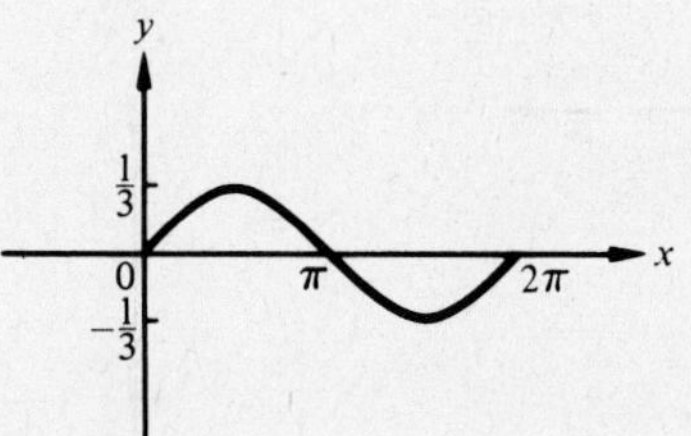

$\frac{1}{3}$, 2π, 0

105 $y = \frac{1}{3} \sin x$

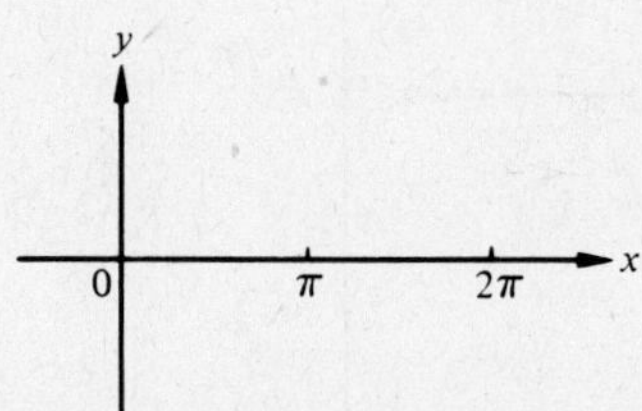

____ (a) ____ (b) ____ (c)

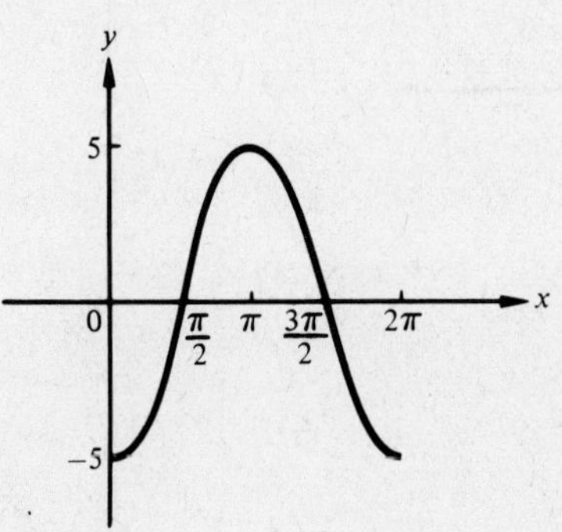

5, 2π, 0

106 $y = -5 \cos x$

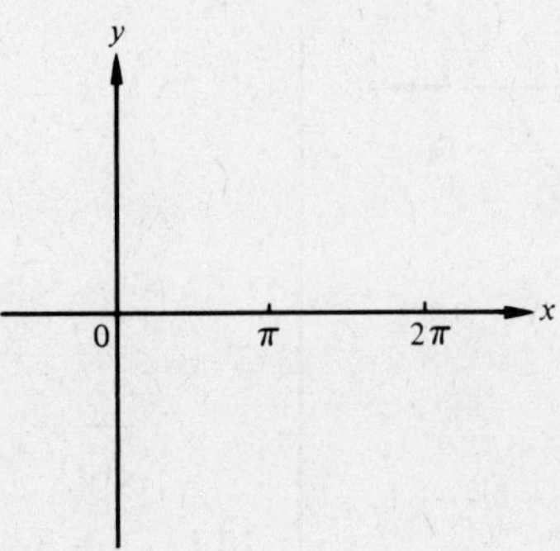

____ (a) ____ (b) ____ (c)

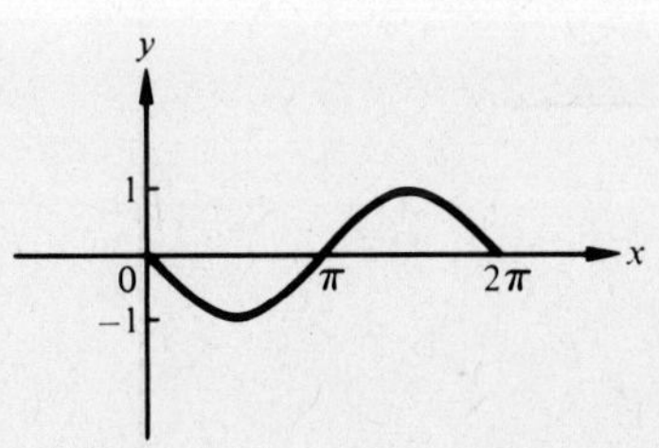

1, 2π, 0

107 $y = -\sin x$

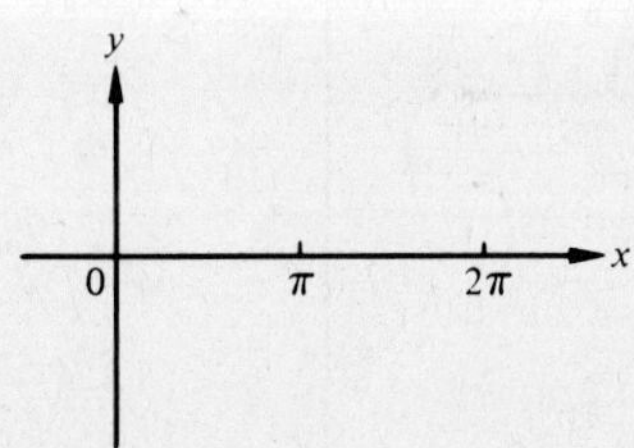

____ (a) ____ (b) ____ (c)

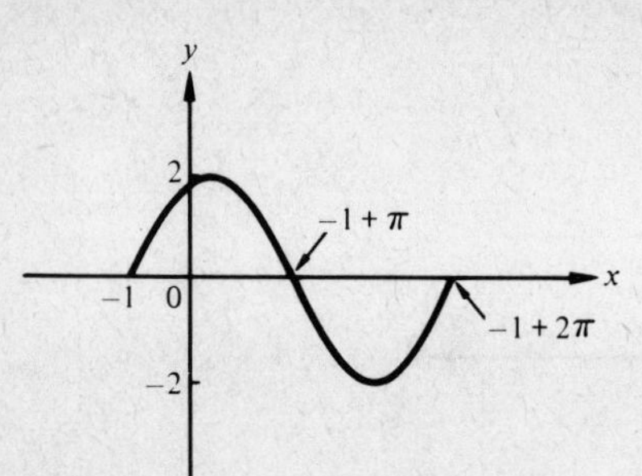

2, 2π, -1

108 $y = 2\sin(x + 1)$

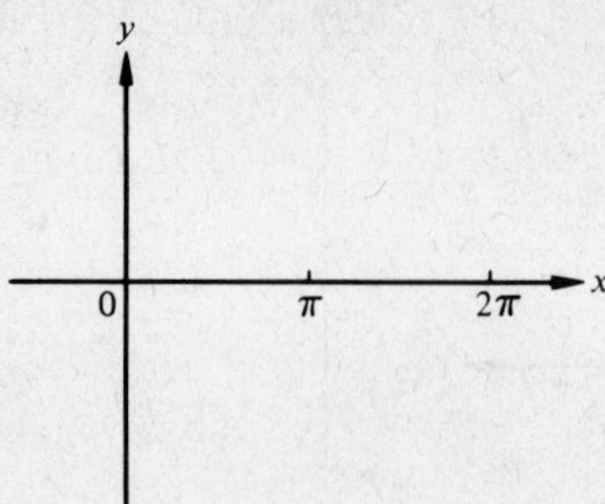

(a) (b) (c)

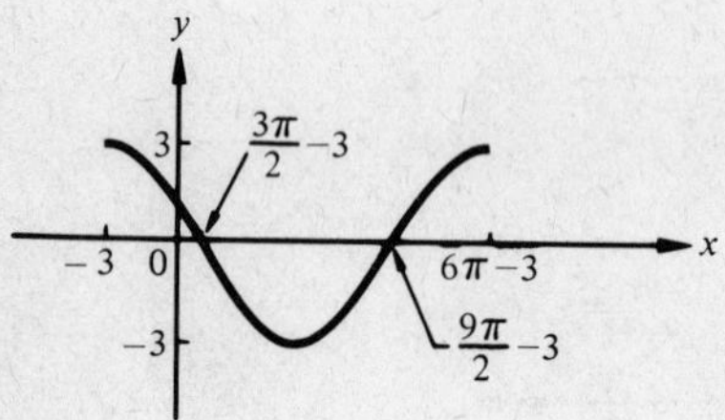

3, 6π, -3

109 $y = 3\cos(\frac{1}{3}x + 1)$

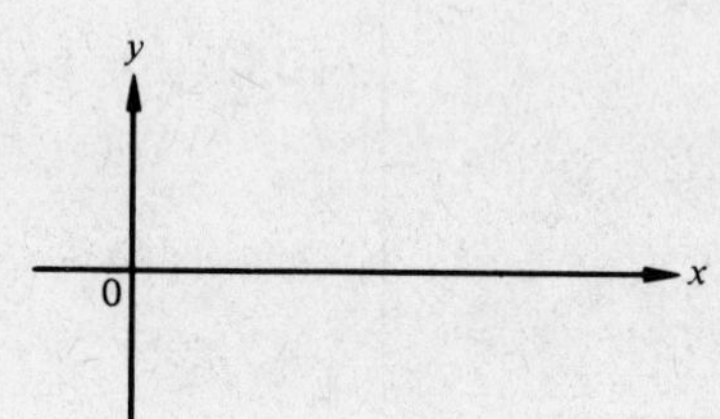

(a) (b) (c)

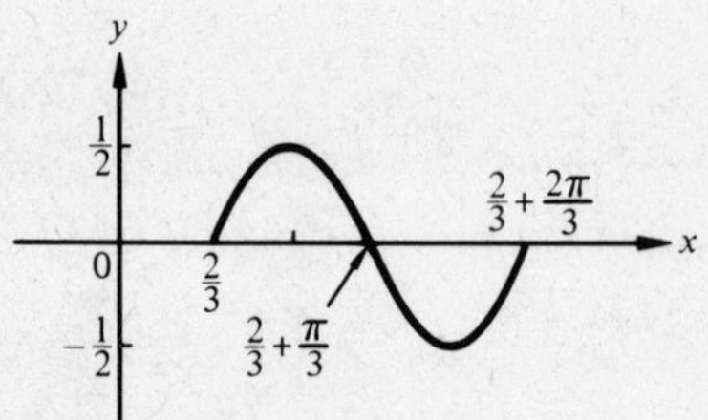

$\frac{1}{2}$, $\frac{2\pi}{3}$, $\frac{2}{3}$

110 $y = \frac{1}{2}\sin(3x - 2)$

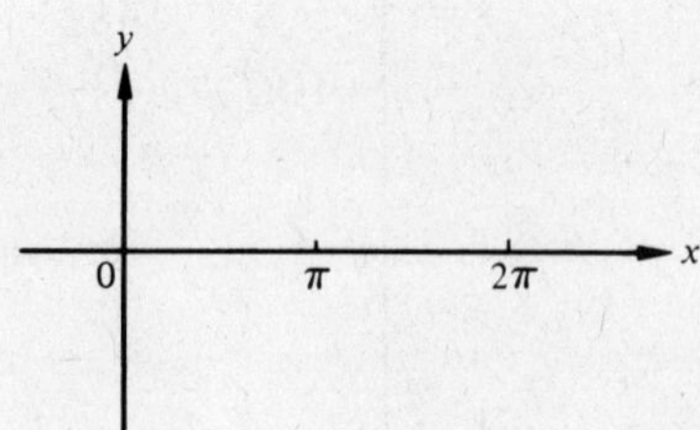

(a) (b) (c)

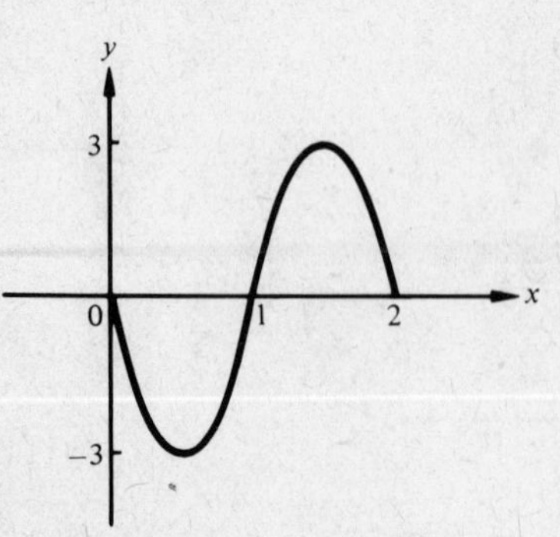

3, 2, 0

111 $y = 3\sin(-\pi x)$
$= -3\sin \pi x$

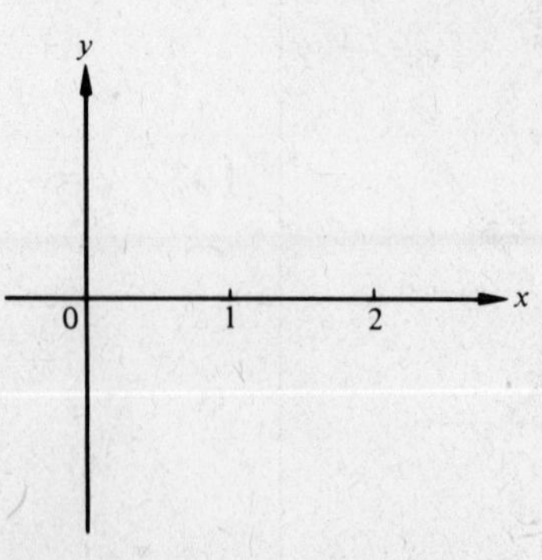

(a) (b) (c)

112 $y = 2\cos(-5x + 1)$
$= 2\cos(5x - 1)$

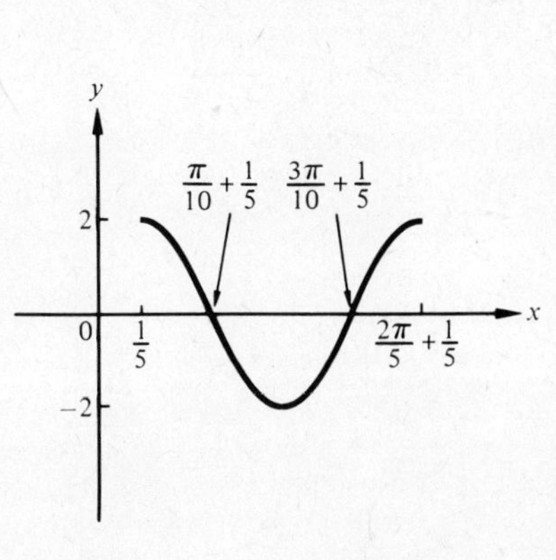

$2, \frac{2\pi}{5}, \frac{1}{5}$

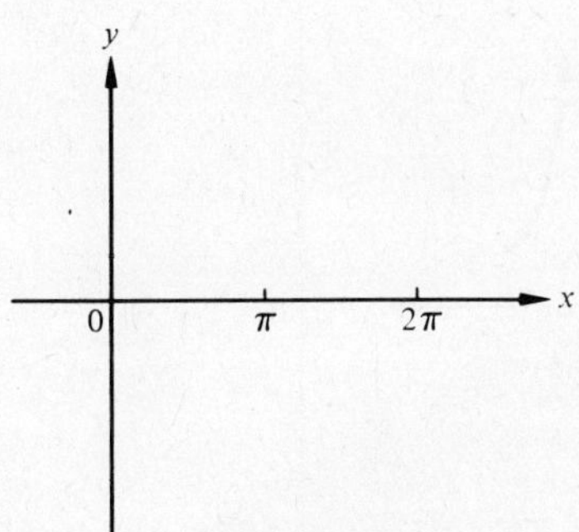

____ (a) ____ (b) ____ (c)

3.2 Graphs of Other Trigonometric Functions

The Tangent

113 The graph of $f(x) = \tan x$ in the interval $[0, 2\pi]$ is

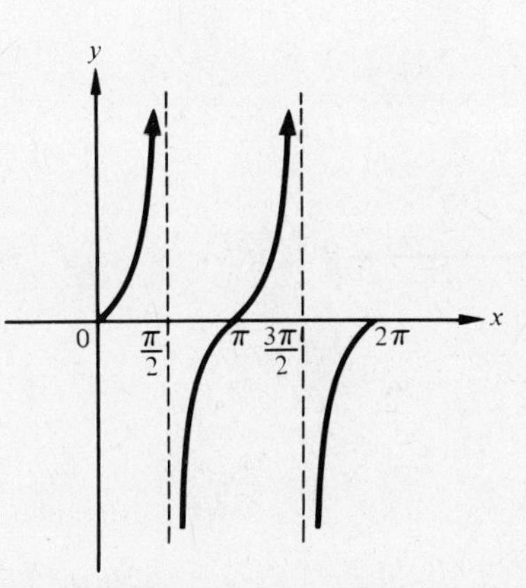

R

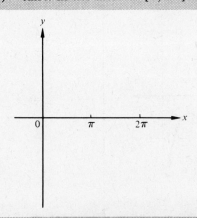

The graph of $f(x) = \tan x$ reveals that the range is the set ____.

The Cotangent, Secant, and Cosecant

114 The graph of $g(x) = \cot x$ in the interval $[0, 2\pi]$ is

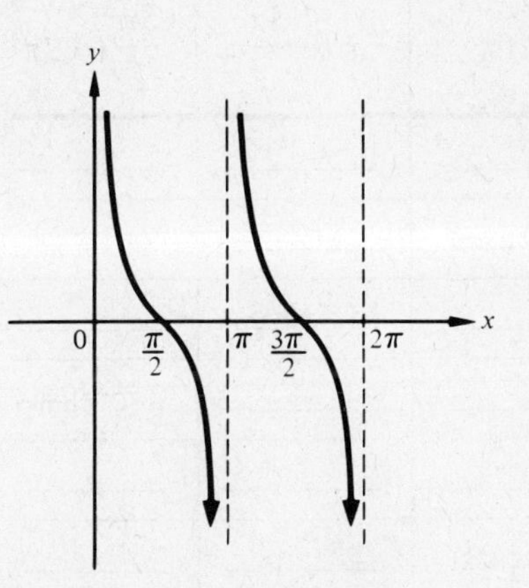

R

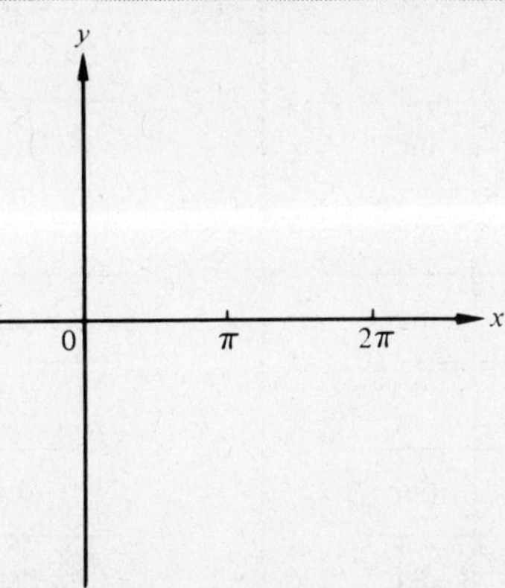

The graph of $g(x) = \cot x$ reveals that the range is the set ____.

115 The graph of $h(x) = \sec x$ in the interval $[0, 2\pi]$ is

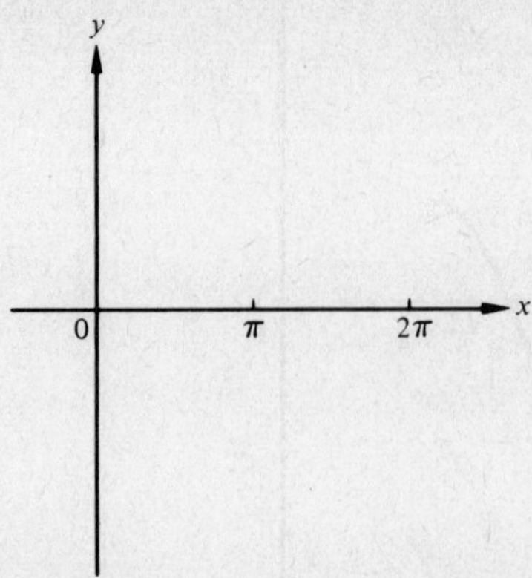

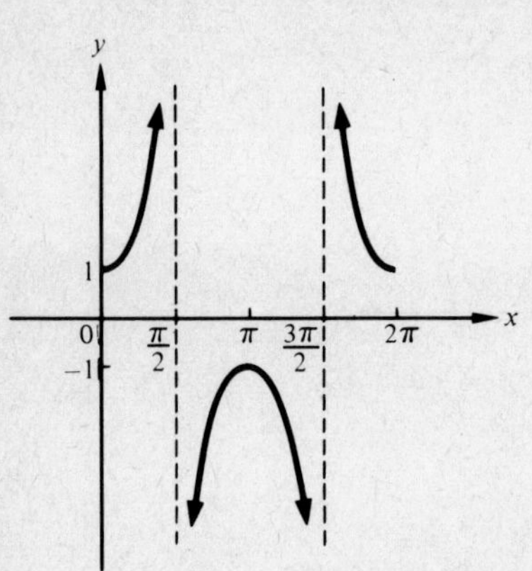

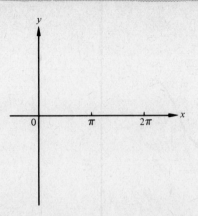

From the graph of $h(x) = \sec x$, we notice that the range is ________________________.

$\{y \mid y \leqslant -1 \text{ or } y \geqslant 1\}$

116 The graph of $k(x) = \csc x$ in the interval $[0, 2\pi]$ is

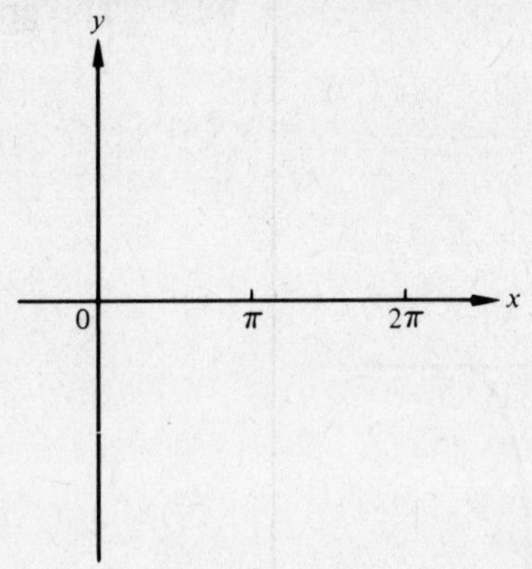

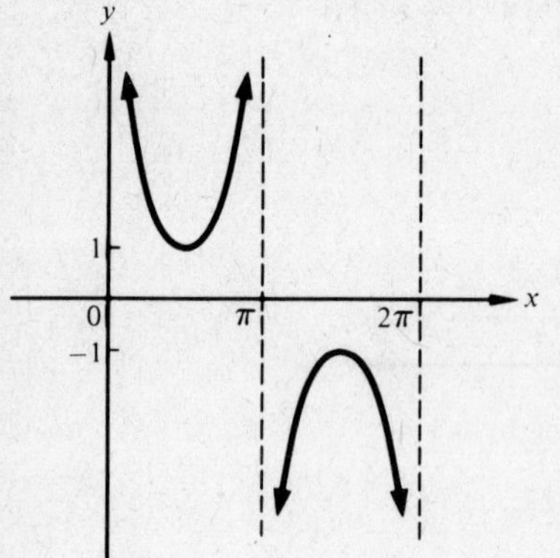

From the graph of $k(x) = \csc x$, we notice that the range is ________________________.

$\{y \mid y \leqslant -1 \text{ or } y \geqslant 1\}$

117

As x increases from:	0 to $\frac{\pi}{2}$	$\frac{\pi}{2}$ to π	π to $\frac{3\pi}{2}$	$\frac{3\pi}{2}$ to 2π
$\sin x$	increases	______	______	______
$\cos x$	______	decreases	______	______
$\tan x$	______	______	increases	______
$\cot x$	______	______	______	decreases
$\sec x$	increases	______	______	______
$\csc x$	______	increases	______	______

	dec	dec	inc
dec		inc	inc
inc	inc		inc
dec	dec	dec	
	inc	dec	dec
dec		inc	dec

In Problems 118–127, graph each function for one period and indicate the extent of the period.

118 $y = \tan 2x$

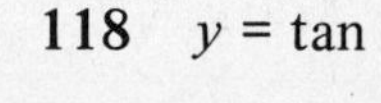

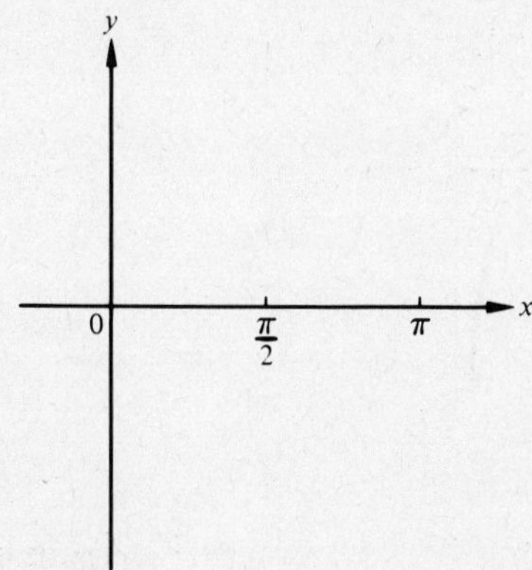

One period of $y = \tan 2x$ is ____.

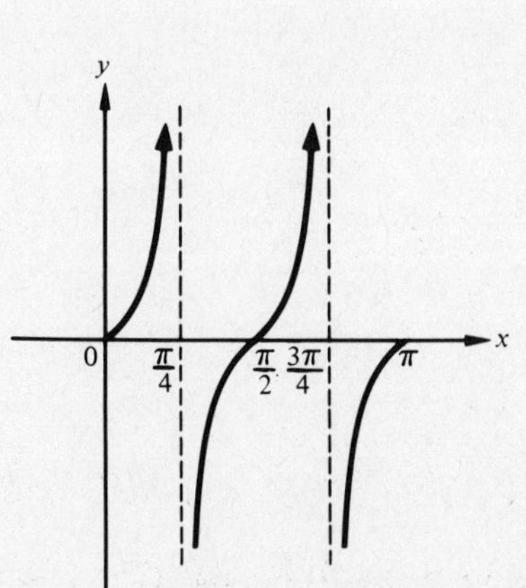

π

119 $y = \tan \dfrac{x}{2}$

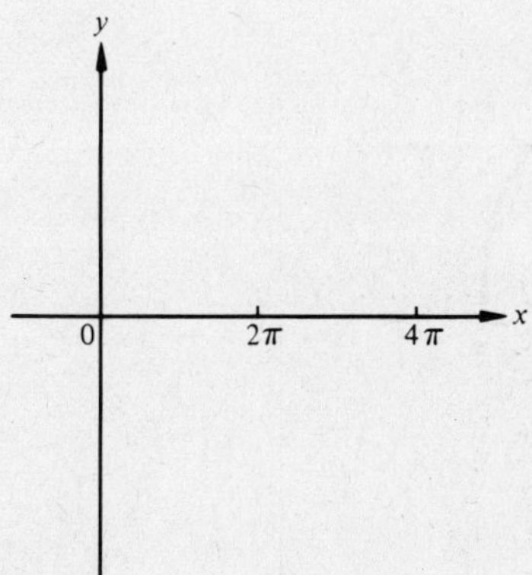

One period of $y = \tan \dfrac{x}{2}$ is ____.

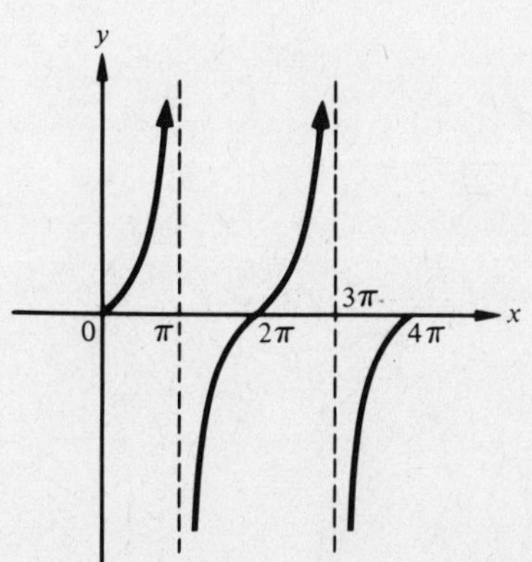

4π

120 $y = \tan(x + 1)$

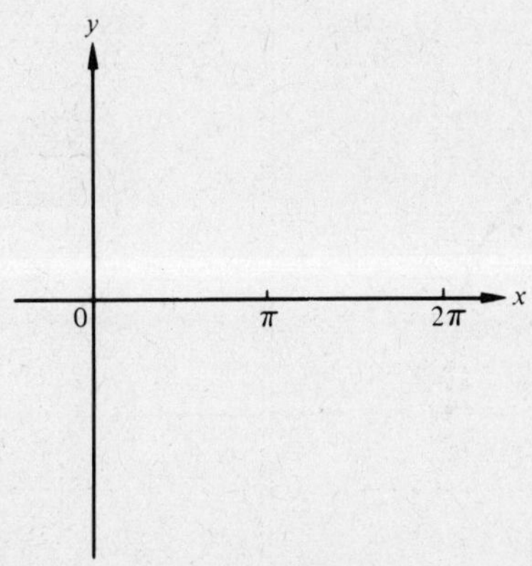

One period of $y = \tan(x + 1)$ is ____.

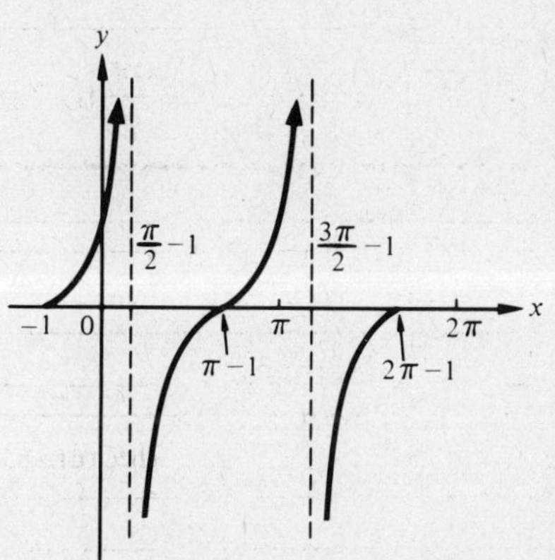

2π

121 $y = 3 \cot x$

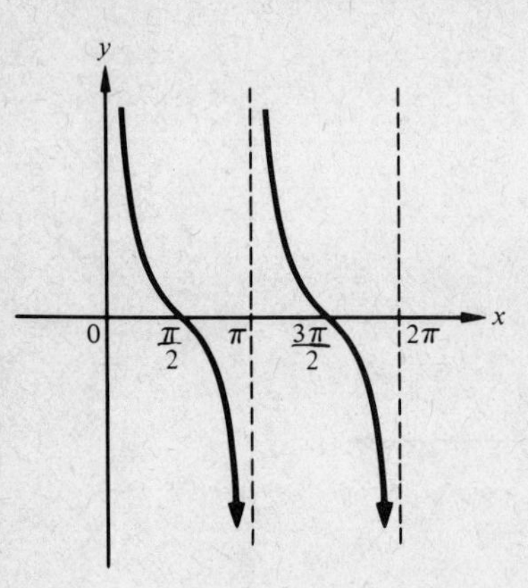

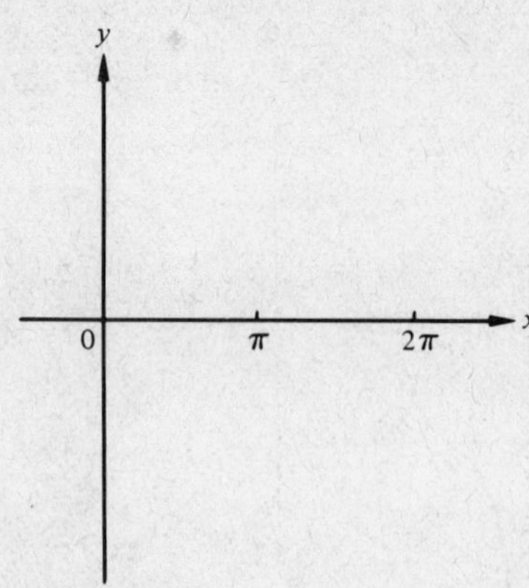

2π

One period of $y = 3 \cot x$ is ____.

122 $y = \cot 5x$

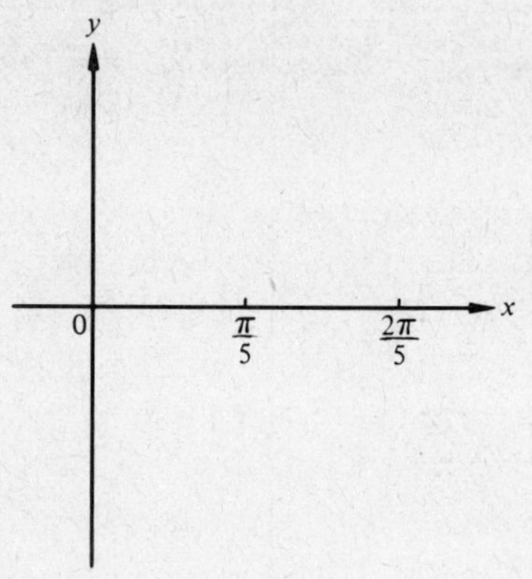

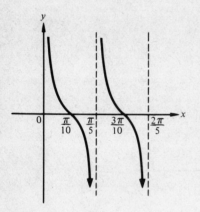

$\frac{2\pi}{5}$

One period of $y = \cot 5x$ is ____.

123 $y = \cot \dfrac{x}{3}$

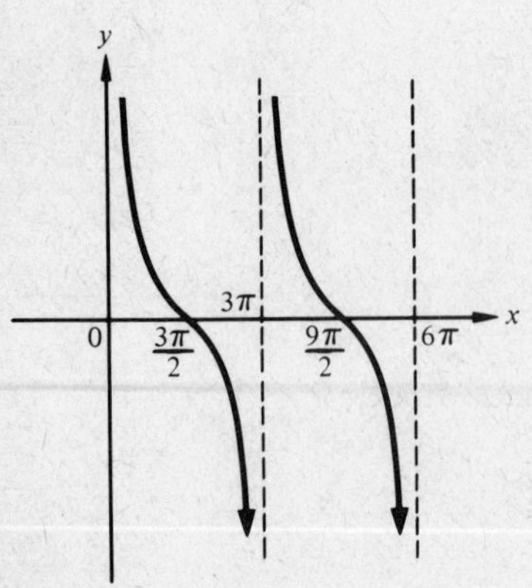

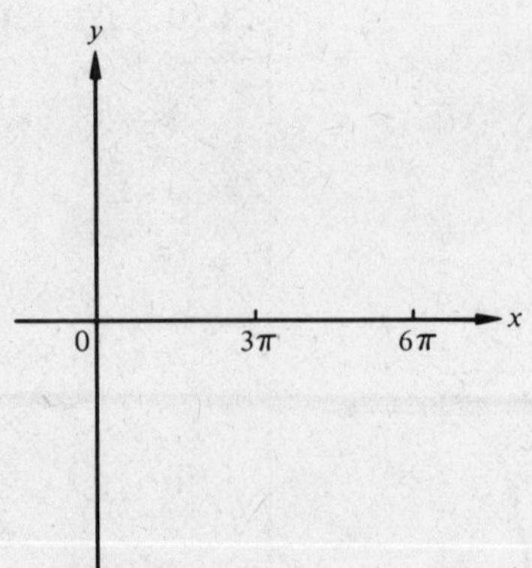

6π

One period of $y = \cot \dfrac{x}{3}$ is ____.

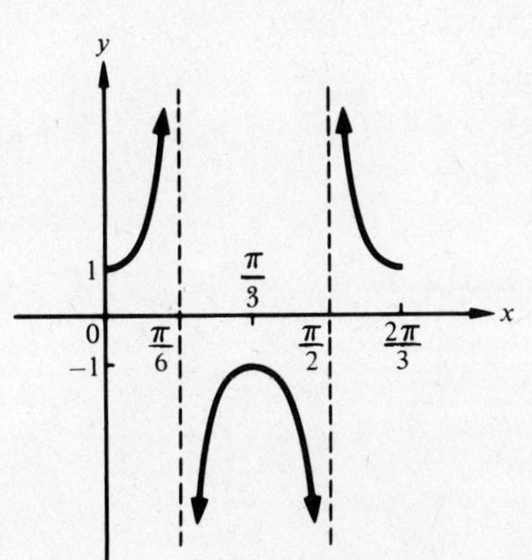

124 $y = \sec 3x$

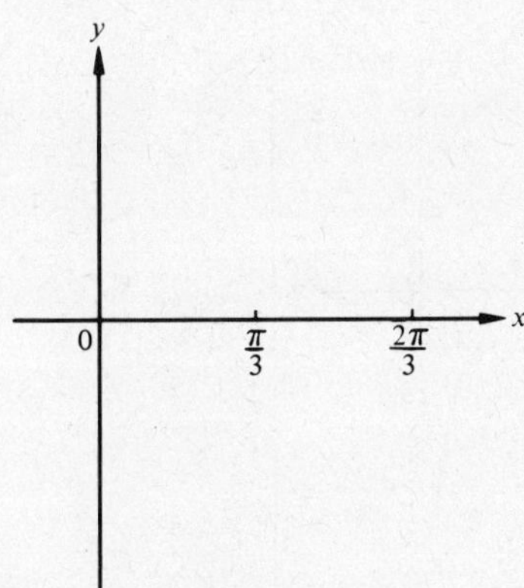

$\frac{2\pi}{3}$

One period of $y = \sec 3x$ is ____.

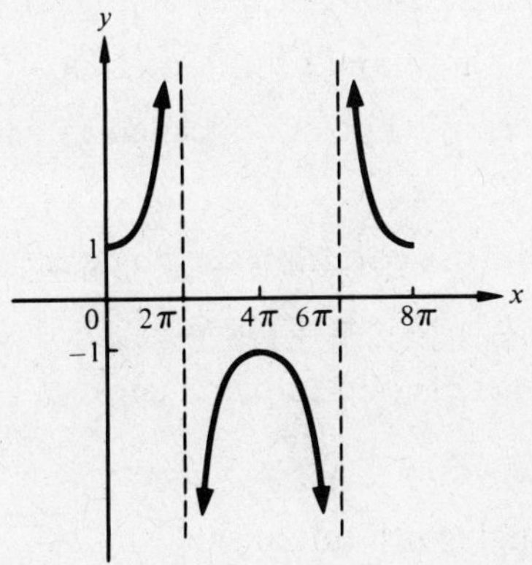

125 $y = \sec \frac{x}{4}$

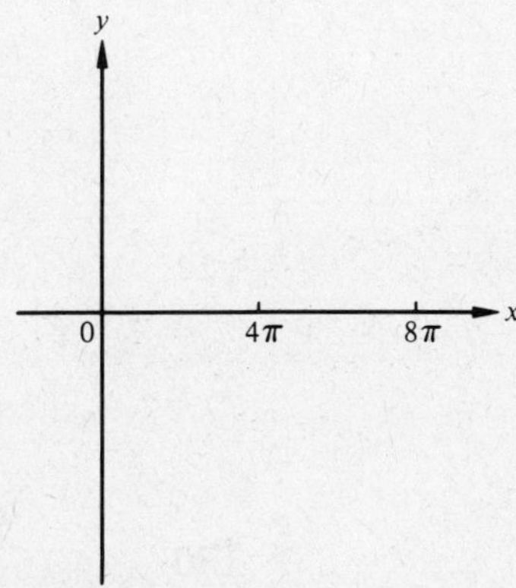

8π

One period of $y = \sec \frac{x}{4}$ is ____.

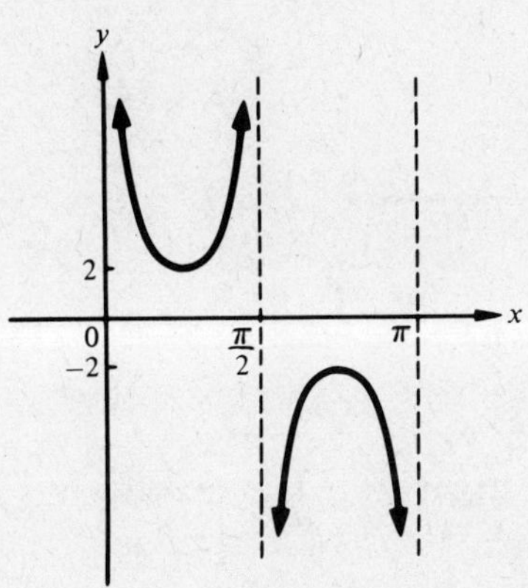

126 $y = 2 \csc 2x$

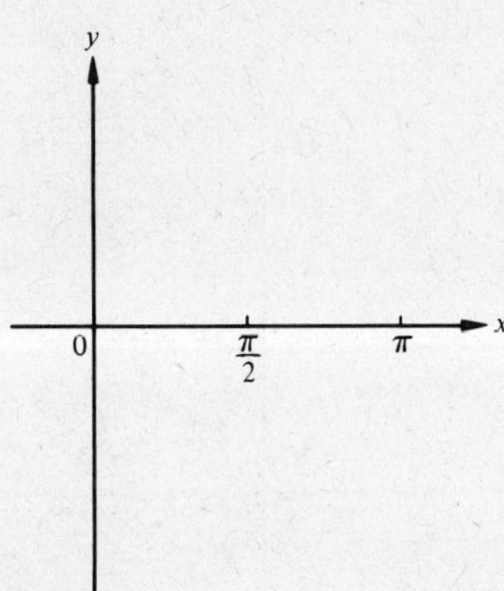

π

One period of $y = 2 \csc 2x$ is ____.

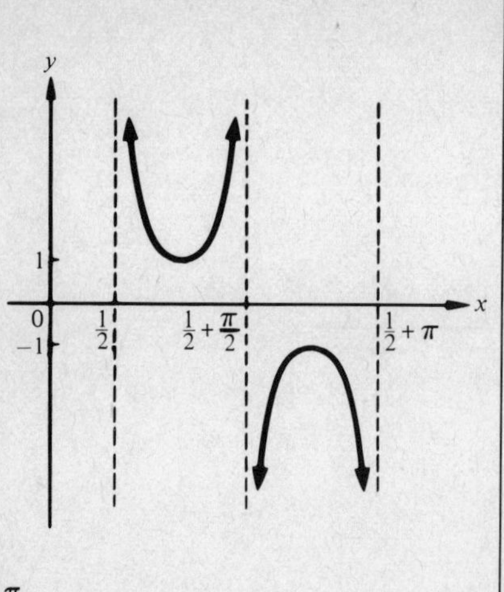

π

127 $y = \csc(2x - 1)$

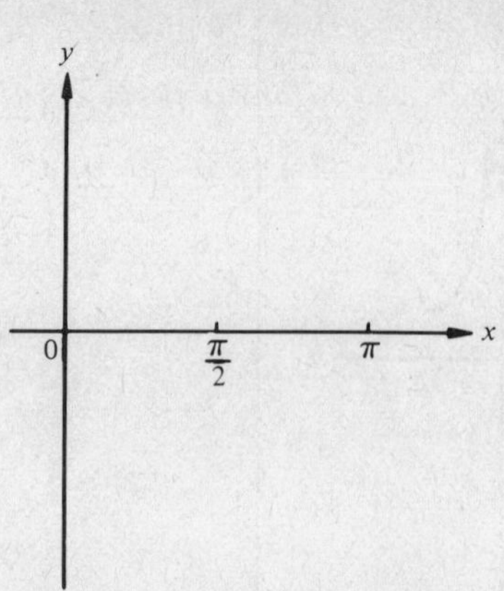

One period of $y = \csc(2x - 1)$ is ____.

4 Trigonometric Functions of Angles

4.1 Angle Measures

ray
vertex
initial
terminal

128 An angle is determined by rotating a ____ about its end point, called the ________ of the angle, from an initial position, called the ________ side of the angle, to a terminal position, called the ________ side of the angle.

positive
negative

129 If an angle is formed by a counterclockwise rotation, the angle is said to be an angle of ________ sense, whereas if the angle is formed by a clockwise rotation, the angle is said to be an angle of ________ sense.

negative, QP
QR, positive
QP, QR, positive
QR, QP

130 Angle α in Figure 1 is a ________ angle with initial side ____ and terminal side ____; β is a ________ angle with initial side ____ and terminal side ____; and γ is a ________ angle with initial side ____ and terminal side ____.

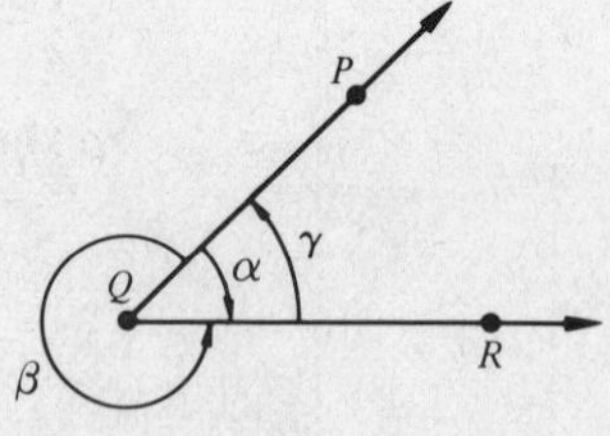

Figure 1

radians
$\frac{1}{360}$
minutes
seconds
60, 3600

131 Angles are measured by using either degrees or ________.

132 One degree (1°) is the measure of a positive angle that is formed by ____ of one complete revolution.

133 A degree can be divided into 60 equal parts called ________ (′). A minute can be divided into 60 equal parts called ________ (″). Thus, an angle of measure 1° is ____′ or ____″.

In Problems 134–135, express the given degree measure in terms of degrees, minutes, and seconds.

Answers	Problem
	134 36.42°
0.42°	36.42° = 36° + ____
60′	= 36° + 0.42 (____)
25.2′	= 36° + ____
60″	= 36° + 25′ + 0.2 (____)
12″	= 36° + 25′ + ____
36°25′12″	= ____
	135 73.37°
0.37°	73.37° = 73° + ____
60′	= 73° + 0.37 (____)
22.2′	= 73° + ____
60″	= 73° + 22′ + 0.2 (____)
12″	= 73° + 22′ + ____
73°22′12″	= ____
1, 1	**136** One radian (1) is the measure of a positive angle that intercepts an arc of length ____ on a circle of radius ____.

In Problems 137–138, indicate both the degree measure and radian measure of an angle formed by the specified motion.

Answers	Problem
	137 One-half of a counterclockwise rotation
360, 2π	Since an angle formed by one complete counterclockwise rotation is either ____° or ____ radians, an angle formed by one-half of a
360, 180	counterclockwise rotation is either $\frac{1}{2}$ (____°) = ____° or
2π, π	$\frac{1}{2}$ (____) = ____ radians.
	138 Two-thirds of a clockwise rotation
-360, -240	An angle formed by a clockwise rotation of two-thirds has a degree measure of $\frac{2}{3}$ (____°) = ____°. The radian measure is
-2π, $-\frac{4\pi}{3}$	$\frac{2}{3}$(____) = ____.

4.2 Conversion of Angle Measures

Answers	Problem
	139 If D represents the degree measure of an angle and R represents the radian measure of the same angle, then the following ratio holds:
$\frac{\pi}{180}$, D, R	$\frac{R}{D} =$ ____, so that $R = \left(\frac{\pi}{180}\right)$ (____) or $D = \left(\frac{180}{\pi}\right)$ (____).

In Problems 140–145, convert the indicated degree measure to radian measure R.

$\frac{\pi}{180}$, $\frac{\pi}{6}$

140 30°

Using Problem 139, we have

$$R = (30)\left(\_\_\_\right) = \_\_\_$$

$\frac{\pi}{180}$, $\frac{\pi}{3}$

141 60°

$$R = (60)\left(\_\_\_\right) = \_\_\_$$

$\frac{\pi}{180}$, $\frac{\pi}{12}$

142 15°

$$R = (15)\left(\_\_\_\right) = \_\_\_$$

$\frac{\pi}{180}$, $\frac{5\pi}{12}$

143 75°

$$R = (75)\left(\_\_\_\right) = \_\_\_$$

$\frac{\pi}{180}$, $-\frac{9\pi}{4}$

144 -405°

$$R = (-405)\left(\_\_\_\right) = \_\_\_$$

$\frac{\pi}{180}$, $-\frac{3\pi}{4}$

145 -135°

$$R = (-135)\left(\_\_\_\right) = \_\_\_$$

In Problems 146–152, convert the given radian measure to degree measure D.

$\frac{180}{\pi}$, 45°

146 $\frac{\pi}{4}$

Using Problem 139, we have

$$D = \left(\frac{\pi}{4}\right)\left(\_\_\_\right) = \_\_\_$$

100°

147 $\frac{5\pi}{9}$

$$D = \left(\frac{5\pi}{9}\right)\left(\frac{180}{\pi}\right) = \_\_\_$$

$\frac{180}{\pi}$, 15°

148 $\frac{\pi}{12}$

$$D = \left(\frac{\pi}{12}\right)\left(\_\_\_\right) = \_\_\_$$

$\frac{180}{\pi}$, -270°

149 $-\frac{3\pi}{2}$

$$D = \left(-\frac{3\pi}{2}\right)\left(\_\_\_\right) = \_\_\_$$

150 π

$\frac{180}{\pi}$, $180°$

$D = (\pi)\left(\_\_\_\right) = \_\_\_$

151 $-\frac{5\pi}{4}$

$\frac{180}{\pi}$, $-225°$

$D = \left(-\frac{5\pi}{4}\right)\left(\_\_\_\right) = \_\_\_$

152 $\frac{11\pi}{6}$

$\frac{180}{\pi}$, $330°$

$D = \left(\_\_\_\right)\left(\frac{11\pi}{6}\right) = \_\_\_$

$\frac{180}{\pi}$, $\frac{180}{\pi}$

larger

153 Since 1 radian corresponds to $D = (1)\left(\_\_\_\right) = \left(\_\_\_\right)^{\circ}$, or approximately 57.3°, an angle of 1 radian is _____ than an angle of 1°.

4.3 Arcs and Sectors of Circles

154 Suppose that θ is the central angle of a circle of radius r and that θ has a radian measure t; then the arc length s determined by θ is given by $s =$ ____ (Figure 2).

tr

θ has radian measure t

Figure 2

In Problems 155–161, find the arc length s determined by the given central angle t or θ of a circle of radius r.

155 $t = \frac{\pi}{6}$ and $r = 6$

Using the formula established in Problem 154, $s =$ ____, so that

tr

$s = \left(\frac{\pi}{6}\right)\_\_\_ = \_\_\_$

6, π

156 $t = \frac{13\pi}{4}$ and $r = 8$

$s = \left(\_\_\_\right)(8) = \_\_\_$

$\frac{13\pi}{4}$, 26π

157 $t = \dfrac{18\pi}{5}$ and $r = 15$

$s = (\_\_\_\_)(15) = \_\_\_\_$

$\dfrac{18\pi}{5}$, 54π

158 $t = 1$ and $r = 7$

$s = (\_\_\_\_)(\_\_\_\_) = \_\_\_\_$

1, 7, 7

159 $\theta = 150°$ and $r = 2$

Before using the formula for arc length, it is necessary to convert 150° to radian measure t. Hence, $t = (150)(\_\_\_\_) = \_\_\_\_$

$\dfrac{\pi}{180}$, $\dfrac{5\pi}{6}$

so $s = (\_\_\_\_)(2) = \_\_\_\_$.

$\dfrac{5\pi}{6}$, $\dfrac{5\pi}{3}$

160 $\theta = 100°$ and $r = 3$

First we convert 100° to radian measure $t = (100)(\_\_\_\_) = \_\_\_\_$

$\dfrac{\pi}{180}$, $\dfrac{5\pi}{9}$

so $s = (\_\_\_\_)(3) = \_\_\_\_$.

$\dfrac{5\pi}{9}$, $\dfrac{5\pi}{3}$

161 $\theta = 270°$ and $r = 2$

The radian measure t of θ is given by $t = (270)(\_\_\_\_) = \_\_\_\_$

$\dfrac{\pi}{180}$, $\dfrac{3\pi}{2}$

so $s = (\_\_\_\_)(\_\_\_\_) = \_\_\_\_$.

$\dfrac{3\pi}{2}$, 2, 3π

162 Suppose that θ is the central angle of a circle of radius r and that θ has a radian measure t; then the area of the sector of the circle A determined by θ is given by either

$A = (\frac{1}{2})r^2 \_\_\_\_$ or $A = (\frac{1}{2})s \_\_\_\_$ (Figure 3)

t, r

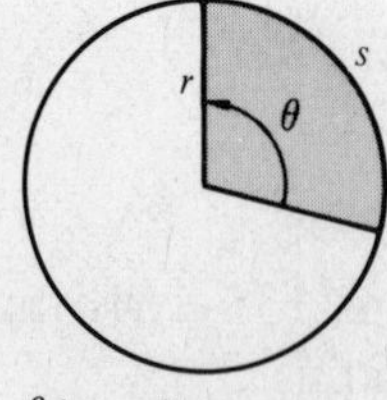

θ has radian measure t

Figure 3

In Problems 163–167, find the area A of a sector determined by the given central angle t or θ of a circle of radius r.

163 $t = \dfrac{\pi}{6}$ and $r = 6$

We know from Problem 162 that $A = \frac{1}{2}r^2 \_\_\_\_$, so

t

$A = \dfrac{1}{2}(\_\_\_\_)^2(\_\_\_\_) = \_\_\_\_$.

6, $\dfrac{\pi}{6}$, 3π

164 $t = \frac{3\pi}{2}$ and $r = 2$

Here we have $A = \frac{1}{2}(\_\_\_)^2(\_\_\_) = \_\_\_.$

$2, \frac{3\pi}{2}, 3\pi$

165 $t = 1$ and $r = 7$

Here we have $A = \frac{1}{2}(\_\_\_)^2(\_\_\_) = \_\_\_.$

$7, 1, \frac{49}{2}$

166 $\theta = 150°$ and $r = 2$

First we convert to radian measure: $t = (150)(\_\_\_) = \_\_\_$

$\frac{\pi}{180}, \frac{5\pi}{6}$

so $A = \frac{1}{2}(\_\_\_)^2(\_\_\_) = \_\_\_.$

$2, \frac{5\pi}{6}, \frac{5\pi}{3}$

167 $\theta = 270°$ and $r = 2$

The radian measure of 270° is given by

$t = (270)(\_\_\_) = \_\_\_$, so that

$\frac{\pi}{180}, \frac{3\pi}{2}$

$A = (\_\_\_)(\_\_\_)^2\left(\frac{3\pi}{2}\right) = \_\_\_$

$\frac{1}{2}, 2, 3\pi$

4.4 Functions Defined on Angles

168 An angle is said to be in *standard position* if it is placed on a cartesian coordinate system so that the vertex corresponds to the ______ and the initial side coincides with the ______.

origin, positive x axis

In Problems 169–176, sketch the indicated angle in standard position. Indicate the quadrant in which the terminal side of θ lies.

169 $\theta = 48°$

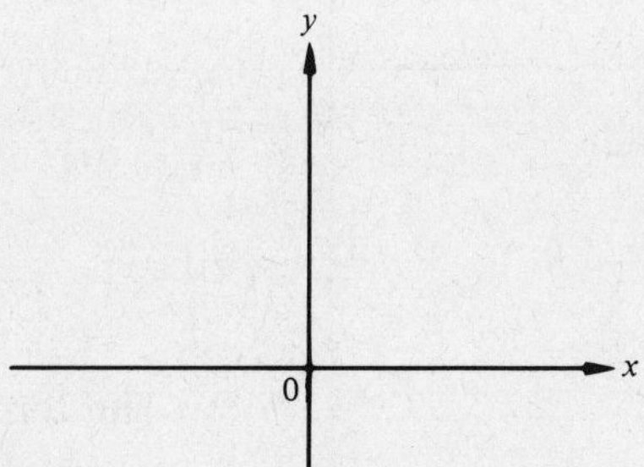

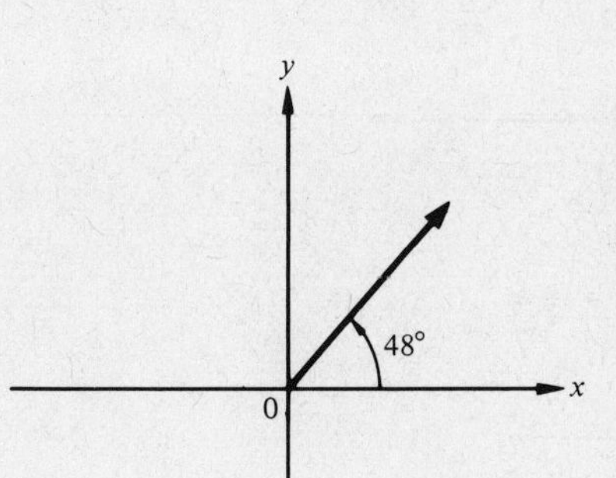

The terminal side of the angle lies in quadrant ____.

I

170 $\theta = 102°$

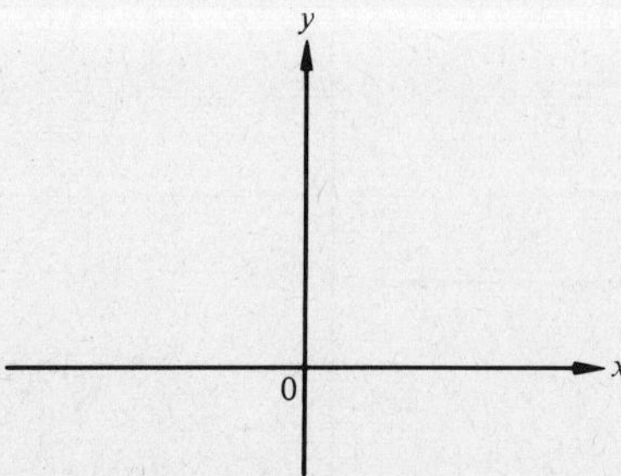

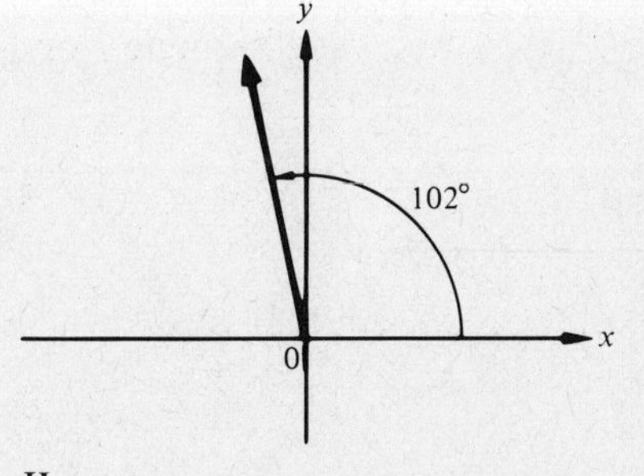

The terminal side lies in quadrant ____.

II

171 $t = -\frac{5\pi}{3}$

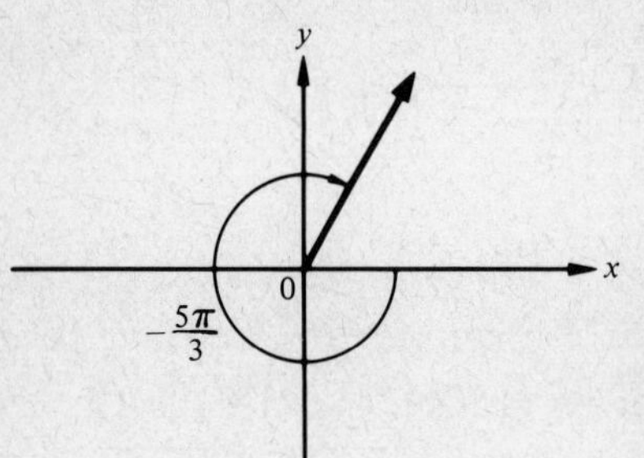

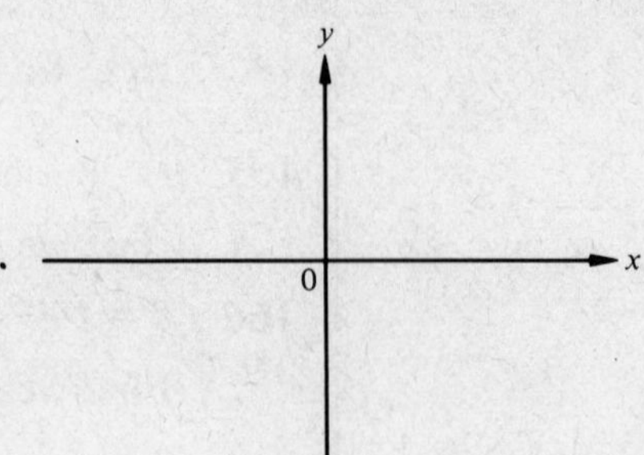

I

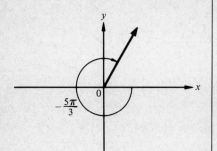

The terminal side lies in quadrant ____.

172 $\theta = 405°$

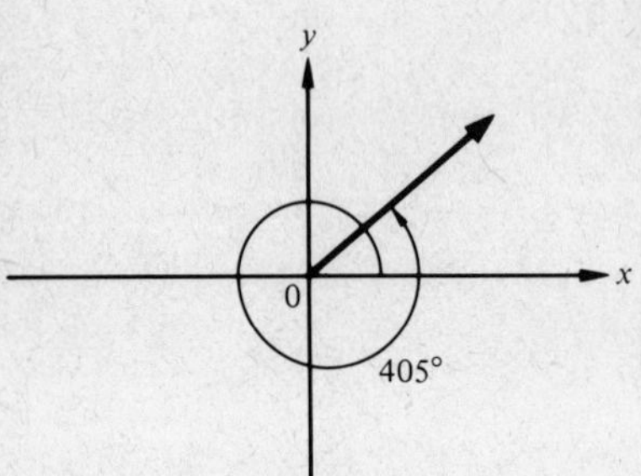

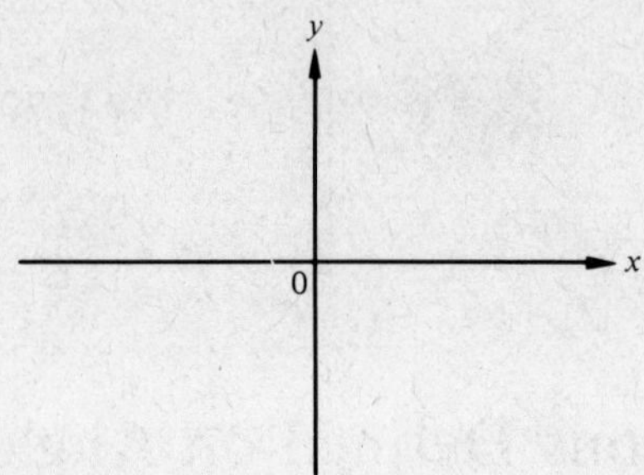

I

The terminal side lies in quadrant ____.

173 $t = \frac{7\pi}{6}$

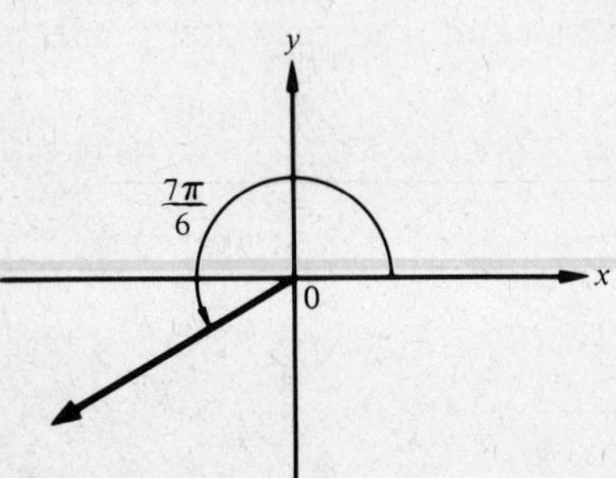

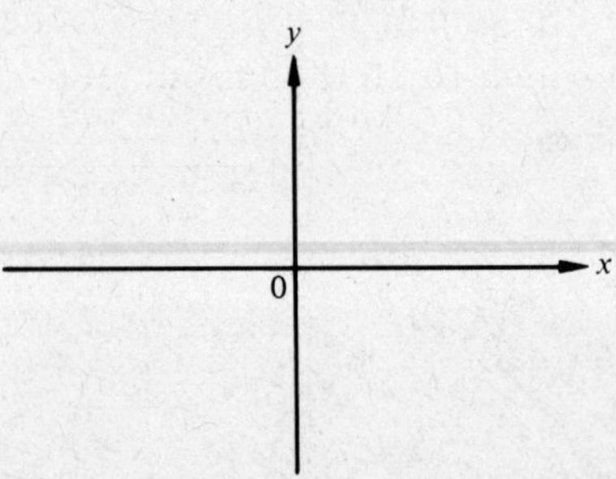

III

The terminal side lies in quadrant ____.

174 $t = 2.1$

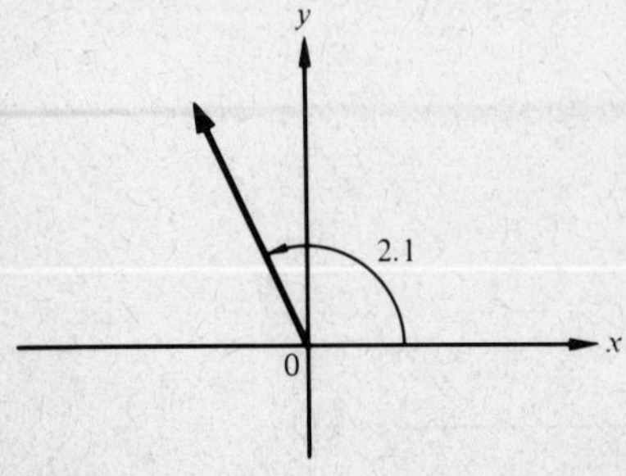

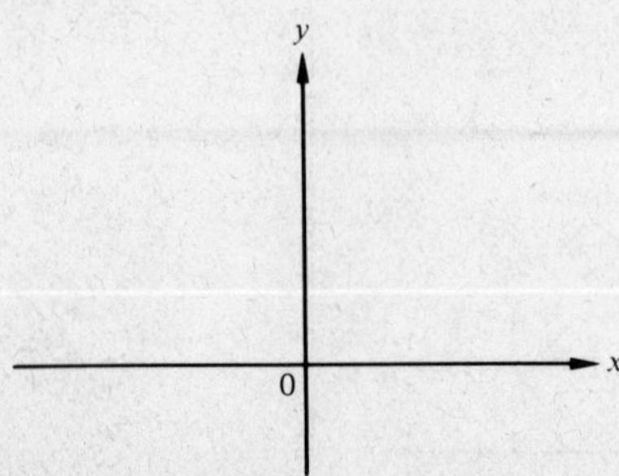

II

The terminal side lies in quadrant ____.

175 $\theta = -260°$

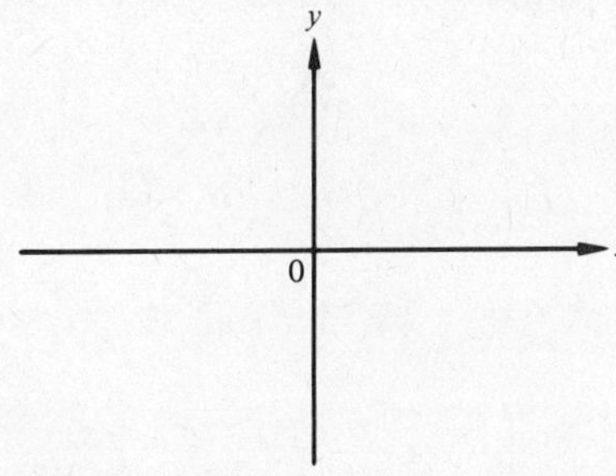

The terminal side lies in quadrant ____.

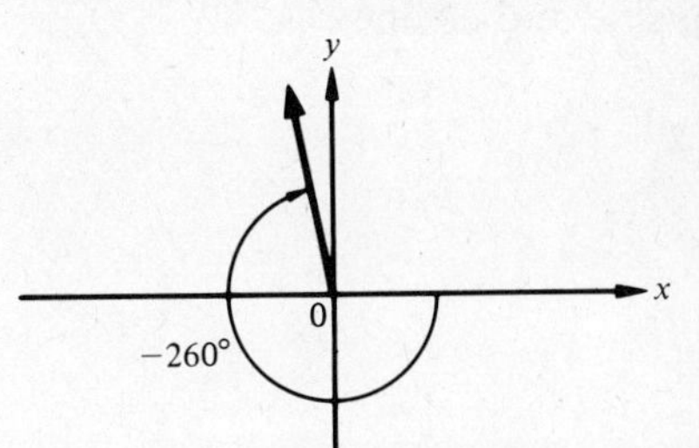

II

176 $t = -7.28$

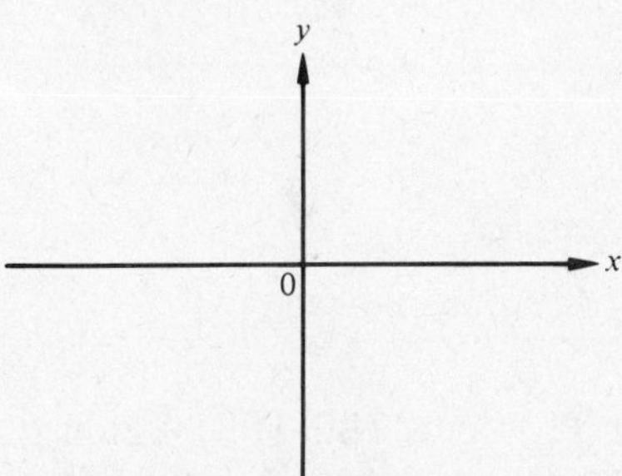

The terminal side lies in quadrant ____.

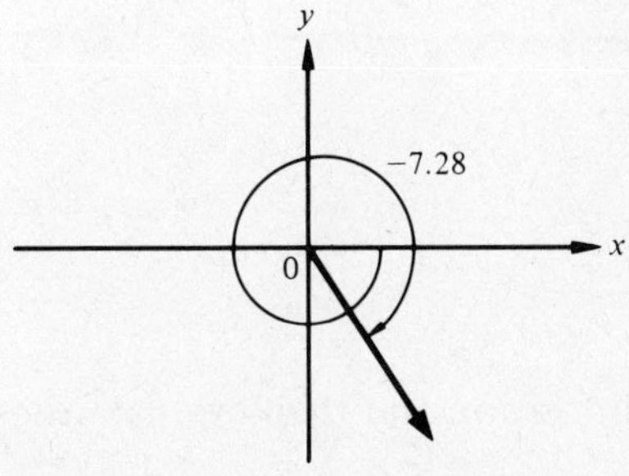

IV

177 Suppose that θ is an angle in standard position with point (x, y) on its terminal side such that $(x, y) \neq (0, 0)$ and r is the distance between (x, y) and $(0, 0)$ (Figure 4).

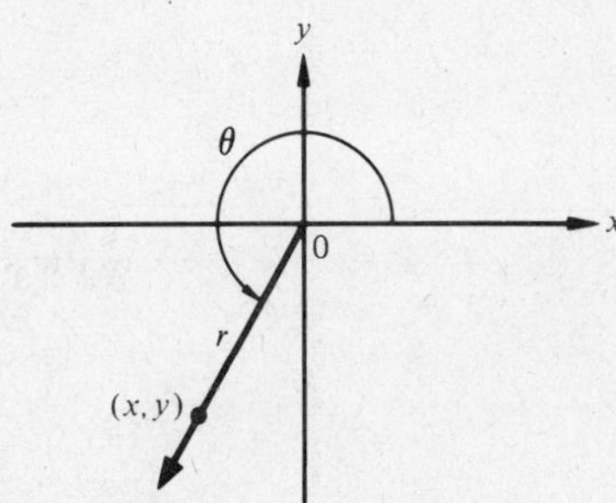

Figure 4

Then, by the definition of the trigonometric functions on θ, we have

$\sin\theta =$ ____ $\qquad \csc\theta =$ ____, $y \neq 0$

$\cos\theta =$ ____ $\qquad \sec\theta =$ ____, $x \neq 0$

$\tan\theta =$ ____, $x \neq 0 \qquad \cot\theta =$ ____, $y \neq 0$

where r can be written in terms of x and y as $r =$ ________.

$\frac{y}{r}, \frac{r}{y}$

$\frac{x}{r}, \frac{r}{x}$

$\frac{y}{x}, \frac{x}{y}$

$\sqrt{x^2 + y^2}$

In Problems 178–181, use the definition of the trigonometric functions on angles to write the tangent, cotangent, secant, and cosecant in terms of the sine and cosine.

$\dfrac{\sin\theta}{\cos\theta}$

178 $\tan\theta = \dfrac{y}{x} = \dfrac{\frac{y}{r}}{\frac{x}{r}} =$ _____

$\dfrac{\cos\theta}{\sin\theta}$

179 $\cot\theta = \dfrac{x}{y} = \dfrac{\frac{x}{r}}{\frac{y}{r}} =$ _____

$\dfrac{x}{r}, \dfrac{1}{\cos\theta}$

180 $\sec\theta = \dfrac{r}{x} = \dfrac{1}{(\_\_\_\_)} =$ _____

$\dfrac{y}{r}, \dfrac{1}{\sin\theta}$

181 $\csc\theta = \dfrac{r}{y} = \dfrac{1}{(\_\_\_\_)} =$ _____

In Problems 182–188, sketch the indicated angle in standard position and then evaluate the six trigonometric functions on the angle.

182 $\theta = -270°$ and $y = 1$, where (x, y) is on the terminal side of θ

0, 1 — We have $x =$ ____ and $y =$ ____, so that

0, 1 — $r = \sqrt{x^2 + y^2} = \sqrt{(\_\_\_\_)^2 + (1)^2} =$ ____. Hence,

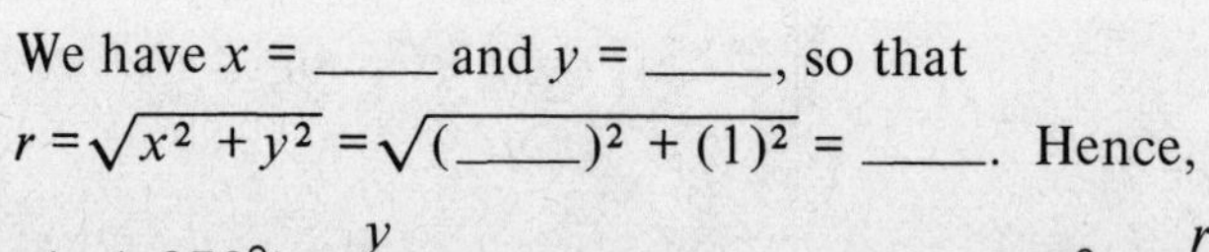

1, 1 — $\sin(-270°) = \dfrac{y}{r} =$ ____ $\qquad \csc(-270°) = \dfrac{r}{y} =$ ____

0, undefined — $\cos(-270°) = \dfrac{x}{r} =$ ____ $\qquad \sec(-270°) = \dfrac{r}{x} =$ ________

undefined, 0 — $\tan(-270°) = \dfrac{y}{x} =$ ________ $\qquad \cot(-270°) = \dfrac{x}{y} =$ ____

183 $t = -\pi$ and $x = -1$, where (x, y) is on the terminal side of t

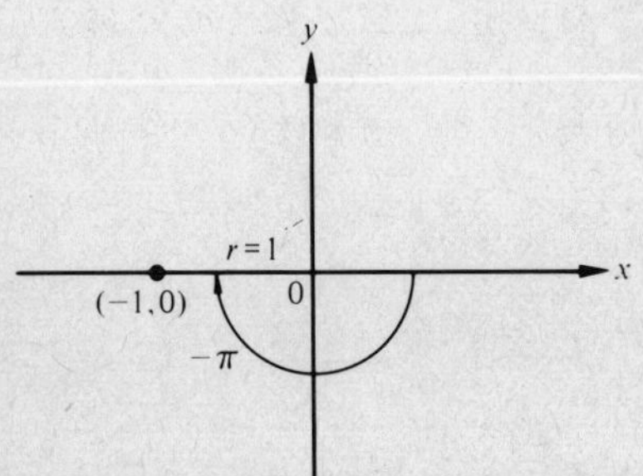

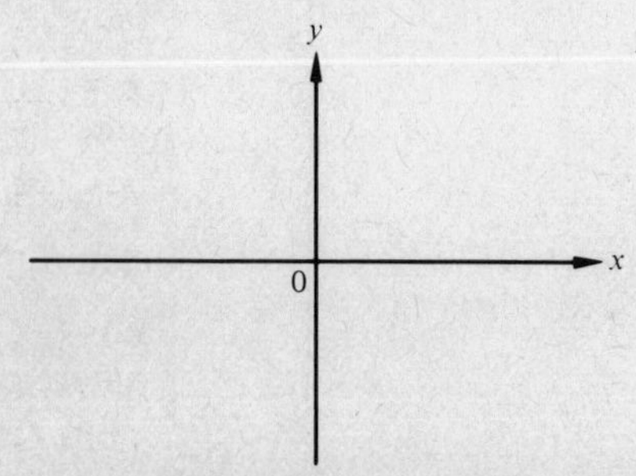

-1, 0

y, 0, 1

y, 0, y, undefined

r, -1, r, -1

x, 0, x, undefined

Here $x =$ ____ and $y =$ ____, so that

$r = \sqrt{x^2 + (\_\_\_\_)^2} = \sqrt{(-1)^2 + (\_\_\_\_)^2} =$ ____. Hence,

$\sin(-\pi) = \frac{\_\_\_\_}{r} =$ ____ $\qquad \csc(-\pi) = \frac{r}{\_\_\_\_} =$ ____

$\cos(-\pi) = \frac{x}{\_\_\_\_} =$ ____ $\qquad \sec(-\pi) = \frac{\_\_\_\_}{x} =$ ____

$\tan(-\pi) = \frac{y}{\_\_\_\_} =$ ____ $\qquad \cot(-\pi) = \frac{\_\_\_\_}{y} =$ ____

184 $t = \frac{\pi}{4}$ and $x = 1$, where (x, y) is on the terminal side of t

$\frac{\pi}{4}$ corresponds to ____°.

45

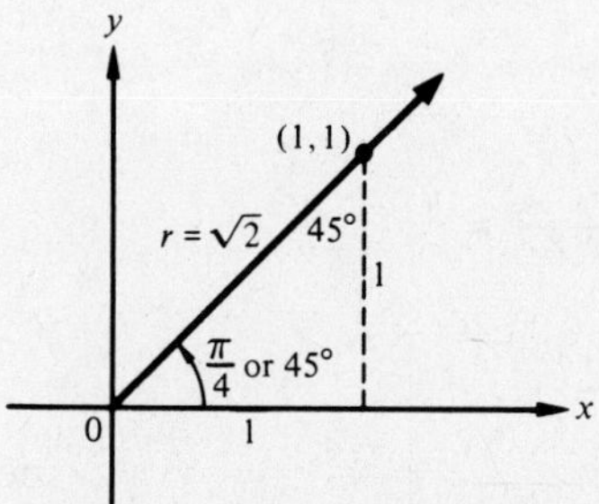

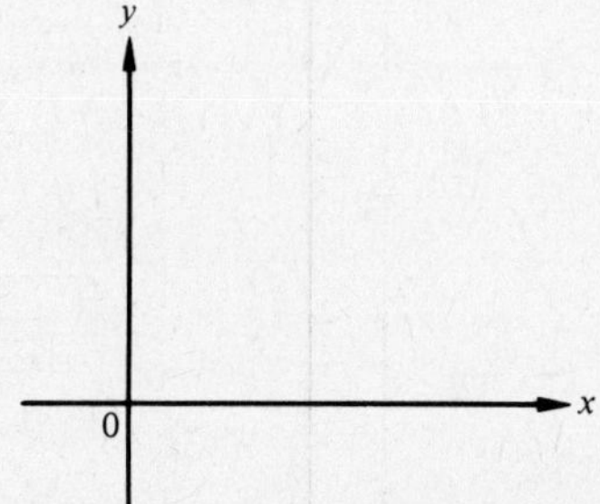

(1, 1)

1, 1, 1, 1, $\sqrt{2}$

In this case, the point ______ is selected on the terminal side of t, so that $x =$ ____, $y =$ ____, and $r = \sqrt{(\_\_\_\_)^2 + (\_\_\_\_)^2} =$ ____. Hence,

1, $\frac{\sqrt{2}}{2}$, 1, $\sqrt{2}$

$\sqrt{2}$, $\frac{\sqrt{2}}{2}$, $\sqrt{2}$, $\sqrt{2}$

1, 1, 1, 1

$\sin 45° = \frac{\_\_\_\_}{\sqrt{2}} =$ ____ $\qquad \csc 45° = \frac{\sqrt{2}}{\_\_\_\_} =$ ____

$\cos 45° = \frac{1}{\_\_\_\_} =$ ____ $\qquad \sec 45° = \frac{\_\_\_\_}{1} =$ ____

$\tan 45° = \frac{1}{\_\_\_\_} =$ ____ $\qquad \cot 45° = \frac{\_\_\_\_}{1} =$ ____

185 $\theta = -135°$ and $x = -1$, where (x, y) is a point on the terminal side of θ.

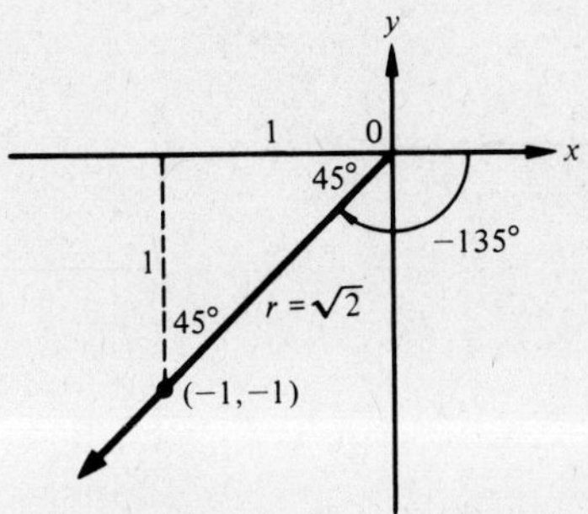

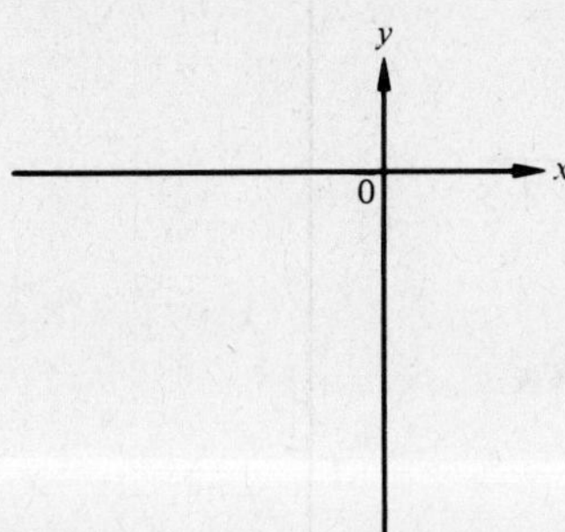

(-1, -1)

-1, -1, $\sqrt{2}$

$-\frac{1}{\sqrt{2}}$, $-\frac{\sqrt{2}}{2}$, $-\sqrt{2}$

$-\frac{1}{\sqrt{2}}$, $-\frac{\sqrt{2}}{2}$, $-\sqrt{2}$

1, 1

Here, the point on the terminal side of θ is ________ so that $x =$ ____, $y =$ ____, and $r =$ ____. Hence

$\sin(-135°) = \frac{\quad}{\_\_\_\_} = \frac{\quad}{\_\_\_\_}$ $\qquad \csc(-135°) =$ ____

$\cos(-135°) = \frac{\quad}{\_\_\_\_} = \frac{\quad}{\_\_\_\_}$ $\qquad \sec(-135°) =$ ____

$\tan(-135°) =$ ____ $\qquad \cot(-135°) =$ ____

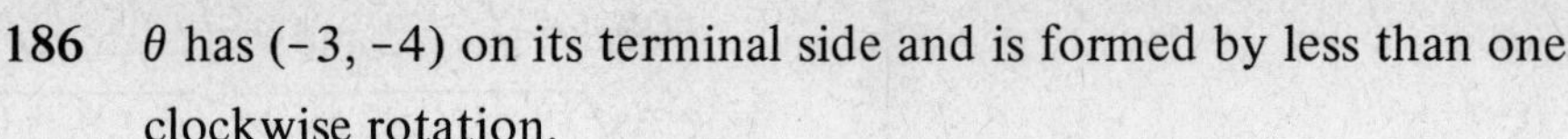

186 θ has $(-3, -4)$ on its terminal side and is formed by less than one clockwise rotation.

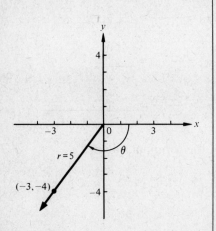

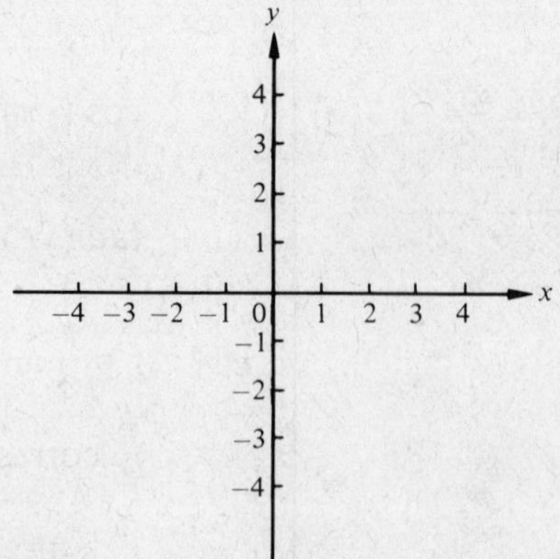

Here $x =$ ____, $y =$ ____, and $r = \sqrt{x^2 + y^2} =$
$\sqrt{(\_\_\_\_)^2 + (\_\_\_\_)^2} =$ ____ so that

$\sin\theta = \dfrac{____}{5} = $ ____ $\qquad \csc\theta = \dfrac{5}{____} = $ ____

$\cos\theta = \dfrac{____}{5} = $ ____ $\qquad \sec\theta = \dfrac{5}{____} = $ ____

$\tan\theta = \dfrac{-\frac{4}{5}}{____} = $ ____ $\qquad \cot\theta = \dfrac{____}{-\frac{4}{5}} = $ ____

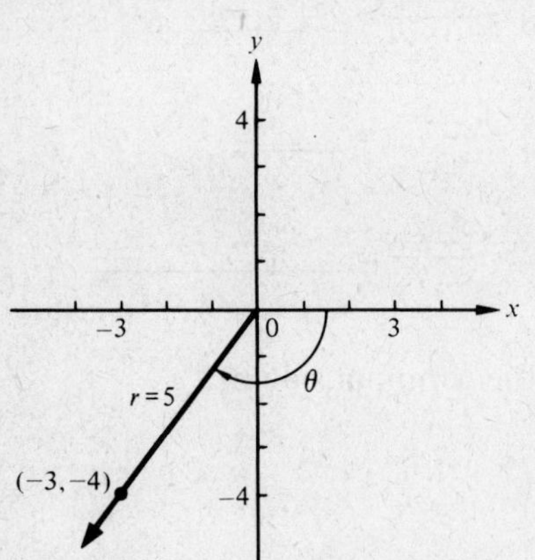

$-3, -4$

$-3, -4, 5$

$-4, -\frac{4}{5}, -4, -\frac{5}{4}$

$-3, -\frac{3}{5}, -3, -\frac{5}{3}$

$-\frac{3}{5}, \frac{4}{3}, -\frac{3}{5}, \frac{3}{4}$

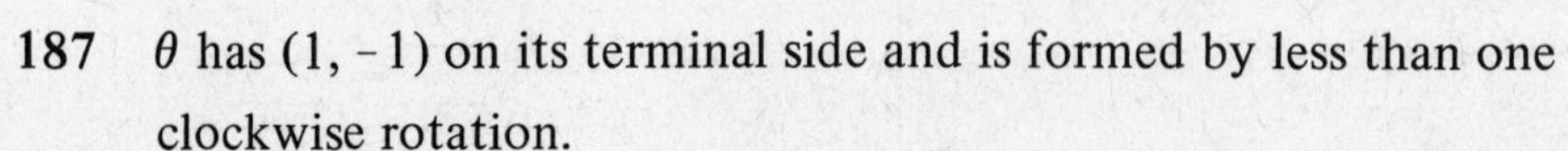

187 θ has $(1, -1)$ on its terminal side and is formed by less than one clockwise rotation.

Here $x =$ ____, $y =$ ____, and $r =$ ____ so that

$\sin\theta = \dfrac{\quad}{____} = $ ____ $\qquad \csc\theta =$ ____

$\cos\theta = \dfrac{\quad}{____} = $ ____ $\qquad \sec\theta =$ ____

$\tan\theta =$ ____ $\qquad \cot\theta =$ ____

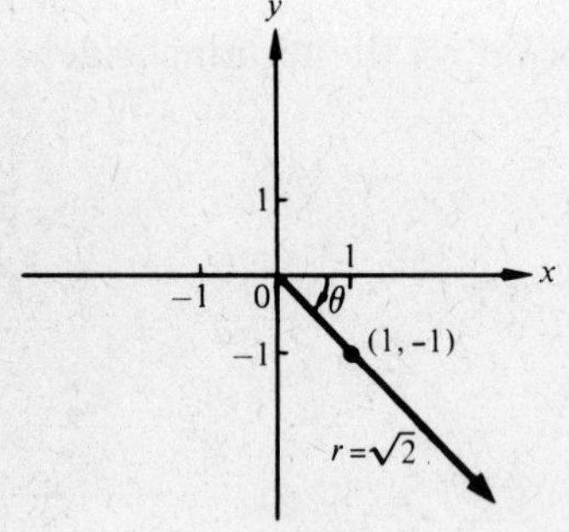

$1, -1, \sqrt{2}$

$-\frac{1}{\sqrt{2}}, -\frac{\sqrt{2}}{2}, -\sqrt{2}$

$\frac{1}{\sqrt{2}}, \frac{\sqrt{2}}{2}, \sqrt{2}$

$-1, -1$

188 θ has (2, -4) on its terminal side and is formed by less than one counterclockwise rotation.

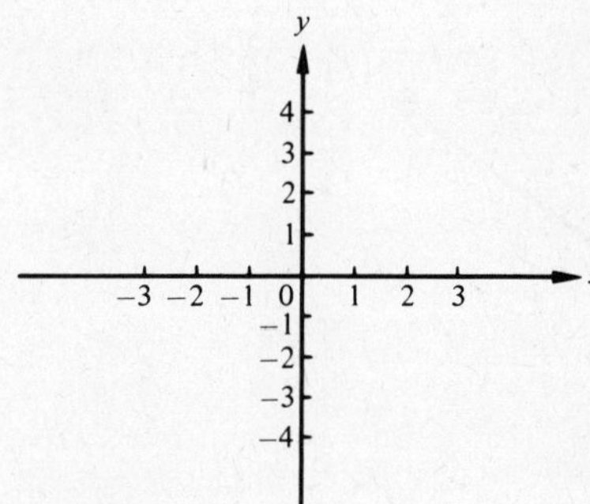

$x =$ ____, $y =$ ____, and $r =$ ____, so that

$\sin\theta = \frac{\_\_\_\_}{\_\_\_\_} = \frac{\_\_\_\_}{\_\_\_\_}$ $\qquad \csc\theta = \frac{\_\_\_\_}{\_\_\_\_}$

$\cos\theta = \frac{\_\_\_\_}{\_\_\_\_} = \frac{\_\_\_\_}{\_\_\_\_}$ $\qquad \sec\theta =$ ____

$\tan\theta =$ ____ $\qquad \cot\theta = \frac{\_\_\_\_}{\_\_\_\_}$

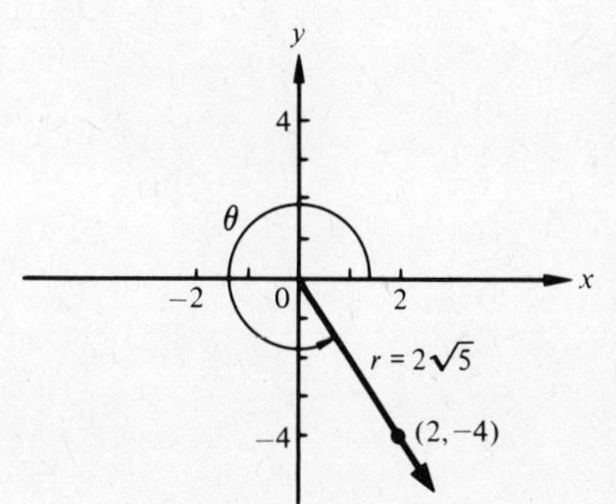

$2, -4, 2\sqrt{5}$

$-\frac{2}{\sqrt{5}}, -\frac{2\sqrt{5}}{5}, -\frac{\sqrt{5}}{2}$

$\frac{1}{\sqrt{5}}, \frac{\sqrt{5}}{5}, \sqrt{5}$

$-2, -\frac{1}{2}$

189 If $(x, -8)$ is 10 units from (0, 0) and is on the terminal side of θ, an angle in standard position with its terminal side in quadrant III (Figure 5),

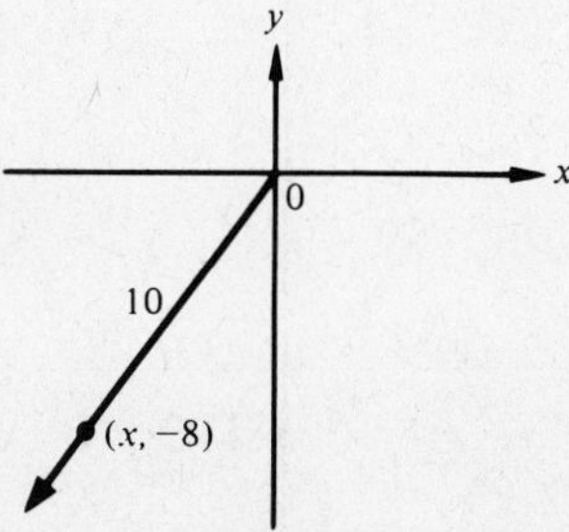

Figure 5

then it follows from the fact that $r = \sqrt{x^2 + y^2}$ that $10 = \sqrt{x^2 + (\_\_\_\_)^2}$, so that $100 = x^2 + (\_\_\_\_)^2$. Hence, $x^2 =$ ____. Since $(x, -8)$ is in quadrant ____, we have $x =$ ____. Consequently,

$\sin\theta = \frac{____}{10} = $ ____ $\qquad \csc\theta =$ ____

$\cos\theta = \frac{____}{10} = $ ____ $\qquad \sec\theta =$ ____

$\tan\theta = \frac{-8}{____} = $ ____ $\qquad \cot\theta =$ ____

$-8, -8, 36$

$\text{III}, -6$

$-8, -\frac{4}{5}, -\frac{5}{4}$

$-6, -\frac{3}{5}, -\frac{5}{3}$

$-6, \frac{4}{3}, \frac{3}{4}$

III

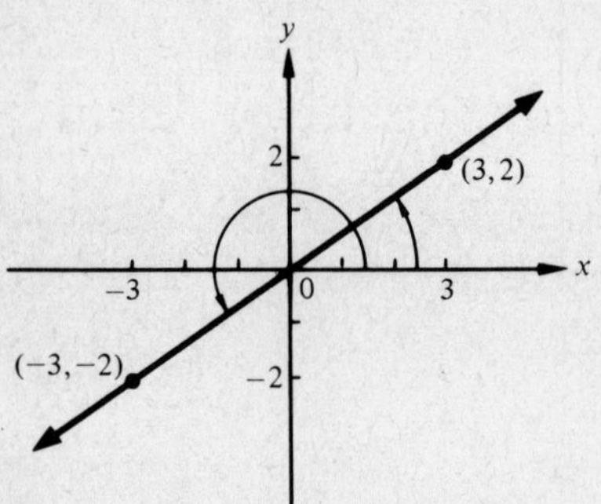

190 An angle θ such that $\tan\theta = \frac{2}{3}$ could have its terminal side in either quadrant I or ____, as illustrated below.

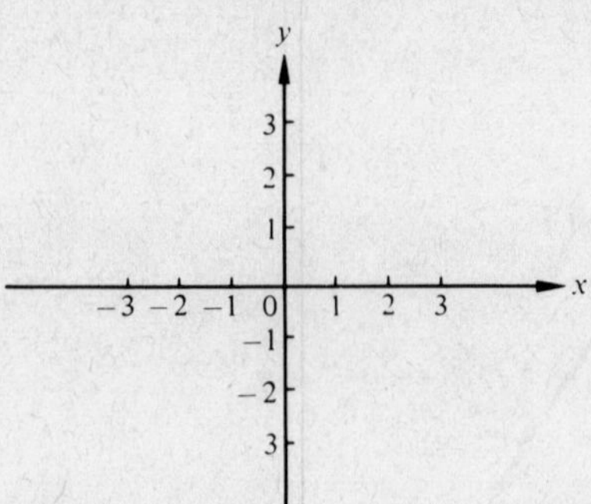

$4,\ \sqrt{13}$

$\frac{3}{\sqrt{13}},\ \frac{3\sqrt{13}}{13}$

III

$-\frac{2\sqrt{13}}{13},\ -\frac{3\sqrt{13}}{13}$

In either case, $r = \sqrt{9 + \_\_\_\_} = \_\_\_\_$, so that if θ is in quadrant I,

$\sin\theta = \frac{2}{\sqrt{13}} = \frac{2\sqrt{13}}{13}$ and $\cos\theta = \frac{}{\_\_\_\_} = \frac{}{\_\_\_\_}$

and if θ is in quadrant ____,

$\sin\theta = \frac{}{\_\_\_\_}$ and $\cos\theta = \frac{}{\_\_\_\_}$

$-x,\ x\sqrt{2}$

$x\sqrt{2},\ -\frac{1}{\sqrt{2}},\ -\frac{\sqrt{2}}{2}$

$x\sqrt{2},\ \frac{1}{\sqrt{2}},\ \frac{\sqrt{2}}{2}$

$-x,\ -1$

IV

191 If $x > 0$ and point $(x, -x)$ is on the terminal side of θ in standard position, then $r = \sqrt{x^2 + (\_\_\_\_)^2} = \_\_\_\_$, so that

$\sin\theta = \frac{-x}{\_\_\_\_} = \frac{}{\_\_\_\_} = \frac{}{\_\_\_\_}$

$\cos\theta = \frac{x}{\_\_\_\_} = \frac{}{\_\_\_\_} = \frac{}{\_\_\_\_}$

$\tan\theta = \frac{\_\_\_\_}{x} = \_\_\_\_$

The terminal side of θ is in quadrant ____.

$\frac{y}{x}$

$\frac{r}{x},\ \frac{y}{x},\ x^2 + y^2$

$r^2,\ \frac{r}{x},\ \sec^2\theta$

192 If (x, y) is on the terminal side of angle θ in standard position, and (x, y) is r units from $(0, 0)$, then by definition $\tan\theta = \frac{}{\_\_\_\_}$ and $\sec\theta = \frac{}{\_\_\_\_}$ so that $\tan^2\theta + 1 = \left(\frac{}{\_\_\_\_}\right)^2 + 1 = \frac{\_\_\_\_}{x^2}$. Since $r^2 = x^2 + y^2$, we have $\tan^2\theta + 1 = \frac{\_\_\_\_}{x^2} = \left(\frac{}{\_\_\_\_}\right)^2 = \_\_\_\_$.

terminal side

193 The values of any one of the trigonometric functions on 100° or 460° would be the same because the angles in standard position have the same ____________.

terminal side

194 The values of any of the trigonometric functions on −200° or 160° would be the same since the two angles in standard position have the same ____________.

4.5 Relationship Between Trigonometric Functions of Angles and Trigonometric Functions of Real Numbers

195 If T is any one of the six trigonometric functions defined on angles and C is the trigonometric function with the same name, defined on real numbers, then $T(\theta) =$ _____, where θ is an angle with radian measure _____.

Answers: $C(t)$; t

196 Since $T(\theta) = C(t)$, it follows that

$T(\theta + 360°) = C(t +$ _____$) = C($_____$) = T($_____$)$

That is, the trigonometric functions are periodic functions of period _____° or _____ radians.

Answers: 2π, t, θ; 360, 2π

197 Since $T(\theta) = C(t)$, the signs of the values of the six trigonometric functions are as follows:

	θ has terminal side in quadrant I	II	III	IV
$\sin\theta$	+	____	____	____
$\cos\theta$	____	____	−	____
$\tan\theta$	____	____	____	−
$\csc\theta$	____	+	____	____
$\sec\theta$	____	−	____	____
$\cot\theta$	+	____	____	____

Answers:
+, −, −
+, −, +
+, −, +
+, −, −
+, −, +
−, +, −

In Problems 198–208, evaluate the indicated trigonometric function by using the information in Problem 195.

198 $\sin 0°$

The radian measure of a 0° angle is _____ so that

$\sin 0° = \sin$ _____ $=$ _____

Answers: 0; 0, 0

199 $\cos(-30°)$

The radian measure of a $-30°$ angle is given by

$t = -30\left(\_\_\_\right) = \_\_\_$, so that $\cos(-30°) = \cos\left(\_\_\_\right) = \_\_\_$.

Answers: $\frac{\pi}{180}$, $-\frac{\pi}{6}$, $-\frac{\pi}{6}$, $\frac{\sqrt{3}}{2}$

200 $\tan 45°$

Here, the radian measure of 45° is $t = 45\left(\_\_\_\right) = \_\_\_$, so that

$$\tan 45° = \tan \_\_\_ = \frac{\sin\frac{\pi}{4}}{\cos \_\_\_} = \frac{\frac{1}{\sqrt{2}}}{\_\_\_} = \_\_\_$$

Answers: $\frac{\pi}{180}$, $\frac{\pi}{4}$; $\frac{\pi}{4}$, $\frac{\pi}{4}$, $\frac{1}{\sqrt{2}}$, 1

201 $\csc(-60°)$

$\frac{\pi}{180}, -\frac{\pi}{3}$

$-60°$ is $-60(\_\_\_) = \_\_\_$ radians, so that

$-\frac{\pi}{3}, -\frac{\pi}{3}, -\frac{\sqrt{3}}{2}, -\frac{2\sqrt{3}}{3}$

$\csc(-60°) = \csc(\_\_\_) = \frac{1}{\sin(\_\_\_)} = \frac{1}{\_\_\_} = \_\_\_$

202 sec 180°

π

The radian measure of 180° is ____. Hence,

$\pi, \pi, -1, -1$

$\sec 180° = \sec \_\_\_ = \frac{1}{\cos \_\_\_} = \frac{1}{\_\_\_} = \_\_\_$

203 csc 270°

$\frac{\pi}{180}, \frac{3\pi}{2}$

$270°$ is $270(\_\_\_) = \_\_\_$ radians, so that

$\frac{3\pi}{2}, \frac{3\pi}{2}, -1, -1$

$\csc 270° = \csc \_\_\_ = \frac{1}{\sin \_\_\_} = \frac{1}{\_\_\_} = \_\_\_$

204 $\sin(-150°)$

$\frac{\pi}{180}, -\frac{5\pi}{6}$

The radian measure of $-150°$ is given by $t = -150(\_\_\_) = \_\_\_$,

$-\frac{5\pi}{6}, -\frac{1}{2}$

so that $\sin(-150°) = \sin(\_\_\_) = \_\_\_$.

205 cos 135°

$\frac{\pi}{180}, \frac{3\pi}{4}$

The radian measure of 135° is given by $t = 135(\_\_\_) = \_\_\_$,

$\frac{3\pi}{4}, -\frac{\sqrt{2}}{2}$

so that $\cos 135° = \cos \_\_\_ = \_\_\_$

206 tan 120°

$\frac{\pi}{180}, \frac{2\pi}{3}$

$120°$ is $120(\_\_\_) = \_\_\_$ radians. Hence,

$\frac{2\pi}{3}, \frac{2\pi}{3}, -\frac{1}{2}, -\sqrt{3}$

$\tan 120° = \tan \_\_\_ = \frac{\sin \frac{2\pi}{3}}{\cos \_\_\_} = \frac{\frac{\sqrt{3}}{2}}{\_\_\_} = \_\_\_$

207 csc 210°

$\frac{\pi}{180}, \frac{7\pi}{6}$

The radian measure for 210° is $t = 210(\_\_\_) = \_\_\_$, so that

$\frac{7\pi}{6}, \frac{7\pi}{6}, -\frac{1}{2}, -2$

$\csc 210° = \csc \_\_\_ = \frac{1}{\sin \_\_\_} = \frac{1}{\_\_\_} = \_\_\_$.

208 $\sec(-300°)$

Since the terminal side of an angle of $-300°$ is the same as the

60

terminal side of the acute angle of ____°, it follows that

60

$\sec(-300°) = \sec \_\_\_°$. However, a 60° angle has a radian

Answers	Frames
$\frac{\pi}{3}$	measure of ____, so that
$\frac{\pi}{3}$, $\cos\frac{\pi}{3}$, 2	$\sec 60° = \sec \underline{\quad} = \frac{1}{\underline{\qquad}} = \underline{\quad}$
2	Thus, $\sec(-300°) = \underline{\quad}$.

5 Evaluation of the Trigonometric Functions

Answers	Frames
	209 To evaluate the trigonometric functions of an angle whose radian measure is less than 0 or greater than 2π (or whose degree measure
0°, 360°	is less than ____ or greater than ____), we add or subtract an
2π, 360°	appropriate multiple of ____ (or ____) in order to reduce the
2π	evaluation of the functions to an angle between 0 and ____
0°, 360°	(or ____ and ____).

In Problems 210–221, reduce the evaluation of the given trigonometric function of the given angle to an evaluation of the same function of an angle between 0 and 2π (if radians are used) or 0° and 360° (if degrees are used). Use $\pi = 3.14$.

Answers	Frames
	210 $\sin\frac{9\pi}{2}$
2, $\frac{\pi}{2}$, $\frac{\pi}{2}$	$\sin\frac{9\pi}{2} = \sin\left(\underline{\quad}\cdot 2\pi + \underline{\quad}\right) = \sin \underline{\quad}$
	211 $\cos\frac{17\pi}{4}$
2, $\frac{\pi}{4}$, $\frac{\pi}{4}$	$\cos\frac{17\pi}{4} = \cos\left(\underline{\quad}\cdot 2\pi + \underline{\quad}\right) = \cos \underline{\quad}$
	212 $\tan\left(-\frac{5\pi}{6}\right)$
$-\frac{5\pi}{6}$, $\frac{7\pi}{6}$	$\tan\left(-\frac{5\pi}{6}\right) = \tan\left[2\pi + \left(\underline{\quad}\right)\right] = \tan \underline{\quad}$ or, alternatively,
	since the tangent is an odd function, we have
$\frac{5\pi}{6}$	$\tan\left(-\frac{5\pi}{6}\right) = -\tan \underline{\quad}$
	213 cos 28.69
4, 3.57, 3.57	cos 28.69 = cos(____ · 6.28 + ____) = cos ____
	214 cot 17.2
2, 4.64, 4.64	cot 17.2 = cot(____ · 6.28 + ____) = cot ____
	215 csc(−8.6)
2, −8.6, 3.96	csc(−8.6) = csc [____ · 6.28 + (____)] = csc ____
	216 sec(−15.4)
3, −15.4, 3.44	sec(−15.4) = sec [____ · 6.28 + (____)] = sec ____

Answers	Frames
	217 cos 420°
360°, 60°	cos 420° = cos(____ + 60°) = cos ____
	218 sin 870°
2, 150°, 150°	sin 870° = sin(____ · 360° + ____) = sin ____
	219 sec(-100°)
360°, 260°	sec(-100°) = sec[____ + (-100°)] = sec ____
	220 tan(-1210°)
4, -1210°, 230°	tan(-1210°) = tan[____ · 360° + (______)] = tan ____
	221 cot(-910°)
3, -910°, 170°	cot(-910°) = cot[____ · 360° + (______)] = cot ____

5.1 Reference Angles

Answers	Frames
	222 The reference angle (θ_R or t_R) associated with a given angle is the
positive acute	______________ angle formed by the terminal side of
x	the given angle and the ____ axis.
	223 The value of a given trigonometric function on an angle is the
	same in absolute value as the value of the trigonometric function
reference	of the __________ angle of that angle.
	224 The sign of a given trigonometric function on an angle is deter-
quadrant	mined by the __________ in which the terminal side of the
	angle lies.

In Problems 225-231, reduce the evaluation of the given trigonometric function to an evaluation of the same function at its reference angle. (Use $\pi = 3.14$.)

Answers	Frames
	225 sin 250°
III	Since the terminal side of 250° is in quadrant ____, the reference
180°, 70°, negative	angle for 250° is θ_R = 250° - ____ = ____. The sine is ________
-sin 70°	in quadrant III, so sin 250° = __________.
	226 cos 3
II	Since the terminal side of 3 is in quadrant ____, the reference
3.14, 3, 0.14	angle is t_R = ____ - ____ = ____. The cosine is negative in
II, -cos 0.14	quadrant ____, so cos 3 = __________.
	227 cot 110°20′
II	Since the terminal side of 110°20′ is in quadrant ____, the
69°40′	reference angle θ_R = ________. The cotangent is negative in
-cot 69°40′	quadrant II. Hence, cot(110°20′) = ______________.
	228 tan 315°
IV	Since the terminal side of 315° lies in quadrant ____, the refer-
315°, 45°	ence angle for 315° is θ_R = 360° - ____ = ____. The tangent is
IV, -tan 45°	negative in quadrant ____, so tan 315° = __________.

229 sin 8.73

sin 8.73 = sin(6.28 + ____) = sin ____

Since the terminal side of __________ lies in quadrant ____, where the sine is __________, we have the reference angle t_R = ____. Thus, sin 8.73 = sin ____.

2.45, 2.45
2.45, II
positive
0.69, 0.69

230 sec 470°

Since 470° = 360° + ____, the terminal side of 470° lies in quadrant ____. The reference angle for 110° is θ_R = 180° - ____ = ____. The secant is negative in quadrant ____, so sec 470° = sec 110° = __________

110°
II, 110°, 70°
II
-sec 70°

231 csc(-15)

Because of periodicity, we have

csc[____ · 6.28 + (____)] = csc ____

The reference angle for 3.84 is given by t_R = ____. Since the cosecant is negative in quadrant ____, we have

csc(-15) = __________

3, -15, 3.84
0.70
III
-csc 0.70

In Problems 232–236, determine the values of the given trigonometric function through the use of its reference angle.

232 cos 225°

Here θ = ____, so θ_R = 225° - ____ = ____. Since the terminal side of 225° lies in quadrant ____, cos 225° is __________.

Hence, cos 225° = - cos 45° = ____.

225°, 180°, 45°
III, negative
$-\frac{\sqrt{2}}{2}$

233 tan 315°

Here θ = ____, so θ_R = ____ - 315° = ____. Since the terminal side of 315° lies in quadrant ____, tan 315° is __________.

Hence, tan 315° = -tan 45° = ____.

315°, 360°, 45°
IV, negative
-1

234 sec 390°

Here θ = ____, so θ_R = 390° - ____ = ____. Since the terminal side of 390° lies in quadrant ____, sec 390° is __________.

Hence, sec 390° = sec 30° = ____.

390°, 360°, 30°
I, positive
$\frac{2\sqrt{3}}{3}$

235 csc 840°

Here θ = ____, which has the same terminal side as 840° - 720° = ____, so θ_R = ____ - 120° = ____. Since the terminal side of 840° lies in quadrant ____, csc 840° is ________.

Hence, csc 840° = csc 60° = ____.

840°
120°, 180°, 60°
II, positive
$\frac{2\sqrt{3}}{3}$

Answer	Problem
	236 sin(-225°)
-225°, 180°, 45°	Here θ = ________, so θ_R = 225° - ____ = ____. Since the ter-
II, positive	minal side of -225° lies in quadrant ____, sin(-225°) is ________.
45°, $\frac{\sqrt{2}}{2}$	Hence, sin(-225°) = sin ____ = ______.

5.2 Values of Trigonometric Functions—Tables

In Problems 237–246, use Appendix Table III for radian measures to determine the value.

Answer	Problem
0.8468	**237** sin 1.01 = ________
0.7317	**238** cos 0.75 = ________
0.2955	**239** sin 0.3 = ________
0.8727	**240** cos 0.51 = ________
0.7602	**241** tan 0.65 = ________
1.315	**242** cot 0.65 = ________
1.851	**243** sec 1 = ________
1.229	**244** csc 0.95 = ________
0.1315	**245** cot 1.44 = ________
1.000	**246** csc 1.55 = ________

In Problems 247–254, use Appendix Table IV for degree measures to determine the value.

Answer	Problem
0.5299	**247** sin 32° = ________
0.7314	**248** cos 43° = ________
1.540	**249** tan 57° = ________
0.9546	**250** cos 17°20′ = ________
3.236	**251** sec 72° = ________
1.093	**252** csc 66°10′ = ________
0.0262	**253** cot 88°30′ = ________
0.7969	**254** sin 52°50′ = ________

In Problems 255–274, use Appendix Table III or IV and reference angles to determine the value of the given expression. (Use π = 3.14.)

Answer	Problem
	255 sec 1.70
-, 1.44, -7.667	sec 1.70 = ____ sec ____________ (reference angle) = ____________
	256 sin 160°
20°, 0.3420	sin 160° = sin ____________ (reference angle) = ____________
	257 sin 3.63
-, 0.49, -0.4706	sin 3.63 = ____ sin ____ = ________
	258 cos 151°
-, 29°, -0.8746	cos 151° = ____ cos ____ = ________

	259 cos(-7)
7, 0.72, 0.7518	cos(-7) = cos ____ = cos ____ = ________
	260 cos 5.71
0.57, 0.8419	cos 5.71 = cos ____ = ________
	261 tan 213°
33°, 0.6494	tan 213° = tan ____ = ________
	262 sec 325°
35°, 1.221	sec 325° = sec ____ = ________
	263 sin(-2)
-2	sin(-2) = sin[6.28 + (____)]
4.28	= sin ____
-, 1.14	= ____ sin ____
-0.9086	= ________
	264 tan 11
1, 4.72	tan 11 = tan (____ · 6.28 + ____)
-, 1.56	= ____ tan ____
-92.620	= ________
	265 csc 171°10′
8°50′	csc 171°10′ = csc ________
6.512	= ________
	266 sin 293°40′
-, 66°20′	sin 293°40′ = ____ sin ________
-0.9159	= ________
	267 cot 14.98
2, 2.42	cot 14.98 = cot(____ · 6.28 + ____)
2.42	= cot ____
-, 0.72	= ____ cot ____
-1.140	= ________
	268 tan 162°30′
-, 17°30′	tan 162°30′ = ____ tan ________
-0.3153	= ________
	269 sec 29.12
4, 4	sec 29.12 = sec(____ · 6.28 + ____)
4	= sec ____
-, 0.86	= ____ sec ____
-1.533	= ________
	270 tan 560°20′
1, 200°20′	tan 560°20′ = tan(____ · 360° + ________)
200°20′	= tan ________
20°20′	= tan ________
0.3706	= ________

Answers	Problems
	271 csc 12.28
1, 6	csc 12.28 = csc(____ · 6.28 + ____)
6	= csc ____
-, 0.28	= ____ csc ____
-3.619	= ________
	272 tan 4.17
1.03, 1.665	tan 4.17 = tan ____ = ______
	273 cos 1033°10′
2, 313°10′	cos 1033°10′ = cos(____ · 360° + ________)
46°50′	= cos ________
0.6841	= ________
	274 cot(-10)
-	cot(-10) = ____ cot 10
-, 1, 3.72	= ____ cot(____ · 6.28 + ____)
-, 0.58, -1.526	= ____ cot ____ = ________

In Problems 275–282, use Appendix Table III or IV and linear interpolation to approximate each value. (Use $\pi = 3.142$.)

Answers	Problems
	275 sin 1.167
0.01, 0.004	$\underline{\quad}\left[\begin{array}{ll} & \begin{array}{cc} t & \sin t \\ 1.16 & 0.9168 \end{array} \\ 0.003 & \left[\begin{array}{cc} 1.167 & ? \\ 1.17 & 0.9208 \end{array}\right] d \end{array}\right]\underline{\qquad}$
0.01, 0.004	$\dfrac{0.003}{\underline{\quad}} = \dfrac{d}{\underline{\quad}}$
0.0012	Hence, d = ________, so that
0.0012, 0.9196	sin 1.167 = 0.9208 - ________ = ________.
	276 cos 0.758
0.002, 0.0069	$0.01\left[\begin{array}{ll} & \begin{array}{cc} t & \cos t \\ 0.75 & 0.7317 \end{array} \\ \underline{\qquad} & \left[\begin{array}{cc} 0.758 & ? \\ 0.76 & 0.7248 \end{array}\right] d \end{array}\right]\underline{\qquad}$
0.002, 0.0069	$\dfrac{\underline{\qquad}}{0.01} = \dfrac{d}{\underline{\qquad}}$
0.0014	Hence, d = ________, so that
0.0014, 0.7262	cos 0.758 = 0.7248 + ________ = ________.
	277 sin 24°43′
3′, 0.0027	$10'\left[\begin{array}{ll} & \begin{array}{cc} \theta & \sin\theta \end{array} \\ \underline{\quad} & \left[\begin{array}{cc} 24°40' & 0.4173 \\ 24°43' & ? \end{array}\right] d \\ & \begin{array}{cc} 24°50' & 0.4200 \end{array} \end{array}\right]\underline{\qquad}$

Answers	
3, 0.0027	$\dfrac{\_\_\_}{10} = \dfrac{d}{\_\_\_\_\_\_}$
0.0008	Hence, d = ________, so that
0.0008, 0.4181	sin 24°43′ = 0.4173 + ________ = ________.
	278 cos 64°26′
4′, 0.0026	$$10'\left[\_\_\_\left[\begin{array}{ll} \theta & \cos\theta \\ 64°20' & 0.4331 \\ 64°26' & ? \\ 64°30' & 0.4305 \end{array}\right]d\right]\_\_\_\_\_\_$$
4, 0.0026	$\dfrac{\_\_\_}{10} = \dfrac{d}{\_\_\_\_\_\_}$
0.0010	Hence, d = ________, so that
0.0010, 0.4315	cos 64°26′ = 0.4305 + ________ = ________.
	279 csc 0.603
0.003, 0.025	$$0.01\left[\_\_\_\left[\begin{array}{ll} t & \csc t \\ 0.60 & 1.771 \\ 0.603 & ? \end{array}\right]d \quad \begin{array}{ll} 0.61 & 1.746 \end{array}\right]\_\_\_\_\_\_$$
0.003, 0.025	$\dfrac{\_\_\_}{0.01} = \dfrac{d}{\_\_\_\_\_\_}$
0.008	d = ________
1.771, 1.763	Hence, csc 0.603 = ________ − 0.008 = ________.
	280 tan 3.048
0.0920	tan 3.048 = −tan(3.14 − 3.048) = −tan(________)
0.008, 0.0101	$$0.01\left[\begin{array}{ll} t & \tan t \\ 0.09 & 0.0902 \end{array} \quad \_\_\_\left[\begin{array}{ll} 0.092 & ? \\ 0.10 & 0.1003 \end{array}\right]d\right]\_\_\_\_\_\_$$
0.0101	$\dfrac{0.008}{0.01} = \dfrac{d}{\_\_\_\_\_\_}$
0.0081	d = ________
0.0081, 0.0922	tan 0.092 = 0.1003 − ________ = ________
0.0920, −0.0922	Thus, tan 3.048 = −tan(________) = ________.
	281 tan 20°25′
5′, 0.0033	$$10'\left[\_\_\_\left[\begin{array}{ll} \theta & \tan\theta \\ 20°20' & 0.3706 \\ 20°25' & ? \end{array}\right]d \quad \begin{array}{ll} 20°30' & 0.3739 \end{array}\right]\_\_\_\_\_\_$$
10, 0.0033	$\dfrac{5}{\_\_\_} = \dfrac{d}{\_\_\_\_\_\_}$
0.0017	d = ________
0.0017, 0.3723	Thus, tan 20°25′ = 0.3706 + ________ = ________.

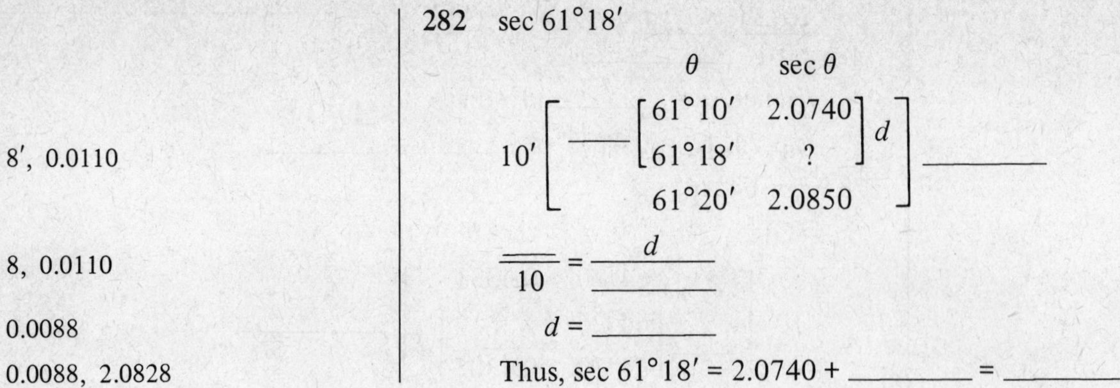

	282 sec 61°18′
	θ $\sec\theta$
8′, 0.0110	$10'\left[\underline{\quad}\left[\begin{matrix}61°10' & 2.0740\\ 61°18' & ?\end{matrix}\right]d \atop \begin{matrix}61°20' & 2.0850\end{matrix}\right]\underline{\qquad}$
8, 0.0110	$\dfrac{\underline{\quad}}{10} = \dfrac{d}{\underline{\qquad}}$
0.0088	$d = \underline{\qquad}$
0.0088, 2.0828	Thus, sec 61°18′ = 2.0740 + ______ = ______.

Chapter Test

1 Find the values of the six trigonometric functions at t if:

(a) $t = \dfrac{3\pi}{2}$ (b) $P(t) = \left(\dfrac{3\sqrt{13}}{13}, -\dfrac{2\sqrt{13}}{13}\right)$

2 Find the values of the following functions.

(a) $\sin\dfrac{11\pi}{6}$ (b) $\cos\dfrac{77\pi}{4}$ (c) $\tan\left(-\dfrac{5\pi}{3}\right)$

3 Find the period of each of the following functions.

(a) $f(x) = 3\cos 4x$ (b) $h(t) = 3\sin 3t$ (c) $g(w) = 4\cos(2w - 1)$

4 Find the values of the other trigonometric functions if $\cos t = \dfrac{3}{4}$, where $\dfrac{3\pi}{2} < t < 2\pi$.

5 Graph one cycle of each of the following functions. Indicate the period, amplitude, and phase shift.

(a) $y = 3\sin x$ (b) $y = \cos(2x - 1)$

6 Graph each of the following functions in the interval $[0, \pi]$.

(a) $f(x) = \tan 2x$ (b) $h(x) = 3\sec 3x$ (c) $g(x) = \csc 2x$

7 (a) Convert $\dfrac{7\pi}{3}$ radians to degree measure.   (b) Convert 405° to radian measure.

(c) What is the arc length and the area of that portion of a circle of radius 4 inches determined by a central angle of 75°?

8 Use the fact that $(-3, 5)$ is on the terminal side of angle θ to evaluate the six trigonometric functions of θ.

9 (a) Construct the angle θ in standard position whose terminal side is in the fourth quadrant and $\cot\theta = -\sqrt{3}$.

(b) Evaluate the remaining trigonometric functions of θ.

10 Evaluate each of the following.

(a) $\sin(-180°)$ (b) $\cot(-120°)$ (c) $\sec 600°$

11 Use Appendix Table III to evaluate each of the following. (Use $\pi = 3.14$.)

(a) $\cos 1.42$ (b) $\sin 6.47$ (c) $\cos 1.363$

12 Use Appendix Table IV to evaluate each of the following.

(a) $\cot 30°10'$ (b) $\tan 570°10'$ (c) $\sec(-120°43')$

Answers

1 (a) $\sin\frac{3\pi}{2} = -1$, $\cos\frac{3\pi}{2} = 0$, $\tan\frac{3\pi}{2}$ and $\sec\frac{3\pi}{2}$ are undefined, $\cot\frac{3\pi}{2} = 0$, $\csc\frac{3\pi}{2} = -1$

(b) $\sin t = -\frac{2\sqrt{13}}{13}$, $\cos t = \frac{3\sqrt{13}}{13}$, $\tan t = -\frac{2}{3}$, $\cot t = -\frac{3}{2}$, $\sec t = \frac{\sqrt{13}}{3}$, $\csc t = -\frac{\sqrt{13}}{2}$

2 (a) $-\frac{1}{2}$ (b) $-\frac{\sqrt{2}}{2}$ (c) $\sqrt{3}$

3 (a) $\frac{\pi}{2}$ (b) $\frac{2\pi}{3}$ (c) π

4 $\sin t = -\frac{\sqrt{7}}{4}$, $\tan t = -\frac{\sqrt{7}}{3}$, $\cot t = -\frac{3\sqrt{7}}{7}$, $\sec t = \frac{4}{3}$, $\csc t = -\frac{4\sqrt{7}}{7}$

5 (a) Period is 2π; amplitude is 3; phase shift is 0 (b) Period is π; amplitude is 1; phase shift is $\frac{1}{2}$

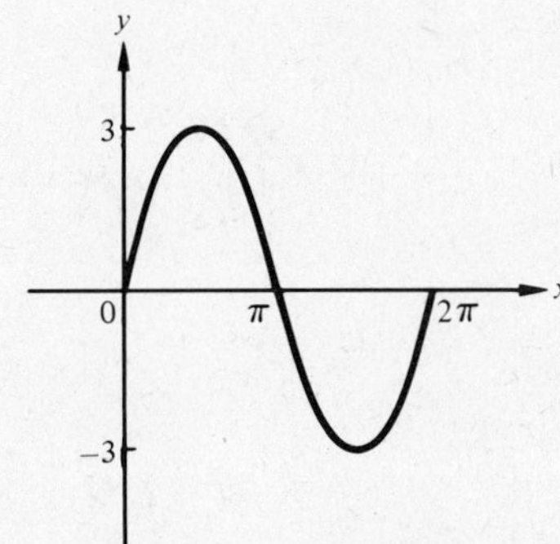

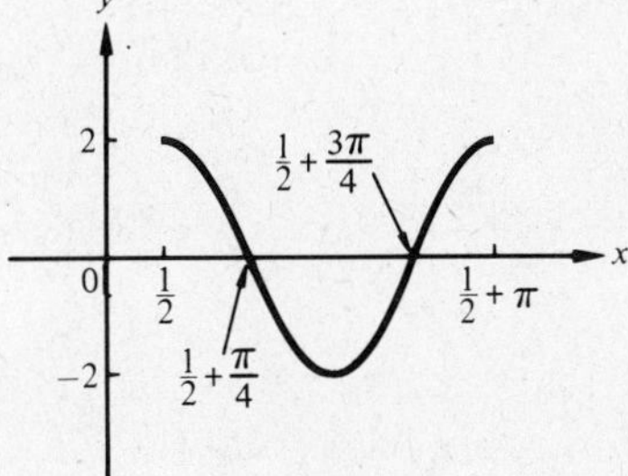

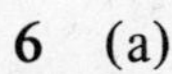
6 (a)

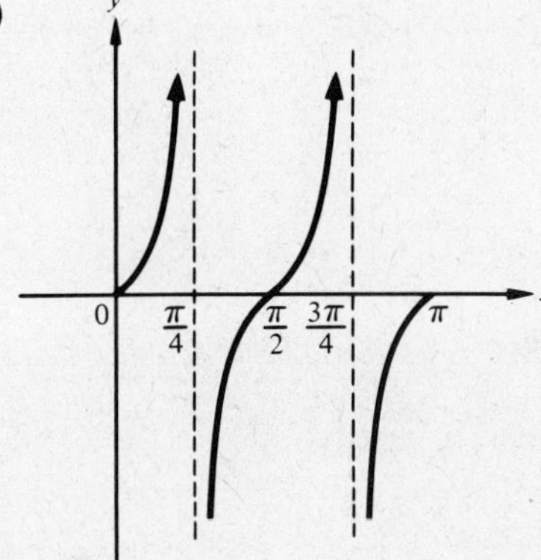

(b)

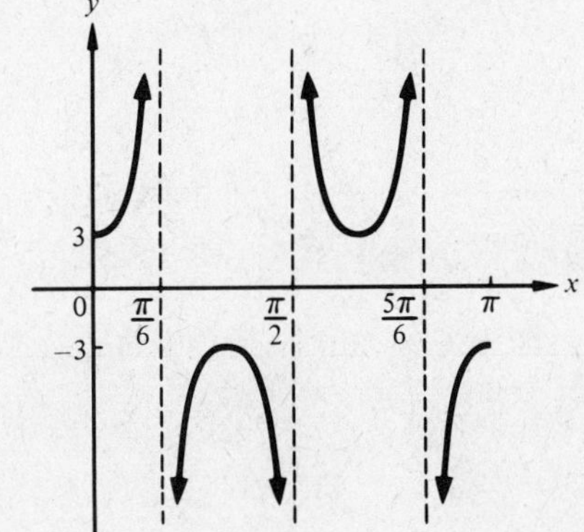

(c)

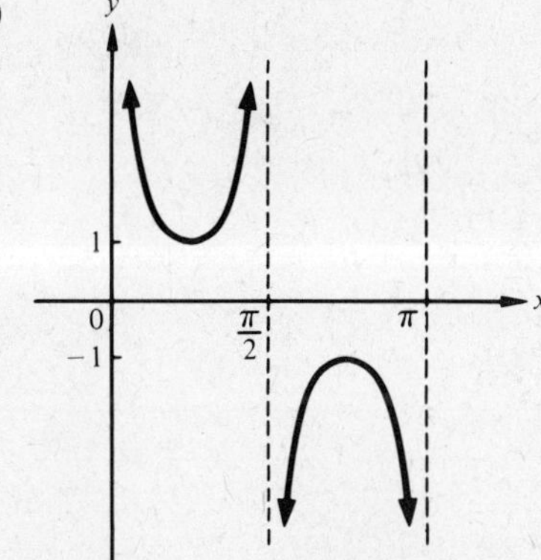

7 (a) 420° (b) $\frac{9\pi}{4}$ (c) $\frac{5\pi}{3}$ inches, $\frac{10\pi}{3}$ square inches

8 $\sin\theta = \frac{5\sqrt{34}}{34}$, $\cos\theta = -\frac{3\sqrt{34}}{34}$, $\tan\theta = -\frac{5}{3}$, $\cot\theta = -\frac{3}{5}$, $\sec\theta = -\frac{\sqrt{34}}{3}$, $\csc\theta = \frac{\sqrt{34}}{5}$

9 (a)

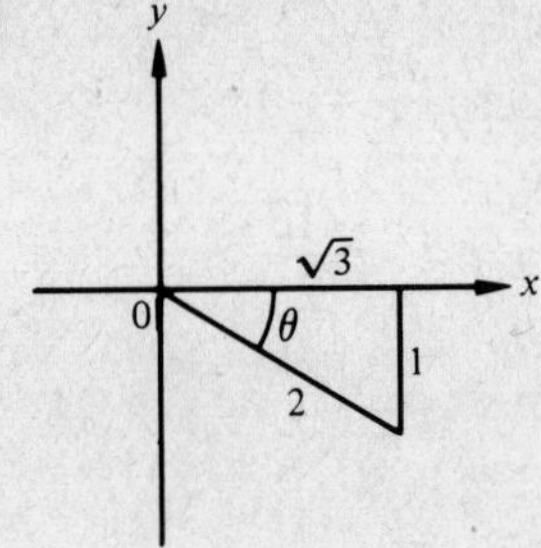

(b) $\sin\theta = -\frac{1}{2}$, $\cos\theta = \frac{\sqrt{3}}{2}$, $\tan\theta = -\frac{\sqrt{3}}{3}$, $\sec\theta = \frac{2\sqrt{3}}{3}$, $\csc\theta = -2$

10 (a) 0 (b) $\frac{\sqrt{3}}{3}$ (c) -2

11 (a) 0.1502 (b) 0.1889 (c) 0.2063

12 (a) 1.720 (b) 0.5812 (c) -1.958

Chapter 6 ANALYTIC TRIGONOMETRY

In this chapter we use the properties of the trigonometric functions to investigate trigonometric identities and trigonometric formulas and the inverses of trigonometric functions. After completing the appropriate sections of this chapter, the student should be able to:

1 Use the fundamental identities to verify other identities.

2 Use trigonometric formulas to verify identities.

3 Define and compute values of the inverse trigonometric functions.

4 Find solution sets of trigonometric equations.

5 Find missing parts of right triangles and convert polar coordinates.

6 Find missing parts of triangles by using the law of sines and law of cosines.

7 Apply trigonometry to solving problems involving vectors.

1 Fundamental Identities

Answer	Problem
$\cos t$	1 $\tan t = \dfrac{\sin t}{\_\_\_\_\_\_}$, $\cos t \neq 0$
$\sin t$	2 $\cot t = \dfrac{\cos t}{\_\_\_\_\_\_}$, $\sin t \neq 0$
$\cos t$	3 $\sec t = \dfrac{1}{\_\_\_\_\_\_}$, $\cos t \neq 0$
$\sin t$	4 $\csc t = \dfrac{1}{\_\_\_\_\_\_}$, $\sin t \neq 0$
$\tan t$	5 $\cot t = \dfrac{1}{\_\_\_\_\_\_}$, $\tan t \neq 0$
$\cos t$	6 $\cos(-t) =$ ______
$-\sin t$	7 $\sin(-t) =$ ______
$-\tan t$	8 $\tan(-t) =$ ______
$-\cot t$	9 $\cot(-t) =$ ______
$\sec t$	10 $\sec(-t) =$ ______
$-\csc t$	11 $\csc(-t) =$ ______
1	12 $\sin^2 t + \cos^2 t =$ ____
$\sec^2 t$	13 $\tan^2 t + 1 =$ ______
$\csc^2 t$	14 $\cot^2 t + 1 =$ ______

In Problems 15–19, use the fundamental identities to express each of the trigonometric function values in terms of x if $\sin t = 1/x$, where $0 < t < \pi/2$.

Answer	Problem
$\dfrac{1}{x}$, $\dfrac{\sqrt{x^2 - 1}}{x}$	15 $\cos t = \sqrt{1 - \sin^2 t} = \sqrt{1 - \left(\_\_\_\right)^2} =$ ______
$\cos t$, $\dfrac{\sqrt{x^2 - 1}}{x}$, $\sqrt{x^2 - 1}$	16 $\tan t = \dfrac{\sin t}{\_\_\_\_\_\_} = \dfrac{\frac{1}{x}}{\_\_\_\_\_\_} = \dfrac{1}{\_\_\_\_\_\_}$
$\sin t$, $\dfrac{\sqrt{x^2 - 1}}{x}$, $\sqrt{x^2 - 1}$	17 $\cot t = \dfrac{\cos t}{\_\_\_\_\_\_} = \dfrac{\_\_\_\_\_\_}{\frac{1}{x}} =$ ______
$\cos t$, $\dfrac{\sqrt{x^2 - 1}}{x}$, $\dfrac{x}{\sqrt{x^2 - 1}}$	18 $\sec t = \dfrac{1}{\_\_\_\_\_\_} = \dfrac{1}{\_\_\_\_\_\_} =$ ______
$\dfrac{x\sqrt{x^2 - 1}}{x^2 - 1}$	$=$ ______
$\sin t$, $\dfrac{1}{x}$, x	19 $\csc t = \dfrac{1}{\_\_\_\_\_\_} = \dfrac{1}{\_\_\_} =$ ___

In Problems 20–41, use the fundamental identities to simplify the given expression.

20 $5 \sin^2 30° + 5 \cos^2 30°$

$5 \sin^2 30° + 5 \cos^2 30° = 5(\sin^2 30° + \underline{\qquad})$ — $\cos^2 30°$

$= 5(\underline{\quad}) = \underline{\quad}$ — 1, 5

21 $\tan 45° \cot 45° + \tan^2 45°$

$\tan 45° \cot 45° + \tan^2 45° = \tan 45° \left(\frac{1}{\underline{\qquad}}\right) + \tan^2 45°$ — $\tan 45°$

$= \underline{\quad} + \tan^2 45°$ — 1

$= \underline{\qquad} = \underline{\quad}$ — $\sec^2 45°$, 2

22 $\left(1 - \sin \frac{\pi}{4}\right)\left(1 + \sin \frac{\pi}{4}\right)$

$\left(1 - \sin \frac{\pi}{4}\right)\left(1 + \sin \frac{\pi}{4}\right) = 1 - \underline{\qquad} = \left(\underline{\qquad}\right)^2$ — $\sin^2 \frac{\pi}{4}$, $\cos \frac{\pi}{4}$

$= \left(\underline{\quad}\right)^2 = \underline{\quad}$ — $\frac{\sqrt{2}}{2}$, $\frac{1}{2}$

23 $\csc^2 73° - \cot^2 73°$

Since $\csc^2 \theta = 1 + \underline{\qquad}$, $\csc^2 \theta - \cot^2 \theta = \underline{\quad}$. Hence, — $\cot^2 \theta$, 1

$\csc^2 73° - \cot^2 73° = \underline{\quad}$. — 1

24 $\sec^2 19° - \tan^2 19°$

Since $\sec^2 \theta = 1 + \underline{\qquad}$, $\sec^2 \theta - \tan^2 \theta = \underline{\quad}$. Hence, — $\tan^2 \theta$, 1

$\sec^2 19° - \tan^2 19° = \underline{\quad}$. — 1

25 $\dfrac{2 \tan(\pi/4)}{1 + \tan^2(\pi/4)}$

Since $1 + \tan^2 \frac{\pi}{4} = \underline{\qquad} = \underline{\quad}$, then — $\sec^2 \frac{\pi}{4}$, 2

$\dfrac{2 \tan(\pi/4)}{1 + \tan^2(\pi/4)} = \dfrac{2(\underline{\quad})}{2} = \underline{\quad}$. — 1, 1

26 $\dfrac{1}{1 + \sin(\pi/6)} + \dfrac{1}{1 - \sin(\pi/6)}$

$$\frac{1}{1 + \sin\frac{\pi}{6}} + \frac{1}{1 - \sin\frac{\pi}{6}} = \frac{1 - \sin\frac{\pi}{6} + 1 + \sin\frac{\pi}{6}}{\underline{\qquad\qquad}}$$

— $1 - \sin^2 \frac{\pi}{6}$

$$= \frac{\underline{\quad}}{\cos^2 \frac{\pi}{6}} = \frac{2}{\underline{\quad}} = \underline{\quad}$$

— 2, $\frac{3}{4}$, $\frac{8}{3}$

27 $\cos 16° \tan 16° \csc 16°$

$\cos 16° \tan 16° \csc 16° = \cos 16° \left(\frac{\sin 16°}{\underline{\qquad}}\right)\left(\frac{1}{\underline{\qquad}}\right)$ — $\cos 16°$, $\sin 16°$

$= \dfrac{\cos 16° \sin 16°}{\underline{\qquad\qquad}} = \underline{\quad}$ — $\cos 16° \sin 16°$, 1

	28 $\cos^2 t(1 + \tan^2 t)$
$\sec^2 t$	$\cos^2 t(1 + \tan^2 t) = \cos^2 t \cdot \underline{\qquad\qquad}$
$\cos^2 t,\ 1$	$= \cos^2 t \left(\dfrac{1}{\underline{\qquad\qquad}}\right) = \underline{\qquad}$
	29 $\dfrac{\sin\theta \sec\theta}{\tan\theta}$
$\cos\theta,\ \tan\theta,\ 1$	$\dfrac{\sin\theta \sec\theta}{\tan\theta} = \dfrac{\sin\theta \left(\dfrac{1}{\underline{\qquad\qquad}}\right)}{\tan\theta} = \dfrac{\underline{\qquad\qquad}}{\tan\theta} = \underline{\qquad}$
	30 $(\sec^2 t - 1)(\csc^2 t - 1)$
$\cot^2 t,\ 1$	$(\sec^2 t - 1)(\csc^2 t - 1) = \tan^2 t(\underline{\qquad\qquad}) = \underline{\qquad}$
	31 $\tan^2 t \csc^2 t - \tan^2 t$
$\csc^2 t - 1$	$\tan^2 t \csc^2 t - \tan^2 t = \tan^2 t(\underline{\qquad\qquad})$
$\cot^2 t,\ 1$	$= \tan^2 t(\underline{\qquad\qquad}) = \underline{\qquad}$
	32 $\csc^2\theta(1 - \cos^2\theta)$
$\sin^2\theta$	$\csc^2\theta(1 - \cos^2\theta) = \csc^2\theta(\underline{\qquad\qquad})$
$\sin^2\theta,\ 1$	$= \left(\dfrac{1}{\underline{\qquad\qquad}}\right)\sin^2\theta = \underline{\qquad}$
	33 $\sec t \cot t$
$\cos t,\ \dfrac{1}{\sin t},\ \csc t$	$\sec t \cot t = \dfrac{1}{\cos t}\left(\dfrac{\underline{\qquad\qquad}}{\sin t}\right) = \dfrac{\underline{\qquad\qquad}}{\underline{\qquad\qquad}} = \underline{\qquad\qquad}$
	34 $\csc\theta \tan\theta$
$\cos\theta,\ \sec\theta$	$\csc\theta \tan\theta = \dfrac{1}{\sin\theta}\left(\dfrac{\sin\theta}{\underline{\qquad\qquad}}\right) = \underline{\qquad\qquad}$
	35 $\cos^2 t(\sec^2 t - 1)$
$\tan^2 t$	$\cos^2 t(\sec^2 t - 1) = \cos^2 t(\underline{\qquad\qquad})$
$\sin^2 t$	$= \cos^2 t \left(\dfrac{\sin^2 t}{\cos^2 t}\right) = \underline{\qquad\qquad}$
	36 $\cot^2 t - \cos^2 t$
$\sin^2 t$	$\cot^2 t - \cos^2 t = \dfrac{\cos^2 t}{\underline{\qquad\qquad}} - \cos^2 t$
$\cos^2 t \sin^2 t$	$= \dfrac{\cos^2 t - \underline{\qquad\qquad\qquad}}{\sin^2 t}$
$1 - \sin^2 t$	$= \dfrac{\cos^2 t(\underline{\qquad\qquad\qquad})}{\sin^2 t}$
$\cos^2 t$	$= \cot^2 t\ \underline{\qquad\qquad}$
	37 $3\sin^2\theta + 4\cos^2\theta$
$1 - \cos^2\theta$	$3\sin^2\theta + 4\cos^2\theta = 3(\underline{\qquad\qquad\qquad}) + 4\cos^2\theta$
$3\cos^2\theta$	$= 3 - \underline{\qquad\qquad\qquad} + 4\cos^2\theta$
$3 + \cos^2\theta$	$= \underline{\qquad\qquad\qquad}$

38 $\sec^4 t - \tan^4 t$

$\sec^4 t - \tan^4 t = (\sec^2 t + \tan^2 t)(\_\_\_\_\_\_\_\_)$

$= (\sec^2 t + \tan^2 t)(\_\_\_\_)$

$= \_\_\_\_\_\_\_\_$

$\sec^2 t - \tan^2 t$

1

$\sec^2 t + \tan^2 t$

39 $\cos^4 \theta - \sin^4 \theta$

$\cos^4 \theta - \sin^4 \theta = (\cos^2 \theta - \sin^2 \theta)(\_\_\_\_\_\_\_\_)$

$= (\cos^2 \theta - \sin^2 \theta)(\_\_\_\_)$

$= \_\_\_\_\_\_\_\_$

$\cos^2 \theta + \sin^2 \theta$

1

$\cos^2 \theta - \sin^2 \theta$

40 $(1 - \cos t)(1 + \sec t)$

$(1 - \cos t)(1 + \sec t) = 1 + \sec t - \_\_\_\_ - \_\_\_\_\_\_$

$= 1 + \sec t - \cos t - \_\_\_\_$

$= \_\_\_\_\_\_\_\_$

$\cos t,\ \cos t \sec t$

1

$\sec t - \cos t$

41 $\dfrac{1}{1 + \sin \theta} + \dfrac{1}{1 - \sin \theta}$

$\dfrac{1}{1 + \sin \theta} + \dfrac{1}{1 - \sin \theta} = \dfrac{1 - \sin \theta + \_\_\_\_\_\_}{(1 + \sin \theta)(1 - \sin \theta)}$

$= \dfrac{2}{\_\_\_\_\_\_}$

$= \dfrac{2}{\_\_\_\_}$

$= \_\_\_\_\_\_$

$1 + \sin \theta$

$1 - \sin^2 \theta$

$\cos^2 \theta$

$2 \sec^2 \theta$

1.1 Verifying Trigonometric Identities

In Problems 42–52, verify the identity.

42 $\cos \theta \tan \theta = \sin \theta$

Proof:

$\cos \theta \tan \theta$

$= \cos \theta \left(\dfrac{\sin \theta}{\cos \theta}\right)$

$= \sin \theta$

43 $\dfrac{\tan t}{\sin t \sec t} = 1$

Proof:

$\dfrac{\tan t}{\sin t \sec t}$

$= \dfrac{\dfrac{\sin t}{\cos t}}{\sin t \left(\dfrac{1}{\cos t}\right)}$

$= \left(\dfrac{\sin t}{\cos t}\right)\left(\dfrac{\cos t}{\sin t}\right)$

$= 1$

44 $\dfrac{\sec t \csc t}{\tan t + \cot t} = 1$

Proof:

$$\frac{\sec t \csc t}{\tan t + \cot t}$$

$$= \frac{\left(\frac{1}{\cos t}\right)\left(\frac{1}{\sin t}\right)}{\frac{\sin t}{\cos t} + \frac{\cos t}{\sin t}}$$

$$= \frac{\frac{1}{\cos t \sin t}}{\frac{\sin^2 t + \cos^2 t}{\cos t \sin t}}$$

$$= \left(\frac{1}{\cos t \sin t}\right)\left(\frac{\cos t \sin t}{1}\right)$$

$$= 1$$

45 $\dfrac{\sin t - \cos t}{\cos t} + 1 = \tan t$

Proof:

$$\frac{\sin t - \cos t}{\cos t} + 1$$

$$= \frac{\sin t - \cos t + \cos t}{\cos t}$$

$$= \tan t$$

46 $\sec t - \sec t \sin^2 t = \cos t$

Proof:

$$\sec t - \sec t \sin^2 t$$

$$= \sec t(1 - \sin^2 t)$$

$$= \left(\frac{1}{\cos t}\right)\cos^2 t$$

$$= \cos t$$

47 $\dfrac{\sin t}{1 - \cos t} - \cot t = \csc t$

Proof:

$$\frac{\sin t}{1 - \cos t} - \cot t$$

$$= \frac{\sin t}{1 - \cos t} - \frac{\cos t}{\sin t}$$

$$= \frac{\sin^2 t - \cos t + \cos^2 t}{(1 - \cos t)\sin t}$$

$$= \frac{1 - \cos t}{(1 - \cos t)\sin t}$$

$$= \frac{1}{\sin t} = \csc t$$

48 $\dfrac{\sin t}{1 - \cos t} + \dfrac{1 + \cos t}{\sin t} = 2 \csc t(1 + \cos t)$

Proof:

$$\frac{\sin t}{1 - \cos t} + \frac{1 + \cos t}{\sin t}$$
$$= \frac{\sin^2 t + 1 - \cos^2 t}{(1 - \cos t)\sin t}$$
$$= \frac{1 - \cos^2 t + 1 - \cos^2 t}{(1 - \cos t)\sin t}$$
$$= \frac{2 - 2\cos^2 t}{(1 - \cos t)\sin t}$$
$$= \frac{2(1 - \cos^2 t)}{(1 - \cos t)\sin t}$$
$$= \frac{2(1 + \cos t)}{\sin t}$$
$$= 2 \csc t(1 + \cos t)$$

49 $\tan\theta \sin\theta + \cos\theta = \sec\theta$

Proof:

$$\tan\theta \sin\theta + \cos\theta$$
$$= \left(\frac{\sin\theta}{\cos\theta}\right)\sin\theta + \cos\theta$$
$$= \frac{\sin^2\theta + \cos^2\theta}{\cos\theta}$$
$$= \frac{1}{\cos\theta}$$
$$= \sec\theta$$

50 $\tan^2\theta - \sin^2\theta = \sin^2\theta \tan^2\theta$

Proof:

$$\tan^2\theta - \sin^2\theta$$
$$= \frac{\sin^2\theta}{\cos^2\theta} - \sin^2\theta$$
$$= \sin^2\theta\left(\frac{1}{\cos^2\theta} - 1\right)$$
$$= \sin^2\theta\left(\frac{1 - \cos^2\theta}{\cos^2\theta}\right)$$
$$= \sin^2\theta\left(\frac{\sin^2\theta}{\cos^2\theta}\right)$$
$$= \sin^2\theta \tan^2\theta$$

51 $\dfrac{\sec\theta + \csc\theta}{\tan\theta + \cot\theta} = \sin\theta + \cos\theta$

Proof:

$$\frac{\sec\theta + \csc\theta}{\tan\theta + \cot\theta}$$
$$= \frac{\dfrac{1}{\cos\theta} + \dfrac{1}{\sin\theta}}{\dfrac{\sin\theta}{\cos\theta} + \dfrac{\cos\theta}{\sin\theta}}$$
$$= \frac{\sin\theta + \cos\theta}{\sin^2\theta + \cos^2\theta}$$
$$= \sin\theta + \cos\theta$$

Proof:

$$\cot\theta + \frac{\sin\theta}{1+\cos\theta}$$
$$= \frac{\cos\theta}{\sin\theta} + \frac{\sin\theta}{1+\cos\theta}$$
$$= \frac{\cos\theta + \cos^2\theta + \sin^2\theta}{\sin\theta\,(1+\cos\theta)}$$
$$= \frac{\cos\theta + 1}{\sin\theta\,(1+\cos\theta)}$$
$$= \csc\theta$$

52 $\cot\theta + \dfrac{\sin\theta}{1+\cos\theta} = \csc\theta$

2 Trigonometric Formulas

2.1 Difference and Sum Formulas

$\cos s$, $\sin s$

53 $\cos(t - s) = \cos t(\underline{\quad\quad}) + \sin t(\underline{\quad\quad})$

In Problems 54–58, use Problem 53 to simplify the expression.

54 $\cos\left(\dfrac{\pi}{2} - t\right)$

$\sin\dfrac{\pi}{2}$ | $\cos\left(\dfrac{\pi}{2} - t\right) = \left(\cos\dfrac{\pi}{2}\right)\cos t + \left(\underline{\quad\quad}\right)\sin t$

0, 1 | $= \underline{\quad} + (\underline{\quad})\sin t$

$\sin t$ | $= \underline{\quad\quad}$

55 $\cos(\pi - t)$

$\cos t$, $\sin\pi$ | $\cos(\pi - t) = \cos\pi(\underline{\quad\quad}) + (\underline{\quad\quad})\sin t$

$\cos t$, 0 | $= -1(\underline{\quad\quad}) + (\underline{\quad})\sin t$

$-\cos t$ | $= \underline{\quad\quad}$

56 $\cos(\theta - 180^\circ)$

$\cos 180^\circ$, $\sin 180^\circ$ | $\cos(\theta - 180^\circ) = \cos\theta(\underline{\quad\quad}) + \sin\theta(\underline{\quad\quad})$

-1, 0 | $= \cos\theta(\underline{\quad}) + \sin\theta(\underline{\quad})$

$-\cos\theta$ | $= \underline{\quad\quad}$

57 $\cos(270^\circ - \theta)$

$\sin 270^\circ$ | $\cos(270^\circ - \theta) = \cos 270^\circ\cos\theta + (\underline{\quad\quad})\sin\theta$

0, -1 | $= (\underline{\quad})\cos\theta + (\underline{\quad})\sin\theta$

$-\sin\theta$ | $= \underline{\quad\quad}$

58 $\cos(\theta - 90^\circ)$

$\sin 90^\circ$ | $\cos(\theta - 90^\circ) = \cos\theta\cos 90^\circ + \sin\theta(\underline{\quad\quad})$

0, 1 | $= \cos\theta(\underline{\quad}) + \sin\theta(\underline{\quad})$

$\sin\theta$ | $= \underline{\quad\quad}$

cos s, sin s

cos$(-s)$, sin$(-s)$

cos s, $-$sin s

cos s, $-$sin s

cos s, sin s

59 $\cos(t+s) = \cos t(\_\_\_\_) - \sin t(\_\_\_\_)$

This identity can be derived from the identity in Problem 53 as follows:

$\cos(t+s) = \cos[t-(-s)]$

$\quad = \cos t[\_\_\_\_\_\_] + \sin t[\_\_\_\_\_\_]$

Since $\cos(-s) =$ ____ and $\sin(-s) =$ ____, it follows that

$\cos t \cos(-s) + \sin t \sin(-s) = \cos t(\_\_\_\_) + \sin t(\_\_\_\_)$

$\quad = \cos t\ \_\_\_\_ - \sin t\ \_\_\_\_$

In Problems 60–63, use Problem 59 to simplify the expression.

sin t

0, 1

$-$sin t

60 $\cos\left(\frac{\pi}{2}+t\right)$

$\cos\left(\frac{\pi}{2}+t\right) = \cos\frac{\pi}{2}\cos t - \sin\frac{\pi}{2}(\_\_\_\_)$

$\quad = \_\_\_\_ \cos t - \_\_\_\_ \sin t$

$\quad = \_\_\_\_$

cos 90°, sin 90°

0, 1

$-$sin θ

61 $\cos(\theta + 90^\circ)$

$\cos(\theta + 90^\circ) = \cos\theta(\_\_\_\_) - \sin\theta(\_\_\_\_)$

$\quad = \cos\theta(\_\_\_\_) - \sin\theta(\_\_\_\_)$

$\quad = \_\_\_\_$

sin θ

0, -1

sin θ

62 $\cos(270^\circ + \theta)$

$\cos(270^\circ + \theta) = \cos 270^\circ \cos\theta - \sin 270^\circ(\_\_\_\_)$

$\quad = (\_\_\_\_)\cos\theta - (\_\_\_\_)\sin\theta$

$\quad = \_\_\_\_$

$\cos\frac{3\pi}{2}$, $\sin\frac{3\pi}{2}$

0, -1

sin t

63 $\cos\left(t+\frac{3\pi}{2}\right)$

$\cos\left(t+\frac{3\pi}{2}\right) = \cos t\left(\_\_\_\_\right) - \sin t\left(\_\_\_\_\right)$

$\quad = \cos t(\_\_\_\_) - \sin t(\_\_\_\_)$

$\quad = \_\_\_\_$

cos s, cos t

64 $\sin(t+s) = \sin t(\_\_\_\_) + \sin s(\_\_\_\_)$

In Problems 65–68, use Problem 64 to simplify the expression.

cos t, $\cos\frac{\pi}{2}$

1, 0

cos t

65 $\sin\left(\frac{\pi}{2}+t\right)$

$\sin\left(\frac{\pi}{2}+t\right) = \sin\frac{\pi}{2}(\_\_\_\_) + \sin t\left(\_\_\_\_\right)$

$\quad = (\_\_\_\_)\cos t + \sin t(\_\_\_\_)$

$\quad = \_\_\_\_$

	66 $\sin(270^\circ + \theta)$
$\cos\theta$, $\cos 270^\circ$	$\sin(270^\circ + \theta) = \sin 270^\circ(\_\_\_\_\_\_) + \sin\theta(\_\_\_\_\_\_\_\_\_\_)$
-1, 0	$= (\_\_\_\_)\cos\theta + \sin\theta(\_\_\_\_)$
$-\cos\theta$	$= \_\_\_\_\_\_\_\_$
	67 $\sin(360^\circ + \theta)$
$\cos\theta$, $\cos 360^\circ$	$\sin(360^\circ + \theta) = \sin 360^\circ(\_\_\_\_\_\_) + \sin\theta(\_\_\_\_\_\_\_\_\_\_)$
0, 1	$= (\_\_\_\_)\cos\theta + \sin\theta(\_\_\_\_)$
$\sin\theta$	$= \_\_\_\_\_\_$
	68 $\sin(\pi + t)$
$\cos t$, $\cos\pi$	$\sin(\pi + t) = \sin\pi(\_\_\_\_\_\_) + \sin t(\_\_\_\_\_\_)$
0, -1	$= (\_\_\_\_)\cos t + \sin t(\_\_\_\_)$
$-\sin t$	$= \_\_\_\_\_\_\_\_$
$\cos s$, $\cos t$	**69** $\sin(t - s) = \sin t(\_\_\_\_\_\_) - \sin s(\_\_\_\_\_\_)$

In Problems 70–73, use Problem 69 to simplify the expression.

	70 $\sin\left(\frac{\pi}{2} - t\right)$
$\cos t$, $\cos\frac{\pi}{2}$	$\sin\left(\frac{\pi}{2} - t\right) = \sin\frac{\pi}{2}(\_\_\_\_\_\_) - \sin t\left(\_\_\_\_\_\_\_\_\right)$
1, 0	$= (\_\_\_\_)\cos t - \sin t(\_\_\_\_)$
$\cos t$	$= \_\_\_\_\_\_$
	71 $\sin(\pi - t)$
$\cos\pi$	$\sin(\pi - t) = \sin\pi\cos t - \sin t(\_\_\_\_\_\_)$
0, -1	$= (\_\_\_\_)\cos t - \sin t(\_\_\_\_)$
$\sin t$	$= \_\_\_\_\_\_$
	72 $\sin(2\pi - t)$
$\cos t$, $\cos 2\pi$	$\sin(2\pi - t) = \sin 2\pi(\_\_\_\_\_\_) - \sin t(\_\_\_\_\_\_\_\_)$
0, 1	$= (\_\_\_\_)\cos t - \sin t(\_\_\_\_)$
$-\sin t$	$= \_\_\_\_\_\_\_\_$
	73 $\sin(\theta - 270^\circ)$
$\sin 270^\circ\cos\theta$	$\sin(\theta - 270^\circ) = \sin\theta\cos 270^\circ - \_\_\_\_\_\_\_\_\_\_\_\_\_\_$
0, -1	$= \sin\theta(\_\_\_\_) - \cos\theta(\_\_\_\_)$
$\cos\theta$	$= \_\_\_\_\_\_$
$\tan s$	**74** $\tan(t + s) = \dfrac{\tan t + \tan s}{1 - \tan t(\_\_\_\_\_\_)}$
$\tan(-t)$	**75** $\tan(t - s) = \dfrac{\tan t + \tan(-s)}{1 - \tan t[\_\_\_\_\_\_\_\_]}$
$\tan s$	$= \dfrac{\tan t - \_\_\_\_}{1 + \tan t\tan s}$

In Problems 76–83, use the formulas for the sum and for the difference to find the value of the given expression.

76 $\sin \frac{7\pi}{12}$

$\frac{\pi}{3}$ — $\frac{7\pi}{12} = \frac{\pi}{4} + \_\_\_\_$, so that

$\sin \frac{7\pi}{12} = \sin\left(\frac{\pi}{4} + \frac{\pi}{3}\right)$

$\cos\frac{\pi}{3}, \cos\frac{\pi}{4}$ — $= \sin\frac{\pi}{4}\left(\_\_\_\_\right) + \sin\frac{\pi}{3}\left(\_\_\_\_\right)$

$\frac{1}{2}, \frac{\sqrt{2}}{2}$ — $= \frac{\sqrt{2}}{2}\left(\_\_\right) + \frac{\sqrt{3}}{2}\left(\_\_\right)$

$\frac{\sqrt{6}}{4}$ — $= \frac{\sqrt{2}}{4} + \_\_\_\_$

$\frac{\sqrt{2}+\sqrt{6}}{4}$ — $= \_\_\_\_\_\_\_\_$

77 $\cos \frac{11\pi}{12}$

$\frac{\pi}{4}$ — $\frac{11\pi}{12} = \frac{2\pi}{3} + \_\_\_\_$, so that

$\cos \frac{11\pi}{12} = \cos\left(\frac{2\pi}{3} + \frac{\pi}{4}\right)$

$\cos\frac{\pi}{4}, \sin\frac{\pi}{4}$ — $= \cos\frac{2\pi}{3}\left(\_\_\_\_\right) - \sin\frac{2\pi}{3}\left(\_\_\_\_\right)$

$\frac{\sqrt{2}}{2}, \frac{\sqrt{2}}{2}$ — $= -\frac{1}{2}\left(\_\_\right) - \frac{\sqrt{3}}{2}\left(\_\_\right)$

$-\frac{\sqrt{2}}{4}, \frac{\sqrt{6}}{4}$ — $= \left(\_\_\right) - \left(\_\_\right)$

$-\frac{\sqrt{2}+\sqrt{6}}{4}$ — $= \_\_\_\_\_\_\_\_$

78 $\cos \frac{5\pi}{12}$

$\frac{\pi}{6}$ — $\frac{5\pi}{12} = \frac{\pi}{4} + \_\_\_\_$, so that

$\cos \frac{5\pi}{12} = \cos\left(\frac{\pi}{4} + \frac{\pi}{6}\right)$

$\cos\frac{\pi}{6}, \sin\frac{\pi}{6}$ — $= \cos\frac{\pi}{4}\left(\_\_\_\_\right) - \sin\frac{\pi}{4}\left(\_\_\_\_\right)$

$\frac{\sqrt{3}}{2}, \frac{1}{2}$ — $= \frac{\sqrt{2}}{2}\left(\_\_\right) - \frac{\sqrt{2}}{2}\left(\_\_\right)$

$\frac{\sqrt{6}}{4}, \frac{\sqrt{2}}{4}$ — $= \left(\_\_\_\_\right) - \left(\_\_\_\_\right)$

$\frac{\sqrt{6}-\sqrt{2}}{4}$ — $= \_\_\_\_\_\_\_\_$

79 sin 285°

$285° = 150° +$ _____, so that

$\sin 285° = \sin(150° + 135°)$

$= \sin 150°(\_\_\_\_\_\_\_\_\_\_) + \sin 135°(\_\_\_\_\_\_\_\_\_\_)$

$= \frac{1}{2}(\_\_\_\_\_) + \frac{\sqrt{2}}{2}(\_\_\_\_\_)$

$= (\_\_\_\_\_) + (\_\_\_\_\_)$

$= \_\_\_\_\_\_\_\_$

Answers: 135°; cos 135°, cos 150°; $-\frac{\sqrt{2}}{2}, -\frac{\sqrt{3}}{2}$; $-\frac{\sqrt{2}}{4}, -\frac{\sqrt{6}}{4}$; $-\frac{\sqrt{2}+\sqrt{6}}{4}$

80 sin 15°

$15° = 45° -$ _____, so that

$\sin 15° = \sin(45° - 30°)$

$= \sin 45° \_\_\_\_\_\_\_\_ - \cos 45° \_\_\_\_\_\_\_\_$

$= (\_\_\_\_)(\_\_\_\_) - (\_\_\_\_)(\_\_\_\_)$

$= (\_\_\_\_\_) - (\_\_\_\_\_)$

$= \_\_\_\_\_\_\_\_$

Answers: 30°; cos 30°, sin 30°; $\frac{\sqrt{2}}{2}, \frac{\sqrt{3}}{2}, \frac{\sqrt{2}}{2}, \frac{1}{2}$; $\frac{\sqrt{6}}{4}, \frac{\sqrt{2}}{4}$; $\frac{\sqrt{6}-\sqrt{2}}{4}$

81 cos 255°

$255° = 210° +$ _____, so that

$\cos 255° = \cos(210° + 45°)$

$= \cos 210°(\_\_\_\_\_\_\_\_\_\_) - \sin 210°(\_\_\_\_\_\_\_\_\_\_)$

$= -\frac{\sqrt{3}}{2}(\_\_\_\_\_) - \left(-\frac{1}{2}\right)(\_\_\_\_\_)$

$= (\_\_\_\_\_) + (\_\_\_\_\_)$

$= \_\_\_\_\_\_\_\_$

Answers: 45°; cos 45°, sin 45°; $\frac{\sqrt{2}}{2}, \frac{\sqrt{2}}{2}$; $-\frac{\sqrt{6}}{4}, \frac{\sqrt{2}}{4}$; $\frac{\sqrt{2}-\sqrt{6}}{4}$

82 $\tan \frac{5\pi}{12}$

$\frac{5\pi}{12} = \frac{2\pi}{3} - \_\_\_\_$, so that

$\tan \frac{5\pi}{12} = \tan\left(\frac{2\pi}{3} - \frac{\pi}{4}\right)$

$= \frac{\tan(2\pi/3) - \_\_\_\_\_\_\_\_}{1 + \tan(2\pi/3)\tan(\pi/4)}$

$= \frac{(\_\_\_\_\_) - (\_\_\_\_)}{1 - (\sqrt{3})(1)}$

$= \_\_\_\_\_\_\_\_ = \_\_\_\_\_\_\_\_$

Answers: $\frac{\pi}{4}$; tan (π/4); $-\sqrt{3}$, 1; $\frac{\sqrt{3}+1}{\sqrt{3}-1}$, $2+\sqrt{3}$

83 tan 255°

Answers	
30°	$255° = 225° + \_\_\_\_$, so that
	$\tan 255° = \tan(225° + 30°)$
tan 30°	$= \dfrac{\tan 225° + \_\_\_\_\_\_}{1 - \tan 225° \tan 30°}$
$\dfrac{1}{\sqrt{3}}, \dfrac{1}{\sqrt{3}}$	$= \dfrac{1 + \_\_\_\_}{1 - 1\left(\_\_\_\_\right)}$
$\dfrac{\sqrt{3}+1}{\sqrt{3}-1}$, $2+\sqrt{3}$	$= \dfrac{\_\_\_\_\_\_}{\_\_\_\_\_\_} = \_\_\_\_\_\_$

In Problems 84–87, write each expression as a single function of one angle and evaluate.

84 sin 27° cos 18° + cos 27° sin 18°

Answers	
18°	$\sin 27° \cos 18° + \cos 27° \sin 18° = \sin(27° + \_\_\_\_)$
45°	$= \sin \_\_\_\_$
$\dfrac{\sqrt{2}}{2}$	$= \_\_\_\_$

85 sin 35° cos 25° + cos 35° sin 25°

Answers	
25°	$\sin 35° \cos 25° + \cos 35° \sin 25° = \sin(35° + \_\_\_\_)$
60°	$= \sin \_\_\_\_$
$\dfrac{\sqrt{3}}{2}$	$= \_\_\_\_$

86 cos 35° cos 55° − sin 35° sin 55°

Answers	
55°	$\cos 35° \cos 55° - \sin 35° \sin 55° = \cos(35° + \_\_\_\_)$
90°	$= \cos \_\_\_\_$
0	$= \_\_\_\_$

87 $\dfrac{\tan 17° + \tan 43°}{1 - \tan 17° \tan 43°}$

Answers	
43°	$\dfrac{\tan 17° + \tan 43°}{1 - \tan 17° \tan 43°} = \tan(17° + \_\_\_\_)$
60°	$= \tan \_\_\_\_$
$\sqrt{3}$	$= \_\_\_\_$

In Problems 88–103, evaluate each expression given that $\cos s = \frac{4}{5}$, for $0 < s < \pi/2$, and that $\sin t = \frac{3}{5}$, for $\pi/2 < t < \pi$.

88 sin s

Answers	
$\cos^2 s$	Since $\sin^2 s + \_\_\_\_\_\_ = 1$, it follows that
$\cos^2 s$	$\sin^2 s = 1 - \_\_\_\_\_\_$
$\frac{4}{5}$	$= 1 - (\_\_\_\_)^2$
$\frac{9}{25}$	$= \_\_\_\_$
I, $\frac{9}{25}$, $\frac{3}{5}$	Since s is in quadrant ____, $\sin s = \sqrt{\_\_\_\_} = \_\_\_\_$.

89 $\cos t$

$\cos^2 t = 1 - \underline{\qquad\qquad}$

$= 1 - (\underline{\qquad})^2$

$= \underline{\qquad}$

Since t is in quadrant ____, where the cosine is ____________, we have $\cos t = -\sqrt{\underline{\qquad}} = \underline{\qquad}$.

$\sin^2 t$

$\frac{3}{5}$

$\frac{16}{25}$

II, negative

$\frac{16}{25}$, $-\frac{4}{5}$

90 $\cos(t + s)$

$\cos(t + s) = \cos t \underline{\qquad\quad} - \underline{\qquad\quad} \sin s$

$= (\underline{\qquad})(\underline{\qquad}) - (\underline{\qquad})(\underline{\qquad})$

$= \underline{\qquad}$

$\cos s$, $\sin t$

$-\frac{4}{5}$, $\frac{4}{5}$, $\frac{3}{5}$, $\frac{3}{5}$

-1

91 $\cos(t - s)$

$\cos(t - s) = \cos t \underline{\qquad\quad} + \underline{\qquad\quad} \sin s$

$= (\underline{\qquad})(\underline{\qquad}) + (\underline{\qquad})(\underline{\qquad})$

$= \underline{\qquad}$

$\cos s$, $\sin t$

$-\frac{4}{5}$, $\frac{4}{5}$, $\frac{3}{5}$, $\frac{3}{5}$

$-\frac{7}{25}$

92 $\sin(s + t)$

$\sin(s + t) = \sin s \cos t + \underline{\qquad\qquad\qquad}$

$= (\underline{\qquad})(\underline{\qquad}) + (\underline{\qquad})(\underline{\qquad})$

$= \underline{\qquad}$

$\sin t \cos s$

$\frac{3}{5}$, $-\frac{4}{5}$, $\frac{3}{5}$, $\frac{4}{5}$

0

93 $\sin(s - t)$

$\sin(s - t) = \underline{\qquad\qquad\qquad} - \sin t \cos s$

$= (\underline{\qquad})(\underline{\qquad}) - (\underline{\qquad})(\underline{\qquad})$

$= \underline{\qquad}$

$\sin s \cos t$

$\frac{3}{5}$, $-\frac{4}{5}$, $\frac{3}{5}$, $\frac{4}{5}$

$-\frac{24}{25}$

94 $\tan s$

$$\tan s = \frac{\sin s}{\underline{\qquad\quad}} = \frac{\frac{3}{5}}{\underline{\qquad}} = \underline{\qquad}$$

$\cos s$, $\frac{4}{5}$, $\frac{3}{4}$

95 $\tan t$

$$\tan t = \frac{\sin t}{\underline{\qquad\quad}} = \frac{\frac{3}{5}}{\underline{\qquad}} = \underline{\qquad}$$

$\cos t$, $-\frac{4}{5}$, $-\frac{3}{4}$

96 $\tan(s + t)$

$$\tan(s + t) = \frac{\tan s + \tan t}{1 - \underline{\qquad\qquad\qquad}} = \frac{\underline{\qquad} + (\underline{\qquad})}{1 + \frac{9}{16}} = \underline{\qquad}$$

$\tan s \tan t$, $\frac{3}{4}$, $-\frac{3}{4}$, 0

97 $\tan(s - t)$

$$\tan(s - t) = \frac{\tan s - \tan t}{\underline{\qquad\qquad\qquad}} = \frac{\underline{\qquad} - (\underline{\qquad})}{1 - \frac{9}{16}} = \underline{\qquad}$$

$1 + \tan s \tan t$, $\frac{3}{4}$, $-\frac{3}{4}$, $\frac{24}{7}$

98 $\cot(t + s)$

$$\cot(t + s) = \frac{\cos(t + s)}{\underline{\qquad\qquad\qquad}} = \frac{-1}{\underline{\qquad}} = \underline{\qquad\qquad\qquad}$$

$\sin(t + s)$, 0, undefined

99 $\cot(t - s)$

$$\cot(t - s) = \frac{\cos(t - s)}{\underline{\qquad\qquad\qquad}} = \frac{\cos(t - s)}{-\sin(s - t)} = \frac{-\frac{7}{25}}{\underline{\qquad}} = \underline{\qquad}$$

$\sin(t - s)$, $\frac{24}{25}$, $-\frac{7}{24}$

100 $\sec(t+s)$

$\sec(t+s) = \dfrac{1}{\rule{3cm}{0.4pt}} = \dfrac{1}{\rule{1cm}{0.4pt}} = \rule{1cm}{0.4pt}$

$\cos(t+s)$, -1, -1

101 $\sec(t-s)$

$\sec(t-s) = \dfrac{1}{\rule{3cm}{0.4pt}} = \dfrac{1}{\rule{1cm}{0.4pt}} = \rule{1cm}{0.4pt}$

$\cos(t-s)$, $-\frac{7}{25}$, $-\frac{25}{7}$

102 $\csc(t+s)$

$\csc(t+s) = \dfrac{1}{\rule{3cm}{0.4pt}} = \dfrac{1}{\rule{1cm}{0.4pt}} = \rule{3cm}{0.4pt}$

$\sin(t+s)$, 0, undefined

103 $\csc(s-t)$

$\csc(s-t) = \dfrac{1}{\rule{3cm}{0.4pt}} = \dfrac{1}{\rule{1cm}{0.4pt}} = \rule{1cm}{0.4pt}$

$\sin(s-t)$, $-\frac{24}{25}$, $-\frac{25}{24}$

In Problems 104–110, prove that the equation is an identity.

104 $\dfrac{\tan(s-t)+\tan t}{1-\tan(s-t)\tan t} = \tan s$

Proof:

$$\frac{\tan(s-t)+\tan t}{1-\tan(s-t)\tan t} = \tan(s-t+t) = \tan s$$

105 $\sin 5\theta \cos 3\theta + \cos 5\theta \sin 3\theta = \sin 8\theta$

Proof:

$$\sin 5\theta \cos 3\theta + \cos 5\theta \sin 3\theta = \sin(5\theta + 3\theta) = \sin 8\theta$$

106 $\tan\left(t+\dfrac{\pi}{3}\right) = \dfrac{\tan t + \sqrt{3}}{1-\sqrt{3}\tan t}$

Proof:

$$\tan\left(t+\frac{\pi}{3}\right) = \frac{\tan t + \tan(\pi/3)}{1-\tan(\pi/3)\tan t} = \frac{\tan t + \sqrt{3}}{1-\sqrt{3}\tan t}$$

107 $\cot\left(t+\dfrac{\pi}{4}\right) + \tan\left(t-\dfrac{\pi}{4}\right) = 0$

Proof:

$$\cot\left(t+\frac{\pi}{4}\right) + \tan\left(t-\frac{\pi}{4}\right) = \frac{1}{\tan(t+\pi/4)} + \tan\left(t-\frac{\pi}{4}\right) = \frac{1}{\dfrac{\tan t + \tan(\pi/4)}{1-\tan t\tan(\pi/4)}}$$

$$+\frac{\tan t - \tan(\pi/4)}{1 + \tan t \tan(\pi/4)}$$

$$= \frac{1 - \tan t \tan(\pi/4)}{\tan t + \tan(\pi/4)}$$

$$+\frac{\tan t - \tan(\pi/4)}{1 + \tan t \tan(\pi/4)}$$

$$= \frac{1 - \tan t}{\tan t + 1} + \frac{\tan t - 1}{1 + \tan t}$$

$$= \frac{1 - \tan t + \tan t - 1}{\tan t + 1}$$

$$= 0$$

108 $\sin\left(t + \frac{\pi}{6}\right) + \sin\left(t - \frac{\pi}{6}\right) = \sqrt{3} \sin t$

Proof:

$$\sin\left(t + \frac{\pi}{6}\right) + \sin\left(t - \frac{\pi}{6}\right)$$

$$= \sin t \cos \frac{\pi}{6} + \sin \frac{\pi}{6} \cos t$$

$$+ \sin t \cos \frac{\pi}{6} - \sin \frac{\pi}{6} \cos t$$

$$= \frac{\sqrt{3}}{2} \sin t + \frac{1}{2} \cos t$$

$$+ \frac{\sqrt{3}}{2} \sin t - \frac{1}{2} \cos t$$

$$= \sqrt{3} \sin t$$

109 $\cos 2\theta \cos \theta - \sin 2\theta \sin \theta = \cos 3\theta$

Proof:

$$\cos 2\theta \cos \theta - \sin 2\theta \sin \theta$$

$$= \cos(2\theta + \theta)$$

$$= \cos 3\theta$$

110 $\dfrac{\tan t + \tan s}{\tan t - \tan s} = \dfrac{\sin(t + s)}{\sin(t - s)}$

Proof:

$$\frac{\tan t + \tan s}{\tan t - \tan s}$$

$$= \frac{\dfrac{\sin t}{\cos t} + \dfrac{\sin s}{\cos s}}{\dfrac{\sin t}{\cos t} - \dfrac{\sin s}{\cos s}}$$

$$= \frac{\dfrac{\sin t \cos s + \cos t \sin s}{\cos t \cos s}}{\dfrac{\sin t \cos s - \sin s \cos t}{\cos t \cos s}}$$

$$= \frac{\sin(t + s)}{\sin(t - s)}$$

2.2 Double and Half Angle Formulas

Answer	Frame
	111 $\sin 2t = \sin(t + t)$
	$= \sin t \cos t + \sin t \cos t$
$2 \sin t \cos t$	$=$ ________

In Problems 112–114, express each function in terms of functions of the indicated value.

Answer	Frame
	112 $\sin 70°$ in terms of functions of $35°$
$2 \sin \theta \cos \theta$	$\sin 2\theta =$ ________, so that
$2 \sin 35° \cos 35°$	$\sin 70° =$ ________.
	113 $\sin 4t$ in terms of functions of $2t$
$2 \sin \theta \cos \theta$	$\sin 2\theta =$ ________, so that
$2 \sin 2t \cos 2t$	$\sin 4t =$ ________.
	114 $\sin \frac{\pi}{4}$ in terms of functions of $\frac{\pi}{8}$
$\cos \theta$	$\sin 2\theta = 2 \sin \theta$ ________, so that
$2 \sin \frac{\pi}{8} \cos \frac{\pi}{8}$	$\sin \frac{\pi}{4} =$ ________.
	115 $\cos 2t = \cos(t + t)$
$\cos t,\ \sin t$	$= \cos t($________$) - \sin t($________$)$
$\cos^2 t,\ \sin^2 t$	$=$ ________ $-$ ________
$\cos^2 t$	**116** $\sin^2 t = 1 -$ ________, so that
	$\cos 2t = \cos^2 t - \sin^2 t$
$1 - \cos^2 t$	$= \cos^2 t - ($________$)$
$2 \cos^2 t$	$=$ ________ $- 1$
$\sin^2 t$	**117** $\cos^2 t = 1 -$ ________, so that
	$\cos 2t = \cos^2 t - \sin^2 t$
$1 - \sin^2 t$	$= ($________$) - \sin^2 t$
$1 - 2 \sin^2 t$	$=$ ________

In Problems 118–122, write each expression as a function of an angle twice as large.

Answer	Frame
	118 $1 - 2 \sin^2 43°$
$\cos 2\theta$	$1 - 2 \sin^2 \theta =$ ________, so that
$\cos 86°$	$1 - 2 \sin^2 43° =$ ________.
	119 $\cos^2 55° - \sin^2 55°$
$\cos 2\theta$	$\cos^2 \theta - \sin^2 \theta =$ ________, so that
$\cos 110°$	$\cos^2 55° - \sin^2 55° =$ ________.
	120 $2 \cos^2 36° - 1$
$\cos 2\theta$	$2 \cos^2 \theta - 1 =$ ________, so that
$\cos 72°$	$2 \cos^2 36° - 1 =$ ________.

121 $\cos^2 t^2 - \sin^2 t^2$

$\cos^2\theta - \sin^2\theta =$ ________, so that

$\cos^2 t^2 - \sin^2 t^2 =$ __________.

Answers: $\cos 2\theta$; $\cos 2t^2$

122 $2\cos^2 3t - 1$

$2\cos^2\theta - 1 =$ ________, so that

$2\cos^2 3t - 1 =$ ________.

Answers: $\cos 2\theta$; $\cos 6t$

123 $\sin^2 \dfrac{s}{2} = \dfrac{1 - \_\_\_\_}{2}$

Answer: $\cos s$

124 $\cos^2 \dfrac{s}{2} = \dfrac{1 + \_\_\_\_}{2}$

Answer: $\cos s$

In Problems 125–128, use the half angle formulas to find the exact value of the expression.

125 $\sin 15°$

$15° = \frac{1}{2}$ (____)

Using Problem 123, we have

$\sin^2 \dfrac{t}{2} =$ ____________, so

$\sin^2 15° = \dfrac{1 - \cos 30°}{2} = \dfrac{1 - \_\_\_\_}{2}$

$=$ __________

Then $\sin 15° = \sqrt{\_\_\_\_\_\_} =$ __________.

Answers: $30°$; $\dfrac{1 - \cos t}{2}$; $\dfrac{\sqrt{3}}{2}$; $\dfrac{2 - \sqrt{3}}{4}$; $\dfrac{2 - \sqrt{3}}{4}$, $\dfrac{\sqrt{2 - \sqrt{3}}}{2}$

126 $\cos \dfrac{\pi}{8}$

$\dfrac{\pi}{8} = \dfrac{1}{2}\left(\_\_\_\_\right)$

Using Problem 124, we have

$\cos^2 \dfrac{t}{2} =$ ____________, so that

$\cos^2 \dfrac{\pi}{8} = \dfrac{1 + \_\_\_\_}{2} = \dfrac{\_\_\_\_}{2}$

$=$ __________

Then $\cos \dfrac{\pi}{8} = \sqrt{\_\_\_\_\_\_} =$ __________.

Answers: $\dfrac{\pi}{4}$; $\dfrac{1 + \cos t}{2}$; $\cos \dfrac{\pi}{4}$, $1 + \dfrac{\sqrt{2}}{2}$; $\dfrac{2 + \sqrt{2}}{4}$; $\dfrac{2 + \sqrt{2}}{4}$, $\dfrac{\sqrt{2 + \sqrt{2}}}{2}$

127 $\sin \frac{5\pi}{12}$

$\frac{5\pi}{12} = \frac{1}{2}(\_\_\_\_)$

$\sin^2 \frac{t}{2} = \_\_\_\_\_\_\_\_$, so that

$\sin^2 \frac{5\pi}{12} = \frac{1 - \_\_\_\_\_\_}{2} = \frac{1 - (\_\_\_\_)}{2}$

$= \_\_\_\_\_\_ = \_\_\_\_\_\_$

Then $\sin \frac{5\pi}{12} = \sqrt{\_\_\_\_\_\_} = \frac{\_\_\_\_\_\_}{2}$.

Answers: $\frac{5\pi}{6}$; $\frac{1 - \cos t}{2}$; $\cos \frac{5\pi}{6}$, $-\frac{\sqrt{3}}{2}$; $\frac{1 + \frac{\sqrt{3}}{2}}{2}$, $\frac{2 + \sqrt{3}}{4}$; $\frac{2 + \sqrt{3}}{4}$, $\sqrt{2 + \sqrt{3}}$

128 $\cos 75^\circ$

$75^\circ = \frac{1}{2}(\_\_\_\_)$

$\cos^2 \frac{t}{2} = \_\_\_\_\_\_\_\_$, so that

$\cos^2 75^\circ = \frac{1 + \_\_\_\_\_\_}{2} = \frac{1 + (\_\_\_\_)}{2}$

$= \_\_\_\_\_\_$

Then $\cos 75^\circ = \sqrt{\_\_\_\_\_\_} = \_\_\_\_\_\_$.

Answers: 150°; $\frac{1 + \cos t}{2}$; $\cos 150^\circ$, $-\frac{\sqrt{3}}{2}$; $\frac{2 - \sqrt{3}}{4}$; $\frac{2 - \sqrt{3}}{4}$, $\frac{\sqrt{2 - \sqrt{3}}}{2}$

129 $\tan 2t = \tan(t + \_\_\_\_) = \frac{\tan t + \tan t}{1 - \_\_\_\_\_\_}$, so

$\tan 2t = \_\_\_\_\_\_\_\_$

Answers: t, $\tan^2 t$; $\frac{2 \tan t}{1 - \tan^2 t}$

In Problems 130–138, assume that $\sin \theta = \frac{3}{5}$ and θ is an angle in standard position, with its terminal side in quadrant I, to evaluate the expression.

130 $\cos \theta$

$\cos \theta = \sqrt{\_\_\_\_\_\_\_\_} = \sqrt{\_\_\_\_\_\_\_\_} = \_\_\_\_$

Answers: $1 - \sin^2 \theta$, $1 - \frac{9}{25}$, $\frac{4}{5}$

131 $\sin 2\theta$

$\sin 2\theta = \_\_\_\_\_\_\_\_\_\_$, so

$\sin 2\theta = 2(\_\_\_\_)(\_\_\_\_) = \_\_\_\_$

Answers: $2 \sin \theta \cos \theta$; $\frac{3}{5}$, $\frac{4}{5}$, $\frac{24}{25}$

132 $\cos 2\theta$

$\cos 2\theta = \cos^2 \theta - \_\_\_\_\_\_$, so

$\cos 2\theta = \_\_\_\_ - \_\_\_\_ = \_\_\_\_$

Answers: $\sin^2 \theta$; $\frac{16}{25}$, $\frac{9}{25}$, $\frac{7}{25}$

133 $\tan 2\theta$

$\tan 2\theta = \frac{\sin 2\theta}{\_\_\_\_\_\_} = \frac{\frac{24}{25}}{\_\_\_\_} = \_\_\_\_$

Answers: $\cos 2\theta$, $\frac{7}{25}$, $\frac{24}{7}$

134 $\cot 2\theta$

$$\cot 2\theta = \frac{1}{\_\_\_\_} = \frac{1}{\_\_\_\_} = \_\_\_\_$$

$\tan 2\theta$, $\frac{24}{7}$, $\frac{7}{24}$

135 $\sec 2\theta$

$$\sec 2\theta = \frac{1}{\_\_\_\_} = \frac{1}{\_\_\_\_} = \_\_\_\_$$

$\cos 2\theta$, $\frac{7}{25}$, $\frac{25}{7}$

136 $\sin \frac{\theta}{2}$

$$\sin \frac{\theta}{2} = \sqrt{\frac{\_\_\_\_}{\_\_\_\_}} = \sqrt{\frac{\_\_\_\_}{\_\_\_\_}} = \frac{\_\_\_\_}{\_\_\_\_} = \frac{\_\_\_\_}{\_\_\_\_}$$

$\frac{1 - \cos\theta}{2}$, $\frac{1 - \frac{4}{5}}{2}$

$\sqrt{\frac{1}{10}}$, $\frac{\sqrt{10}}{10}$

137 $\cos \frac{\theta}{2}$

$$\cos \frac{\theta}{2} = \sqrt{\frac{\_\_\_\_}{\_\_\_\_}} = \sqrt{\frac{\_\_\_\_}{\_\_\_\_}} = \frac{\_\_\_\_}{\_\_\_\_} = \frac{\_\_\_\_}{\_\_\_\_}$$

$\frac{1 + \cos\theta}{2}$, $\frac{1 - \frac{4}{5}}{2}$

$\sqrt{\frac{9}{10}}$, $\frac{3\sqrt{10}}{10}$

138 $\tan \frac{\theta}{2}$

$$\tan \frac{\theta}{2} = \frac{\sin \frac{\theta}{2}}{\_\_\_\_} = \frac{\frac{\sqrt{10}}{10}}{\_\_\_\_} = \frac{\_\_\_\_}{}$$

$\cos \frac{\theta}{2}$, $\frac{3\sqrt{10}}{10}$, $\frac{1}{3}$

In Problems 139–145, prove that the equation is an identity.

139 $\cot\theta \sin 2\theta = 1 + \cos 2\theta$

Proof:

$\cot\theta \sin 2\theta$

$= \left(\frac{\cos\theta}{\sin\theta}\right) 2 \sin\theta \cos\theta$

$= 2\cos^2\theta$

$= 2\left(\frac{1 + \cos 2\theta}{2}\right)$

$= 1 + \cos 2\theta$

140 $\left(\sin \frac{t}{2} - \cos \frac{t}{2}\right)^2 = 1 - \sin t$

Proof:

$\left(\sin \frac{t}{2} - \cos \frac{t}{2}\right)^2$

$= \sin^2 \frac{t}{2} - 2 \sin \frac{t}{2} \cos \frac{t}{2} + \cos^2 \frac{t}{2}$

$= 1 - 2 \sin \frac{t}{2} \cos \frac{t}{2}$

$= 1 - \sin t$

141 $\tan\theta \sin 2\theta = 2\sin^2\theta$

Proof:

$\tan\theta \sin 2\theta$

$= \left(\dfrac{\sin\theta}{\cos\theta}\right) 2\sin\theta\cos\theta$

$= 2\sin^2\theta$

142 $\dfrac{1 - \cos 2t}{\sin 2t} = \tan t$

Proof:

$\dfrac{1 - \cos 2t}{\sin 2t}$

$= \dfrac{1 - (1 - 2\sin^2 t)}{2\sin t\cos t}$

$= \dfrac{2\sin^2 t}{2\sin t\cos t}$

$= \dfrac{\sin t}{\cos t}$

$= \tan t$

143 $\csc 2\theta - \cot 2\theta = \tan\theta$

Proof:

$\csc 2\theta - \cot 2\theta$

$= \dfrac{1}{\sin 2\theta} - \dfrac{\cos 2\theta}{\sin 2\theta}$

$= \dfrac{1 - \cos 2\theta}{\sin 2\theta}$

$= \dfrac{2\sin^2\theta}{2\sin\theta\cos\theta}$

$= \dfrac{\sin\theta}{\cos\theta}$

$= \tan\theta$

144 $\dfrac{\sin 3\theta}{\sin\theta} - \dfrac{\cos 3\theta}{\cos\theta} = 2$

Proof:

$\dfrac{\sin 3\theta}{\sin\theta} - \dfrac{\cos 3\theta}{\cos\theta}$

$= \dfrac{\sin 3\theta\cos\theta - \cos 3\theta\sin\theta}{\sin\theta\cos\theta}$

$= \dfrac{\sin(3\theta - \theta)}{\sin\theta\cos\theta}$

$= \dfrac{\sin 2\theta}{\sin\theta\cos\theta}$

$= \dfrac{2\sin\theta\cos\theta}{\sin\theta\cos\theta}$

$= 2$

Proof:

$$\frac{\sin 2\theta}{\sin\theta} - \frac{\cos 2\theta}{\cos\theta}$$
$$= \frac{\sin 2\theta \cos\theta - \cos 2\theta \sin\theta}{\sin\theta\cos\theta}$$
$$= \frac{\sin(2\theta - \theta)}{\sin\theta\cos\theta}$$
$$= \frac{\sin\theta}{\sin\theta\cos\theta}$$
$$= \frac{1}{\cos\theta}$$
$$= \sec\theta$$

145 $\dfrac{\sin 2\theta}{\sin\theta} - \dfrac{\cos 2\theta}{\cos\theta} = \sec\theta$

3 Inverse Trigonometric Functions

3.1 Inverses of Sine and Cosine Functions

$y,\ x$

$-1,\ 1,\ -\frac{\pi}{2},\ \frac{\pi}{2}$

146 $y = \sin^{-1} x$ is equivalent to sin ____ = ____, where ____ $\leqslant x \leqslant$ ____ and ____ $\leqslant y \leqslant$ ____.

$y,\ x$

$-1,\ 1,\ 0,\ \pi$

147 $y = \cos^{-1} x$ is equivalent to cos ____ = ____, where ____ $\leqslant x \leqslant$ ____ and ____ $\leqslant y \leqslant$ ____.

$\sin t,\ \frac{\pi}{2}$

$\frac{\pi}{6},\ \frac{\pi}{6}$

148 $\sin^{-1} \frac{1}{2} = t$ is equivalent to ________ $= \frac{1}{2}$, where $-\frac{\pi}{2} \leqslant t \leqslant$ ____, so that $t =$ ____; that is, $\sin^{-1} \frac{1}{2} =$ ____.

-1

$0,\ \pi,\ \pi,\ \pi$

149 $\cos^{-1}(-1) = t$ is equivalent to $\cos t =$ ____, where ____ $\leqslant t \leqslant$ ____, so that $t =$ ____; that is, $\cos^{-1}(-1) =$ ____.

$1,\ \frac{\pi}{2}$

$\frac{\pi}{2}$

150 If $y = \sin^{-1} 1$, then $\sin y =$ ____, so $y =$ ____.

Hence, $\sin^{-1} 1 =$ ____.

$-1,\ 1$

151 $\cos^{-1} 2$ is not defined because ____ $\leqslant \cos t \leqslant$ ____.

$\frac{\sqrt{3}}{2},\ \frac{\pi}{3}$

$\frac{\pi}{3}$

152 If $y = \text{Arcsin}\,\frac{\sqrt{3}}{2}$, then $\sin y =$ ____, so $y =$ ____.

Hence, $\text{Arcsin}\,\frac{\sqrt{3}}{2} =$ ____.

$t,\ -\frac{1}{2}$

$\frac{\pi}{3},\ \frac{\pi}{3}$

II, $\frac{\pi}{3},\ \frac{2\pi}{3}$

$\frac{2\pi}{3}$

153 If $\cos^{-1}\left(-\frac{1}{2}\right) = t$, then cos ____ = ____. We know that cos ____ $= \frac{1}{2}$, so that ____ is the reference angle for $P(t)$ in quadrant ____. Hence, $t = \pi -$ ____ = ____, so that $\cos^{-1}\left(-\frac{1}{2}\right) =$ ____

0, 0	**154** $\sin^{-1}(\sin \pi) = \sin^{-1}$ ____ = ____
$-1, \pi$	**155** $\cos^{-1}\left[\sin\left(-\frac{\pi}{2}\right)\right] = \cos^{-1}($____$) =$ ____

In Problems 156–170, evaluate the given expression.

	156 $\text{Arcsin}\left(-\frac{\sqrt{3}}{2}\right)$
$-\frac{\pi}{3}$	Let $y = \text{Arcsin}\left(-\frac{\sqrt{3}}{2}\right)$; then $\sin y = -\frac{\sqrt{3}}{2}$, so $y =$ ____.
$-\frac{\pi}{3}$	Hence, $\text{Arcsin}\left(-\frac{\sqrt{3}}{2}\right) =$ ____.
$-\frac{\pi}{6}$	**157** $\sin^{-1}\left(-\frac{1}{2}\right) =$ ____
$\frac{\pi}{4}$	**158** $\cos^{-1}\frac{\sqrt{2}}{2} =$ ____
0	**159** Arcsin 0 = ____
0.85	**160** $\sin^{-1} 0.7513 =$ ____
1.55	**161** $\cos^{-1} 0.0208 =$ ____
-1.2	**162** $\sin^{-1}(-0.9320) =$ ____
3	**163** $\cos^{-1}(-0.9902) =$ ____
$\frac{5\pi}{6}$	**164** $\cos^{-1}\left(-\frac{\sqrt{3}}{2}\right) =$ ____
$\frac{\pi}{4}, \frac{\sqrt{2}}{2}$	**165** $\sin\left(\cos^{-1}\frac{\sqrt{2}}{2}\right) = \sin$ ____ = ____
$\frac{\pi}{2}$, 0	**166** $\cos(\sin^{-1} 1) = \cos$ ____ = ____
$\frac{\pi}{2}$, 1	**167** $\sin(\cos^{-1} 0) = \sin$ ____ = ____
	168 $\sin(\cos^{-1}\frac{3}{5})$
$\frac{3}{5}$	Let $y = \cos^{-1}\frac{3}{5}$; $\cos y =$ ____, so
$\cos^2 y$, $\frac{9}{25}$, $\frac{4}{5}$	$\sin y = \sqrt{1 - \text{\_\_\_\_}} = \sqrt{1 - \text{\_\_\_\_}} =$ ____.
$\frac{4}{5}$	Hence, $\sin(\cos^{-1}\frac{3}{5}) =$ ____.
	169 $\sin(\cos^{-1} 0.73)$
0.73	Let $y = \cos^{-1} 0.73$; then $\cos y =$ ____, so
$\cos^2 y$, 0.5329, 0.68	$\sin y = \sqrt{1 - \text{\_\_\_\_}} = \sqrt{1 - \text{\_\_\_\_}} =$ ____.
0.68	Hence, $\sin(\cos^{-1} 0.73) =$ ____.
	170 $\cos(\sin^{-1}\frac{5}{13})$
$\frac{5}{13}$	Let $y = \sin^{-1}\frac{5}{13}$; then $\sin y =$ ____, so
$\sin^2 y$, $\frac{12}{13}$	$\cos y = \sqrt{1 - \text{\_\_\_\_}} =$ ____.
$\frac{12}{13}$	Hence, $\cos(\sin^{-1}\frac{5}{13}) =$ ____.
$-\frac{y}{2}$, $\sin\left(-\frac{y}{2}\right)$	**171** If $y = -2\sin^{-1} x$, then $\sin^{-1} x =$ ____, so that $x =$ ____.

$y - 7$

$3y - 21$, $\cos(3y - 21)$

172 If $y = 7 + \frac{1}{3}\cos^{-1} x$, then $\frac{1}{3}\cos^{-1} x =$ ________, so that $\cos^{-1} x =$ ________ and $x =$ ________________.

173 The graph of $f(x) = \sin^{-1} x$ is

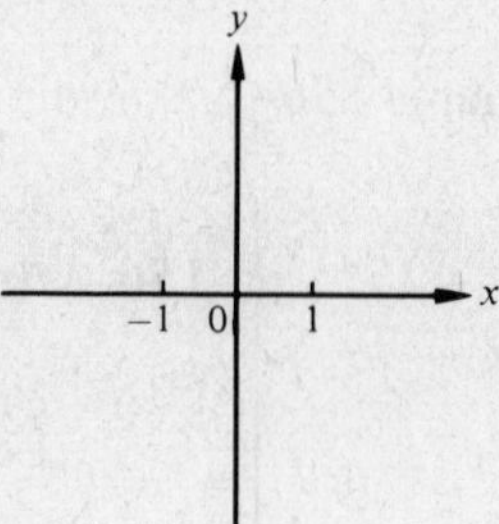

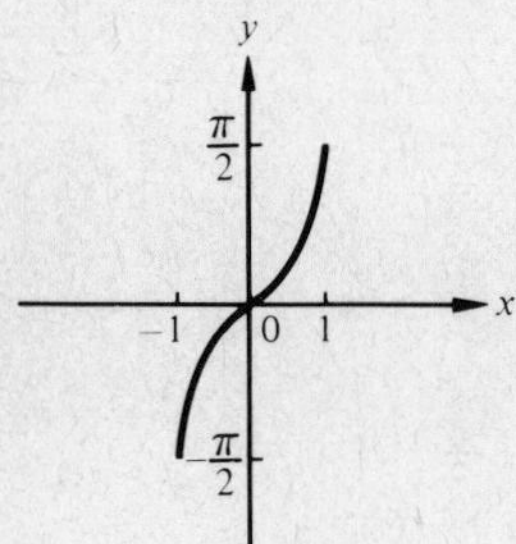

174 The graph of $g(x) = \cos^{-1} x$ is

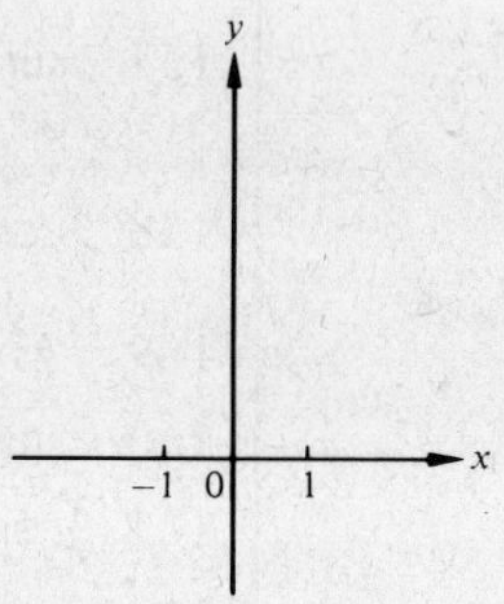

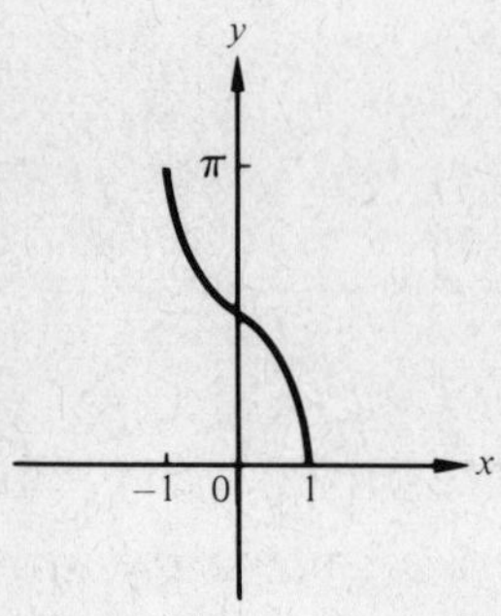

175 From the graph of $f(x) = \sin^{-1} x$ in Problem 173, we notice that the domain of f is $\{x \mid$ ____________$\}$ and the range of f is $\left\{y \mid \right.$ ____________$\left.\right\}$.

$-1 \leqslant x \leqslant 1$

$-\frac{\pi}{2} \leqslant y \leqslant \frac{\pi}{2}$

176 From the graph of $g(x) = \cos^{-1} x$ in Problem 174, we notice that the domain of g is $\{x \mid$ ____________$\}$ and the range of g is $\{y \mid$ ____________$\}$.

$-1 \leqslant x \leqslant 1$

$0 \leqslant y \leqslant \pi$

3.2 Other Inverse Trigonometric Functions

177 $y = \text{Arctan}\, x = \tan^{-1} x$ is the inverse function of $f(x) = \tan x$ if and only if $x =$ _____ and ____ $< y <$ ____.

$\tan y$, $-\frac{\pi}{2}$, $\frac{\pi}{2}$

In Problems 178–187, evaluate the expression.

178 $\text{Arctan}\sqrt{3}$

$y = \text{Arctan}\sqrt{3}$, so $\tan y =$ ____. Hence, $y =$ ____.

$\sqrt{3}$, $\frac{\pi}{3}$

179 $\tan^{-1}\frac{1}{\sqrt{3}}$

$y = \tan^{-1}\frac{1}{\sqrt{3}}$, so $\tan y =$ ____ $=$ ____ and $y =$ ____.

$\frac{1}{\sqrt{3}}$, $\frac{\sqrt{3}}{3}$, $\frac{\pi}{6}$

Answer	Problem
$\frac{\pi}{4}$	**180** $\tan^{-1} 1 =$ ____
$\frac{\pi}{3}$, $\sqrt{3}$	**181** $\tan(\tan^{-1}\sqrt{3}) = \tan$ ____ $=$ ____
1.21	**182** $\tan^{-1} 2.650 =$ ____
0, 0	**183** $\tan(\sin^{-1} 0) = \tan$ ____ $=$ ____
-5.798	**184** $\tan[\cos^{-1}(-0.17)] =$ ____
$\frac{\sqrt{3}}{2}$	**185** $\sin(\tan^{-1}\sqrt{3}) =$ ____
$-\frac{\pi}{3}$, $\frac{1}{2}$	**186** $\cos[\tan^{-1}(-\sqrt{3})] = \cos($ ____ $) =$ ____
-1, $-\frac{\pi}{4}$	**187** $\tan^{-1}\left(\cot\frac{3\pi}{4}\right) = \tan^{-1}($ ____ $) =$ ____

4 Trigonometric Equations

In Problems 188–203, assume that t represents a radian measure such that $0 \leqslant t < 2\pi$, or that θ represents a degree measure such that $0° \leqslant \theta < 360°$. Find the solution set of the given equation.

Answer	Problem
	188 $\sin t = \frac{\sqrt{2}}{2}$
$\frac{\pi}{2}$	First, it is necessary to find the reference angle t_R; that is, we must find t_R, for $0 < t_R <$ ____, such that $\sin t_R = \frac{\sqrt{2}}{2}$. Here,
$\frac{\pi}{4}$, II	$t_R =$ ____. Since the sine is positive in quadrants I and ____, we
$\frac{\pi}{4}$	use the reference angle $t_R =$ ____ to get a solution in quadrant I
$\frac{\pi}{4}$, II	of $t =$ ____ and a solution in quadrant ____ of
$\frac{\pi}{4}$, $\frac{3\pi}{4}$	$t = \pi -$ ____ $=$ ____
$\frac{\pi}{4}$, $\frac{3\pi}{4}$	Hence, the solution set is $\{$ ____, ____ $\}$.
	189 $\cos t = -\frac{1}{2}$
$\frac{1}{2}$	Here, the reference angle t_R must satisfy $\cos t_R =$ ____, for
$\frac{\pi}{2}$, $\frac{\pi}{3}$	$0 < t_R <$ ____, so that $t_R =$ ____. Since the cosine is negative in
III, $\frac{\pi}{3}$	quadrants II and ____, we use the reference angle $t_R =$ ____ to

$\frac{\pi}{3}, \frac{2\pi}{3}$	get a solution in quadrant II of $t = \pi -$ ___ = ___ and a solution
III, $\frac{\pi}{3}, \frac{4\pi}{3}$	in quadrant ___ of $t = \pi +$ ___ = ___.
$\frac{2\pi}{3}, \frac{4\pi}{3}$	Hence, the solution set is $\{$___, ___$\}$.

190 $\tan t = -1$

1	Here, the reference angle t_R must satisfy $\tan t_R$ = ___, for
$\frac{\pi}{2}$	$0 < t_R <$ ___, so that $t_R = \frac{\pi}{4}$. Since the tangent is negative in
II	quadrants ___ and IV, we get a solution in quadrant II of
$\frac{\pi}{4}, \frac{3\pi}{4}$, IV	$t = \pi -$ ___ = ___ and a solution in quadrant ___ of
$\frac{\pi}{4}, \frac{7\pi}{4}, \frac{3\pi}{4}, \frac{7\pi}{4}$	$t = 2\pi -$ ___ = ___. Hence, the solution set is $\{$___, ___$\}$.

191 $\cos t = \frac{\sqrt{3}}{2}$

t_R, $\frac{\sqrt{3}}{2}$	Here, the reference angle t_R must satisfy cos ___ = ___, for
$\frac{\pi}{6}$	$0 < t_R < \frac{\pi}{2}$, so that t_R = ___. Since the cosine is positive in
IV, $\frac{\pi}{6}$	quadrants I and ___, we have a solution in quadrant I of t = ___
IV, $\frac{\pi}{6}, \frac{11\pi}{6}$	and a solution in quadrant ___ of $t = 2\pi -$ ___ = ___.
$\frac{\pi}{6}, \frac{11\pi}{6}$	Hence, the solution set is $\{$___, ___$\}$.

192 $\tan \theta = 3.732$

θ_R, 3.732	Here, the reference angle θ_R must satisfy tan ___ = ______,
90°	where $0° < \theta_R <$ ___. From Appendix Table IV, we find that
75°, III	θ_R = ___. Since the tangent is positive in quadrants I and ___,
75°	we have a solution in quadrant I of θ = ___ and a solution in
III, 75°, 255°	quadrant ___ of $\theta = 180° +$ ___ = ___. Hence, the solution
75°, 255°	set is {___, ___}.

193 $\sec \theta = -1.126$

θ_R, 1.126	Here, the reference angle θ_R must satisfy sec ___ = ___, where
90°	$0° < \theta_R <$ ___. From Appendix Table IV, we find that
27°20′	θ_R = ______. Since the secant is negative in quadrants II and
III, 27°20′	___, we have a solution in quadrant II of $\theta = 180° -$ ______ =
152°40′, III, 27°20′	______ and a solution in quadrant ___ of $\theta = 180° +$ ______ =
207°20′, 152°40′, 207°20′	______. Hence, the solution set is {______, ______}.

θ_R, 0.7133
0°, 90°
45°30′
IV, 45°30′
225°30′, IV, 45°30′
314°30′, 225°30′, 314°30′

194 $\sin\theta = -0.7133$

Here, the reference angle θ_R must satisfy sin ____ = ________, where ____ $< \theta_R <$ ____. From Appendix Table IV we find that θ_R = ________. Since the sine is negative in quadrants III and ____, we have a solution in quadrant III of θ = 180° + ________ = ________ and a solution in quadrant ____ of θ = 360° - ____ = ________. Hence, the solution set is {________, ________}.

1
360°
1, 810°, 1
1170°, 1,
1170°, 22°30′, 112°30′, 202°30′
292°30′
22°30′, 112°30′, 202°30′, 292°30′

195 $\sin 4\theta = 1$

We know that sin 90° = ____. Since the sine is a periodic function of period _____, it follows that sin (90° + 360°) = sin (450°) = ____, sin (450° + 360°) = sin (_____) = ____, and sin (810° + 360°) = sin (_____) = ____, so that 4θ = 90°, 450°, 810°, or _____. Hence, θ = ________, ________, ________, or ________, and the solution set is given by {________, ________, ________, ________}.

π
$\frac{3\pi}{2}$
$\pi, \frac{3\pi}{2}$

196 $\sin t \cos t = 0$

This equation is equivalent to $\sin t = 0$ or $\cos t = 0$. The solution set for $\sin t = 0$ is {0, ____} and the solution set for $\cos t = 0$ is $\left\{\frac{\pi}{2}, \text{\_\_\_\_}\right\}$, so that the solution set for $\sin t \cos t = 0$ is $\left\{0, \frac{\pi}{2}, \text{\_\_\_\_}, \text{\_\_\_\_}\right\}$.

2 sin t - 1
2 sin t - 1, $\frac{1}{2}$
0, π
$\frac{\pi}{6}, \frac{5\pi}{6}$
$\frac{\pi}{6}, \frac{5\pi}{6}, \pi$

197 $2 \tan t \sin t - \tan t = 0$

First we can rewrite the equation by factoring the left-hand side: tan t(____________) = 0 so that tan t = 0 or ____________ = 0, that is, tan t = 0 or sin t = ____. The solution set for tan t = 0 is {____, ____} and the solution set for $\sin t = \frac{1}{2}$ is {____, ____}, so the solution set of the given equation is {0, ____, ____, ____}.

cos θ
cos θ
0, $-\frac{1}{2}$
270°, 330°

90°, 210°, 270°, 330°

198 $\sin 2\theta + \cos\theta = 0$

Using the identity $\sin 2\theta = 2 \sin\theta$ _____, we can rewrite the equation as $2\sin\theta\cos\theta + \cos\theta = 0$, so ____ $(2\sin\theta + 1) = 0$. Hence, $\cos\theta$ = ____ or $\sin\theta$ = ____. The solution set of $\cos\theta = 0$ is {90°, ____} and the solution set of $\sin\theta = -\frac{1}{2}$ is {210°, ____}, so that the solution set of the given equation is {_____, _____, _____, _____}.

199 $\cos^2\theta - \sin^2\theta = 0$

Since $\cos 2\theta = \cos^2\theta - \sin^2\theta$, we can rewrite the equation as ________ = 0. We know that $\cos 90° = \cos 270° = \cos 450° = \cos 630°$ = ____, so that 2θ = ____, ____, ____, or ____. Hence, θ = ____, ____, ____, or ____, and the solution set is {____, ____, ____, ____}.

$\cos 2\theta$

0, 90°, 270°, 450°, 630°

45°, 135°, 225°, 315°

45°, 135°, 225°, 315°

200 $\sin^2 t = 4$

$\sin^2 t = 4$ implies that $\sin t$ = ____; but since $|\sin t| \leqslant$ ____, we conclude that this equation does not have a solution. Hence the solution set is ____.

±2, 1

$\emptyset$

201 $\sin^2 t = \cos^2 t$

Since $\sin^2 t + \cos^2 t$ = ____, we have $\sin^2 t + \sin^2 t$ = ____, so that $2\sin^2 t = 1$, that is, $\sin t = \frac{1}{\sqrt{2}}$ or $\sin t$ = ____. $\sin t = \frac{1}{\sqrt{2}}$ yields $t = \frac{\pi}{4}$ or ____ and $\sin t = -\frac{1}{\sqrt{2}}$ yields $t = \frac{5\pi}{4}$ or ____. Hence the solution set is {____, ____, ____, ____}.

1, 1

$-\frac{1}{\sqrt{2}}$

$\frac{3\pi}{4}, \frac{7\pi}{4}$

$\frac{\pi}{4}, \frac{3\pi}{4}, \frac{5\pi}{4}, \frac{7\pi}{4}$

202 $\sin\left(\frac{\theta}{2}\right) + \cos\theta = 1$

From the identity in Problem 123, we have $\sin^2\left(\frac{\theta}{2}\right)$ = ________ so that $2\sin^2\left(\frac{\theta}{2}\right)$ = ______. The given equation can be rewritten as $\sin\left(\frac{\theta}{2}\right) = 1 - \cos\theta$. Hence, $2\sin^2\left(\frac{\theta}{2}\right)$ = __________, that is, $2\sin^2\left(\frac{\theta}{2}\right) - \sin\left(\frac{\theta}{2}\right)$ = ____, so that

$\sin\left(\frac{\theta}{2}\right)\cdot\left[\_\_\_\_\_\_\_\_\_\_\_\_\right] = 0$

Thus, $\sin\left(\frac{\theta}{2}\right)$ = ____ or $\sin\left(\frac{\theta}{2}\right)$ = ____.

$\sin\left(\frac{\theta}{2}\right) = 0$ yields $\frac{\theta}{2} = 0°$

$\sin\left(\frac{\theta}{2}\right) = \frac{1}{2}$ yields $\frac{\theta}{2} = 30°$ or ____

Hence, $\theta = 0°, 60°$, or ____.

The solution set is {____, ____, ____}.

$\frac{1-\cos\theta}{2}$

$1 - \cos\theta$

$\sin\left(\frac{\theta}{2}\right)$

0

$2\sin\left(\frac{\theta}{2}\right) - 1$

0, ½

150°

300°

0°, 60°, 300°

203 $\tan\theta - \cot\theta = 0$

Since $\cot\theta = \frac{1}{\tan\theta}$, we can rewrite the equation as $\tan\theta$ = ____, so that $\tan^2\theta$ = ____. Hence, $\tan\theta = 1$ or $\tan\theta$ = ____, so the solution set is {____, ____, ____, ____}.

$\frac{1}{\tan\theta}$

1, −1

45°, 135°, 225°, 315°

5 Right Triangle Trigonometry and Polar Coordinates

5.1 Right Triangle Trigonometry

In Problems 204–209, express the value of the trigonometric functions in terms of the sides of the right triangle shown in Figure 1.

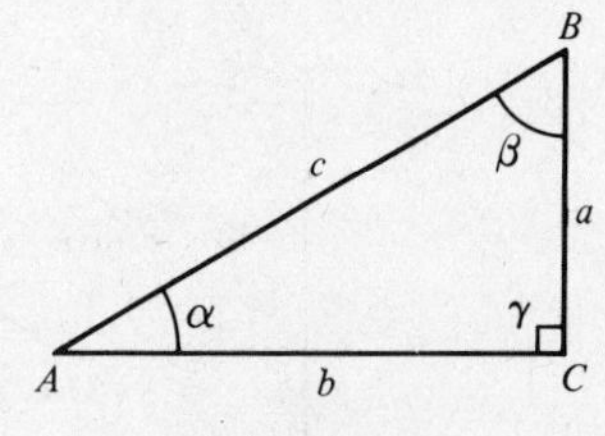

Figure 1

Answer	Problem
a, length of side opposite α	**204** $\sin \alpha = \dfrac{\_\_\_\_}{c} = \dfrac{\_\_\_\_\_\_\_\_\_\_\_\_\_\_\_\_}{\text{length of hypotenuse}}$
c, length of hypotenuse	**205** $\cos \alpha = \dfrac{b}{\_\_\_\_} = \dfrac{\text{length of side adjacent to } \alpha}{\_\_\_\_\_\_\_\_\_\_\_\_\_\_\_\_}$
b, length of side adjacent to α	**206** $\tan \alpha = \dfrac{a}{\_\_\_\_} = \dfrac{\text{length of side opposite } \alpha}{\_\_\_\_\_\_\_\_\_\_\_\_\_\_\_\_}$
c, length of side opposite α	**207** $\csc \alpha = \dfrac{\_\_\_\_}{a} = \dfrac{\text{length of hypotenuse}}{\_\_\_\_\_\_\_\_\_\_\_\_\_\_\_\_}$
b, length of side adjacent to α	**208** $\sec \alpha = \dfrac{c}{\_\_\_\_} = \dfrac{\text{length of hypotenuse}}{\_\_\_\_\_\_\_\_\_\_\_\_\_\_\_\_}$
b, length of side opposite α	**209** $\cot \alpha = \dfrac{\_\_\_\_}{a} = \dfrac{\text{length of side adjacent to } \alpha}{\_\_\_\_\_\_\_\_\_\_\_\_\_\_\_\_}$

In Problems 210–215, find the unknown parts of the given right triangle. Give lengths to the nearest tenth and angle measures to the nearest ten minutes (10′).

210 $\alpha = 30°$ and $a = 1$ (Figure 2)

Figure 2

Answer	Problem
90°, 60°	Since $\beta + 30° =$ ____, we have $\beta =$ ____.
b, $\sqrt{3}$	$\cot 30° = \dfrac{\_\_\_\_}{1}$, and since $\cot 30° =$ ____, it follows that
$\sqrt{3}$ or 1.7	$b =$ ____________. Finally,
	$c^2 = 1^2 + b^2$
$\sqrt{3}$	$= 1^2 + (\_\_\_\_)^2$
3	$= 1 +$ ____
4	$=$ ____
	so that
4, 2	$c = \sqrt{\_\_\_\_} =$ ____

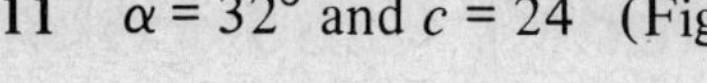

211 $\alpha = 32°$ and $c = 24$ (Figure 3)

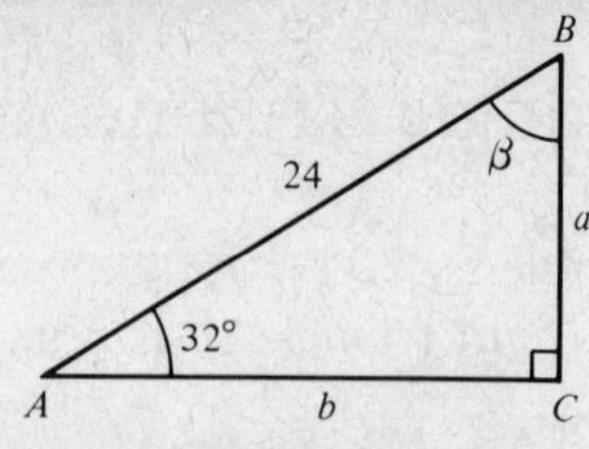

Figure 3

Answers	
90°, 58°	Since $\beta + 32° =$ ____, we have $\beta =$ ____.
c, 24	$\sin 32° = \dfrac{a}{\_\_\_} = \dfrac{a}{\_\_\_}$, so that
0.5299, 12.7	$a = 24 \sin 32° = 24$ (________) = ____
c, 24	$\cos 32° = \dfrac{b}{\_\_\_} = \dfrac{b}{\_\_\_}$, so that
0.8480, 20.4	$b = 24 \cos 32° = 24$(________) = _____

212 $b = 4$ and $\beta = 60°$ (Figure 4)

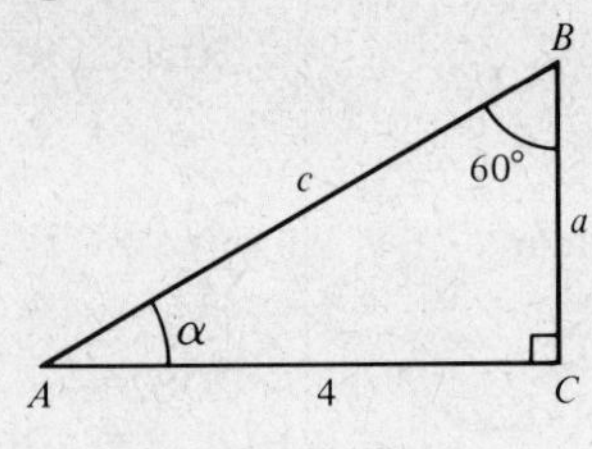

Figure 4

Answers	
90°, 30°	Since $\alpha + 60° =$ ____, $\alpha =$ ____.
a	$\tan 30° = \dfrac{\_\_\_}{4}$, so that
$\tan 30°, \dfrac{1}{\sqrt{3}}, \dfrac{4}{\sqrt{3}}, \dfrac{4\sqrt{3}}{3}$	$a = 4$(________) $= 4\left(\dfrac{\quad}{\_\_\_\_}\right) = \dfrac{\quad}{\_\_\_\_} = \dfrac{\quad}{\_\_\_\_}$
$\dfrac{8}{\sqrt{3}}, \dfrac{8\sqrt{3}}{3}$	Therefore, $c = \sqrt{a^2 + b^2} = \sqrt{\dfrac{16}{3} + 16} = \dfrac{\quad}{\_\_\_\_} = \dfrac{\quad}{\_\_\_\_}$.

213 $\alpha = 41°$ and $a = 21.7$ (Figure 5)

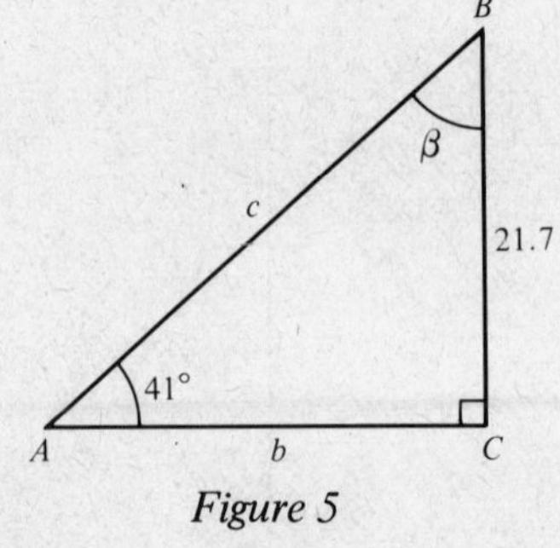

Figure 5

Answers	
41°, 49°	$\beta = 90° -$ ____ = ____
b, tan 41°, 0.8693, 25	$\tan 41° = \dfrac{21.7}{\_\_\_}$, so that $b = \dfrac{21.7}{\_\_\_\_\_\_} = \dfrac{21.7}{\_\_\_\_\_\_} =$ ____.
25	$\sec 41° = \dfrac{c}{b} = \dfrac{c}{\_\_\_}$, so that
sec 41°, 1.3250, 33.1	$c = 25$(________) = 25(________) = _____

214 $a = 3$ and $b = 4$ (Figure 6)

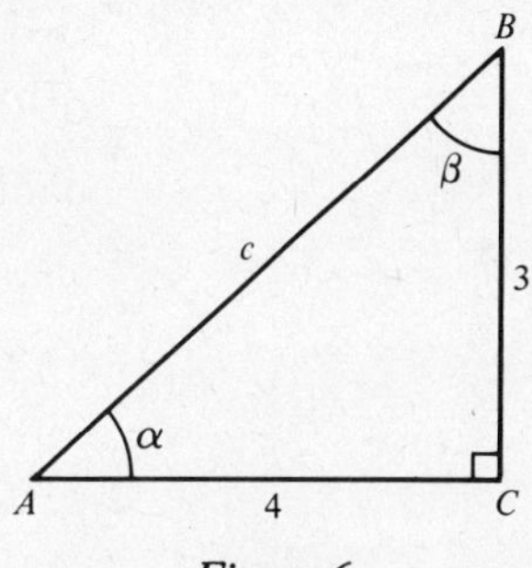

Figure 6

3, 9, 25, 5 — Here, $c^2 = 4^2 + (\_\_\_\_)^2 = 16 + \_\_\_\_ = \_\_\_\_$, so that $c = \_\_\_\_$.

0.6, 36°50′ — $\sin \alpha = \_\_\_\_$, so from Appendix Table IV, we have $\alpha = \_\_\_\_\_\_\_\_$.

36°50′, 53°10′ — $\beta = 90° - \_\_\_\_\_\_\_\_ = \_\_\_\_\_\_\_\_$

215 $a = 2$ and $c = 10$ (Figure 7)

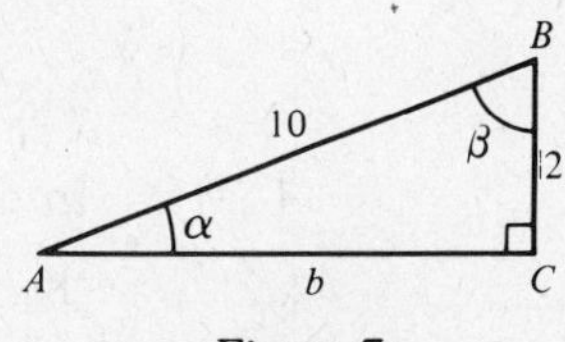

Figure 7

100, 4, 96, $4\sqrt{6}$ or 9.8 — Here, $b^2 = \_\_\_\_ - \_\_\_\_ = \_\_\_\_$, so $b = \_\_\_\_\_\_\_\_$.

0.2, 11°30′ — $\sin \alpha = \_\_\_\_$, so from Appendix Table IV we have $\alpha = \_\_\_\_$.

11°30′, 78°30′ — $\beta = 90° - \_\_\_\_\_\_\_\_ = \_\_\_\_\_\_\_\_$

216 The acute angle $\angle HAB$ in Figure 8, formed by the line of sight $\overline{AB}$ and the horizontal line $\overline{AH}$ through the eye of the observer, is called the angle of ________.

elevation

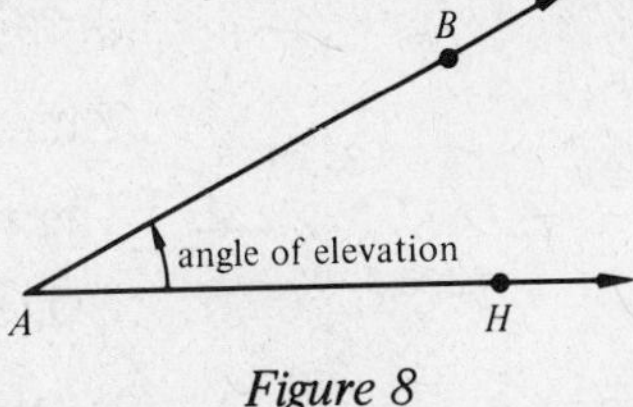

Figure 8

217 The acute angle $\angle POB$ in Figure 9, formed by the line of sight $\overline{OB}$ and the horizontal line $\overline{OP}$ through the eye of the observer, is called the angle of ____________.

depression

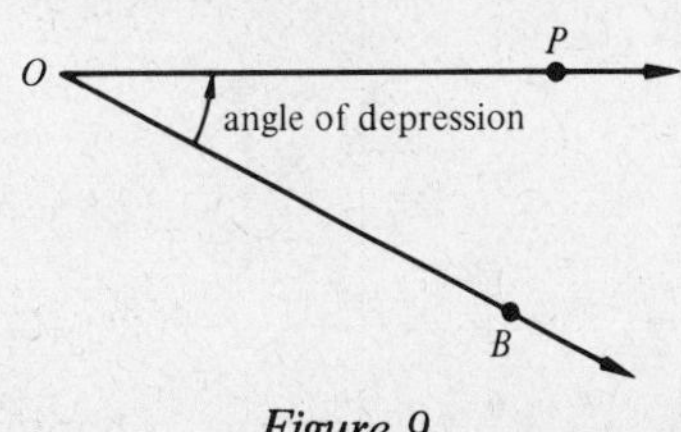

Figure 9

In Problems 218–221, use right triangle trigonometry to find the missing information.

218 A tower 40 feet high stands on the bank of a river. From the top of the tower the angle of depression of a point C on the opposite bank is 14° (Figure 10). What is the distance from the base of the tower B to C across the river?

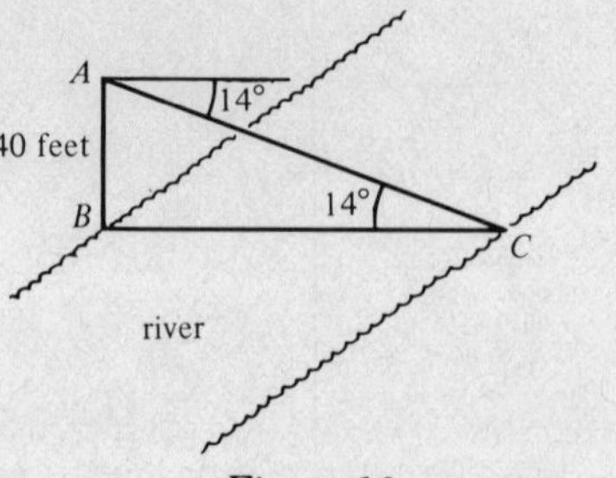

Figure 10

$\overline{BC}$, 14°	The desired distance is ____. In ΔABC, the angle ACB is ____.
40, 4.0110, 160.4	$\cot 14° = \frac{\overline{BC}}{\_\_}$, so that $\overline{BC} = 40 \cot 14° = 40(\_\_\_\_) = \_\_\_\_$.
160.4 feet	Therefore, the width of the river is ________.

219 In order to measure the height of a pine tree, a stake is driven into the ground 200 feet from the tree. If the tree is assumed to be vertical and the angle of elevation from the stake to the top of the tree is 30° (Figure 11), how tall is the tree?

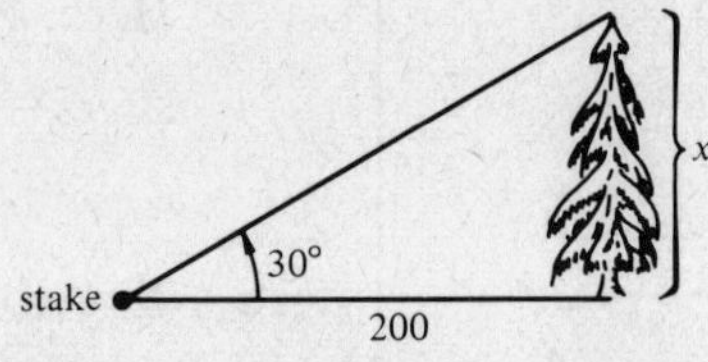

Figure 11

200	Let x be the height of the tree. Then, $\tan 30° = \frac{x}{\_\_}$,
tan 30°, 0.5774, 115.5	so that $x = 200(\_\_\_\_) = 200(\_\_\_\_) = \_\_\_\_$.
115.5 feet	Therefore, the height of the tree is ________.

220 The angle of depression of a boat 210 feet from the bottom of Niagara Falls, as seen from Goat Island at the top of the Falls, is 40° (Figure 12). How high is Niagara Falls?

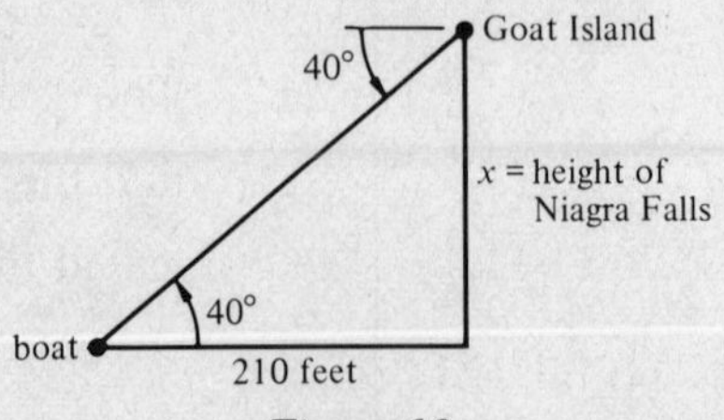

Figure 12

210	Let x be the height of Niagara Falls. Then, $\tan 40° = \frac{x}{\_\_}$,
tan 40°, 0.8391, 176.2	so that $x = 210(\_\_\_\_) = 210(\_\_\_\_) = \_\_\_\_$.
176.2 feet	Therefore, the height of Niagara Falls is ________.

221 If a man who is 6 feet tall casts a shadow that is 5 feet in length (Figure 13), what is the angle of elevation of the sun?

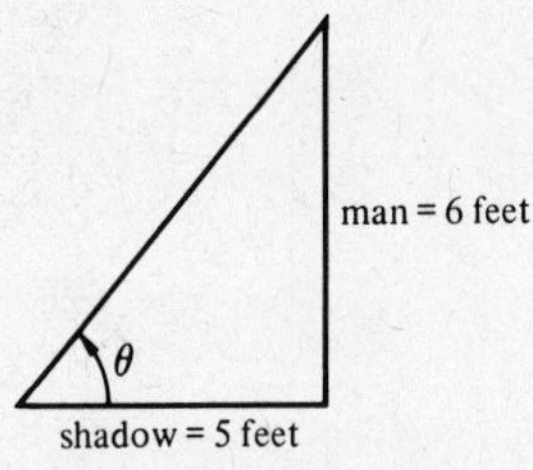

Figure 13

5, 1.2 — Let the angle of elevation be θ. Then $\tan\theta = \dfrac{6}{\_\_\_\_} = \_\_\_\_$, so

50°10′, 50°10′ — that $\theta = \_\_\_\_$. Therefore, the angle of elevation is ____.

In Problems 222–223, use the indicated right triangle to evaluate the given expression.

222 $\cos(\sin^{-1}\frac{3}{5})$

$\frac{3}{5}$ — Let $t = \sin^{-1}\frac{3}{5}$. Then $\sin t = \_\_\_\_$ (Figure 14).

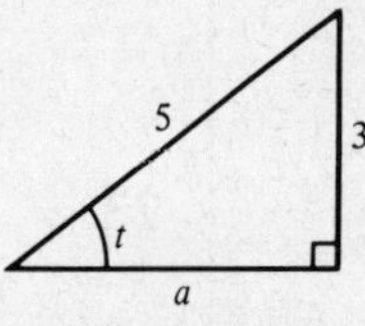

Figure 14

25, 4 — By the Pythagorean theorem, $a^2 + 9 = \_\_\_\_$, so that $a = \_\_\_\_$.

t, $\frac{4}{5}$ — Hence, $\cos(\sin^{-1}\frac{3}{5}) = \cos \_\_\_\_ = \_\_\_\_$.

223 $\tan\left(\cos^{-1}\dfrac{5}{\sqrt{26}}\right)$

$\dfrac{5}{\sqrt{26}}$ — Let $t = \cos^{-1}\dfrac{5}{\sqrt{26}}$. Then $\cos t = \dfrac{}{\_\_\_\_}$ (Figure 15).

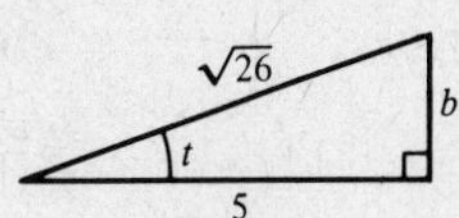

Figure 15

25, 1 — $b^2 + \_\_\_\_ = 26$, so that $b = \_\_\_\_$.

t, $\frac{1}{5}$ — Hence, $\tan\left(\cos^{-1}\dfrac{5}{\sqrt{26}}\right) = \tan \_\_\_\_ = \_\_\_\_$.

5.2 Polar Coordinates

224 A coordinate system that associates pairs of numbers with points in the plane based on a "grid" composed of concentric circles and rays emanating from the common center of the circles is called a ________ system.

polar coordinate

Answer	Frame
polar axis pole	**225** The frame of reference for the polar coordinate system consists of a fixed ray, called the ________________, and a fixed point, called the ______.
angle initial distance	**226** The position of a point P is determined uniquely by r and θ, where θ is any __________ in standard position having the polar axis as its __________ side and with its terminal side containing P, and r is the directed ______________ along the terminal side of θ between the pole and P.
one-to-one	**227** A polar coordinate system does not establish a ______________ correspondence between points in a plane and ordered pairs (r, θ).
point	**228** The ordered pairs (3, 30°) and (3, 390°) represent the same _______.
2	**229** The point represented by $\left(2, \frac{7\pi}{4}\right)$ can also be represented by $\left(\_\_\_\_, -\frac{\pi}{4}\right)$.
180° π	**230** If $r < 0$, the point (r, θ) is determined by plotting $(\lvert r\rvert, \theta + \_\_\_\_)$ or $(\lvert r\rvert, \theta + \_\_\_\_)$, depending on whether θ is measured in degrees or radians.
200°	**231** The point represented by (−2, 20°) can be determined by plotting (2, ______).
5	**232** The point represented by $\left(-5, \frac{3\pi}{2}\right)$ can be determined by plotting $\left(\_\_\_\_, \frac{\pi}{2}\right)$.

Conversion of Polar Coordinates to Rectangular Coordinates

Answer	Frame
$r\cos\theta$, $r\sin\theta$	**233** If (r, θ) is a polar representation of a point P, then the rectangular coordinates (x, y) of P can be found by the transformation or conversion formulas $x =$ __________ and $y =$ __________.

In Problems 234–239, find the rectangular coordinates of the point with the given polar coordinates.

Answer	Frame
	234 (4, 30°)
4, 30°	Since $r =$ _____ and $\theta =$ _____, then
sin 30°, cos 30°	$y = 4(\_\_\_\_\_\_\_\_)$ and $x = 4(\_\_\_\_\_\_\_\_)$
$\frac{1}{2}$, $\frac{\sqrt{3}}{2}$	$= 4\left(\_\_\_\right)$ $\qquad$ $= 4\left(\_\_\_\right)$
2, $2\sqrt{3}$	$=$ ____ $\qquad$ $=$ ____
$(2\sqrt{3}, 2)$	Hence, $(x, y) =$ __________.

235 $\left(-6, \frac{3\pi}{4}\right)$

Since $r =$ ____ and $\theta =$ ____, then

$y = -6\left(\sin \underline{\quad}\right)$ and $x = \underline{\quad}\left(\cos \frac{3\pi}{4}\right)$

$= -6\left(\underline{\quad}\right)$ $= -6\left(\underline{\quad}\right)$

$=$ ________ $=$ ________

Hence, $(x, y) =$ ______________.

Answers: $-6, \frac{3\pi}{4}$; $\frac{3\pi}{4}, -6$; $\frac{\sqrt{2}}{2}, -\frac{\sqrt{2}}{2}$; $-3\sqrt{2}, 3\sqrt{2}$; $(3\sqrt{2}, -3\sqrt{2})$

236 $(9, 90°)$

$r =$ ____ and $\theta =$ ____, so that

$y = 9($________$)$ and $x = 9($________$)$

$= 9($____$)$ $= 9($____$)$

$=$ ____ $=$ ____

Hence, $(x, y) =$ ________.

Answers: 9, 90°; sin 90°, cos 90°; 1, 0; 9, 0; (0, 9)

237 $(8, -420°)$

$r =$ ____ and $\theta =$ ____, so that

$y = 8[$____________$]$ and $x = 8[$____________$]$

$= 8\left(\underline{\quad}\right)$ $= 8\left(\underline{\quad}\right)$

$=$ ________ $=$ ____

Hence, $(x, y) =$ ____________.

Answers: $8, -420°$; $\sin(-420°), \cos(-420°)$; $-\frac{\sqrt{3}}{2}, \frac{1}{2}$; $-4\sqrt{3}, 4$; $(4, -4\sqrt{3})$

238 (π, π)

$r =$ ____ and $\theta =$ ____, so that

$y = \pi(\sin$ ____$)$ and $x =$ ____ $(\cos \pi)$

$=$ ____ $=$ ____

Hence, $(x, y) =$ ________.

Answers: π, π; π, π; $0, -\pi$; $(-\pi, 0)$

239 $\left(-1, \frac{13\pi}{6}\right)$

Here $r =$ ____ and $\theta =$ ____ so that

$y = -1\left(\sin \underline{\quad}\right)$ and $x = -1\left(\cos \underline{\quad}\right)$

$= -\left(\underline{\quad}\right)$ $= -\left(\underline{\quad}\right)$

$=$ ____ $=$ ____

Hence, $(x, y) =$ ______________.

Answers: $-1, \frac{13\pi}{6}$; $\frac{13\pi}{6}, \frac{13\pi}{6}$; $\frac{1}{2}, \frac{\sqrt{3}}{2}$; $-\frac{1}{2}, -\frac{\sqrt{3}}{2}$; $\left(-\frac{\sqrt{3}}{2}, -\frac{1}{2}\right)$

Conversion of Rectangular Coordinates to Polar Coordinates

$x^2 + y^2$, $\frac{y}{x}$

240 If (x, y) is a rectangular representation of a point P, then (r, θ), the polar coordinates of P, can be found by the formulas

$r = \sqrt{\_\_\_\_}$ and $\tan \theta = \_\_\_\_$.

In Problems 241–246, find the polar coordinates of the point with the given rectangular coordinates.

241 $(-4, 0)$

-4, 0

Since $x =$ ____ and $y =$ ____, then

-4, 0, 4, 0

$r = \sqrt{x^2 + y^2} = \sqrt{(\_\_\_\_)^2 + (\_\_\_\_)^2} =$ ____ and $\tan \theta = \frac{y}{x} =$ ____.

π or $180°$

Since $(-4, 0)$ lies on the negative x axis, $\theta =$ ____.

$(4, \pi)$ or $(4, 180°)$

Hence, $(r, \theta) =$ ____.

242 $(-3\sqrt{3}, -3)$

$-3\sqrt{3}$, -3

$x =$ ____ and $y =$ ____, so that

$-3\sqrt{3}$, -3, 6, $\frac{1}{\sqrt{3}}$, $\frac{\sqrt{3}}{3}$

$r = \sqrt{(\_\_\_\_)^2 + (\_\_\_\_)^2} =$ ____ and $\tan \theta = \_\_\_\_ = \_\_\_\_$.

III, $210°$ or $\frac{7\pi}{6}$

Since $(-3\sqrt{3}, -3)$ lies in quadrant ____, $\theta =$ ____.

$(6, 210°)$ or $\left(6, \frac{7\pi}{6}\right)$

Hence, $(r, \theta) =$ ____.

243 $(-6, 6\sqrt{3})$

-6, $6\sqrt{3}$

$x =$ ____ and $y =$ ____, so that

-6, $6\sqrt{3}$, 12, $-\sqrt{3}$

$r = \sqrt{(\_\_\_\_)^2 + (\_\_\_\_)^2} =$ ____ and $\tan \theta =$ ____.

II, $120°$ or $\frac{2\pi}{3}$

Since $(-6, 6\sqrt{3})$ lies in quadrant ____, $\theta =$ ____.

$(12, 120°)$ or $\left(12, \frac{2\pi}{3}\right)$

Hence, $(r, \theta) =$ ____.

244 (π, π)

π, π

$x =$ ____ and $y =$ ____, so that

π, π, $\sqrt{2}\pi$, 1

$r = \sqrt{(\_\_\_\_)^2 + (\_\_\_\_)^2} =$ ____ and $\tan \theta =$ ____.

I, $45°$ or $\frac{\pi}{4}$

Since (π, π) lies in quadrant ____, $\theta =$ ____.

$(\sqrt{2}\pi, 45°)$ or $\left(\sqrt{2}\pi, \frac{\pi}{4}\right)$

Hence, $(r, \theta) =$ ____.

245 $(0, -27)$

0, -27

$x =$ ____ and $y =$ ____, so that

0, -27, 27, undefined

$r = \sqrt{(\_\_\_\_)^2 + (\_\_\_\_)^2} =$ ____ and $\tan \theta =$ ____.

negative, $\frac{3\pi}{2}$ or $270°$

Since $(0, -27)$ lies on the ____ y axis, $\theta =$ ____.

$\left(27, \frac{3\pi}{2}\right)$ or $(27, 270°)$

Hence, $(r, \theta) =$ ____.

−4, 3

−4, 3, 5, $-\frac{3}{4}$

II, 143°8′

(5, 143°8′)

246 (−4, 3)

$x =$ ____ and $y =$ ____, so that

$r = \sqrt{(\_\_\_\_)^2 + (\_\_\_\_)^2} =$ ____ and $\tan\theta =$ ____.

Since (−4, 3) lies in quadrant ____, $\theta =$ ____.

Hence, $(r, \theta) =$ ________.

6 Law of Sines and Law of Cosines

6.1 Law of Sines

247 The law of sines states that in any ΔABC (Figure 1),

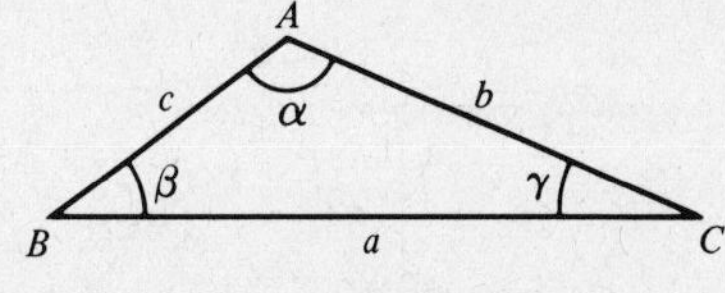

Figure 1

a, b, γ

$\sin\alpha$, b, $\sin\gamma$

$$\frac{\sin(\_\_\_\_)}{a} = \frac{\sin\beta}{\_\_\_\_} = \frac{\sin(\_\_\_\_)}{c}$$

or, equivalently,

$$\frac{a}{\_\_\_\_} = \frac{\_\_\_\_}{\sin\beta} = \frac{c}{\_\_\_\_}$$

In Problems 248–252, use the law of sines to find the missing parts of the illustrated triangles.

248 If $\alpha = 45°$, $\beta = 60°$, and $a = 5$ (Figure 2), find b, c, and γ.

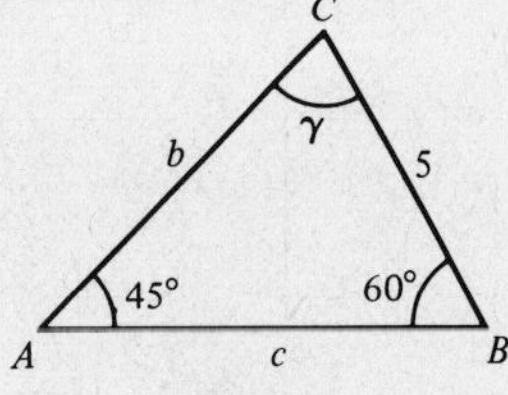

Figure 2

180°, 75°

sines

60°

60°, $\frac{\sqrt{3}}{2}$, $\frac{5\sqrt{6}}{2}$, 6.1

sines

45°

Since $45° + 60° + \gamma =$ ____, it follows that $\gamma =$ ____.

Next we can solve for b by using the law of ________.

$$\frac{b}{\sin \_\_\_\_} = \frac{5}{\sin 45°}$$

so that

$$b = \frac{5\,(\sin \_\_\_\_)}{\sin 45°} = \frac{5(\_\_\_\_)}{\frac{\sqrt{2}}{2}} = \_\_\_\_\_\_\_\_ = \_\_\_\_$$

Similarly, we can solve for c by using the law of ________.

$$\frac{c}{\sin 75°} = \frac{5}{\sin \_\_\_\_}$$

so that

$45°, \frac{\sqrt{2}}{2}, 6.8$

$$c = \frac{5(\sin 75°)}{\sin \_\_\_\_} = \frac{5(0.9659)}{\_\_\_\_} = \_\_\_\_$$

249 If $a = 12$, $\alpha = 45°$, and $\beta = 105°$ (Figure 3), find c.

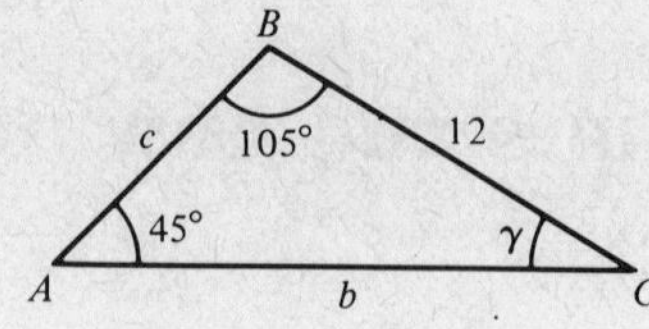

Figure 3

$180°, 30°$

Since $\alpha + \beta + \gamma = \_\_\_\_$, then $\gamma = \_\_\_\_$, so that

$\sin 45°, \sin 30°$

$$\frac{12}{\_\_\_\_} = \frac{c}{\_\_\_\_}$$

Then

$\sin 45°, \frac{\sqrt{2}}{2}, 6\sqrt{2}, 8.5$

$$c = \frac{12(\sin 30°)}{\_\_\_\_} = \frac{12\left(\frac{1}{2}\right)}{\_\_\_\_} = \_\_\_\_ = \_\_\_\_$$

250 If $a = 20$, $\gamma = 51°$, and $\beta = 42°$ (Figure 4), find b.

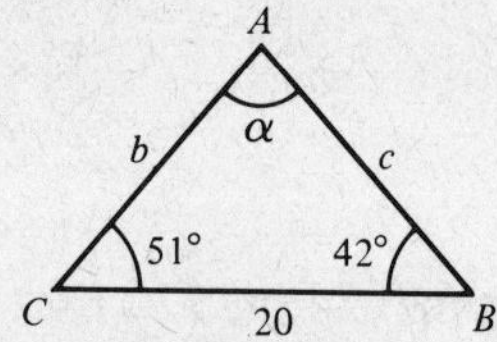

Figure 4

$180°, 87°$

Since $\alpha + \beta + \gamma = \_\_\_\_$, then $\alpha = \_\_\_\_$, so that

$b, \sin 87°$

$$\frac{\_\_\_\_}{\sin 42°} = \frac{20}{\_\_\_\_}$$

Then

$\sin 87°, 0.6691, 13.4$

$$b = \frac{20(\sin 42°)}{\_\_\_\_} = \frac{20(\_\_\_\_)}{0.9986} = \_\_\_\_$$

251 If $c = 25$, $\gamma = 77°$, and $\beta = 68°$ (Figure 5), find b.

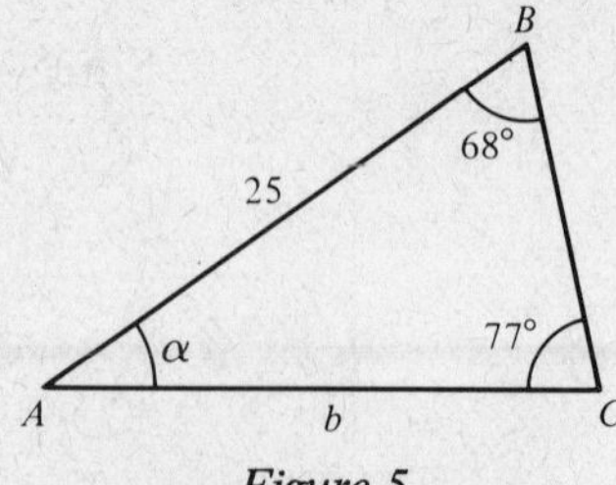

Figure 5

$\sin 77°$

$$\frac{b}{\sin 68°} = \frac{25}{\_\_\_\_}$$

so that

$\sin 68°, 0.9272, 23.8$

$$b = \frac{25(\_\_\_\_)}{\sin 77°} = \frac{25(\_\_\_\_)}{0.9744} = \_\_\_\_$$

252 If $\alpha = 36°$, $b = 10$, and $\beta = 40°$ (Figure 6), find a.

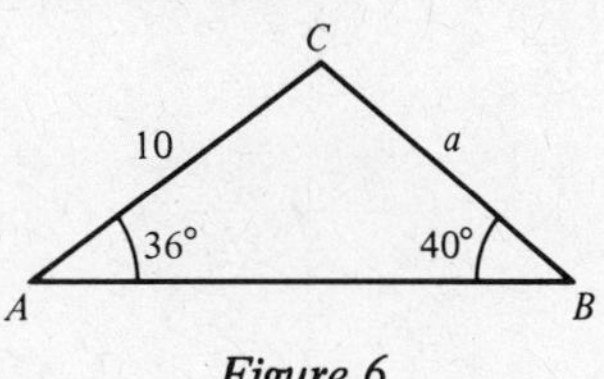

Figure 6

40° — $\dfrac{a}{\sin 36°} = \dfrac{10}{\sin \_\_\_\_}$

so that

40°, 0.6428, 9.1 — $a = \dfrac{10(\sin 36°)}{\sin \_\_\_\_} = \dfrac{10(0.5878)}{\_\_\_\_\_\_} = \_\_\_\_$

253 Find the lengths of the sides of a parallelogram to the nearest tenth if one of the diagonals is 5 inches long and forms angles of 33° and 25° with the sides at each end (Figure 7).

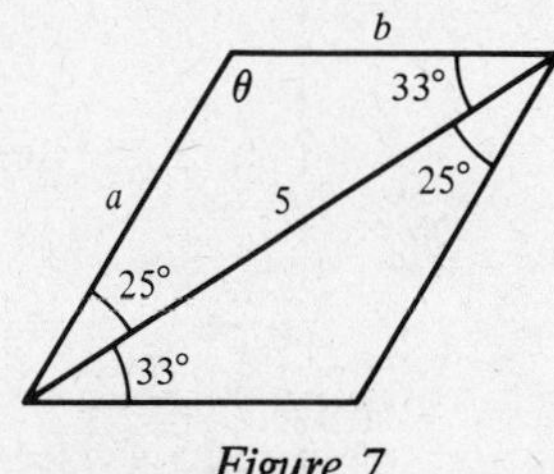

Figure 7

180°, 122° — Since $\theta + 25° + 33° = \_\_\_\_$, $\theta = \_\_\_\_$.

By the law of sines,

122° — $\dfrac{a}{\sin 33°} = \dfrac{5}{\sin \_\_\_\_}$

so that

122°, 0.8480 — $a = \dfrac{5(\sin 33°)}{\sin \_\_\_\_} = \dfrac{5(0.5446)}{(\_\_\_\_\_\_)}$

3.2 inches — $= \_\_\_\_\_\_\_\_$

Similarly,

25° — $\dfrac{b}{\sin \_\_\_\_} = \dfrac{5}{\sin 122°}$

so that

25°, 0.4226 — $b = \dfrac{5(\sin \_\_\_\_)}{\sin 122°} = \dfrac{5(\_\_\_\_\_\_)}{0.8480}$

2.5 inches — $= \_\_\_\_\_\_\_\_$

254 An airplane P is sighted simultaneously from two towns, A and B, that are 50,000 feet apart. The angle of sight from town A is 35° and that from town B is 60°, as is shown in Figure 8. How high is the airplane?

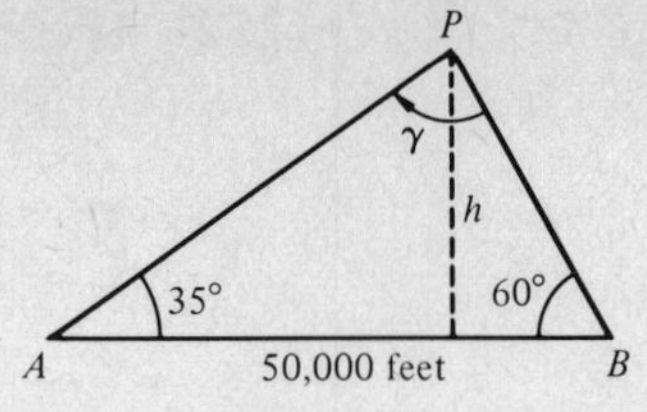

Figure 8

95°, 85°

$\gamma = 180° -$ ____ = ____. By the law of sines,

85°

$$\frac{\overline{PB}}{\sin 35°} = \frac{50{,}000}{\sin \text{\_\_\_\_}}$$

so that

85°, 0.9962

$$\overline{PB} = \frac{50{,}000(\sin 35°)}{\sin \text{\_\_\_\_}} = \frac{50{,}000\ (0.5736)}{\text{\_\_\_\_\_\_}}$$

28,789 feet

Hence, $\overline{PB}$ is approximately ____________. From right triangle trigonometry, we have $\sin 60° = \frac{h}{\overline{PB}}$, so that

28,789; $\frac{\sqrt{3}}{2}$; 24,931

$$h = \overline{PB} \sin 60° = \text{\_\_\_\_\_\_}\left(\text{\_\_\_\_}\right) = \text{\_\_\_\_\_\_}$$

24,931 feet

Therefore, the airplane is ____________ high.

The Ambiguous Case

opposite

one, two

255 The ambiguous case of the law of sines occurs when we are given two sides and an angle ____________ one of them. In this case there may be no possible triangle, ____ triangle, or ____ different triangles with the given parts.

no possible

256 In Figure 9, if α is an acute angle and $a < h = b \sin \alpha$, then there is ____________ triangle.

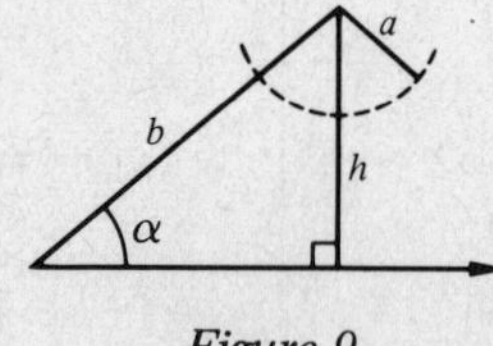

Figure 9

one possible

257 In Figure 10, if α is an acute angle and $a > b$, then there is ____________ triangle.

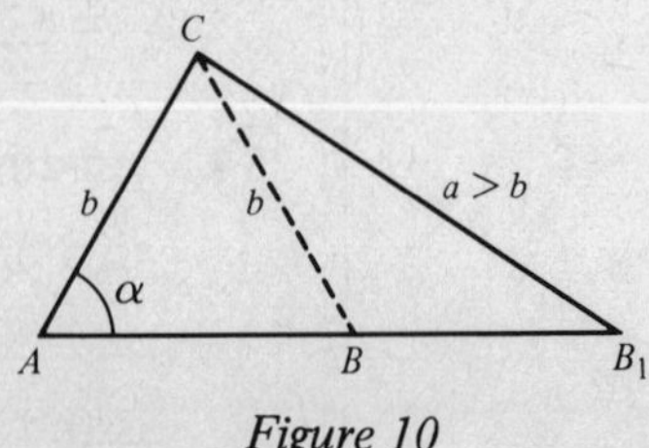

Figure 10

two possible

258 In Figure 11, if α is an acute angle and $b > a > h = b \sin \alpha$, then there are ______ triangles.

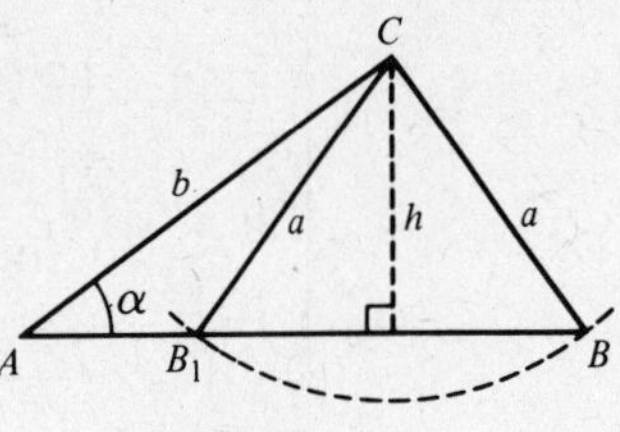

Figure 11

In Problems 259–263, determine if the given parts define a triangle and then use the law of sines to determine the specified unknown parts of all possible triangles. Where required, find degree measures to the nearest 10 minutes (10′).

259 If $c = 24$, $b = 67$, and $\gamma = 121°$, find β.

Using the law of sines, we have

sin 121°

$$\frac{\sin \beta}{67} = \frac{\_\_\_\_}{24}$$

so that

24, sin 59°, 0.8572

$$\sin \beta = \frac{67(\sin 121°)}{\_\_\_\_} = \frac{67(\_\_\_\_)}{24} = \frac{67(\_\_\_\_)}{24}$$

2.4, impossible

Hence, $\sin \beta =$ ____. Therefore, it is ______ to construct a triangle with the given data, since $|\sin \beta| \leq 1$.

260 If $a = 3$, $b = 4$, and $\alpha = 30°$, find all unknown parts of all possible triangles.

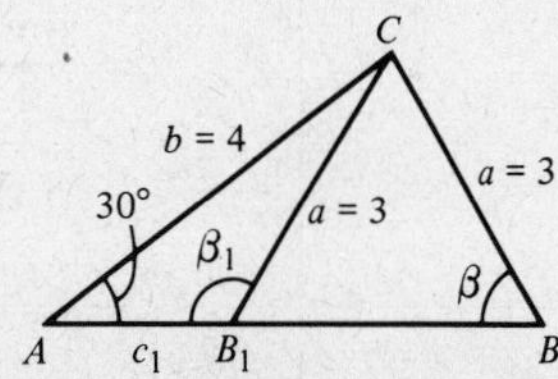

Figure 12

b sin α

From Figure 12, we observe that $a >$ ______ and $a < b$.

two

Therefore, there are ____ triangles. Solving ΔABC by using the law of sines, we obtain

sin β

$$\frac{\sin 30°}{3} = \frac{\_\_\_\_}{4}$$

so that

sin 30°, 0.5

$$\sin \beta = \tfrac{4}{3}(\_\_\_\_) = \tfrac{4}{3}(\_\_\_\_) = 0.6667$$

41°50′

Hence, $\beta =$ ______.

108°10′

$\gamma = 180° - 30° - 41°50' =$ ______

To find c, we have

c, 3

$$\frac{\_\_\_\_}{\sin 108°10'} = \frac{\_\_\_\_}{\sin 30°}$$

so that

sin 30°, 0.9502, 5.7

$$c = \frac{3(\sin 108°10')}{\_\_\_\_\_\_} = \frac{3(\_\_\_\_\_\_)}{0.500} = \_\_\_\_$$

138°10'

To solve for ΔAB_1C, we have $\beta_1 = 180° - 41°50' =$ ______.

138°10', 11°50'

$\gamma_1 = 180° - 30° -$ ______ $=$ ______

Using the law of sines,

sin 11°50'

$$\frac{c_1}{\_\_\_\_\_\_} = \frac{3}{\sin 30°}$$

so

sin 30°, 0.5000, 1.2

$$c_1 = \frac{3(\sin 11°50')}{\_\_\_\_\_\_} = \frac{3(0.2051)}{\_\_\_\_\_\_} = \_\_\_\_$$

Thus, the solutions for ΔABC are

108°10'

$\beta = 41°50'$, $\gamma =$ ______, $\alpha = 30°$, $a = 3$, $b = 4$, and

5.7, 138°10'

$c =$ ____, and for ΔAB_1C, we have $\beta_1 =$ ______, $\gamma_1 = 11°50'$,

1.2

$\alpha = 30°$, $a = 3$, $b = 4$, and $c_1 =$ ____.

261 If $\alpha = 30°$, $a = 4$, and $b = 10$ (Figure 13), find β.

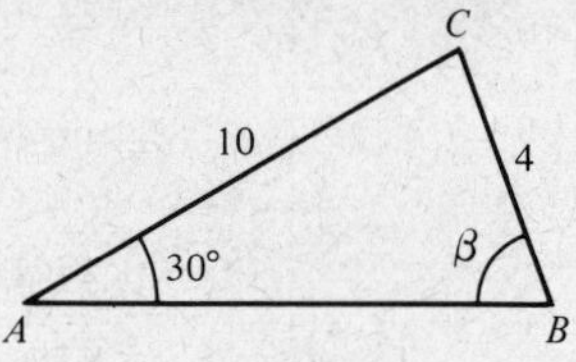

Figure 13

sines

We begin to solve for β by using the law of ______. Hence,

10, 4

$$\frac{\sin \beta}{\_\_\_\_} = \frac{\sin 30°}{\_\_\_\_}$$

so that

10

$$\sin \beta = \frac{\_\_\_\_(\sin 30°)}{4}$$

10, $\frac{1}{2}$

$$= \frac{(\_\_\_\_)(\_\_\_\_)}{4}$$

$\frac{5}{4}$

$$= \_\_\_\_$$

1

But for *any* β, $\sin \beta <$ ____, so we conclude that $\sin \beta = \frac{5}{4}$ does *not* have a solution, hence there is no possible triangle with the indicated parts.

262 If $\alpha = 30°$, $a = 2$, and $c = 3$ (Figure 14), find γ.

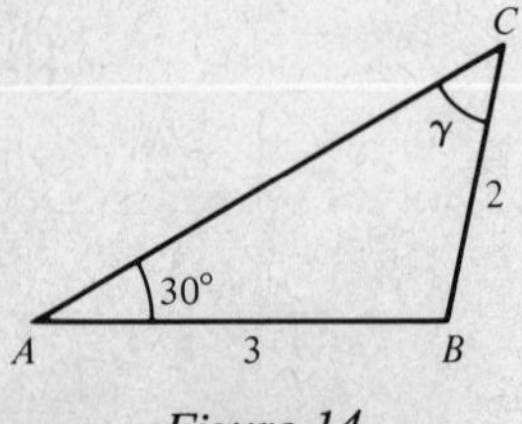

Figure 14

We can begin to solve for γ by using the law of ________. Hence,

$$\frac{\sin \gamma}{3} = \frac{\sin 30^\circ}{\_\_\_\_}$$

so

$$\sin \gamma = \frac{3(\sin 30^\circ)}{\_\_\_\_} = \frac{3(\frac{1}{2})}{\_\_\_} = \_\_\_\_ = \_\_\_\_\_$$

From Appendix Table IV we get $\gamma =$ ________. However, since the sine is positive in quadrant II, $\gamma = 180^\circ -$ ______ = ________ is also a solution of the equation and a possible angle in the given triangle because its sum with 30° does not exceed _____. Hence, since γ can equal ________ or ________, there are two possible triangles.

sines
2
2, 2, $\frac{3}{4}$, 0.75
48°40′
48°40, 131°20′
180°
48°40′, 131°20′

263 If $a = 20$, $b = 15$, and $\alpha = 30^\circ$, find β.

Using the law of sines, we have

$$\frac{\sin \beta}{\_\_\_\_} = \frac{\sin 30^\circ}{\_\_\_\_\_}$$

so that

$$\sin \beta = \frac{15(\sin 30^\circ)}{\_\_\_\_\_} = \_\_\_\_ = \_\_\_\_\_\_\_$$

If β is an acute angle, then $\beta =$ ____; if β is an obtuse angle, then $\beta =$ ______. However, it is not possible to have both a 158° angle and a 30° angle in the same triangle, since the sum of the angle measures in a triangle is ______. Thus $\beta =$ ____ (Figure 15).

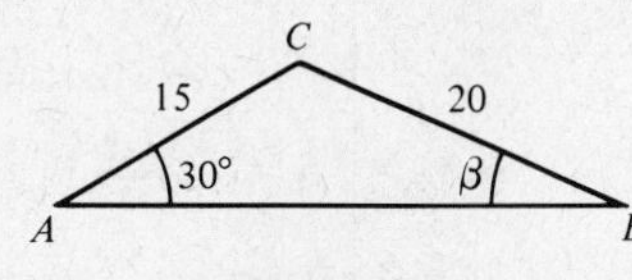

Figure 15

15, 20
20, $\frac{3}{8}$, 0.3750
22°
158°
180°, 22°

6.2 Law of Cosines

264 The law of cosines states that in any ΔABC (Figure 16),

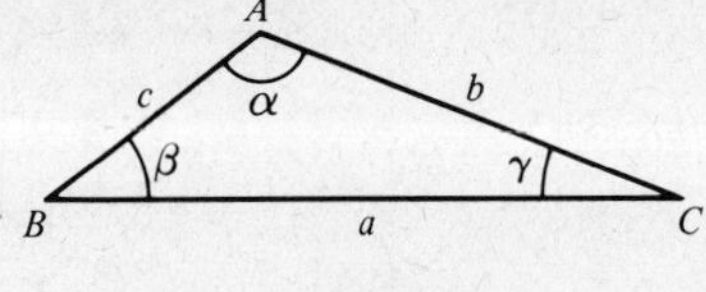

Figure 16

$c^2 =$ ______________________

$b^2 =$ ______________________

$a^2 =$ ______________________

$a^2 + b^2 - 2ab \cos \gamma$
$a^2 + c^2 - 2ac \cos \beta$
$b^2 + c^2 - 2bc \cos \alpha$

In Problems 265–271, use the law of cosines to find the indicated missing part of the given triangle. Find angle measures to the nearest 10 minutes (10′).

265 If $\alpha = 150°$, $b = 6$, and $c = 8$ (Figure 17), find a.

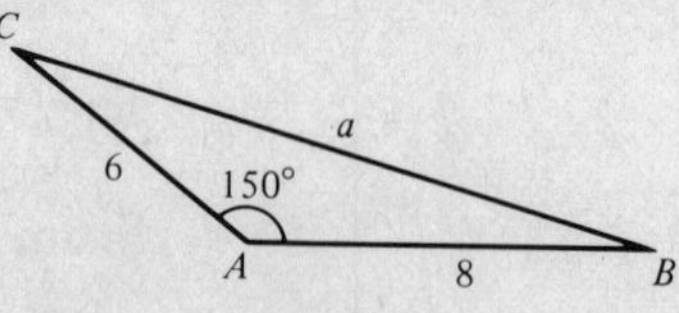

Figure 17

$2bc \cos \alpha$	$a^2 = b^2 + c^2 -$ ________
	so that
$\cos 150°$	$a^2 = 36 + 64 - 2(6)(8)$ ________
-0.8660	$= 36 + 64 - 2(6)(8)($________$)$
83.136	$= 100 +$ ________
183.136	$=$ ________
183.136, 13.5	Hence, $a = \sqrt{\_\_\_\_\_\_} =$ ____

266 If $\alpha = 100°$, $b = 10$, and $c = 12$ (Figure 18), find a.

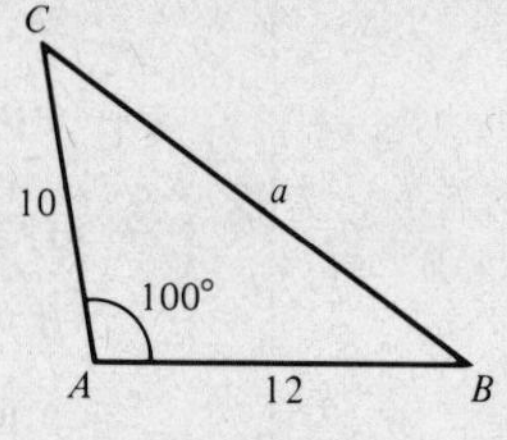

Figure 18

$\cos \alpha$	$a^2 = b^2 + c^2 - 2bc$ ________
	so that
$\cos 100°$	$a^2 = 100 + 144 - 2(10)(12)$ ________
-0.1736	$= 100 + 144 - 2(10)(12)($________$)$
41.664	$= 100 + 144 +$ ________
285.664	$=$ ________
16.9	Hence, $a =$ ____.

267 If $a = 8$, $b = 6$, and $\gamma = 60°$ (Figure 19), find c.

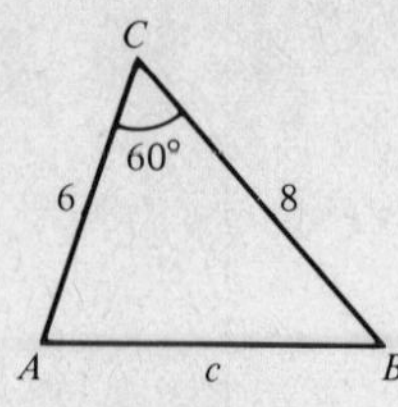

Figure 19

$2ab \cos \gamma$	$c^2 = a^2 + b^2 -$ ________
	so that
$\frac{1}{2}$	$c^2 = 64 + 36 - 2(8)(6)($____$)$
52	$=$ ____
7.2	Therefore, $c =$ ____.

268 If $a = 2.9$, $c = 2.3$, and $\beta = 120°$ (Figure 20), find b.

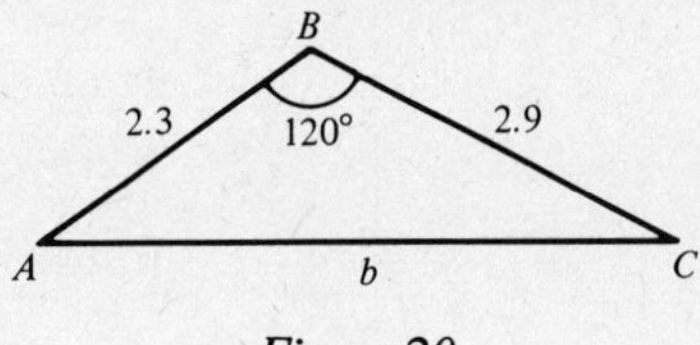

Figure 20

Answer	
$2ac \cos \beta$	$b^2 = a^2 + c^2 -$ __________
	so that
$-\frac{1}{2}$	$b^2 = 8.41 + 5.29 - 2(2.3)(2.9)($____$)$
6.67	$= 8.41 + 5.29 +$ ____
20.37	$=$ ______
4.51	Therefore, $b =$ ____.

269 If $a = 5$, $b = 4$, and $c = 7$ (Figure 21), find α.

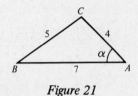

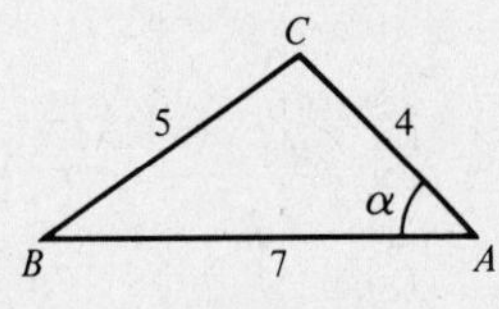

Figure 21

Answer	
$\cos \alpha$	$a^2 = b^2 + c^2 - 2bc$ ______
	so that
$\cos \alpha$	$25 = 16 + 49 - 2(4)(7)$ ______
$56 \cos \alpha$	$= 65 -$ ________
0.7143, 44°20′	Then $\cos \alpha =$ ______. Hence, $\alpha =$ ______.

270 If $a = 9$, $b = 12$, and $c = 15$ (Figure 22), find γ.

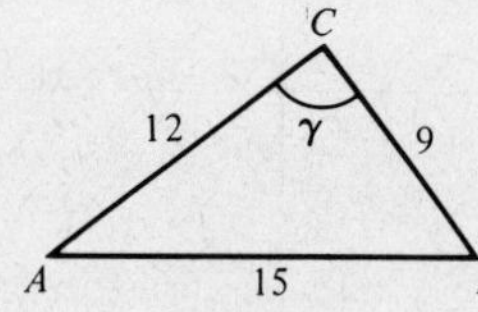

Figure 22

Answer	
$2ab \cos \gamma$	$c^2 = a^2 + b^2 -$ __________, so
12, 12	$15^2 = 9^2 + ($____$)^2 - 2(9)($____$) \cos \gamma$
225, 225, 216	____ $=$ ____ $-$ ____ $\cos \gamma$
0	$\cos \gamma =$ ____
90°	Hence, $\gamma =$ ____.

271 If $a = 10$, $b = 5$, and $c = 8$ (Figure 23), find β.

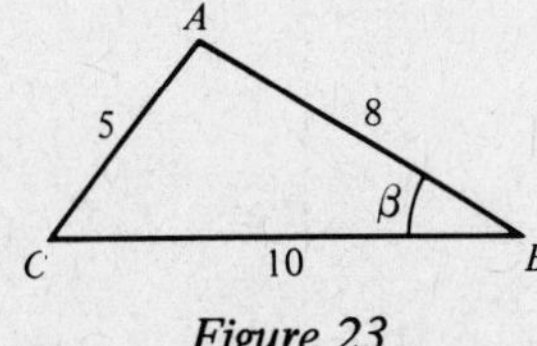

Figure 23

$\cos\beta$	$b^2 = a^2 + c^2 - 2ac$ ________, so
$\cos\beta$	$25 = 100 + 64 - 2(10)(8)$ ________
$\frac{139}{160}$, 0.8688, 29°40′	Then, $\cos\beta =$ ____ = ________. Hence, $\beta =$ ________.

272 A motorboat is propelled west at a rate of 30 feet per second. Find the strength of the tide, which alters the course of the boat 20° south of west and reduces the speed to 25 feet per second (Figure 24).

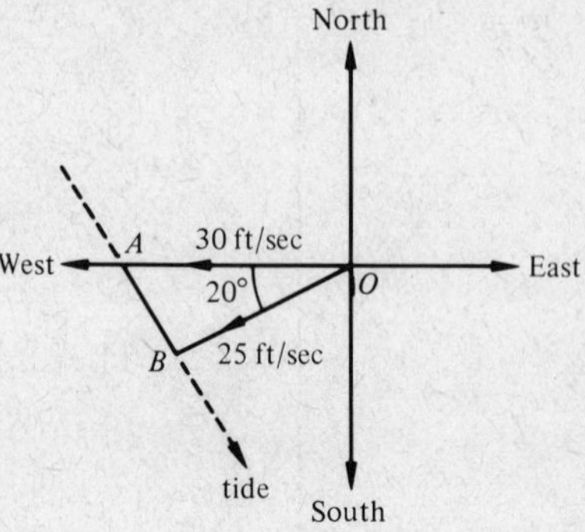

Figure 24

Let $|\overline{AB}|$ represent the strength of the tide in feet per second. Then, by the law of cosines,

cos 20°	$|\overline{AB}|^2 = |\overline{OA}|^2 + |\overline{OB}|^2 - 2|\overline{OA}| \cdot |\overline{OB}|$ (________)
0.9397	$= 900 + 625 - 2(30)(25)$ (________)
115.4	= ______
10.7	Hence, $|\overline{AB}|$ is approximately _____.
10.7 feet per second	Therefore, the strength of the tide is about ____________.

273 Two trains depart simultaneously from the same depot, D. The angle determined by the tracks on which they leave is 118°40′ (Figure 25). If one travels at 40 miles per hour and the other at 90 miles per hour, how far apart are the trains after 3 hours?

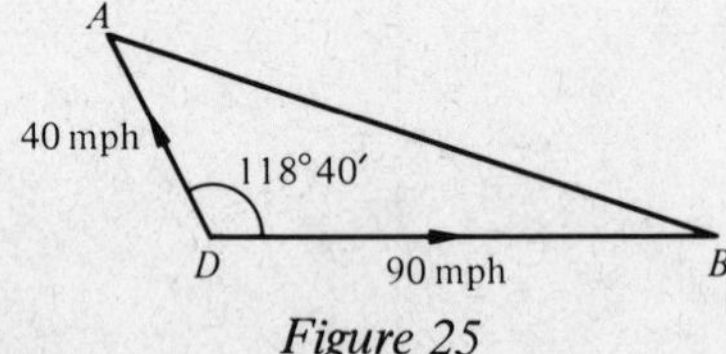

Figure 25

At the end of 3 hours,

40, 120, 90, 270	$|\overline{AD}| = 3($____$) =$ ____ and $|\overline{DB}| = 3($____$) =$ ____
	so that, by the law of cosines, we have
270, 270	$|\overline{AB}|^2 = 120^2 + ($____$)^2 - 2(120)($____$) \cos 118°40'$
72,900; 64,800	$= 14{,}400 +$ ________ $-$ (________) $\cos 118°40'$
87,300; 64,800	= ________ $-$ (________)$(-\cos 61°20')$
87,300; 64,800	= ________ + ________ $\cos 61°20'$
118,385	= ________
344	so that, $|\overline{AB}|$ is approximately __________.
344 miles	Therefore, the trains are about ____ apart after 3 hours.

7 Vectors and Trigonometry

7.1 Geometry of Vectors in the Plane

Answers		Frame
line segment direction terminal	274	A vector in the plane is a ________________ with ________. A vector has an initial point and a ________ point.
AB, |**AB**|	275	A vector having initial point A and terminal point B is denoted by ____. Its length or magnitude is denoted by ______.
zero	276	The vector whose initial and terminal points coincide is called the ______ vector.
0	277	|**0**| = ____.
magnitudes	278	Two vectors are considered to be equal if they have equal ____________ and the same directions.
coincide	279	If a is a scalar (real) number and **u** is a vector, then a**u** is the vector with magnitude $\|a\|$ times that of **u** and in the direction of ____ if $a > 0$, and in the direction ________ of ____ if $a < 0$.

v

u

u + v

Answers		Frame
u opposite, **u**	280	If a is a scalar (real) number and **u** is a vector, then a**u** is the vector with magnitude $\|a\|$ times that of **u** and in the direction of ____ if $a > 0$, and in the direction ________ of ____ if $a < 0$.
six times	281	6**u** is a vector in the direction of **u** with magnitude ________ the magnitude of **u**.
opposite $\frac{1}{3}$|**v**|	282	$-\frac{1}{3}$**v** is a vector that has its direction ________ that of **v** and $\|-\frac{1}{3}$**v**$\|$ = ______.
(−**v**)	283	**u** − **v** = **u** + ______

7.2 Analytic Representation of Vectors in the Plane

Answers		Frame
radius position **u** = ⟨a, b⟩, x y	284	Suppose that vector **u** is positioned in a plane with a cartesian coordinate system so that the initial point of **u** is the origin and the terminal point is at point (a, b). Then **u** is called a ______ vector or ________ vector. We identify such a vector **u** as ________; a is called the ____ component and b is called the ____ component of **u**.
3, 5	285	**u** = ⟨3, 5⟩ has x component ____ and y component ____.
−2, −1	286	**u** = ⟨−2, −1⟩ has x component ____ and y component ____.

0, 4

287 **u** = ⟨0, 4⟩ has x component ____ and y component ____.

$\sqrt{a^2 + b^2}$

288 If **u** = ⟨a, b⟩, then the magnitude of **u** is given by |**u**| = ________.

$(-3)^2 + (-4)^2$, 5

289 If **u** = ⟨−3, −4⟩, then |**u**| = $\sqrt{\qquad\qquad}$ = ____.

$\sqrt{4^2 + 3^2}$, 5, 5, 25

290 If **u** = ⟨4, 3⟩ and **v** = ⟨−3, 4⟩, then
|**u**||**v**| = ________ $\sqrt{(-3)^2 + 4^2}$ = (____)(____) = ____

c, d

291 If **u** = ⟨a, b⟩ and **v** = ⟨c, d⟩, then **u** = **v** whenever
a = ____ and b = ____.

3, 2

292 ⟨−3, y⟩ = ⟨−x, 2⟩ if and only if x = ____ and y = ____.

$\frac{5}{3}$, −1

293 ⟨5, y⟩ = ⟨3x, −1⟩ if and only if x = ____ and y = ____.

5

3, 1, 2

294 ⟨3x + 2, 2y − 1⟩ = ⟨5, 3⟩ if and only if 3x + 2 = ____ and
2y − 1 = ____, so x = ____ and y = ____.

cos 60°, 3

sin 60°, $3\sqrt{3}$

295 Assume that |**u**| = 6 and that the direction angle for **u** is 60°.
Then the x component of **u** is given by x = 6(________) = ____,
and the y component is given by y = 6(________) = ____.

In Problems 296–297, assume that vector **u** has the first set of coordinates as its initial point and the second set of coordinates as its terminal point. Represent **u** as a radius vector and then find |**u**| and θ, the direction angle for **u**.

6, 5, 4, 5, ⟨1, −1⟩

−1, $\sqrt{2}$

−1, −45°

296 (5, 5) and (6, 4)
u = ⟨____ − ____, ____ − ____⟩ = ________.
Thus |**u**| = $\sqrt{1^2 + (\quad)^2}$ = ____. Also
tan θ = ____ and so θ = ____.

⟨3, 1⟩, $\sqrt{10}$

$\frac{1}{3}$, 18°30′

297 (−1, −3) and (2, −2)
u = ________ and |**u**| = ____. Also,
tan θ = ____ and so θ = ________.

7.3 Vector Algebra in Terms of Components

⟨$a + c, b + d$⟩

⟨$a - c, b - d$⟩

⟨ca, cb⟩

298 If **u** = ⟨a, b⟩ and **v** = ⟨c, d⟩, then
u + **v** = ⟨a, b⟩ + ⟨c, d⟩ = ____________
and
u − **v** = ____________
If **u** = ⟨a, b⟩ and c is a scalar (real) number, then
c**u** = c⟨a, b⟩ = ________.

In Problems 299–306, use **u** = ⟨2, 3⟩ and **v** = ⟨4, −1⟩ to find each expression.

2 + 4, 3 − 1; ⟨6, 2⟩

299 **u** + **v** = ⟨____________⟩ = ________

2 − 4, 3 + 1; ⟨−2, 4⟩

300 **u** − **v** = ⟨____________⟩ = ________

6, 9; −12, 3

⟨−6, 12⟩

301 3**u** − 3**v** = 3⟨2, 3⟩ − 3⟨4, −1⟩
= ⟨________⟩ + ⟨________⟩
= ________

302 $|\mathbf{u}||\mathbf{v}| = \sqrt{2^2+3^2}\ \sqrt{4^2+(-1)^2}$

$= \_\_\_\_ \cdot \_\_\_\_$ — $\sqrt{13},\ \sqrt{17}$

$= \_\_\_\_\_\_\_\_ = \_\_\_\_\_\_$ — $\sqrt{(13)(17)},\ \sqrt{221}$

303 $|\mathbf{u}+\mathbf{v}| = |\langle 2, 3\rangle + \langle 4, -1\rangle|$

$= |\_\_\_\_\_\_|$ — $\langle 6, 2\rangle$

$= \sqrt{6^2+2^2} = \sqrt{\_\_\_} = \_\_\_\_\_\_$ — $40,\ 2\sqrt{10}$

304 $|\mathbf{u}| + |\mathbf{v}| = |\langle 2, 3\rangle| + |\langle 4, -1\rangle|$

$= \sqrt{4+9} + \sqrt{16+1}$

$= \sqrt{\_\_\_} + \sqrt{\_\_\_}$ — $13,\ 17$

$= \_\_\_\_\_\_\_\_$ — $\sqrt{13}+\sqrt{17}$

305 $|3\mathbf{u} - 2\mathbf{v}| = |3\langle 2, 3\rangle - 2\langle 4, -1\rangle|$

$= |\langle 6, 9\rangle + \langle -8, 2\rangle|$

$= |\_\_\_\_\_\_\_\_|$ — $\langle -2, 11\rangle$

$= \_\_\_\_\_\_$ — $5\sqrt{5}$

306 $|5\mathbf{u} + 2\mathbf{v}| = |5\langle 2, 3\rangle + 2\langle 4, -1\rangle|$

$= |\_\_\_\_\_\_ + \_\_\_\_\_\_|$ — $\langle 10, 15\rangle,\ \langle 8, -2\rangle$

$= |\_\_\_\_\_\_\_\_|$ — $\langle 18, 13\rangle$

$= \_\_\_\_\_\_$ — $\sqrt{493}$

7.4 Applications of Vectors

307 A boat, with a still-water speed of 10 miles per hour, heading northeast, is pushed off course by a current of 3 miles per hour moving in the direction of 60° S of W. What is the resulting speed and direction (to the nearest 10 minutes) of the boat's path? The situation is illustrated in Figure 1.

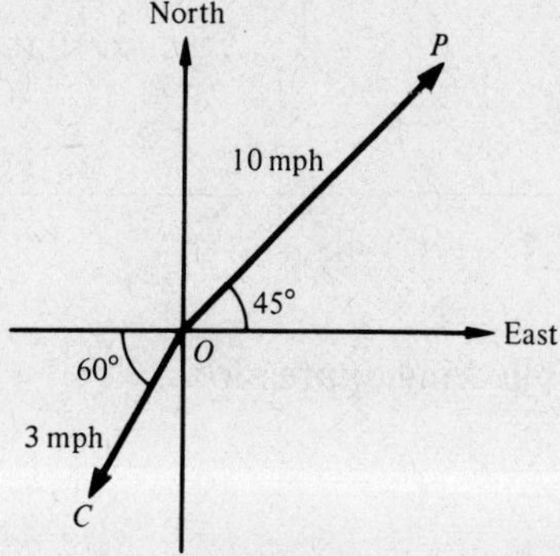

Figure 1

The path of the boat in still water is defined by the vector

$\boldsymbol{OP} = \langle 10 \cos 45°, \_\_\_\_\_\_\rangle$ — 10 sin 45°

$= \langle 5\sqrt{2}, \_\_\_\_\rangle$ — $5\sqrt{2}$

$= \langle 7.07, \_\_\_\rangle$ — 7.07

and the current is defined by the vector

$\cos 240°$, $\sin 240°$

$OC = \langle 3(\_\_\_\_\_\_), 3(\_\_\_\_\_\_)\rangle$

-2.60

$= \langle -1.50, \_\_\_\_\rangle$

The velocity vector of the boat's path is given by the resultant

$\langle 7.07, 7.07\rangle$, $\langle -1.50, -2.60\rangle$

vector $OP + OC = \_\_\_\_\_\_\_\_ + \_\_\_\_\_\_\_\_$

$\langle 5.57, 4.47\rangle$

$= \_\_\_\_\_\_\_\_$

$\langle 5.57, 4.47\rangle$, 51, 7.14

Thus the resulting speed is $|\_\_\_\_\_\_\_\_| = \sqrt{\_\_\_\_} = \_\_\_\_$.

0.80, 38°40′

The direction angle θ satisfies $\tan\theta = \_\_\_\_$, so that $\theta = \_\_\_\_$.

308 Two forces, one of 63 pounds and the other of 45 pounds, yield a resultant force of 75 pounds. Find the angle (to the nearest 10 minutes) between the resultant force and the force of 63 pounds. Assume that the force vectors are **u**, **v**, and **w** as shown in Figure 2, where θ is the desired angle.

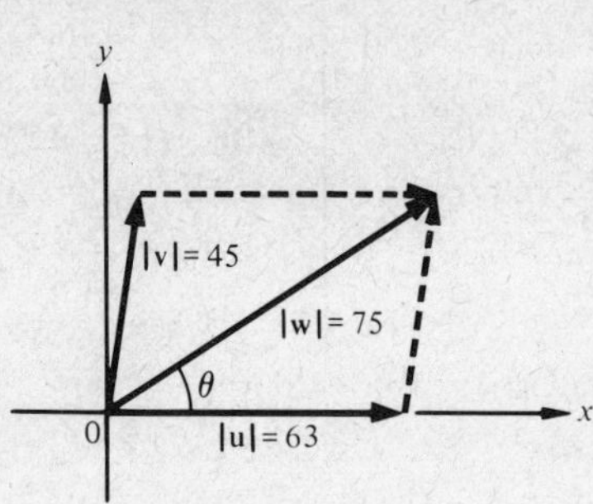

Figure 2

By the law of cosines,

$|\mathbf{w}|^2$, $\cos\theta$

$|\mathbf{v}|^2 = |\mathbf{u}|^2 + \_\_\_\_ - 2|\mathbf{u}||\mathbf{w}|\_\_\_\_$

so that

75^2, 75

$45^2 = 63^2 + \_\_\_\_ - 2(63)(\_\_\_\_)\cos\theta$

7569, 0.8010

Thus $\cos\theta = \dfrac{\_\_\_\_}{9450} = \_\_\_\_$

36°50′

so that $\theta = \_\_\_\_$.

Chapter Test

1 Simplify the following expressions.

(a) $\dfrac{\sec\theta\csc\theta}{\tan\theta + \cot\theta}$ (b) $\dfrac{\cos t + \sin^2 t\sec t}{\sec t}$

(c) $(\sin\theta + \cos\theta)^2 - \sin 2\theta$ (d) $\dfrac{\sin^2 2t}{(1 + \cos 2t)^2} + 1$

2 Write each of the following expressions in terms of the cosine or sine of t or θ.

(a) $\cos(\pi + t)$ (b) $\cos(2\pi - t)$ (c) $\sin(\theta + 180°)$ (d) $\sin(270° - \theta)$

3 (a) Use the fact that $\dfrac{\pi}{8} = \dfrac{1}{2}\left(\dfrac{\pi}{4}\right)$, together with the half angle formula for the cosine, to find $\cos\left(\dfrac{\pi}{8}\right)$.

(b) Use the formula for $\sin(t - s)$, together with the fact that $15° = 45° - 30°$, to evaluate $\sin 15°$.

4 If $\sin t = \frac{3}{5}, \frac{\pi}{2} < t < \pi$, find

(a) $\cos t$ (b) $\sin 2t$ (c) $\cos 2t$ (d) $\tan 2t$

5 Prove that $\frac{\sin 2t}{1 + \cos 2t} = \tan t$

6 Evaluate each of the following expressions.

(a) $\sin^{-1}\left(\frac{1}{2}\right)$ (b) $\cos^{-1}\left(-\frac{\sqrt{2}}{2}\right)$ (c) $\tan^{-1}(-1)$ (d) $\sin\left[\cos^{-1}\left(-\frac{3}{4}\right)\right]$

7 Find the solution set of the following equations.

(a) $2 \sin \theta - 1 = 0$, for $0 \leqslant \theta < 360°$ (b) $4 \cos^2 t - 3 = 0$, for $0 \leqslant t < 2\pi$

(c) $2 \sin^2 \theta - \sqrt{3} \sin \theta = 0$, for $0 \leqslant \theta < 360°$

8 Convert the given polar coordinates to rectangular coordinates.

(a) $(2, 120°)$ (b) $(-3, 420°)$ (c) $\left(5, \frac{\pi}{6}\right)$ (d) $\left(4, \frac{17\pi}{4}\right)$

9 Convert the given rectangular coordinates to polar coordinates, where $0 \leqslant \theta < 2\pi$ and $r > 0$.

(a) $(1, -1)$ (b) $(1, \sqrt{3})$ (c) $(-\sqrt{3}, -1)$ (d) $(-2, 2)$

10 Find a in ΔABC if $\alpha = 90°$, $\gamma = 45°$, and $c = 4$.

11 Find c in ΔABC if $\alpha = 45°$, $\beta = 105°$, and $a = 12$.

12 Find b in ΔABC if $\beta = 60°$, $a = 8$, and $c = 6$.

13 Let A and B be two points on opposite sides of a river. From A, a line $AC = 275$ feet is laid off and the angles $CAB = 125°40'$ and $ACB = 48°50'$ are measured. Find the length of $\overline{AB}$.

14 An airplane, with a still-air speed of 710 miles per hour, heading east, is pushed off course by a wind blowing from the south at 65 miles per hour. Find the velocity vector of the path of the airplane. What is the speed and direction (to the nearest 10 minutes) of the airplane?

Answers

1 (a) 1 (b) 1 (c) 1 (d) $\sec^2 t$

2 (a) $-\cos t$ (b) $\cos t$ (c) $-\sin \theta$ (d) $-\cos \theta$

3 (a) $\frac{\sqrt{2+\sqrt{2}}}{2}$ (b) $\frac{\sqrt{6}-\sqrt{2}}{4}$

4 (a) $-\frac{4}{5}$ (b) $-\frac{24}{25}$ (c) $\frac{7}{25}$ (d) $-\frac{24}{7}$

5 *Proof:* $\frac{\sin 2t}{1 + \cos 2t} = \frac{2 \sin t \cos t}{1 + \cos^2 t - \sin^2 t} = \frac{2 \sin t \cos t}{2 \cos^2 t} = \frac{\sin t}{\cos t} = \tan t$

6 (a) $\frac{\pi}{6}$ (b) $\frac{3\pi}{4}$ (c) $-\frac{\pi}{4}$ (d) $\frac{\sqrt{7}}{4}$

7 (a) $\{30°, 150°\}$ (b) $\left\{\frac{\pi}{6}, \frac{5\pi}{6}, \frac{7\pi}{6}, \frac{11\pi}{6}\right\}$ (c) $\{0, 60°, 180°, 120°\}$

8 (a) $(-1, \sqrt{3})$ (b) $\left(-\frac{3}{2}, -\frac{3\sqrt{3}}{2}\right)$ (c) $\left(\frac{5\sqrt{3}}{2}, \frac{5}{2}\right)$ (d) $(2\sqrt{2}, 2\sqrt{2})$

9 (a) $\left(\sqrt{2}, \frac{7\pi}{4}\right)$ (b) $\left(2, \frac{\pi}{3}\right)$ (c) $\left(2, \frac{7\pi}{6}\right)$ (d) $\left(2\sqrt{2}, \frac{3\pi}{4}\right)$

10 $4\sqrt{2}$

11 $6\sqrt{2}$

12 $2\sqrt{13}$

13 2161 feet

14 $\langle 710, 65\rangle$, 713 miles per hour, $5°10'$ N of E

Chapter 7 ANALYTIC GEOMETRY

In this chapter we use the cartesian coordinate system to relate the geometry of certain graphs, called conics, to the algebraic representations of the graphs. After working the problems of this chapter, the student should be able to:

1 Determine the equation of a circle with a given radius and center.

2 Determine the equation of an ellipse and sketch its graph.

3 Determine the equation of a hyperbola and sketch its graph.

4 Determine the equation of a parabola and sketch its graph.

5 Use the definition of a conic, in terms of the directrix, eccentricity, and focus, to determine the equation of that conic.

1 Circle

circle	**1** Geometrically, a ________ in a plane can be defined as the set of
radius	all points that are at a fixed distance r, called the ________,
center	from a fixed point C, called the ________.
	2 Let (h, k) be the center of a circle radius r. Then an equation of
$y - k$	the circle is $(x - h)^2 +$ (________)$^2 = r^2$.

In Problems 3–6, find an equation of the circle with the given center and radius.

	3 Center $(2, -3)$ and radius 3
2, −3, 3	Here $h =$ ____, $k =$ ____, and $r =$ ____, so that
2, −3, 3	$(x -$ ____$)^2 + [y -$ (____)$]^2 =$ (____)^2 or
$(x - 2)^2 + (y + 3)^2 = 9$	________________________________.
	4 Center $(-1, 4)$ and radius $\sqrt{5}$
$-1, 4, \sqrt{5}$	Here $h =$ ____, $k =$ ____, and $r =$ ____, so that
$-1, 4, \sqrt{5}$	$[x -$ (____)$]^2 + (y -$ ____$)^2 =$ (____)^2 or
$(x + 1)^2 + (y - 4)^2 = 5$	________________________________.
	5 Center $(0, 0)$ and radius 5
0, 0, 5	Here $h =$ ____, $k =$ ____, and $r =$ ____, so that
0, 0, 5, $x^2 + y^2 = 25$	$(x -$ ____$)^2 + (y -$ ____$)^2 =$ (____)^2 or ________________.
	6 Center $(0, 0)$ and radius 1
0, 0, 1	Here $h =$ ____, $k =$ ____, and $r =$ ____, so that
0, 0, 1, $x^2 + y^2 = 1$	$(x -$ ____$)^2 + (y -$ ____$)^2 =$ (____)^2 or ________________.

In Problems 7–11, find the center (h, k) and the radius r of the circle with the given equation.

	7 $(x - 2)^2 + (y + 3)^2 = 16$
	Writing this equation in standard form, we have
−3, 4	$(x - 2)^2 + [y -$ (____)$]^2 =$ (____)^2, so that
$(2, -3)$, 4	$(h, k) =$ ________ and $r =$ ____.
	8 $(x + 3)^2 + (y + 1)^2 = 5$
	First, rewrite the equation as
$-3, -1, \sqrt{5}$	$[x -$ (____)$]^2 + [y -$ (____)$]^2 =$ (____)^2
$(-3, -1)$, $\sqrt{5}$	so that $(h, k) =$ ________ and $r =$ ____.
	9 $x^2 + y^2 - 2x + 4y = 4$
	First, rewrite the equation in standard form by completing the square. Thus,
$y^2 + 4y$	$(x^2 - 2x) +$ (________) $= 4$
1, $y^2 + 4y + 4$, 5	$(x^2 - 2x +$ ____$) +$ (________________) $= 4 +$ ____
$y + 2$	$(x - 1)^2 +$ (________)$^2 = 9$
−2, 3	$(x - 1)^2 + [y -$ (____)$]^2 =$ (____)^2
$(1, -2)$, 3	so that $(h, k) =$ ________ and $r =$ ____.

Answers	Problems
	10 $x^2 + y^2 + 8y = 9$
	Completing the square, we get
16, 16, 25	$x^2 + y^2 + 8y +$ ____ $= 9 +$ ____ $=$ ____
0, −4, 5	or $(x -$ ____$)^2 + [y - ($____$)]^2 = ($____$)^2$
(0, −4), 5	so that $(h, k) =$ ________ and $r =$ ____.
	11 $x^2 + y^2 + 4x - 10y = 3$
	Completing the square, we get
4, 25, 25, 32	$(x^2 + 4x +$ ____$) + (y^2 - 10y +$ ____$) = 3 + 4 +$ ____ $=$ ____
2, 5, 32	or $(x +$ ____$)^2 + (y -$ ____$)^2 =$ ____
(−2, 5), $4\sqrt{2}$	so that $(h, k) =$ ________ and $r =$ ____.

1.1 Translation of Axes

Answers	Problems
parallel	**12** If two cartesian coordinate systems have corresponding axes that are ____________ and have the same positive directions, then we say that these systems are obtained from one another by
translation	________________.
coordinate axes	**13** If P is a point in the xy plane, the coordinates of P depend on how the ____________________ are placed in the plane.
(0, 0)	**14** If the axes are placed so that the origin coincides with P, then P has the coordinates ______.
(3, 4)	**15** If another pair of axes are formed by "shifting" or "translating" the y axis 3 units to the left to get a $\bar{y}$ axis and the x axis 4 units down to get a $\bar{x}$ axis (Figure 1), the point P, from Problem 14, has coordinates ______ in this new coordinate system, the $\bar{x}\bar{y}$ system.

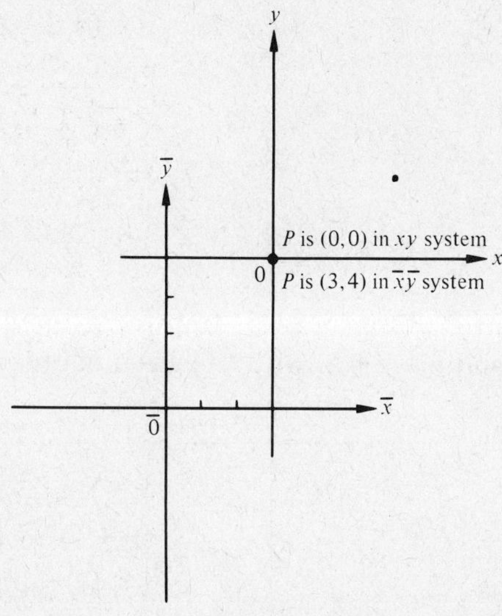

Figure 1

16 Suppose that the coordinate axes xy are translated to coordinate axes $\bar{x}\bar{y}$ so that the origin $\bar{O}$ of the $\bar{x}\bar{y}$ system has coordinates (h, k) in the xy system. If P has coordinates (x, y) in the xy system and coordinates $(\bar{x}, \bar{y})$ in the $\bar{x}\bar{y}$ system, then

$x =$ ________ and $y =$ ________ or, equivalently,

$\bar{x} =$ ________ and $\bar{y} =$ ________.

$\bar{x} + h,\ \bar{y} + k$

$x - h,\ y - k$

In Problems 17–21, find the coordinates of P in the $\bar{x}\bar{y}$ system, obtained by translation, if $\bar{O}$ has coordinates (h, k) in the xy system and P has the given coordinates in the xy system.

17 $P = (2, 3); (h, k) = (1, 2)$

$\bar{x} = x - h$ and $\bar{y} = y - k$

$\bar{x} =$ (____) - (____) and $\bar{y} =$ (____) - (____), so that

$(\bar{x}, \bar{y}) =$ ________

2, 1, 3, 2

(1, 1)

18 $P = (-3, 4); (h, k) = (5, -7)$

$\bar{x} = x - h$ and $\bar{y} = y - k$

$\bar{x} =$ (____) - (____) and $\bar{y} =$ (____) - (____), so that

$(\bar{x}, \bar{y}) =$ ________

-3, 5, 4, -7

(-8, 11)

19 $P = (-4, -2); (h, k) = (8, 0)$

$\bar{x} = x - h$ and $\bar{y} =$ ________

$\bar{x} =$ (____) - (____) and $\bar{y} =$ (____) - (____), so that

$(\bar{x}, \bar{y}) =$ ________

$y - k$

-4, 8, -2, 0

(-12, -2)

20 $P = (5, -7); (h, k) = (-3, -1)$

$\bar{x} =$ ______ and $\bar{y} =$ ______

$\bar{x} =$ (____) - (____) and $\bar{y} =$ (____) - (____), so that

$(\bar{x}, \bar{y}) =$ ________

$x - h,\ y - k$

5, -3, -7, -1

(8, -6)

21 $P = (4, 0); (h, k) = (-3, 5)$

$\bar{x} =$ ______ and $\bar{y} =$ ______

$\bar{x} =$ (____) - (____) and $\bar{y} =$ (____) - (____), so that

$(\bar{x}, \bar{y}) =$ ________

$x - h,\ y - k$

4, -3, 0, 5

(7, -5)

In Problems 22–25, transform the given equation to the form $\bar{x}^2 + \bar{y}^2 = r^2$. (Use $x = \bar{x} + h$ and $y = \bar{y} + k$, with the given coordinates (h, k) in Problems 22–23.)

22 $x^2 + y^2 - 2x + 4y + 1 = 0; (h, k) = (1, -2)$

$x = \bar{x} +$ _____ and $y = \bar{y} +$ _____. Upon substitution,

$(\bar{x} + 1)^2 + (\bar{y} - 2)^2 - 2($________$) + 4($________$) + 1 = 0$, so that

$\bar{x}^2 + 2\bar{x} + 1 +$ ____________ $- 2\bar{x} - 2 +$ ________ $+ 1 = 0$.

Hence, $\bar{x}^2 + \bar{y}^2 =$ _____.

1, -2

$\bar{x} + 1,\ \bar{y} - 2$

$\bar{y}^2 - 4\bar{y} + 4,\ 4\bar{y} - 8$

4

$\bar{x} - 3,\ \bar{y} + 2$

$\bar{x} - 3,\ \bar{y} + 2$

$\bar{x}^2 - 6\bar{x} + 9,\ 6\bar{x} - 18$

25

23 $x^2 + y^2 + 6x - 4y - 12 = 0; (h, k) = (-3, 2)$

$x =$ _____ and $y =$ _____. Upon substitution,

$(\bar{x} - 3)^2 + (\bar{y} + 2)^2 + 6($_____$) - 4($_____$) - 12 = 0$, so that

________ $+ \bar{y}^2 + 4\bar{y} + 4 +$ _______ $- 4\bar{y} - 8 - 12 = 0$.

Hence, $\bar{x}^2 + \bar{y}^2 =$ ____.

4, 9, 13

$y - 3,\ 1$

$y - 3,\ \bar{y}^2$

24 $x^2 + y^2 + 4x - 6y + 12 = 0$

First, complete the square. Thus,

$x^2 + 4x +$ ____ $+ y^2 - 6y +$ ____ $= -12 +$ ____ or, equivalently,

$(x + 2)^2 + ($_____$)^2 =$ ____.

Letting $\bar{x} = x + 2$ and $\bar{y} =$ _____, we have $\bar{x}^2 +$ ____ $= 1$.

1, 1, 2

$x - 1,\ y + 1,\ 16$

$x - 1,\ y + 1,\ \bar{x}^2 + \bar{y}^2 = 16$

25 $x^2 + y^2 - 2x + 2y - 14 = 0$

Completing the square,

$x^2 - 2x +$ ____ $+ y^2 + 2y +$ ____ $= 14 +$ ____, or

$($_____$)^2 + ($_____$)^2 =$ ____.

Letting $\bar{x} =$ _____ and $\bar{y} =$ _____ we have __________.

2 Ellipse

sum

foci

26 Geometrically, an ellipse is defined to be the set of all points P in the plane such that the ____ of the distances from P to two fixed points, called _____, is a constant.

In Problems 27–31, refer to Figure 1.

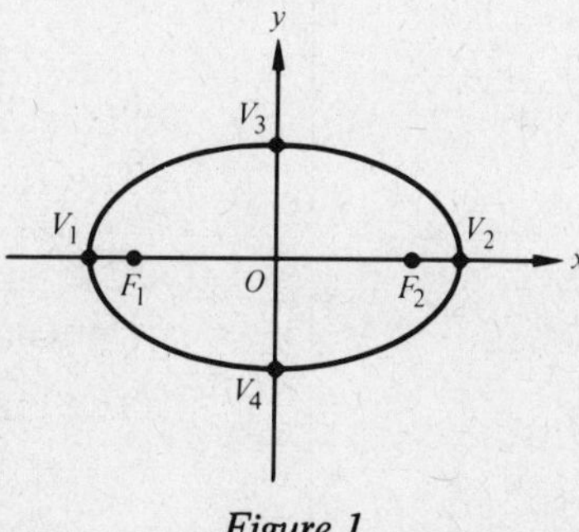

Figure 1

foci

27 If F_1 and F_2 are the two fixed points referred to in Problem 26, then F_1 and F_2 are the _____ of the ellipse.

center

center

28 The midpoint of the line segment $\overline{F_1F_2}$ is called the _______ of the ellipse. The ellipse is symmetrical with respect to each of two perpendicular lines that intersect at its _______.

vertices

29 The four points V_1, V_2, V_3, and V_4 are called the _________ of the ellipse.

major axis

30 The line segment $\overline{V_1V_2}$ is called the ___________ of the ellipse.

minor axis

31 The line segment $\overline{V_3 V_4}$ is called the ______________ of the ellipse.

c^2

32 An equation for the ellipse with foci at $F_2 = (c, 0)$ and $F_1 = (-c, 0)$ is $\frac{x^2}{a^2} + \frac{y^2}{b^2} = 1$, where $b^2 = a^2 - $ ____ and $a > b$ (Figure 2).

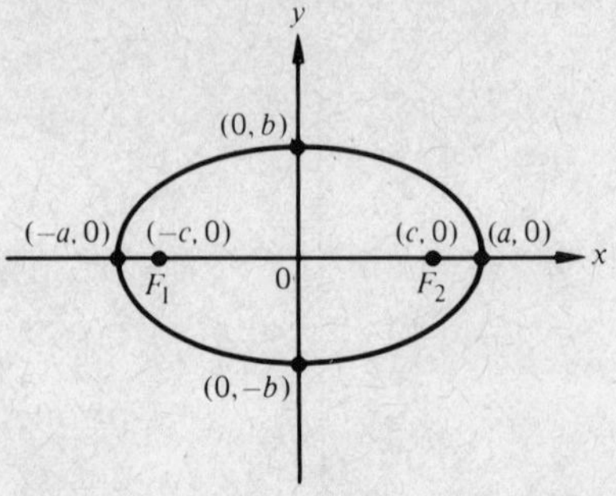

Figure 2

major

2b, minor

33 Figure 2 also shows that $2a$ is the length of the ______ axis and ____ is the length of the ______ axis.

In Problems 34–36, find the equation of the ellipse presented in Problem 32 whose foci are on the x axis and whose center is at the origin. Sketch the graph.

34 $a = 2, b = 1$

$\frac{x^2}{a^2} + \frac{y^2}{b^2} = 1$, so that, by substitution,

2, 1, 4, 1

$\frac{x^2}{(\_\_\_\_)^2} + \frac{y^2}{(\_\_\_\_)^2} = 1$ or $\frac{x^2}{\_\_\_\_} + \frac{y^2}{\_\_\_\_} = 1$

The foci are determined by the equation $b^2 = a^2 - c^2$

$a^2 - b^2$

or $c^2 = $ __________.

1, 3, $\sqrt{3}$

Hence, $c^2 = 2^2 - (\_\_\_\_)^2$ or $c^2 = $ ____ so that $c = $ ____.

The graph is

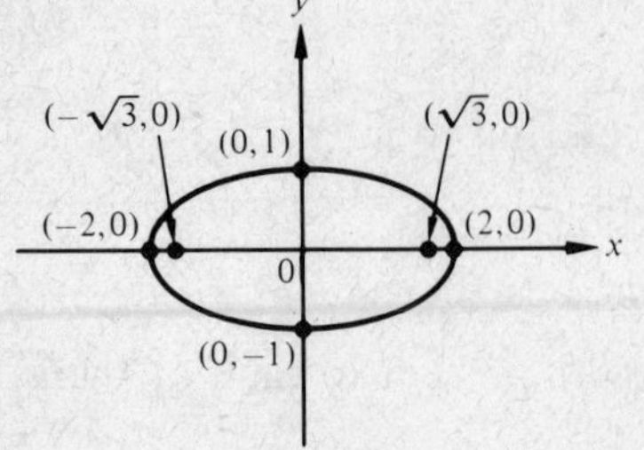

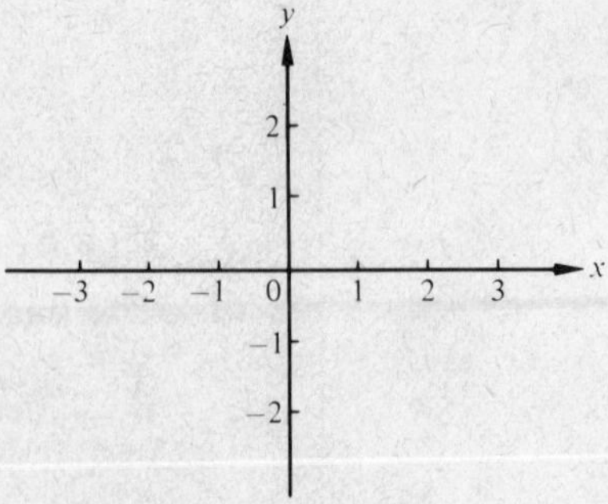

35 $a = 5, c = 4$

First, find b^2 by the equation $b^2 = a^2 - c^2$, so that

4, 9

$b^2 = 5^2 - (____)^2 = $ ____

$\frac{y^2}{b^2}$

Next, substituting in the equation $\frac{x^2}{a^2} + \frac{\quad}{\quad} = 1$, we have

9

$\frac{x^2}{25} + \frac{y^2}{\quad} = 1$. The graph is

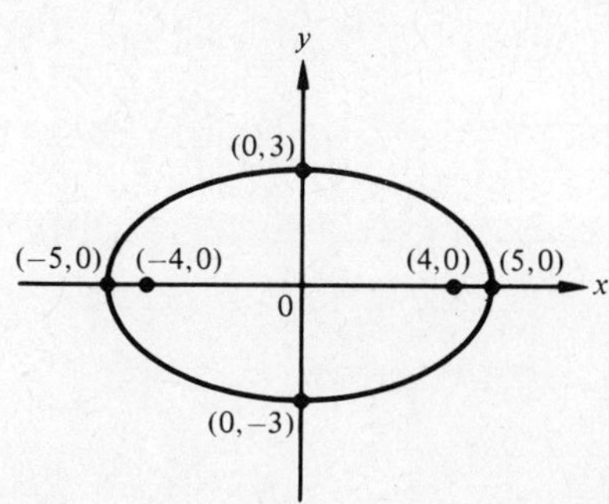

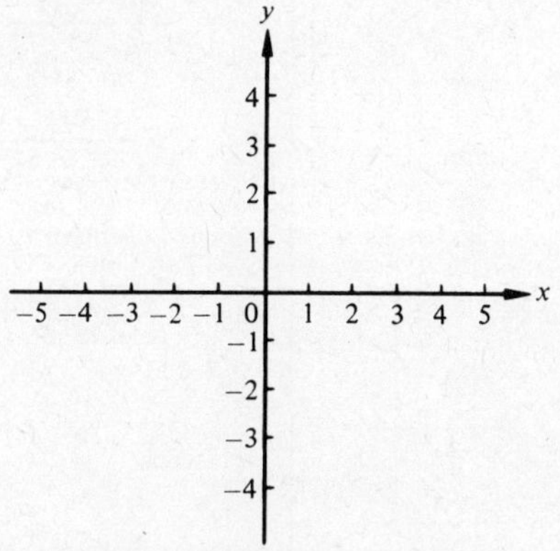

36 $b = 12, c = 5$

$b^2 + c^2$

First, find a^2 by the equation $a^2 =$ __________, so that

12, 5, 169

$a^2 = (\_\_\_)^2 + (\_\_\_)^2 = \_\_\_$

$\frac{x^2}{a^2} + \frac{y^2}{b^2} = 1$

Next, substituting in the equation __________, we have

169, 144

$\frac{x^2}{\quad} + \frac{y^2}{\quad} = 1$. The graph is

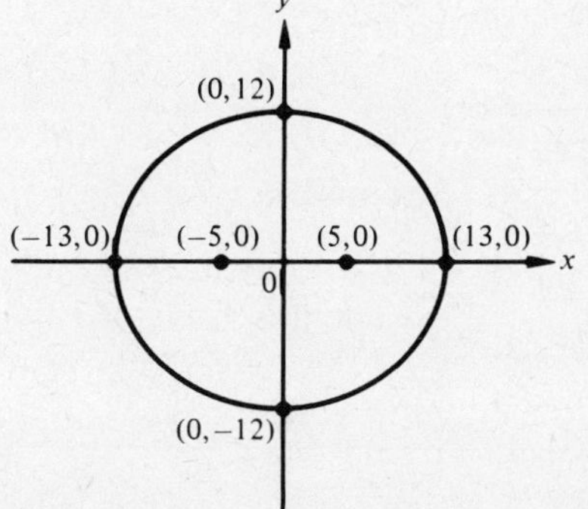

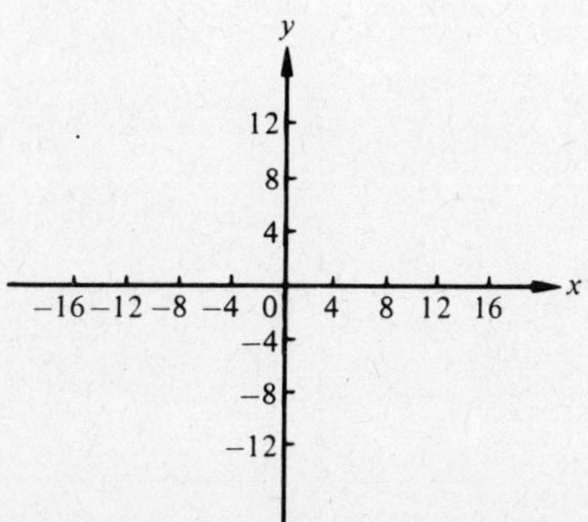

b^2

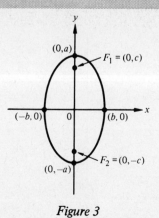

37 An equation for the ellipse with foci at $F_1 = (0, c)$ and $F_2 = (0, -c)$ is $\frac{x^2}{b^2} + \frac{y^2}{a^2} = 1$, where $c^2 = a^2 -$ ____ and $a > b$ (Figure 3).

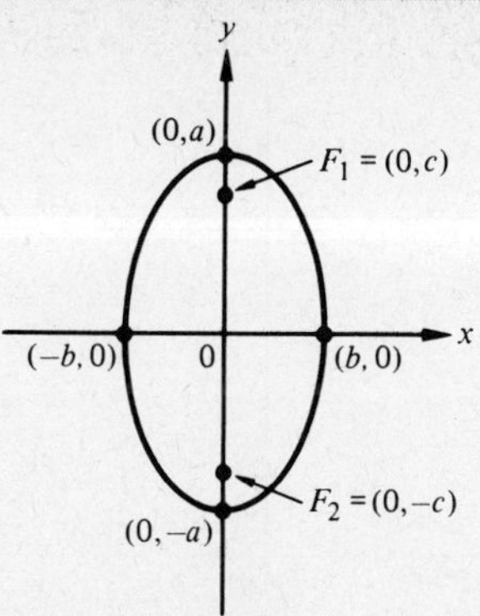

Figure 3

In Problems 38–40, find the equation of the ellipse presented in Problem 37 whose foci are on the y axis and whose center is at the origin. Sketch the graph.

38 $a = 5, b = 4$

$\frac{x^2}{b^2} + \frac{y^2}{a^2} = 1$, so that, by substitution,

$\frac{x^2}{(\_\_\_)^2} + \frac{y^2}{(\_\_\_)^2} = 1$ or $\frac{x^2}{\_\_\_} + \frac{y^2}{\_\_\_} = 1$.

Also $c^2 = a^2 - b^2 = \_\_\_ - \_\_\_ = \_\_\_$ or $c = \_\_\_$.

The graph is

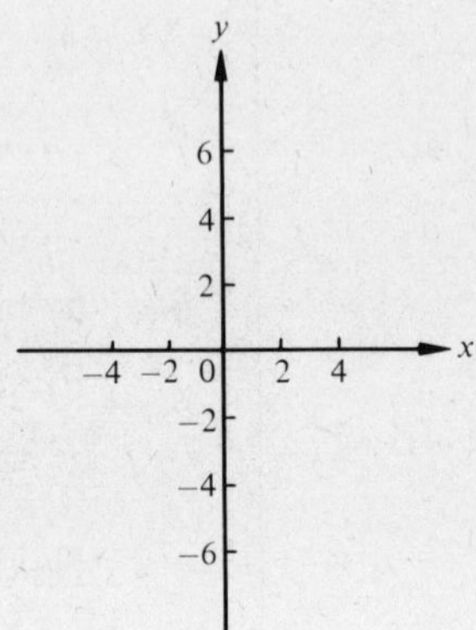

Answers: 4, 5, 16, 25

25, 16, 9, 3

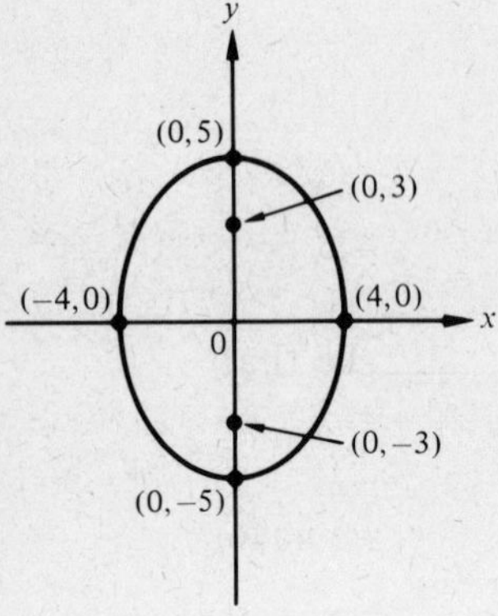

39 $a = 7, c = 4$

First, find b by the equation $b^2 = a^2 - c^2$, so that

$b^2 = (\_\_\_)^2 - (\_\_\_)^2$ or $b^2 = \_\_\_$, so $b = \_\_\_$. Next,

substituting in the equation $\frac{x^2}{b^2} + \frac{y^2}{\_\_\_} = 1$, we have

$\frac{x^2}{(\_\_\_)^2} + \frac{y^2}{(\_\_\_)^2} = 1$ or ________. The graph is

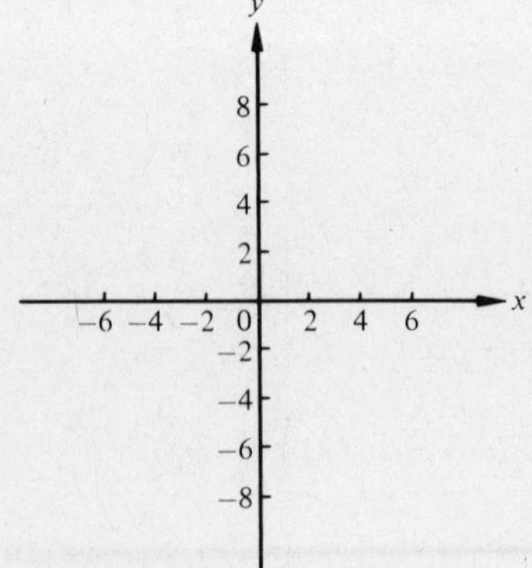

Answers: 7, 4, 33, $\sqrt{33}$

a^2

$\sqrt{33}$, 7, $\frac{x^2}{33} + \frac{y^2}{49} = 1$

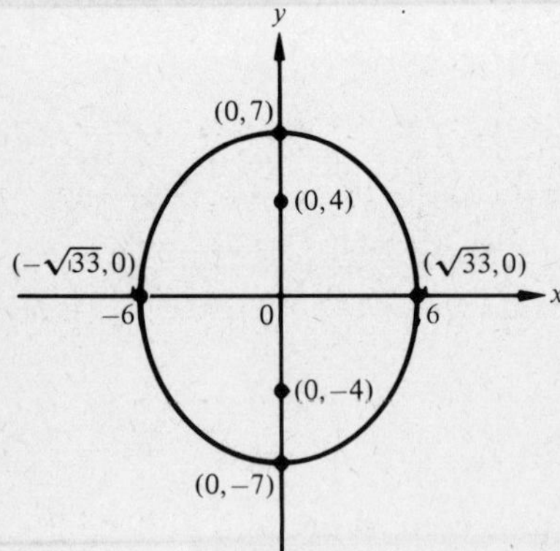

40 $b = 2, c = 3$

First, find a by the equation $a^2 = \_\_\_\_\_\_$, so that

$a^2 = (\_\_\_)^2 + (\_\_\_)^2$ or $a^2 = \_\_\_$, so $a = \_\_\_$. Next, substituting in the equation $\frac{x^2}{\_\_\_} + \frac{y^2}{\_\_\_} = 1$, we have ________.

Answers: $b^2 + c^2$

2, 3, 13, $\sqrt{13}$

b^2, a^2, $\frac{x^2}{4} + \frac{y^2}{13} = 1$

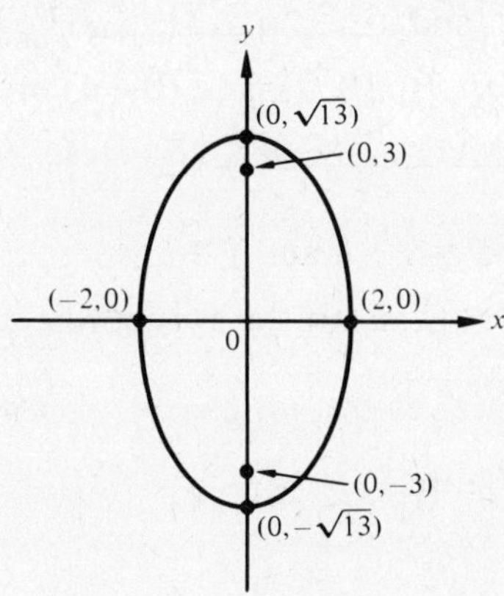

The graph is

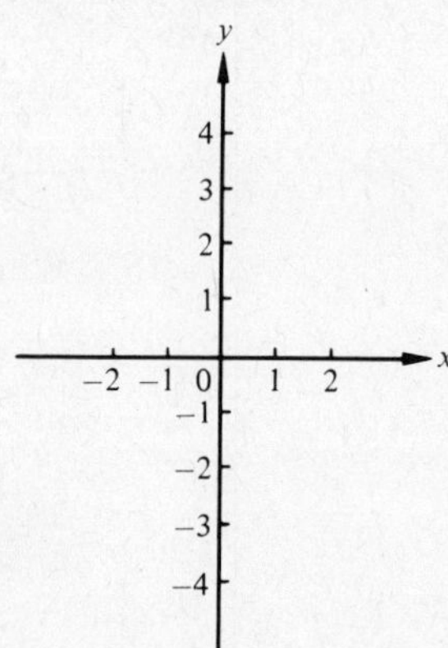

x

41 If an equation of an ellipse is of the form $\frac{x^2}{a^2} + \frac{y^2}{b^2} = 1$, for $a > b$, then the major axis lies on the ____ axis.

y

42 If an equation of an ellipse is of the form $\frac{x^2}{b^2} + \frac{y^2}{a^2} = 1$, for $a > b$, then the major axis lies on the ____ axis.

In Problems 43–44, find the coordinates of the foci and the vertices of the given ellipse.

43 $9x^2 + 16y^2 = 144$

144

First, divide both members of the equation by ____ to transform it to the standard form. Thus,

144, 144, 144, 16, 9

$$\frac{9x^2}{\_\_\_\_} + \frac{16y^2}{\_\_\_\_} = \frac{144}{\_\_\_\_} \quad \text{or} \quad \frac{x^2}{\_\_\_\_} + \frac{y^2}{\_\_\_\_} = 1$$

16, 9, 4, 3

Hence, $a^2 =$ ____ and $b^2 =$ ____ or $a =$ ____ and $b =$ ____.

From the equation $c^2 = a^2 - b^2$, we have

16, 9, 7, $\sqrt{7}$

$c^2 =$ ____ − ____ or $c^2 =$ ____, so $c =$ ____.

x^2

Since a^2 is associated with the ____ term in this equation, the

x

major axis lies on the ____ axis. Hence, the coordinates of the

$(-\sqrt{7}, 0)$, $(\sqrt{7}, 0)$

foci are $(-c, 0)$ and $(c, 0)$ or ________ and ________. The coordinates of the vertices are $(-a, 0)$, $(a, 0)$, $(0, b)$, and $(0, -b)$ or

(-4, 0), (4, 0), (0, 3), (0, -3)

__________, __________, __________, and __________.

44 $9x^2 + y^2 = 9$

9

Dividing both members by ____, the standard form of this ellipse

1, 9

is $\frac{x^2}{\_\_\_\_} + \frac{y^2}{\_\_\_\_} = 1$.

9, 1, 3, 1

Hence, $a^2 =$ ____ and $b^2 =$ ____ or $a =$ ____ and $b =$ ____.

$a^2 - b^2$

Since $c^2 =$ ____________, by substitution,

9, 1, 8, $2\sqrt{2}$

$c^2 =$ ____ − ____ or $c^2 =$ ____, so $c =$ ____.

y^2

Since a^2 is associated with the ____ term in this equation, the

Answers	
y	major axis lies on the ____ axis. Thus, the coordinates of the foci
$(0, 2\sqrt{2})$, $(0, -2\sqrt{2})$	are $(0, c)$ and $(0, -c)$ or ________ and ________. The coordinates of the vertices are $(-b, 0)$, $(b, 0)$, $(0, a)$, and $(0, -a)$ or
$(-1, 0)$, $(1, 0)$, $(0, 3)$, $(0, -3)$	______, ______, ______, and ______.

45 When we use the translation equations $\overline{x} = x - h$ and $\overline{y} =$ ______, an equation for the ellipse with center at (h, k) in the xy coordinate system and with a horizontal major axis is

$$\frac{(x-h)^2}{a^2} + \frac{(y-k)^2}{b^2} = 1, \quad \text{for } a > b \quad \text{(Figure 4).}$$

Answer: $y - k$

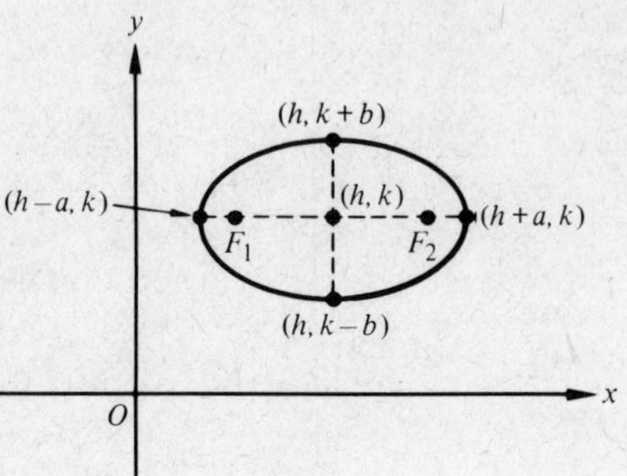

Figure 4

46 An equation for the ellipse with center at (h, k) and with a vertical major axis is $\frac{(x-h)^2}{\_\_\_\_} + \frac{(y-k)^2}{\_\_\_\_} = 1$, for $a > b$ (Figure 5).

Answer: b^2, a^2

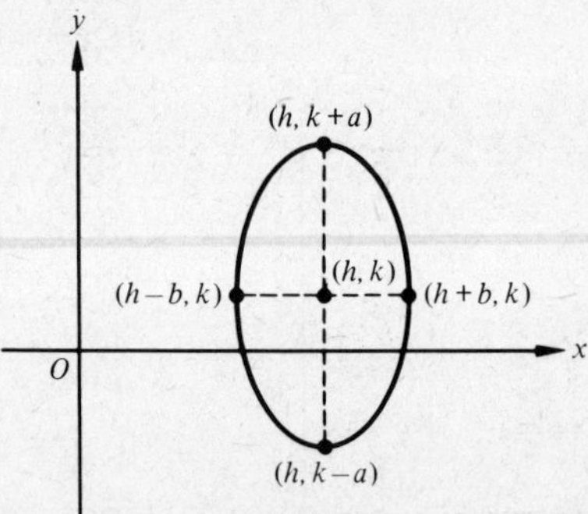

Figure 5

In Problems 47–49, find the coordinates of the center, foci, and vertices of the ellipse with the given equation and sketch the graph.

47 $\frac{(x-1)^2}{4} + \frac{(y+2)^2}{9} = 1$

Answers	
1, −2	Since $h =$ ____ and $k =$ ____, the coordinates of the center are
$(1, -2)$, 9, 4, 3	______. Also, $a^2 =$ ____ and $b^2 =$ ____ or $a =$ ____ and
2	$b =$ ____. The coordinates of the vertices are $(h - b, k)$, $(h + b, k)$,
$(-1, -2)$, $(3, -2)$	$(h, k + a)$, and $(h, k - a)$ or ________, ________,

(1, 1), (1, -5)
5, $\sqrt{5}$
vertical
$(1, -2+\sqrt{5})$, $(1, -2-\sqrt{5})$

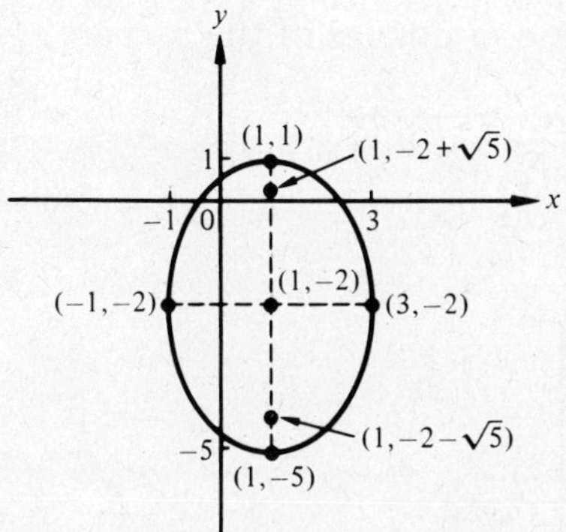

-2, 3
(-2, 3), 25, 16, 5
4, horizontal

(-7, 3), (3, 3), (-2, 7), (-2, -1)
9, 3

(-5, 3), (1, 3)

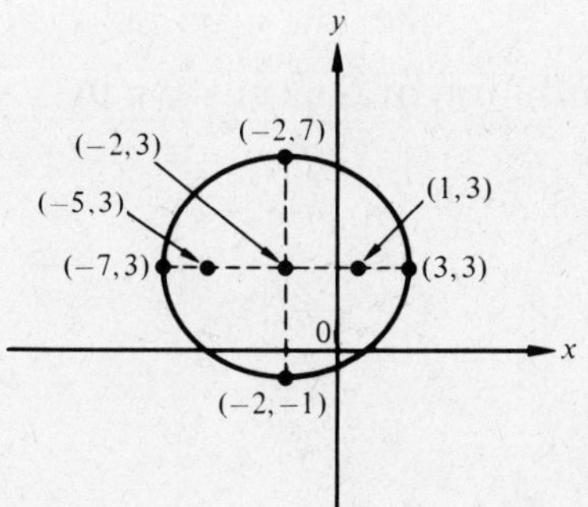

1, 4, 16
1, 2, 36, 36

4, 9
-1, -2

_________, and ____________. Since $c^2 = a^2 - b^2$, by substitution, $c^2 =$ ____ or $c =$ ____. The major axis of this ellipse is __________, so that the coordinates of the foci are $(h, k + c)$ and $(h, k - c)$ or ____________ and ____________. The graph is

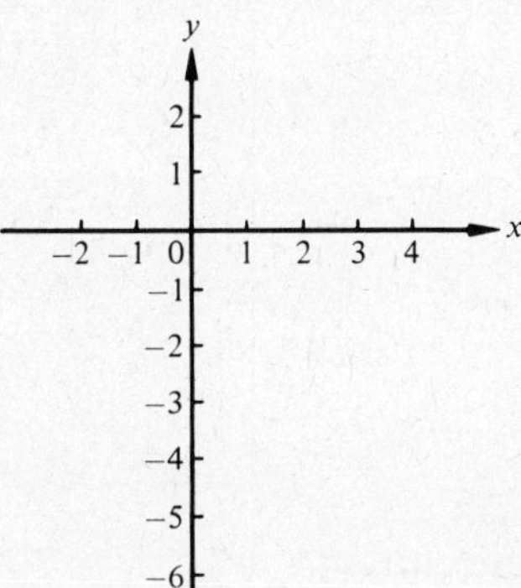

48 $\dfrac{(x+2)^2}{25} + \dfrac{(y-3)^2}{16} = 1$

Since $h =$ ____ and $k =$ ____, the coordinates of the center are _________. Also, $a^2 =$ ____ and $b^2 =$ ____ or $a =$ ____ and $b =$ ____. The major axis is _____________. The coordinates of the vertices are $(h - a, k)$, $(h + a, k)$, $(h, k + b)$, and $(h, k - b)$ or ____________, ____________, ____________, and ____________. From the equation $c^2 = a^2 - b^2$, we have $c^2 =$ ____ or $c =$ ____. Hence, the coordinates of the foci are $(h - c, k)$ and $(h + c, k)$ or ____________ and ____________. The graph is

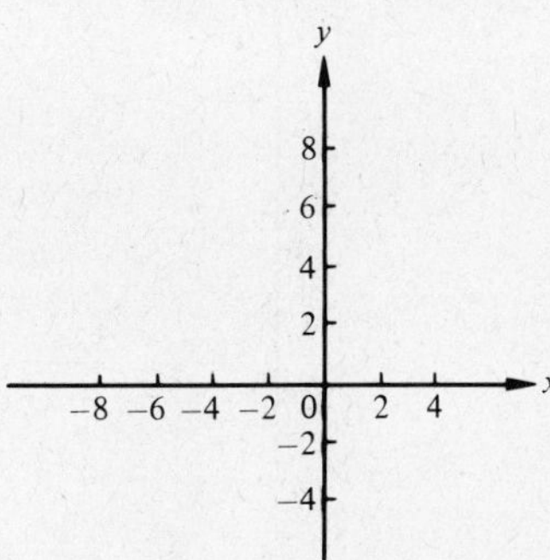

49 $9x^2 + 4y^2 + 18x + 16y - 11 = 0$

First, complete the square to get $9(x^2 + 2x +$ ____$) + 4(y^2 + 4y +$ ____$) = 11 + 9 +$ ____ or, equivalently, $9(x +$ ____$)^2 + 4(y +$ ____$)^2 =$ ____. Dividing by ____, the standard form of this ellipse is $\dfrac{(x+1)^2}{\_\_\_\_} + \dfrac{(y+2)^2}{\_\_\_\_} = 1$. Since $h =$ ____ and $k =$ ____, the coordinates of the center are

$(-1, -2)$, 9, 4 3
2,
y
$(-3, -2)$
$(1, -2)$, $(-1, 1)$, $(-1, -5)$
5, $\sqrt{5}$
$(-1, -2 + \sqrt{5})$
$(-1, -2 - \sqrt{5})$

__________. Also, $a^2 =$ ____ and $b^2 =$ ____ or $a =$ ____ and $b =$ ____. Since a^2 is associated with the y term, the major axis is parallel to the ____ axis. Hence, the coordinates of the vertices are $(h - b, k)$, $(h + b, k)$, $(h, k + a)$, and $(h, k - a)$ or __________, __________, __________, and __________. Also $c^2 = a^2 - b^2$, so that $c^2 =$ ____ or $c =$ ____. The coordinates of the foci are $(h, k + c)$ and $(h, k - c)$ or ______________ and ______________. The graph is

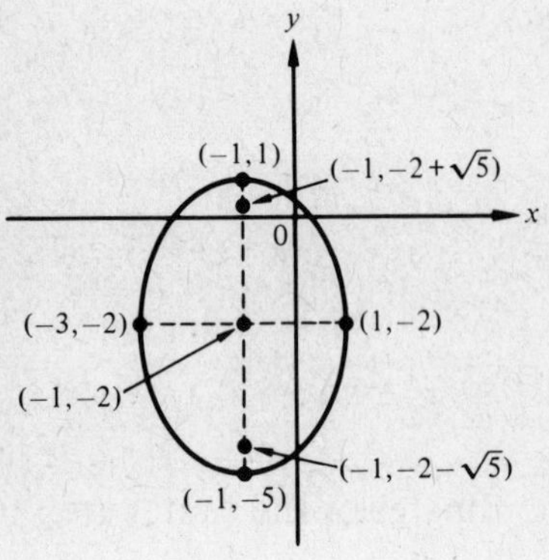

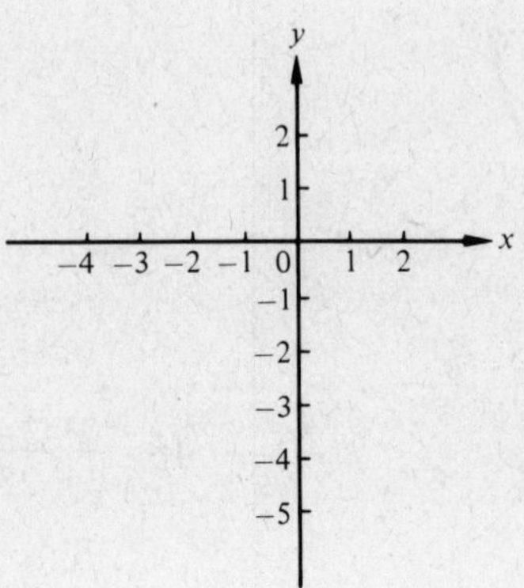

In Problems 50–53, find an equation of the ellipse that satisfies the given conditions.

vertices, $(2, 3)$
-3, 10, 5
5, 4, 2
x
$\frac{(y - k)^2}{b^2}$
2, 2
$\frac{(x - 2)^2}{25} + \frac{(y - 3)^2}{4} = 1$

50 The vertices are $(-3, 3)$, $(7, 3)$, $(2, 1)$, and $(2, 5)$.

The center is the intersection of the two line segments determined by the __________. Hence, $(h, k) =$ _____. The major axis is $2a = 7 - ($____$) =$ ____, so that $a =$ ____. The minor axis is $2b =$ ____ $- 1 =$ ____, so that $b =$ ____. Since the major axis is parallel to the ____ axis, the standard form of the equation is

$$\frac{(x - h)^2}{a^2} + \frac{\quad}{\_\_\_\_\_\_} = 1.$$ By substitution,

$$\frac{(x - \_\_\_)^2}{5^2} + \frac{(y - 3)^2}{(\_\_\_)^2} = 1 \text{ or}$$

____

b^2, a^2
6, 5, 11, $\sqrt{11}$
11

51 Center at $(-3, 2)$, with $a = 6$ and $c = 5$, and major axis parallel to the y axis

The standard form of the equation for this ellipse is

$$\frac{(x - h)^2}{\_\_\_\_} + \frac{(y - k)^2}{\_\_\_\_} = 1,$$ where $b^2 = a^2 - c^2$. Hence,

$b^2 = ($____$)^2 - ($____$)^2$ or $b^2 =$ ____, so that $b =$ ____. By substitution, we have $\frac{(x + 3)^2}{\_\_\_\_} + \frac{(y - 2)^2}{36} = 1.$

52 Center at $(-4, -2)$, $c = \sqrt{7}$, and one vertex of the major axis at $(0, -2)$

Since the major axis contains the center $(-4, -2)$ and one vertex is $(0, -2)$, the major axis is parallel to the ____ axis. Hence, the standard form of this equation is ____________, where $c^2 =$ ________. The value of a is the distance between the points $(-4, -2)$ and ________. Thus, $a = 0 - (\_\_\_\_) = \_\_\_\_$, $b^2 = a^2 - \_\_\_\_ = 16 - (\_\_\_\_)^2 = \_\_\_\_$.

By substitution, we have $\frac{[x - (\_\_\_\_)]^2}{16} + \frac{(y+2)^2}{\_\_\_\_} = 1$, or ________________

x

$\frac{(x-h)^2}{a^2} + \frac{(y-k)^2}{b^2} = 1$

$a^2 - b^2$

$(0, -2)$

$-4, 4, c^2, \sqrt{7}, 9$

$-4, 9$

$\frac{(x+4)^2}{16} + \frac{(y+2)^2}{9} = 1$

53 Center at the origin, with $a = 2b$ and $c = 3$, and major axis parallel to the y axis

Since $c^2 = a^2 - b^2$, by substitution, $3^2 = (2b)^2 - b^2$ or $9 = (\_\_\_\_)b^2$, so that $b^2 = \_\_\_\_$ or $b = \_\_\_\_$. Hence, $a = 2b = 2(\_\_\_\_) = \_\_\_\_$. The standard form of the equation of this ellipse is ____________, so that, by substitution, we have

$\frac{x^2}{(\_\_\_\_)^2} + \frac{y^2}{(\_\_\_\_)^2} = 1$ or ____________

$3, 3, \sqrt{3}$

$\sqrt{3}, 2\sqrt{3}$

$\frac{x^2}{b^2} + \frac{y^2}{a^2} = 1$

$\sqrt{3}, 2\sqrt{3}, \frac{x^2}{3} + \frac{y^2}{12} = 1$

3 Hyperbola

54 Geometrically, a hyperbola is defined in a plane to be the set of points P such that the absolute value of the ____________ of distances from P to two distinct fixed points, called ______, is a constant.

difference

foci

In Problems 55–59, refer to Figure 1.

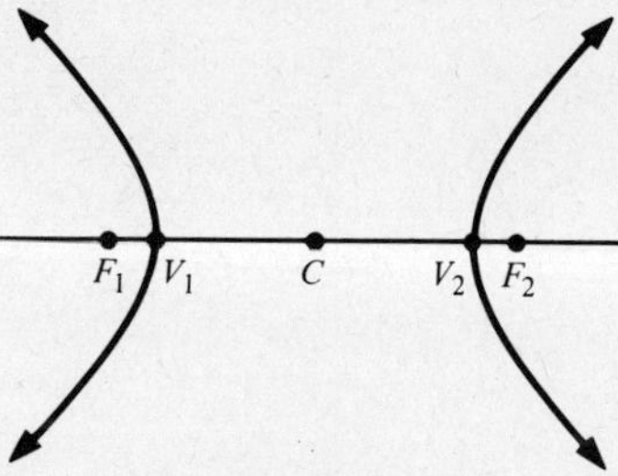

Figure 1

55 The points ____ and ____ are called the foci of the hyperbola.

56 The points ____ and ____ are called the vertices of the hyperbola.

F_1, F_2

V_1, V_2

symmetry

57 The line determined by the foci is a line of ________ of the hyperbola.

center

58 The midpoint C of the line segment $\overline{F_1F_2}$ is called the ________ of the hyperbola.

transverse

59 The line determined by F_1 and F_2 is the ________ axis of the hyperbola.

$c^2 - a^2$

60 An equation of the hyperbola with foci $F_1 = (-c, 0)$ and $F_2 = (c, 0)$ and the constant difference $k = 2a$ is $\frac{x^2}{a^2} - \frac{y^2}{b^2} = 1$, where $b^2 =$ ________.

3.1 Properties of the Hyperbola

symmetric

y axis

61 The graph of the hyperbola $\frac{x^2}{a^2} - \frac{y^2}{b^2} = 1$ is ________ with respect to both the x axis and the ________.

y, $(-a, 0)$, $(a, 0)$

62 The x intercepts of the hyperbola $\frac{x^2}{a^2} - \frac{y^2}{b^2} = 1$ are found by letting ____ $= 0$ and solving for x to get ________ and ________.

y

63 The graph of the hyperbola $\frac{x^2}{a^2} - \frac{y^2}{b^2} = 1$ does not intersect the ____ axis.

$\frac{a^2}{x^2}$

64 Solving $\frac{x^2}{a^2} - \frac{y^2}{b^2} = 1$ for y, we get the equation

$$y = \pm\frac{b}{a}x\sqrt{1 - \frac{\quad}{\quad}}$$

large

$\pm\frac{b}{a}x$

65 Since the expression $1 - \frac{a^2}{x^2}$ approaches 1 as $|x|$ gets very ________, the graph of the hyperbola approaches the lines whose equations are $y =$ ____.

asymptotes

66 The lines $y = \pm\frac{b}{a}x$ are called the ________ of the hyperbola $\frac{x^2}{a^2} - \frac{y^2}{b^2} = 1$ (Figure 2)

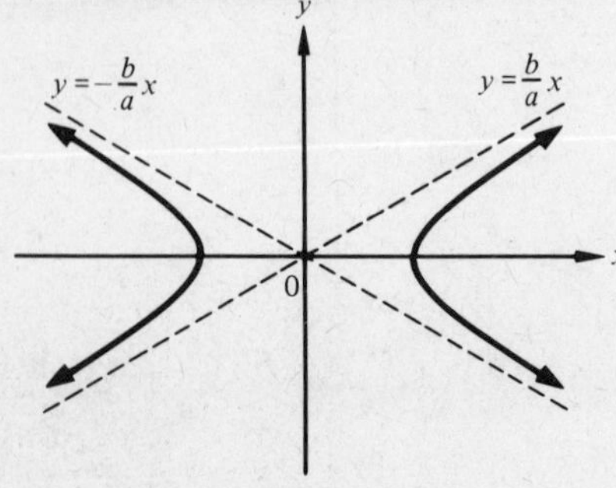

Figure 2

$2a$, $2b$

67 The asymptotes of the hyperbola can be drawn by constructing the rectangle of dimensions ____ by ____, as shown in Figure 3.

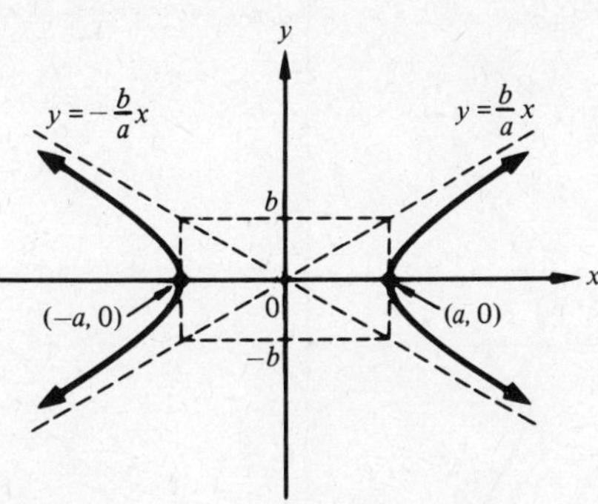

Figure 3

In Problems 68–69, find the coordinates of the foci and vertices, and the equations of the asymptotes for the given hyperbola; also, sketch the graph.

68 $\dfrac{x^2}{25} - \dfrac{y^2}{16} = 1$

25, 16, 5, 4

Here $a^2 =$ ____ and $b^2 =$ ____ or $a =$ ____ and $b =$ ____. From

25, 16, 41

the equation $c^2 = a^2 + b^2$, we have $c^2 =$ ____ + ____ = ____, so

$\sqrt{41}$

$c =$ ____. The coordinates of the foci are $(-c, 0)$ and $(c, 0)$ or

$(-\sqrt{41}, 0)$, $(\sqrt{41}, 0)$

________ and ________. The coordinates of the vertices are $(-a, 0)$ and $(a, 0)$ or ________ and ________. The

$(-5, 0)$, $(5, 0)$

$-\dfrac{b}{a}x$

equations of the asymptotes are $y = \dfrac{b}{a}x$ and $y =$ ____ or

$\dfrac{4}{5}x$, $-\dfrac{4}{5}x$

$y =$ ____ and $y =$ ____. The graph is

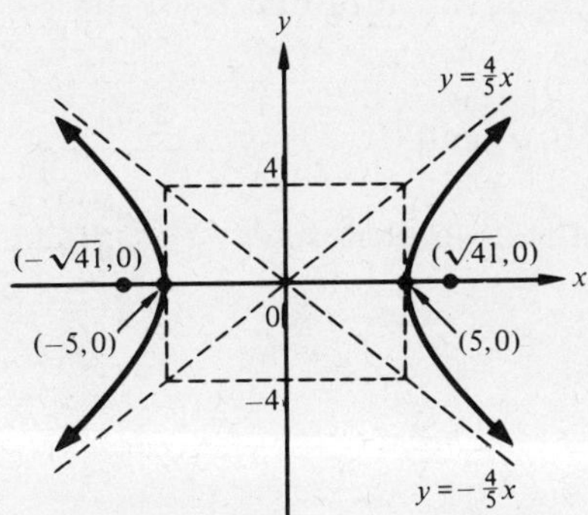

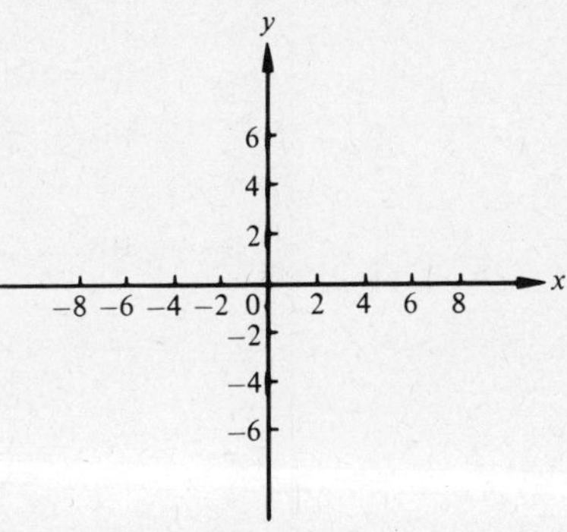

69 $\dfrac{x^2}{9} - \dfrac{y^2}{16} = 1$

9, 16, 3, 4

Here $a^2 =$ ____ and $b^2 =$ ____ or $a =$ ____ and $b =$ ____. Also,

25, 5

$c^2 = a^2 + b^2 =$ ____, so that $c =$ ____. The coordinates of the foci

$(-5, 0)$, $(5, 0)$

are $(-c, 0)$ and $(c, 0)$ or ________ and ________. The coordi-

$(-3, 0)$

nates of the vertices are $(-a, 0)$ and $(a, 0)$ or ________ and

(3, 0)

$\pm\frac{4}{3}x$

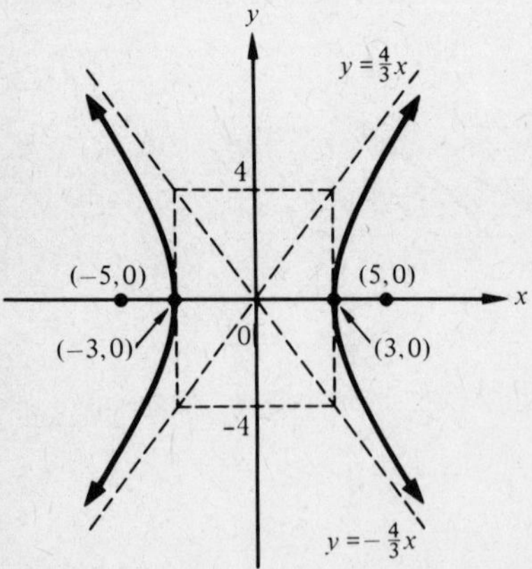

$a^2 + b^2$, $(0, c)$

$(0, -c)$

$\frac{a}{b}$

________. The equations of the asymptotes are $y = \pm\frac{b}{a}x$ or

$y =$ ________. The graph is

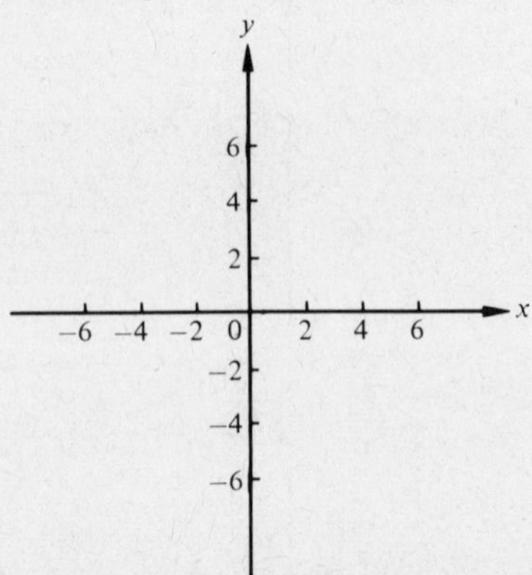

70 An equation of the hyperbola whose vertices are $(0, a)$ and $(0, -a)$ is $\frac{y^2}{a^2} - \frac{x^2}{b^2} = 1$, where $c^2 =$ __________ and whose foci are ________ and __________.

71 The equations of the asymptotes of the hyperbola $\frac{y^2}{a^2} - \frac{x^2}{b^2} = 1$ are $y = \pm$ _____ x

In Problems 72–73, find the coordinates of the foci and vertices, and the equations of the asymptotes for the given hyperbola; also, sketch the graph.

36, 9, 6, 3

45, $3\sqrt{5}$

$(0, 3\sqrt{5})$, $(0, -3\sqrt{5})$

(0, 6)

(0, −6)

$\pm 2x$

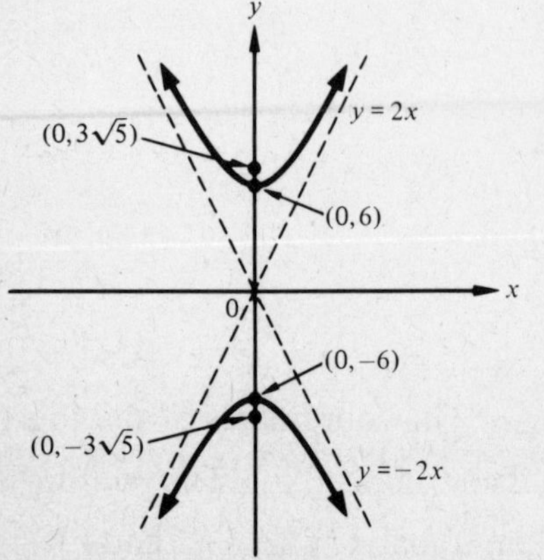

72 $\frac{y^2}{36} - \frac{x^2}{9} = 1$

Here $a^2 =$ ____ and $b^2 =$ ____ or $a =$ ____ and $b =$ ____. Also, $c^2 = a^2 + b^2 =$ ____, so that $c =$ ____. The coordinates of the foci are $(0, c)$ and $(0, -c)$ or __________ and __________. The coordinates of the vertices are $(0, a)$ and $(0, -a)$ or ________ and ________. The equations of the asymptotes are $y = \pm\frac{a}{b}x$ or $y =$ ____. The graph is

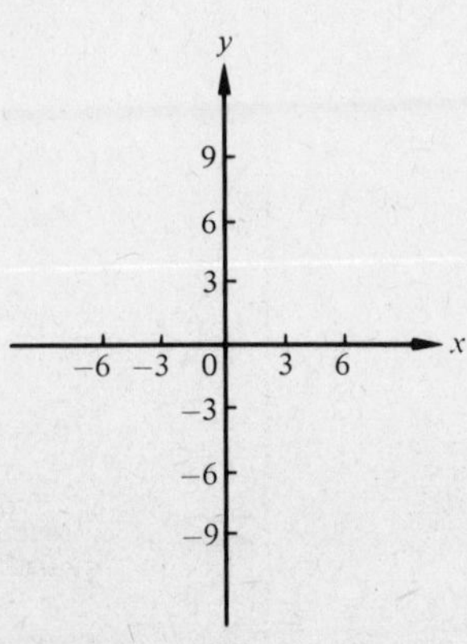

4, 5, 2, $\sqrt{5}$

b^2, 9, 3

(0, 3), (0, −3)

(0, 2)

(0, −2)

$\pm\frac{2\sqrt{5}}{5}x$

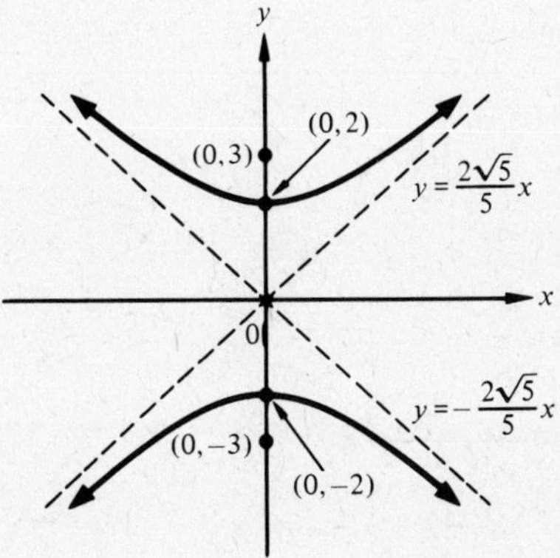

$x - h$

$\frac{a}{b}$

73 $\frac{y^2}{4} - \frac{x^2}{5} = 1$

Here a^2 = ____ and b^2 = ____, or $a \doteq$ ____ and b = ____. Also, $c^2 = a^2 +$ ____ = ____, so that c = ____. The coordinates of the foci are $(0, c)$ and $(0, -c)$ or ______ and ______. The coordinates of the vertices are $(0, a)$ and $(0, -a)$ or ______ and ______. The equations of the asymptotes are $y = \pm\frac{a}{b}x$ or y = ______. The graph is

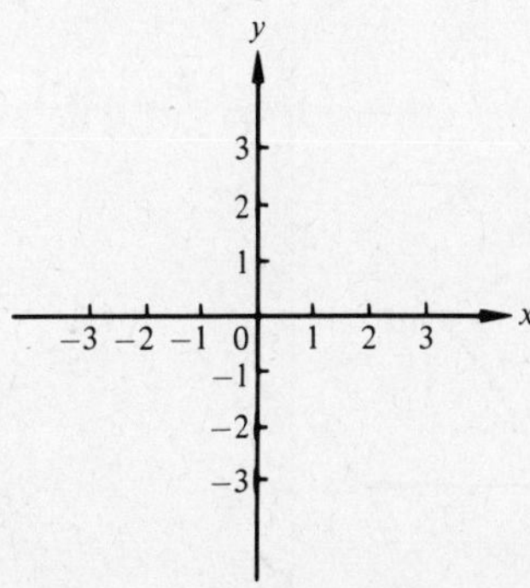

74 If the center of a hyperbola is translated so that the point (h, k) is the center, the forms of the equations are

(i) $\frac{(x-h)^2}{a^2} - \frac{(y-k)^2}{b^2} = 1$, with asymptotes

$y - k = \pm\frac{b}{a}$(______) (Figure 4)

(ii) $\frac{(y-k)^2}{a^2} - \frac{(x-h)^2}{b^2} = 1$, with asymptotes

$y - k = \pm$ ___ $(x - h)$ (Figure 5)

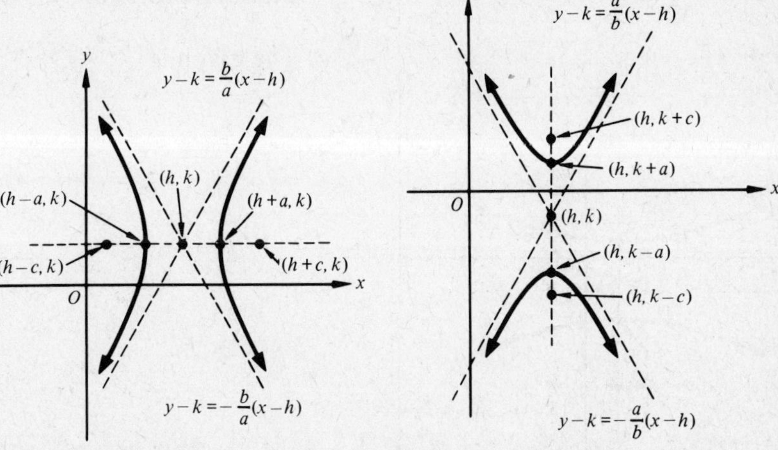

Figure 4 *Figure 5*

In Problems 75–78, find the coordinates of the center, foci, and vertices, and the equations of the asymptotes for the given hyperbola; also, sketch the graph.

75 $\dfrac{(x-2)^2}{4} - \dfrac{(y+1)^2}{5} = 1$

Since $h =$ ____ and $k =$ ____, the coordinates of the center $(h, k) =$ ________. Also, $a^2 =$ ____ and $b^2 =$ ____ or $a =$ ____ and $b =$ ____. Since $c^2 = a^2 + b^2 =$ ____, then $c =$ ____. The coordinates of the foci are $(h - c, k)$ and $(h + c, k)$ or ________ and ________. The coordinates of the vertices are $(h - a, k)$ and $(h + a, k)$ or ________ and ________. The equations of the asymptotes are $y - k = \pm\dfrac{b}{a}(x - h)$ or ____________.

The graph is

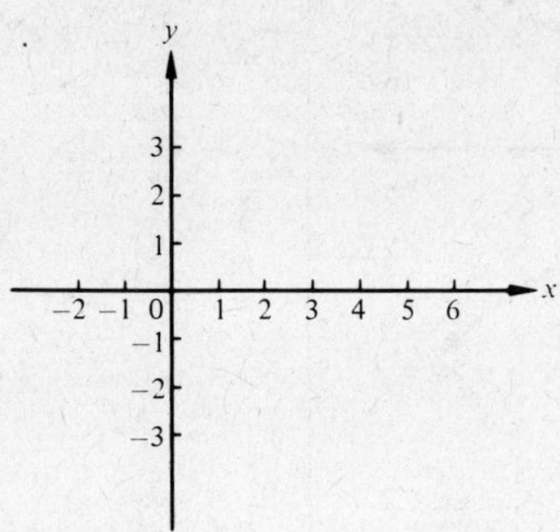

2, −1

(2, −1), 4, 5, 2

$\sqrt{5}$, 9, 3

(−1, −1)

(5, −1)

(0, −1), (4, −1)

$y + 1 = \pm\dfrac{\sqrt{5}}{2}(x - 2)$

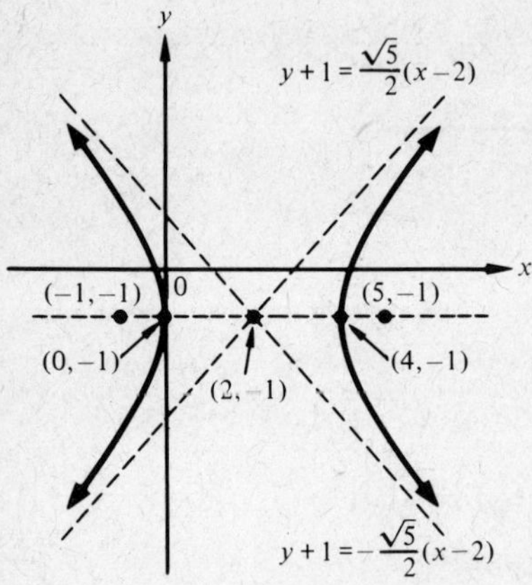

76 $\dfrac{(y+3)^2}{25} - \dfrac{(x+4)^2}{144} = 1$

Here $h =$ ____ and $k =$ ____, so that the coordinates of the center $(h, k) =$ ________. Also, $a^2 =$ ____ and $b^2 =$ ____ or $a =$ ____ and $b =$ ____. Since $c^2 = a^2 + b^2 =$ ____, then $c =$ ____. The coordinates of the foci are $(h, k + c)$ and $(h, k - c)$ or ________ and ________. The coordinates of the vertices are $(h, k + a)$ and $(h, k - a)$ or ________ and ________. The equations of the asymptotes are $y - k = \pm\dfrac{a}{b}(x - h)$ or ____________.

The graph is

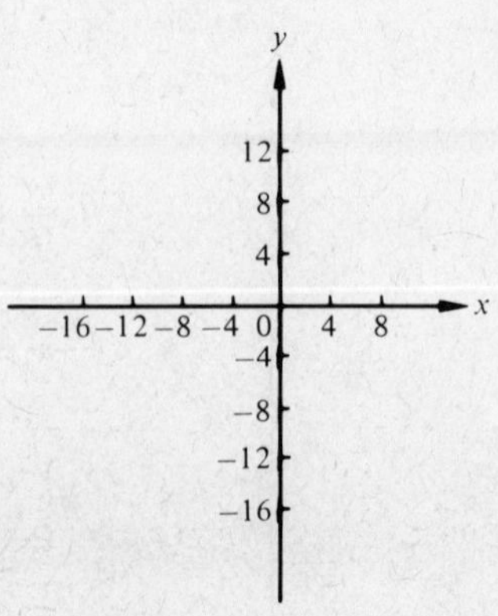

−4, −3

(−4, −3), 25, 144, 5

12, 169, 13

(−4, 10)

(−4, −16)

(−4, 2), (−4, −8)

$y + 3 = \pm\dfrac{5}{12}(x + 4)$

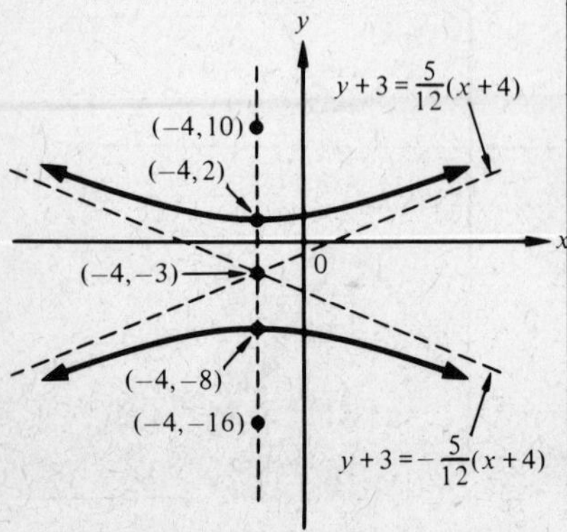

77 $16x^2 - 9y^2 - 32x - 36y - 164 = 0$

First, complete the square to get

$16(x^2 - 2x +$ ____ $) - 9(y^2 + 4y +$ ____ $) = 164 +$ ____ $+$ ____

or $16($ ________ $)^2 - 9($ ________ $)^2 =$ ____.

Dividing both members by ____, we obtain $\dfrac{(x-1)^2}{9} - \dfrac{(y+2)^2}{\_\_\_\_} = 1$.

Here $h =$ ____ and $k =$ ____, so that the coordinates of the center $(h, k) =$ ________. Also, $a^2 =$ ____ and $b^2 =$ ____ or $a =$ ____ and $b =$ ____. Since $c^2 = a^2 + b^2 =$ ____, then $c =$ ____. The coordinates of the foci are $(h - c, k)$ and $(h + c, k)$ or ________ and ________. The coordinates of the vertices are $(h - a, k)$ and $(h + a, k)$ or ________ and ________. The equations of the asymptotes are $y - k = \pm \dfrac{b}{a}(x - h)$ or ________________.

The graph is

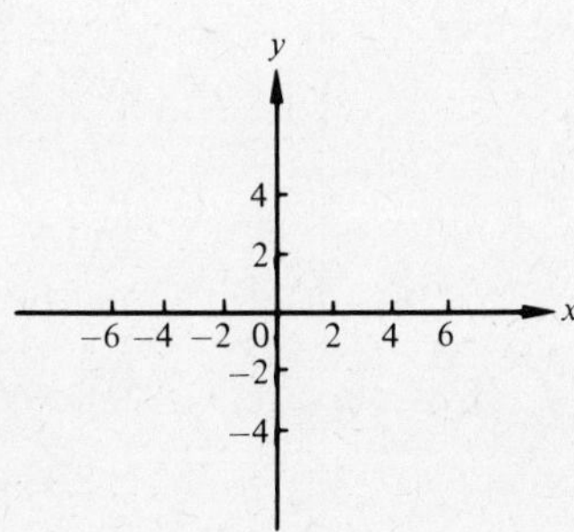

1, 4, 16, −36

$x - 1$, $y + 2$, 144

144, 16

1, −2

(1, −2), 9, 16, 3

4, 25, 5

(−4, −2)

(6, −2)

(−2, −2), (4, −2)

$y + 2 = \pm\frac{4}{3}(x - 1)$

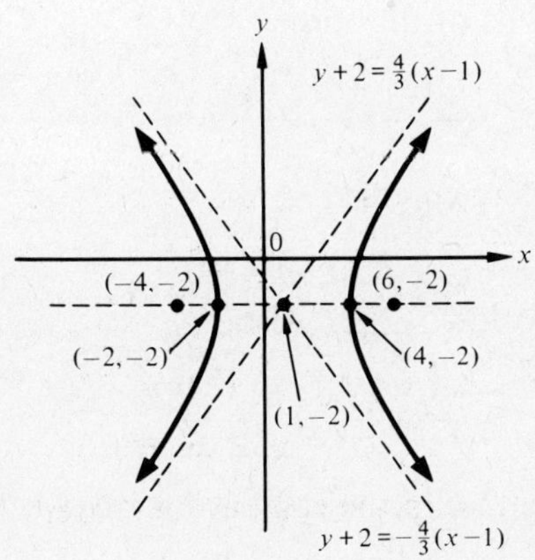

78 $y^2 - x^2 - 6y - 10x - 20 = 0$

Completing the square, we have

$(y^2 - 6y +$ ____ $) - (x^2 + 10x +$ ____ $) = 20 +$ ____ or

$($ ________ $)^2 - ($ ________ $)^2 =$ ____. Dividing both members by ____, we obtain $\dfrac{(y-3)^2}{\_\_\_\_} - \dfrac{(x+5)^2}{\_\_\_\_} = 1$. Here $h =$ ____ and $k =$ ____, so that the coordinates of the center $(h, k) =$ ________. Also, $a^2 =$ ____ and $b^2 =$ ____ or $a =$ ____ and $b =$ ____. Since $c^2 = a^2 + b^2 =$ ____, $c =$ ______. The coordinates of the foci are $(h, k - c)$ and $(h, k + c)$ or ________________ and ________________. The coordinates of the vertices are $(h, k - a)$ and $(h, k + a)$ or ________ and ________. The equations of the asymptotes are $y - k = \pm \dfrac{a}{b}(x - h)$ or ____________.

9, 25, −16

$y - 3$, $x + 5$, 4

4, 4, 4, −5

3, (−5, 3)

4, 4, 2, 2

8, $2\sqrt{2}$

$(-5, 3 - 2\sqrt{2})$

$(-5, 3 + 2\sqrt{2})$

(−5, 1), (−5, 5)

$y - 3 = \pm(x + 5)$

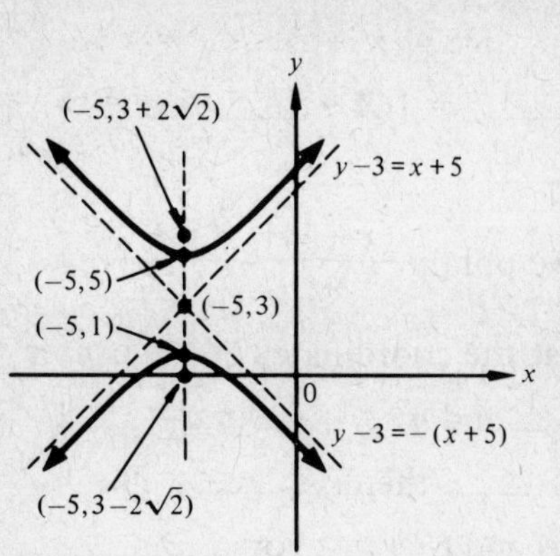

The graph is

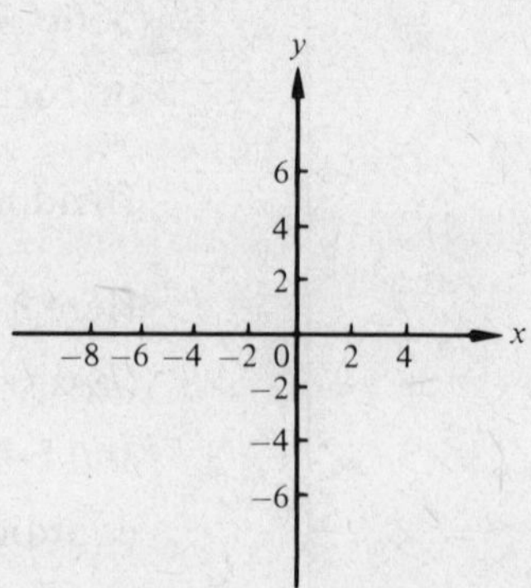

In Problems 79–81, find an equation of the hyperbola that satisfies the given conditions.

79 Center at (3, 4), one focus at (8, 4), one vertex at (6, 4)

The given points indicate that the transverse axis is parallel to the ____ axis. Hence, the equation will be of the form $\frac{(x-h)^2}{a^2} - \frac{(y-k)^2}{b^2} = 1$. Also, $(h, k) =$ ________, so that $h =$ ____ and $k =$ ____. The given focus can be expressed as $(8, 4) = (h + c, k)$, so that $h + c =$ ____. Since $h =$ ____, then ____ $+ c = 8$, or $c =$ ____. Also, $(6, 4) = (h + a, k)$ or $h + a =$ ____ or ____ $+ a = 6$, so that $a =$ ____. Since $c^2 = a^2 + b^2$, then $b = \sqrt{c^2 - a^2} = \sqrt{\_\_\_\_} =$ ____. Substituting these values of a, b, h, and k in the standard form, we have $\frac{(x - \_\_\_\_)^2}{3^2} - \frac{(y-4)^2}{(\_\_\_\_)^2} = 1$ or ________________.

Answers: x; $(3, 4)$, 3; 4; 8; 3, 3, 5; 6, 3; 3, 16, 4; 3, 4, $\frac{(x-3)^2}{9} - \frac{(y-4)^2}{16} = 1$

80 Vertices at (4, 1) and (4, 5), one focus at (4, 7)

The center (h, k) is the midpoint of the line segment determined by the two ____________. Hence, $h =$ ____ and $k = \frac{\_\_\_\_ + \_\_\_\_}{2} =$ ____. The value of a is the distance from the center to the vertex. Hence, $a = 5 -$ ____ $=$ ____. The value of c is the distance from the center to the focus. Hence, $c =$ ____ $- 3 =$ ____. Then $b = \sqrt{c^2 - a^2} = \sqrt{\_\_\_\_} =$ ____. The given points indicate that the transverse axis is parallel to the ____ axis, so that the standard form of the equation is $\frac{(y-k)^2}{a^2} - \frac{(x-h)^2}{b^2} = 1$. By substitution, $\frac{(y - \_\_\_\_)^2}{2^2} - \frac{(x-4)^2}{(\_\_\_\_)^2} = 1$ or ________________.

Answers: vertices, 4; 1, 5, 3; 3, 2; 7, 4, 12, $2\sqrt{3}$; y; 3, $2\sqrt{3}$; $\frac{(y-3)^2}{4} - \frac{(x-4)^2}{12} = 1$

81 Vertices at (−3, 0) and (3, 0), asymptotes $y = \pm\frac{2}{3}x$

The center (h, k) is the midpoint of the line segment determined by the two vertices. Hence $h =$ ____ and $k =$ ____. The value of a is the distance from the center to a vertex, hence $a =$ ____. The standard form of the equations of the asymptotes is $y = \pm\frac{b}{a}x$, since the transverse axis is parallel to the ____ axis. Thus $\frac{b}{a} = \frac{2}{3}$, and since $a =$ ____, we have $\frac{b}{\_\_\_\_} = \frac{2}{3}$, so $b =$ ____. Since the center is at the origin, the standard form of the equation is $\frac{x^2}{a^2} - \frac{y^2}{b^2} = 1$. Substituting the values of a and b, we have $\frac{x^2}{(\_\_\_\_)^2} - \frac{y^2}{(\_\_\_\_)^2} = 1$ or ____________.

Answers: 0, 0; 3; x; 3, 3, 2; 3, 2, $\frac{x^2}{9} - \frac{y^2}{4} = 1$

4 Parabola

82 Geometrically, a parabola is defined to be the set of all points P in the plane such that the distance from P to a fixed point, called the ________, is equal to the distance from P to a fixed line, called the ____________.

Answers: focus; directrix

In Problems 83–87, refer to Figure 1.

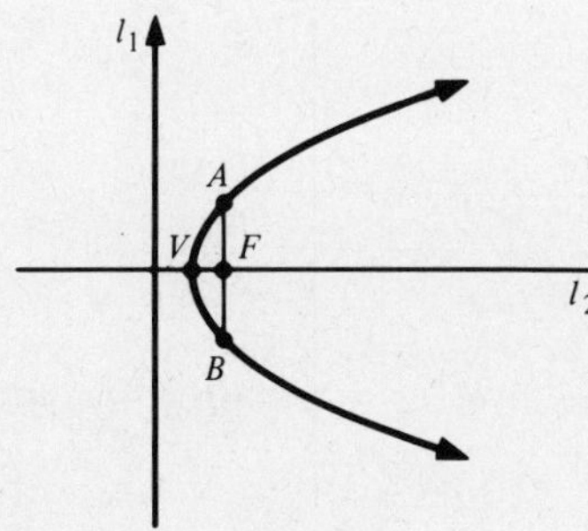

Figure 1

83 The point F is called the ________ of the parabola.

84 The line l_1 is called the ____________ of the parabola.

85 The point V is called the ________ of the parabola.

86 The line l_2, containing the focus and the vertex, is called the ________________ of the parabola.

87 The line segment $\overline{AB}$, which has its end points on the parabola and is perpendicular to the axis of symmetry at the focus, is called the focal chord or ____________ of the parabola.

Answers: focus; directrix; vertex; axis of symmetry; latus rectum

$4cx$

88 An equation of the parabola with focus $(c, 0)$ and directrix $x = -c$ is $y^2 =$ _____, where $c > 0$ (Figure 2).

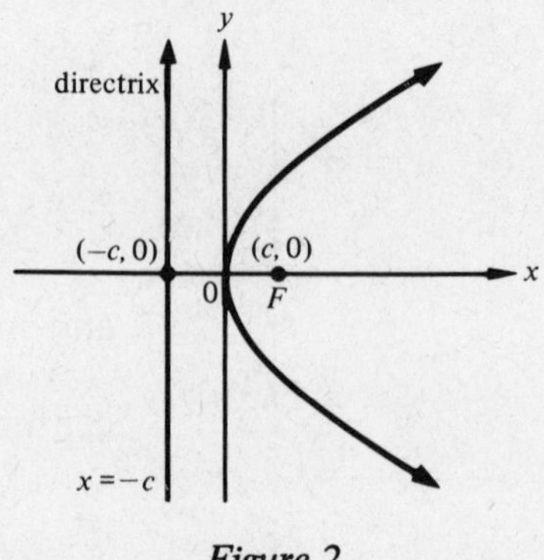

Figure 2

$-4cx$

89 An equation of the parabola with focus $(-c, 0)$ and directrix $x = c$ is $y^2 =$ _____, where $c > 0$ (Figure 3).

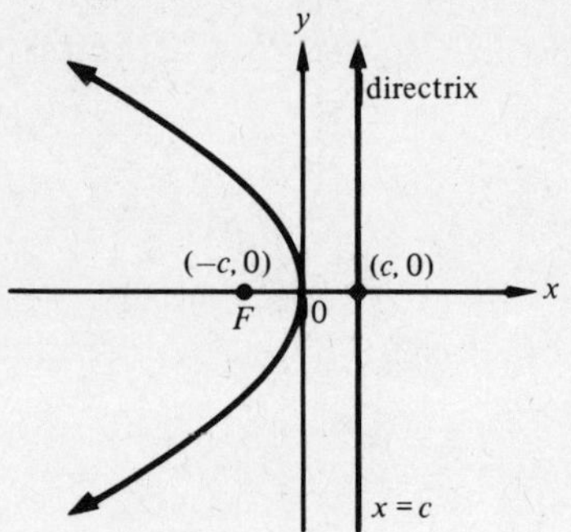

Figure 3

$4cy$

90 An equation of the parabola with focus $(0, c)$ and directrix $y = -c$ is $x^2 =$ _____, where $c > 0$ (Figure 4).

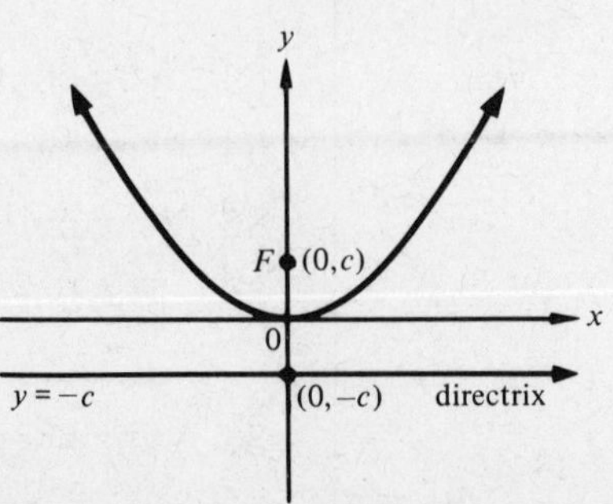

Figure 4

$-4cy$

91 An equation of the parabola with focus $(0, -c)$ and directrix $y = c$ is $x^2 =$ ______, where $c > 0$ (Figure 5).

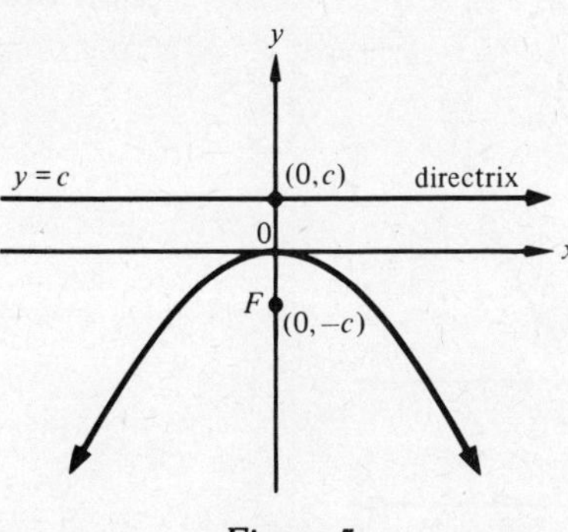

Figure 5

In Problems 92–95, find the focus and directrix of the given parabola and sketch the graph.

16, 4

(4, 0), -4

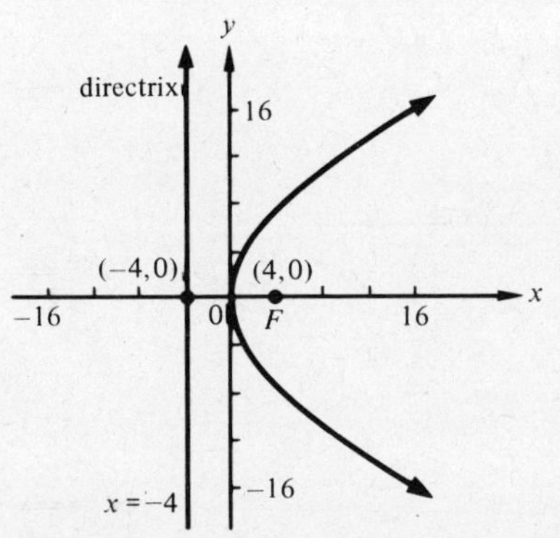

92 $y^2 = 16x$

Here $4c =$ ____, so that $c =$ ____. Hence, the focus is $(c, 0) =$ ________ and the directrix is $x = -c$ or $x =$ ____.

The graph is

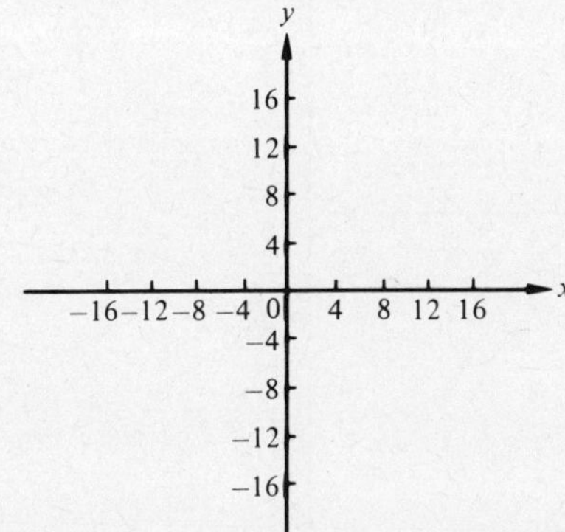

-8, 2

(-2, 0), 2

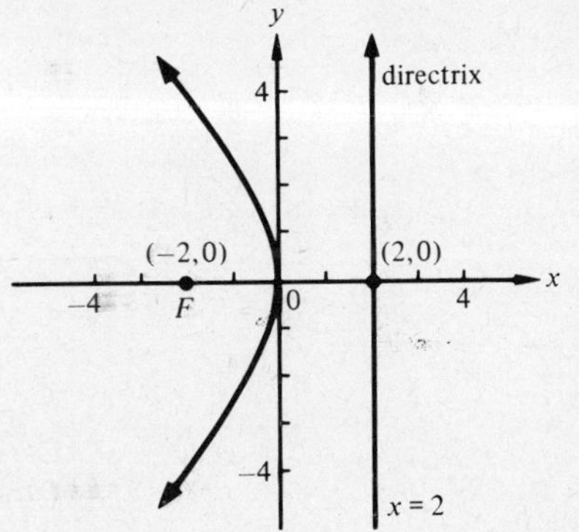

93 $y^2 = -8x$

Here $-4c =$ ____, so that $c =$ ____. Hence, the focus is $(-c, 0) =$ __________ and the directrix is $x = c$ or $x =$ ____.

The graph is

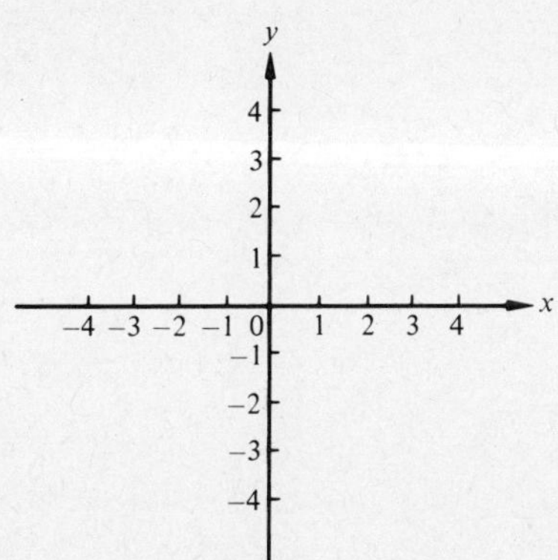

94 $x^2 = 10y$

Here $4c =$ ____, so that $c =$ ____. Hence, the focus is ________ and the directrix is $y =$ ____. The graph is

10, $\frac{5}{2}$, $(0, \frac{5}{2})$
$-\frac{5}{2}$

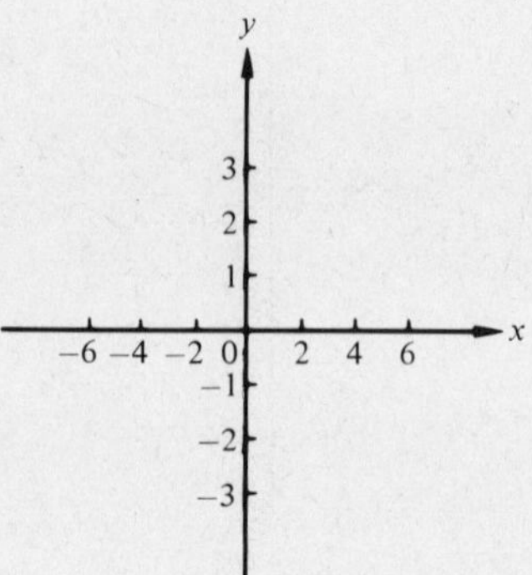

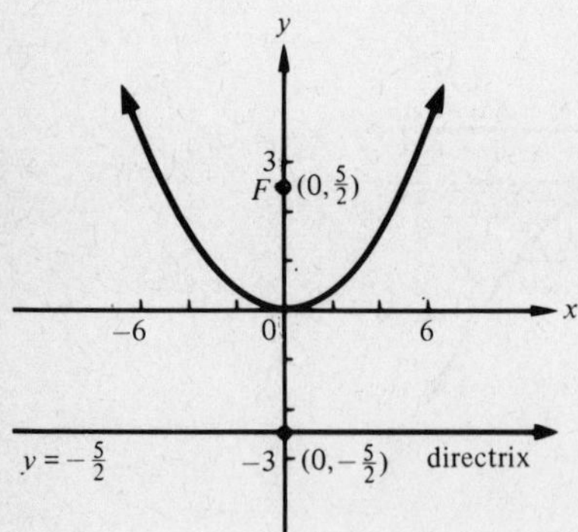

95 $x^2 = -4y$

Here $-4c =$ ____, so that $c =$ ____. Hence, the focus is ________ and the directrix is $y =$ ____. The graph is

−4, 1, (0, −1)
1

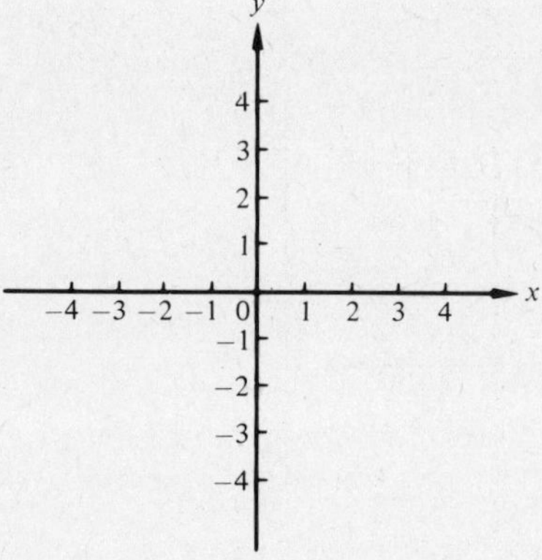

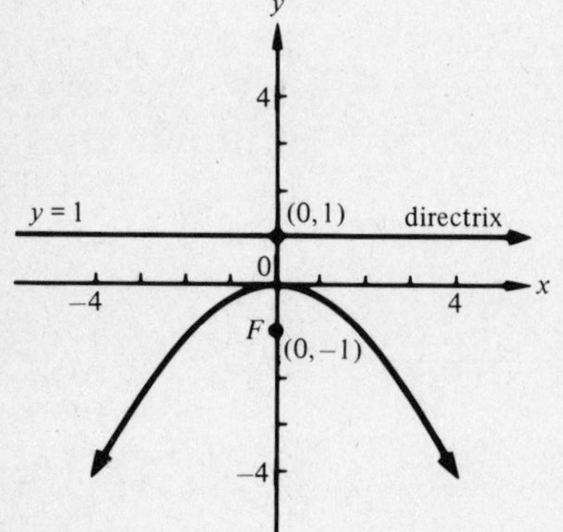

96 If the vertex of a parabola is translated to the point (h, k), then an equation of the parabola, assuming that $c > 0$, is

(i) $(y - k)^2 = 4c(x - h)$, with directrix $x =$ _______ $h - c$
(ii) $(y - k)^2 = -4c(x - h)$, with directrix $x =$ _______ $h + c$
(iii) $(x - h)^2 = 4c(y - k)$, with directrix $y =$ _______ $k - c$
(iv) $(x - h)^2 = -4c(y - k)$, with directrix $y =$ _______ $k + c$

In Problems 97–102, find the focus, vertex, and directrix of the given parabola; then sketch the graph.

97 $(y - 3)^2 = 12(x + 1)$

Here $h = -1$ and $k = 3$, so that the vertex $(h, k) =$ _________.   (−1, 3)
Also, $4c =$ ____, so that $c =$ ____. Hence, the focus   12, 3
$(h + c, k) =$ ________. The directrix is $x =$ ______ = ____.   (2, 3), $h - c$, −4

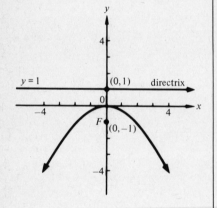

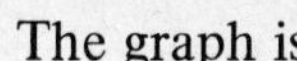
The graph is

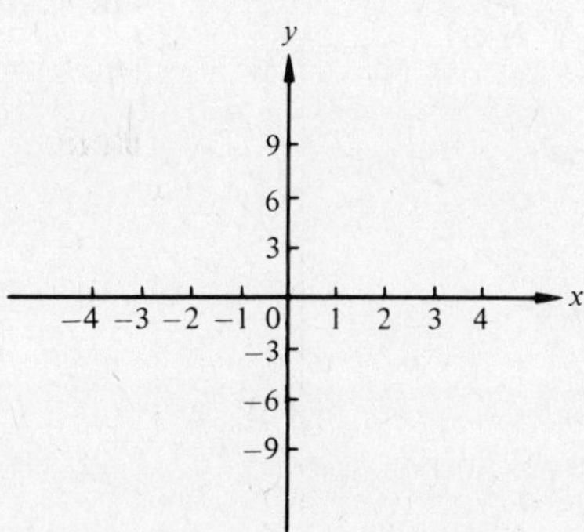

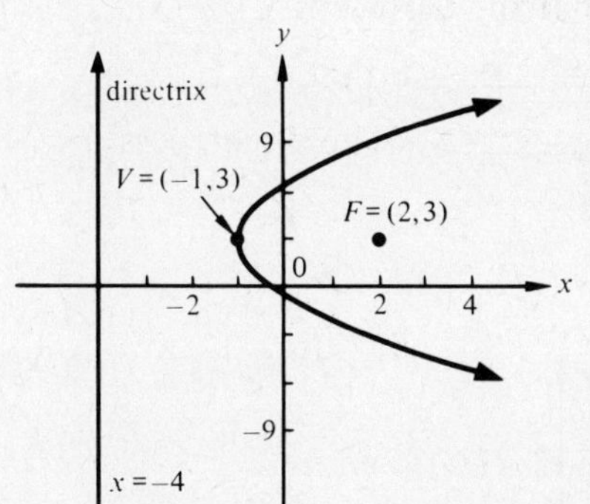

98 $(y+2)^2 = -16(x-3)$

Here, $h = 3$ and $k =$ ____, so that the vertex $(h, k) =$ __________.
Also, $-4c =$ ____, so that $c =$ ____. Hence, the focus
$(h - c, k) =$ __________. The directrix is $x =$ _______ $=$ ____.
The graph is

−2, (3, −2)
−16, 4
(−1, −2), $h + c$, 7

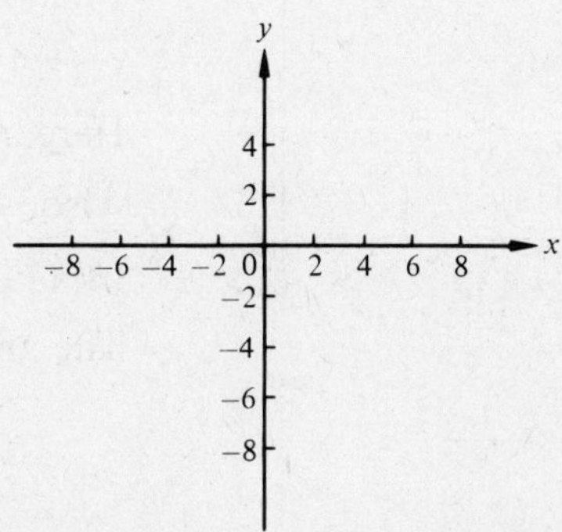

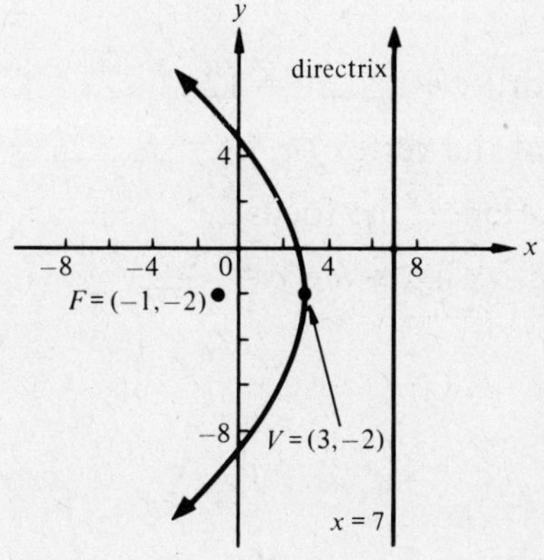

99 $(x-2)^2 = 8(y+2)$

Here, $h =$ ____ and $k =$ ____, so that the vertex $(h, k) =$ ________.
Also, $4c =$ ____, so that $c =$ ____. Hence, the focus
$(h, k + c) =$ ________. The directrix is $y =$ ________ $=$ ____.
The graph is

2, −2, (2, −2)
8, 2
(2, 0), $k - c$, −4

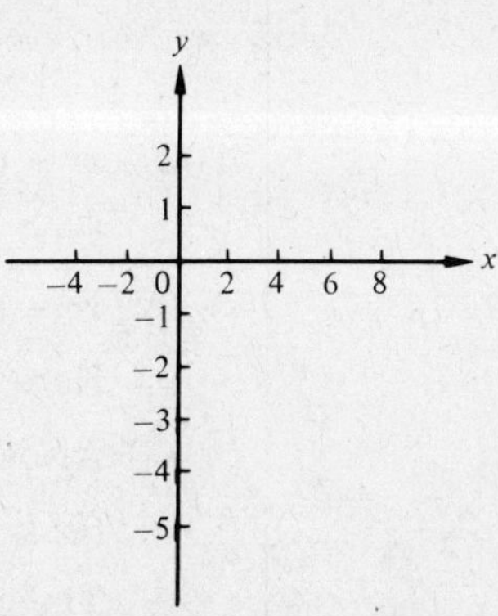

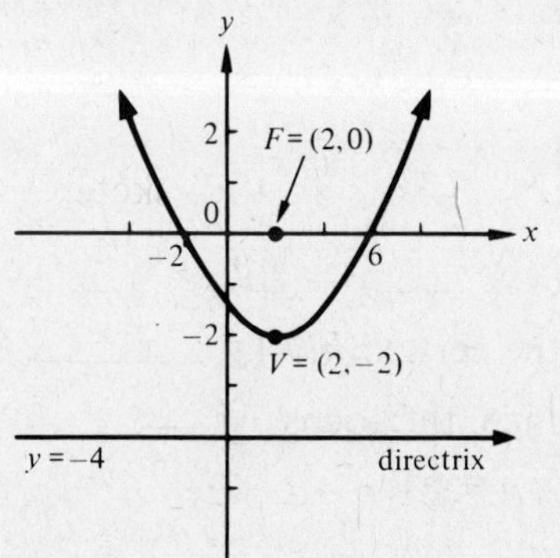

100 $(x+3)^2 = -4(y+2)$

Here, $h =$ ____ and $k =$ ____, so that the vertex $(h, k) =$ ____________. Also, $-4c =$ ____, so that $c =$ ____. Hence, the focus $(h, k - c) =$ ____________. The directrix is $y =$ ________ $=$ ____. The graph is

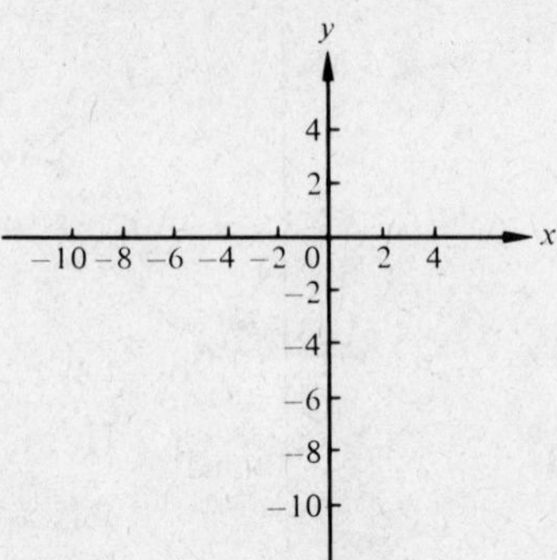

-3, -2

(-3, -2), -4, 1

(-3, -3)

$k + c$, -1

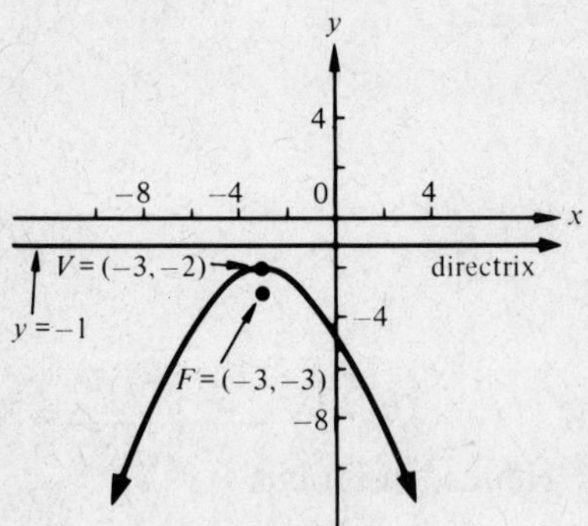

101 $y^2 + 4y - 12x + 16 = 0$

First, completing the square, we have

$y^2 + 4y +$ ____ $= 12x - 16 +$ ____ or $(y +$ ____$)^2 = 12(x -$ ____$)$.

Here, $h =$ ____ and $k =$ ____, so that the vertex $(h, k) =$ ________.

Also, $4c =$ ____, so that $c =$ ____. Hence, the focus $(h + c, k) =$ ___________. The directrix is $x = h - c =$ ____.

The graph is

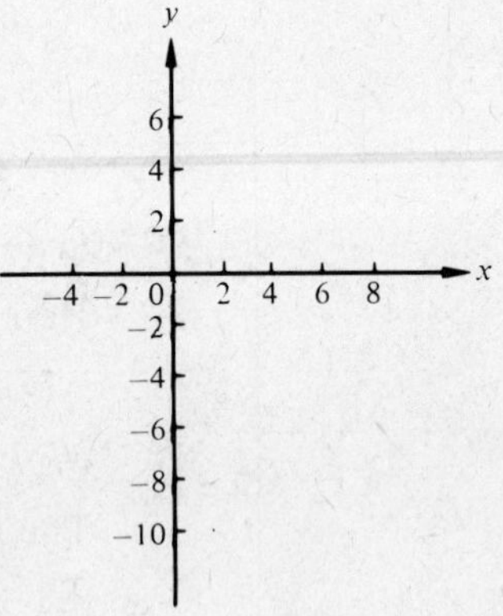

4, 4, 2, 1

1, -2, (1, -2)

12, 3

(4, -2), -2

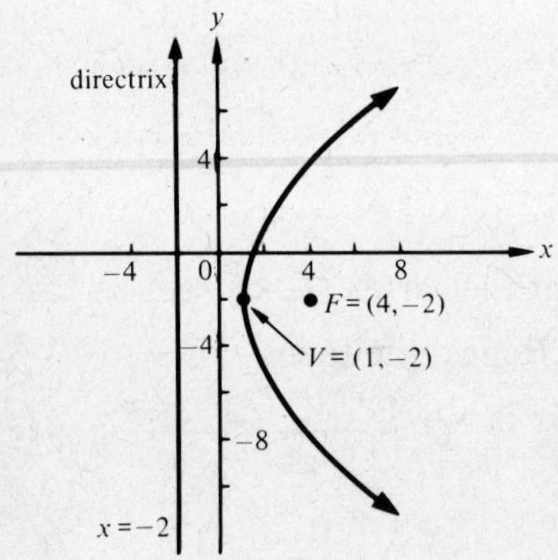

102 $x^2 - 6x - 20y + 49 = 0.$

First, completing the square, we obtain

$x^2 - 6x +$ ____ $= 20y - 49 +$ ____ or $(x -$ ____$)^2 = 20(y -$ ____$)$.

Here, $h =$ ____ and $k =$ ____, so that the vertex $(h, k) =$ ________.

Also, $4c =$ ____, so that $c =$ ____. Hence, the focus $(h, k + c) =$ _________. The directrix is $y = k - c =$ ____.

The graph is

9, 9, 3, 2

3, 2, (3, 2)

20, 5

(3, 7), -3

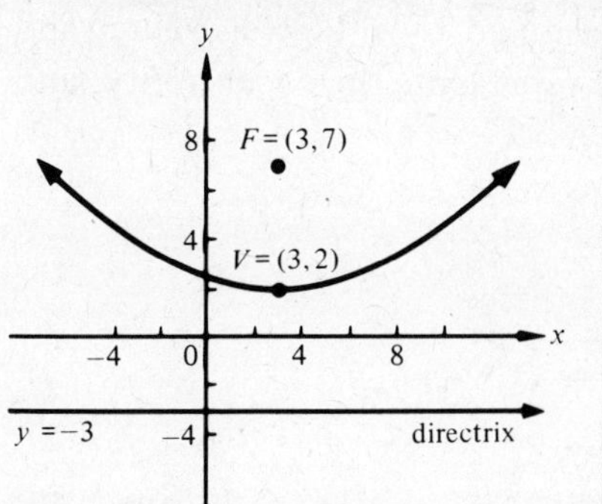

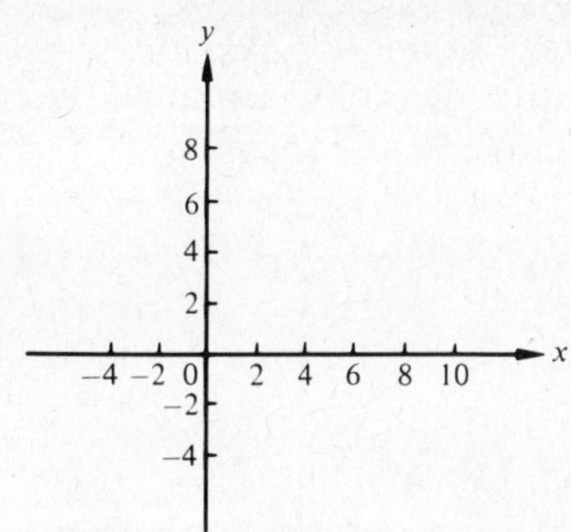

In Problems 103–105, find an equation of the parabola that satisfies the given conditions.

103 The vertex is the point (2, 0) and the focus is the origin. The value of c is the distance from the vertex to the focus. Hence, $c = 2 -$ ____ = ____. Also, $h =$ ____ and $k =$ ____. The standard form of the equation of this parabola is $(y - k)^2 =$ ________. By substitution, $(y -$ ____$)^2 = -4($____$)(x -$ ____$)$ or ____________

0, 2, 2, 0
$-4c(x - h)$
0, 2, 2
$y^2 = -8(x - 2)$

104 The focus is the point $(-3, -5)$ and the directrix is $y = -1$. The vertex lies midway between the focus and the directrix. Hence, the vertex $(h, k) =$ ________. The value of c is the distance from the vertex to the focus or $c =$ ____. Hence, the equation is of the form $(x - h)^2 = -4c(y - k)$, so that $[x - ($____$)]^2 = -4($____$)[y - ($____$)]$ or ____________

$(-3, -3)$
2
$-3, 2, -3$
$(x + 3)^2 = -8(y + 3)$

105 The vertex is the point $(3, -4)$ and the directrix is $y = -8$. The value of c is the distance between the vertex and the directrix. Hence, $c = -4 - ($____$) =$ ____. The standard form of the equation is $(x - h)^2 = ($____$)(y - k)$. Since $(h, k) =$ ________, we have $(x -$ ____$)^2 = 4($____$)[y - ($____$)]$ or ____________

$-8, 4$
$4c, (3, -4)$
$3, 4, -4$
$(x - 3)^2 = 16(y + 4)$

5 Conics

106 The graphs of the ellipse, the hyperbola, and the parabola are called ________.

conics

107 In general, a conic is determined by a given point F, called the ________, a given line d not containing F, called the ________ associated with F, and a positive number e, called the ________.

focus, directrix
eccentricity

108 A conic contains a point P if and only if $\frac{|\overline{FP}|}{|\overline{PD}|} =$ ____, where D is the foot of the perpendicular from P to D, a point on the directrix.

e

In Problems 109–111, find an equation of the conic determined by the given values and identify the conic. Assume that d represents the directrix, e represents the eccentricity, and F represents the focus.

109 $e = \frac{1}{2}$, $F = (1, 1)$, and d is the line $x = -1$

If $P = (x, y)$ is a point on the conic, then

1, 1

$r_1 = |\overline{FP}| = \sqrt{(x - \_\_\_)^2 + (y - \_\_\_)^2}$ or

$(x-1)^2 + (y-1)^2$

$r_1{}^2 =$ __________.

-1, $(x+1)^2$

Also, $r_2 = |\overline{PD}| = x - (\_\_\_)$, so that $r_2{}^2 =$ ________. Since, by definition, $\dfrac{|\overline{FP}|}{|\overline{PD}|} = e$ or $\dfrac{r_1}{r_2} = e$, then $\dfrac{r_1{}^2}{r_2{}^2} = e^2$. Hence,

$\frac{1}{2}$, $(x+1)^2$

$\dfrac{(x-1)^2 + (y-1)^2}{(x+1)^2} = \left(\_\_\_\right)^2$ or $4(x-1)^2 + 4(y-1)^2 =$ ______.

Simplifying, and completing the square, we have

$\frac{16}{3}$, $\frac{16}{9}$, $\frac{4}{3}$

$3(x - \frac{5}{3})^2 + 4(y-1)^2 =$ ____ or $\dfrac{(x-\frac{5}{3})^2}{\_\_\_} + \dfrac{(y-1)^2}{\_\_\_} = 1$.

an ellipse

Hence, the conic is __________.

110 $e = 1$, $F = (-1, 1)$, and d is the line $y = -2$

If $P = (x, y)$ is a point on the conic, then

$y - 1$

$r_1 = |\overline{FP}| = \sqrt{(x+1)^2 + (\_\_\_\_\_)^2}$ or

$(x+1)^2 + (y-1)^2$

$r_1{}^2 =$ __________

-2, $(y+2)^2$

Also, $r_2 = |\overline{PD}| = y - (\_\_\_)$, so that $r_2{}^2 =$ ________. Since

e^2

$\dfrac{r_1}{r_2} = \dfrac{|\overline{FP}|}{|\overline{PD}|} = e$, then $r_1{}^2 = r_2{}^2(\_\_\_\_)$. Hence,

1

$(x+1)^2 + (y-1)^2 = (\_\_\_\_)(y+2)^2$. After simplifying, we

3, $\frac{1}{2}$

obtain $(x+1)^2 = 6y +$ ____ or $(x+1)^2 = 6(y + \_\_\_\_)$.

a parabola

Hence, the conic is __________.

111 $e = 2$, $F = (-3, -2)$, and d is the line $x = -1$

If $P = (x, y)$ is a point on the conic, then

-3, -1

$r_1{}^2 = |\overline{FP}|^2 = (x - \_\_\_\_)^2 + (y+2)^2$. Also, $r_2{}^2 = (x - \_\_\_\_)^2$.

Since $r_1{}^2 = r_2{}^2 e^2$, by substitution, we obtain

$(y+2)^2$, 4

$(x+3)^2 +$ ________ $= (x+1)^2(\_\_\_\_)$

After simplifying and completing the square, we obtain

$(y+2)^2$, $\frac{16}{3}$

$3(x + \frac{1}{3})^2 -$ ________ $= \frac{16}{3}$ or $\dfrac{(x+\frac{1}{3})^2}{\frac{16}{9}} - \dfrac{(y+2)^2}{\_\_\_\_} = 1$.

a hyperbola

Hence, the conic is __________.

112 If the eccentricity of a conic is 1, that is, $e = 1$, then the conic is a __________.

parabola

113 If the eccentricity of a conic is less than 1, that is, $e < 1$, then the conic is an __________.

ellipse

114 If the eccentricity of a conic is greater than 1, that is, $e > 1$, then the conic is a __________.

hyperbola

115 For ellipses and hyperbolas of the form $\frac{(x-h)^2}{a^2} \pm \frac{(y-k)^2}{b^2} = 1$, where $c^2 = a^2 - b^2$ for ellipses and $c^2 =$ ________ for hyperbolas, the eccentricity $e = \frac{}{\_\_\_\_}$ and the directrices are $x = \pm \frac{}{\_\_\_\_}$.

$a^2 + b^2$

$\frac{c}{a}, \frac{a^2}{c}$

In Problems 116–119, find the eccentricity and equations of the directrices.

116 $\frac{x^2}{9} - \frac{y^2}{16} = 1$

9, 16, 25 — Here, $a^2 =$ ____ and $b^2 =$ ____, so that $c^2 = a^2 + b^2 =$ ____ or

5, 3 — $c =$ ____. Hence, $e = \frac{c}{a} = \frac{5}{\_\_\_\_}$ and the directrices are

$\frac{9}{5}$ — $x = \pm \frac{a^2}{c}$ or $x = \pm \frac{}{\_\_\_\_}$.

117 $\frac{(x-1)^2}{4} + \frac{(y+2)^2}{3} = 1$

4, 3, 1 — Here, $a^2 =$ ____ and $b^2 =$ ____, so that $c^2 = a^2 - b^2 =$ ____ or

1, $\frac{1}{2}$ — $c =$ ____. Hence, $e = \frac{c}{a} = \frac{}{\_\_\_\_}$ and the directrices are $x = \pm \frac{a^2}{c}$

4 — or $x = \pm$____.

118 $\frac{(x+3)^2}{7} - \frac{(y+2)^2}{5} = 1$

7, 5, 12 — Here, $a^2 =$ ____ and $b^2 =$ ____, so that $c^2 = a^2 + b^2 =$ ____ or

$2\sqrt{3}$, $\sqrt{7}$, $\frac{2\sqrt{21}}{7}$ — $c =$ ____. Hence, $e = \frac{c}{a} = \frac{2\sqrt{3}}{\_\_\_\_} = \frac{}{\_\_\_\_}$ and the directrices are

$\frac{7}{2\sqrt{3}}$, $\frac{7\sqrt{3}}{6}$ — $x = \pm \frac{a^2}{c}$ or $x = \pm \frac{}{\_\_\_\_} = \pm \frac{}{\_\_\_\_}$.

119 $\frac{(x+7)^2}{10} + \frac{y^2}{6} = 1$

10, 6, 4 — Here, $a^2 =$ ____ and $b^2 =$ ____, so that $c^2 = a^2 - b^2 =$ ____ or

2, $\frac{\sqrt{10}}{5}$ — $c =$ ____. Hence, $e = \frac{c}{a} = \frac{}{\_\_\_\_}$ and the directrices are

5 — $x = \pm \frac{a^2}{c}$ or $x = \pm$____.

Chapter Test

1 Find the center and radius of each of the following circles:
(a) $(x-3)^2 + (y+4)^2 = 81$ (b) $x^2 + y^2 - 6x + 8y = 11$

2 Find a translation of axes that reduces the equation $x^2 + y^2 + 4x - 2y - 9 = 0$ to the form $\bar{x}^2 + \bar{y}^2 = r^2$.

3 Find the coordinates of the foci, center, and vertices of the following ellipses:

(a) $\dfrac{(x-1)^2}{4}+\dfrac{(y+2)^2}{3}=1$ (b) $2x^2+3y^2-4x+18y-7=0$

4 Find an equation of the ellipse with center at $(-3, 4)$, with foci at $(-3, 6)$ and $(-3, 2)$, and with one vertex at $(-2, 4)$.

5 Find the coordinates of the foci, center, and vertices, and equations of the asymptotes of the following hyperbolas:

(a) $\dfrac{(x+2)^2}{16}-\dfrac{(y-4)^2}{9}=1$ (b) $x^2-2y^2+2x+8y-5=0$

6 Find an equation of the hyperbola whose vertices are $(-1, 3)$ and $(3, 3)$ and one of whose asymptotes is $y=\frac{3}{2}x+\frac{3}{2}$.

7 Find the coordinates of the vertex, focus, and an equation of the directrix of the following parabolas:

(a) $(y+2)^2=-12(x-3)$ (b) $x^2-4x-16y-12=0$

8 Find an equation of the parabola whose vertex is the point $(-1, -3)$ and whose directrix is the line $y=-1$.

9 Find an equation of the conic whose eccentricity $e=\frac{2}{3}$, whose focus is the origin, and whose directrix is $y=-2$.

10 Find the eccentricity and the directrices of the following conics:

(a) $\dfrac{(x-3)^2}{14}+\dfrac{(y+2)^2}{9}=1$ (b) $\dfrac{(x+1)^2}{16}-\dfrac{(y+2)^2}{9}=1$

Answers

1 (a) center: $(3, -4)$; radius: 9 (b) center: $(3, -4)$; radius: 6

2 $\bar{x}=x+2, \bar{y}=y-1$

3 (a) foci: $(0, -2), (2, -2)$
center: $(1, -2)$
vertices: $(-1, -2), (3, -2), (1, -2+\sqrt{3}), (1, -2-\sqrt{3})$

(b) foci: $(1-\sqrt{6}, -3), (1+\sqrt{6}, -3)$
center: $(1, -3)$
vertices: $(1+3\sqrt{2}, -3), (1-3\sqrt{2}, -3), (1, -3+2\sqrt{3}), (1, -3-2\sqrt{3})$

4 $5(x+3)^2+(y-4)^2=5$

5 (a) foci: $(-7, 4), (3, 4)$; center: $(-2, 4)$; vertices: $(-6, 4), (2, 4)$; asymptotes: $y-4=\pm\frac{3}{4}(x+2)$

(b) foci: $(-1, 2+\sqrt{3}), (-1, 2-\sqrt{3})$; center: $(-1, 2)$; vertices: $(-1, 3), (-1, 1)$;
asymptotes: $y-2=\pm\dfrac{\sqrt{2}}{2}(x+1)$

6 $\dfrac{(x-1)^2}{4}-\dfrac{(y-3)^2}{9}=1$

7 (a) vertex: $(3, -2)$; focus: $(0, -2)$; directrix: $x=6$
(b) vertex: $(2, -1)$; focus: $(2, 3)$; directrix: $y=-5$

8 $(x+1)^2=-8(y+3)$

9 $9x^2+5y^2-16y-16=0$

10 (a) $e=\dfrac{\sqrt{70}}{14}$; $x=\pm\dfrac{14\sqrt{5}}{5}$ (b) $e=\dfrac{5}{4}$; $x=\pm\dfrac{16}{5}$

Chapter 8 SYSTEMS OF EQUATIONS AND LINEAR PROGRAMMING

The main objectives of the chapter are to investigate methods of solving systems of linear equations and inequalities, along with systems that involve quadratic equations. In working the problems of the chapter, the student will be able to:

1 Solve systems of linear equations by the substitution method and by the elimination method.

2 Solve systems of linear equations by using matrices and row reduction.

3 Solve systems of linear equations by using determinants (Cramer's rule).

4 Solve systems of linear inequalities and apply the results to linear programming.

5 Solve systems with quadratic equations.

1 Systems of Linear Equations

system

1 A set of two or more linear equations is called a ________ of linear equations.

ordered pairs

2 The solution set of a linear system containing two variables is the set of all ________________ of numbers that satisfy each of the equations in the system simultaneously.

points

3 Geometrically, the solution set of a linear system containing two variables is the set of ordered pairs of numbers corresponding to the ________ of intersection of the graphs of the linear equations, provided that such points exist.

independent

4 If the graphs of two equations intersect at one point, the system is said to be ________________.

inconsistent

5 If the graphs of two linear equations are parallel lines, the intersection is the null set, and the system is said to be ________________.

dependent

6 If the graphs of two linear equations coincide, the intersection is every point of each line, and the system is said to be ____________.

In Problems 7–9, sketch the graphs of each system on the same coordinate system. Use the graphs to determine whether the system is independent, inconsistent, or dependent.

7 $\begin{cases} x + y = 2 & (1) \\ -2x - 2y = 3 & (2) \end{cases}$

Rewrite the system by putting each equation in slope-intercept form.

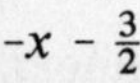

$\begin{cases} y = -x + 2 & (1) \\ y = \_\_\_\_\_\_\_\_ & (2) \end{cases}$

parallel

Since the slopes of equations (1) and (2) are equal and the y intercepts are different, the lines are ________, so that the system does not have a solution. The graph is

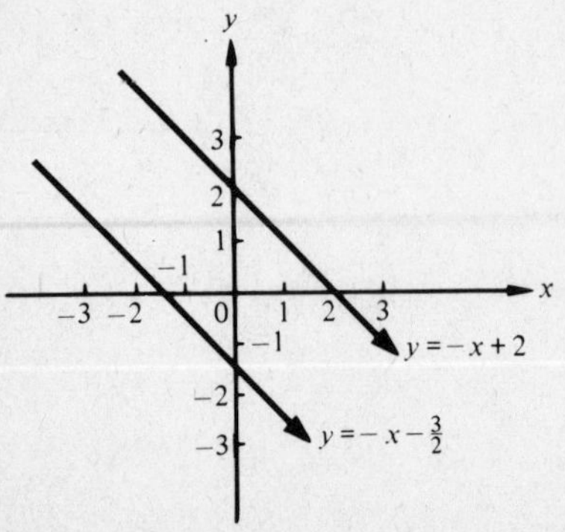

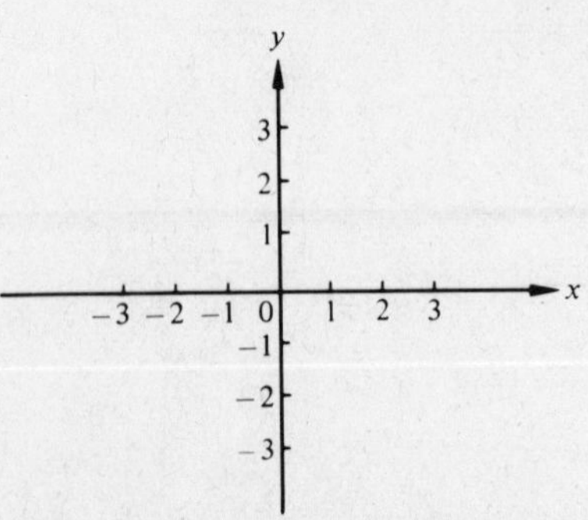

inconsistent

Thus the system is ________________.

8 $\begin{cases} 3x - y = 12 & (1) \\ 5x + 3y = 34 & (2) \end{cases}$

Rewrite the system by putting each equation in slope-intercept form.

$\begin{cases} y = 3x - 12 & (1) \\ y = \_\_\_\_\_\_\_\_ & (2) \end{cases}$

$-\frac{5}{3}x + \frac{34}{3}$

Since the slopes of equations (1) and (2) are different, the lines ________ at one point. The graph is

intersect

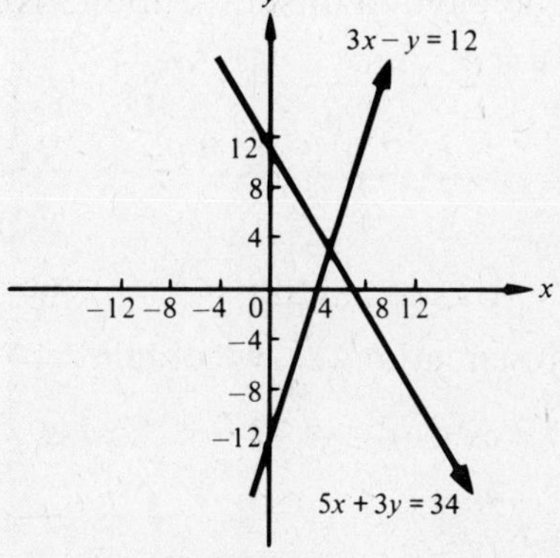

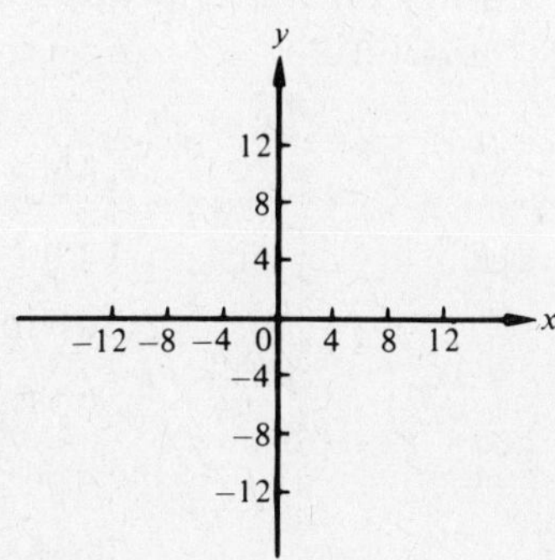

Thus the system is ________________.

independent

9 $\begin{cases} 2x + y = 3 & (1) \\ 4x + 2y = 6 & (2) \end{cases}$

The system can be rewritten in the slope-intercept form.

$\begin{cases} y = -2x + 3 & (1) \\ y = \_\_\_\_\_\_\_\_ & (2) \end{cases}$

$-2x + 3$

Since equations (1) and (2) have the same slope and the same y intercept, each equation represents the same _____. The graph is

line

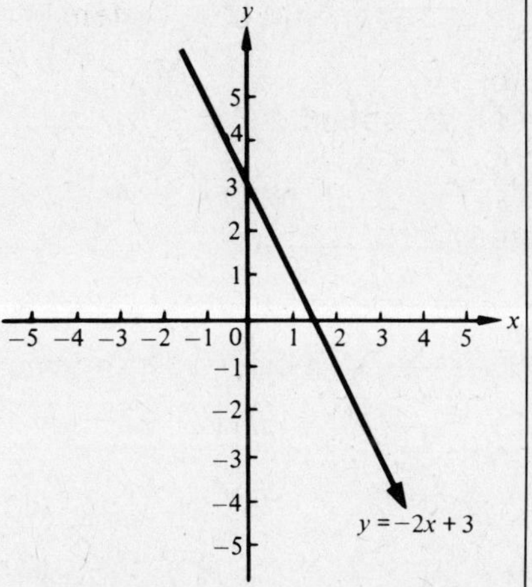

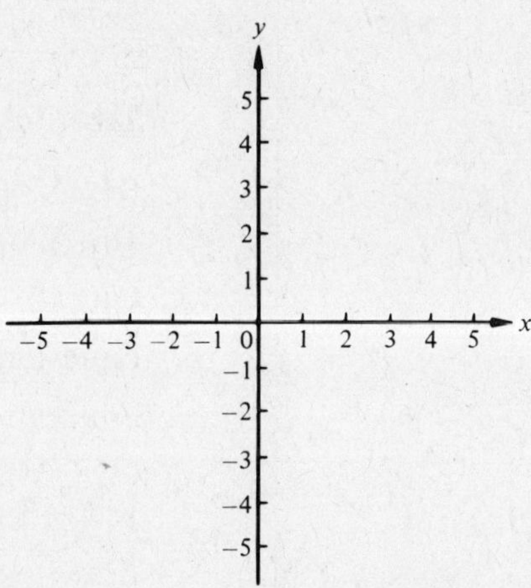

Thus the system is ______________.

dependent

1.1 The Substitution Method

substitute one	10	To solve a linear system containing two variables by the substitution method, we use one of the equations to express one of the variables in terms of the other, and then __________ this value into the remaining equation to obtain an equation in ____ variable.

In Problems 11–16, find the solution set of each system by the method of substitution and check the solution.

11
$$\begin{cases} 3x - y = 12 & (1) \\ 5x + 3y = 34 & (2) \end{cases}$$

Answer	
$3x - 12$	Solving equation (1) for y, we have $y =$ ________.
	Substituting y from equation (1) into equation (2), we obtain
$3x - 12$, 36	$5x + 3($________$) = 34$ or $5x + 9x -$ ____ $= 34$
70, 5	Then $14x =$ ____ or $x =$ ____.
	Substituting 5 for x in equation (1), we obtain
3	$3(5) - y = 12$, so $y =$ ____.
{(5, 3)}	The solution set of the given system is ________.
	Check:
15, 12	Equation (1): $3(5) - (3) =$ ____ $- 3 =$ ____
34	Equation (2): $5(5) + 3(3) = 25 + 9 =$ ____

12
$$\begin{cases} x - 9y = 11 & (1) \\ 7x + 2y = 12 & (2) \end{cases}$$

Answer	
$9y + 11$	Solving equation (1) for x, we have $x =$ ________.
	Substituting x from equation (1) into equation (2), we obtain
$9y + 11$, 77	$7($________$) + 2y = 12$ or $63y +$ ____ $+ 2y = 12$
-65, -1	so $65y =$ ____ or $y =$ ____.
	Substituting -1 for y in equation (1), we obtain
2	$x - 9(-1) = 11$ or $x =$ ____
{(2, −1)}	The solution set of the given system is ________.
	Check:
11	Equation (1): $(2) - 9(-1) = 2 + 9 =$ ____
12	Equation (2): $7(2) + 2(-1) = 14 - 2 =$ ____

13
$$\begin{cases} 2x - 3y = -6 & (1) \\ 2x - y = -4 & (2) \end{cases}$$

Answer	
$2x + 4$	Solving equation (2) for y, we obtain $y =$ ________.
	Substituting y from equation (2) into equation (1), we obtain

$2x + 4$, -6

6, $-\frac{3}{2}$

1

$\{(-\frac{3}{2}, 1)\}$

-6

-4

$x - 2$

$x - 2$,

$3x - 4$, 12, 4

2

$\{(4, 2)\}$

2

8

satisfies

$6 - x - y$

0

7

2

1

3

$\{(2, 1, 3)\}$

6

0

7

$2x - 3($__________$) = -6$, so $-4x - 12 =$ ____.

Therefore, $-4x =$ ____ or $x =$ ____.

Substituting $-\frac{3}{2}$ for x in equation (2), we obtain

$2(-\frac{3}{2}) - y = -4$ or $y =$ ____

The solution set of the given system is __________.

Check:

Equation (1): $2(-\frac{3}{2}) - 3(1) = -3 - 3 =$ ____

Equation (2): $2(-\frac{3}{2}) - (1) = -3 - 1 =$ ____

14
$$\begin{cases} x - y = 2 & (1) \\ x + 2y = 8 & (2) \end{cases}$$

Solving equation (1) for y, we obtain $y =$ _____.

Substituting y from equation (1) into equation (2), we obtain

$x + 2($_________$) = 8$. Simplifying this equation, we get

______ $= 8$ or $3x =$ ____. Therefore, $x =$ ____.

Substituting 4 for x in equation (1), we obtain

$(4) - y = 2$, so that $y =$ ____.

The solution set of the given system is ______________.

Check:

Equation (1): $(4) - (2) =$ ____

Equation (2): $(4) + 2(2) =$ ____

15
$$\begin{cases} x + y + z = 6 & (1) \\ 2x - y - z = 0 & (2) \\ x - y + 2z = 7 & (3) \end{cases}$$

We are seeking an ordered triple of numbers (x, y, z) that ________ all three equations simultaneously. Solving for z in equation (1), we get $z =$ __________. Then, substituting for z from equation (1) into equations (2) and (3), respectively, we obtain

$$\begin{cases} 2x - y - (6 - x - y) = \_\_\_\_ & (4) \\ x - y + 2(6 - x - y) = \_\_\_\_ & (5) \end{cases}$$

Equations (4) and (5) can be solved as a linear system containing two variables; that is,

$3x - 6 = 0$ and $-x - 3y + 12 = 7$

so that $x =$ ____. Substituting 2 for x in equation (5), we obtain $y =$ ____. Now we substitute 2 for x and 1 for y in equation (1) to obtain $z =$ ____.

The solution set of the given system is _______________.

Check:

Equation (1): $2 + 1 + 3 =$ ____

Equation (2): $4 - 1 - 3 =$ ____

Equation (3): $2 - 1 + 6 =$ ____

16 $\begin{cases} 2x + 3y + z = 6 & (1) \\ x - 2y + 3z = -3 & (2) \\ 3x + y - z = 8 & (3) \end{cases}$

Solving for x in equation (2), we get $x =$ ______________.

Substituting for x in equation (1), we obtain

$2($______________$) + 3y + z = 6$ or

$7y - 5z = 12$ (4)

Substituting for x in equation (3), we obtain

$3(-3 + 2y - 3z) + y - z =$ ____ or

$7y - 10z = 17$ (5)

Solving the new linear system, equations (4) and (5), we obtain $z =$ ____ and $y =$ ____. Substituting 1 for y and -1 for z in equation (2), we obtain $x =$ ____.

The solution set of the given system is ______________.

Check:

Equation (1): $4 + 3 - 1 =$ ____

Equation (2): $2 - 2 - 3 =$ ____

Equation (3): $6 + 1 + 1 =$ ____

Answers: $-3 + 2y - 3z$; $-3 + 2y - 3z$; 8; -1, 1; 2; $\{(2, 1, -1)\}$; 6; -3; 8

17 A man said to his son, "I am now four times as old as you, but in 20 years I shall be only twice as old as you." Find the age of each. Let x be the man's age and y be the son's age. The values of x and y are found by solving the system

$\begin{cases} \text{\_\_\_\_\_\_\_\_} \\ \text{\_\_\_\_\_\_\_\_\_\_\_\_} \end{cases}$

The system is written

$\begin{cases} x = 4y & (1) \\ x - 2y = 20 & (2) \end{cases}$

Substituting $4y$ for x in equation (2), we obtain

$4y - 2y = 20$ or ____ $= 20$, so $y =$ ____.

Substituting ____ for y in equation (1), we have

$x = 4($____$)$ or $x =$ ____

Thus the man is ____ years old and the son is ____ years old.

Answers: $x = 4y$; $x + 20 = 2(y + 20)$; $2y$, 10; 10; 10, 40; 40, 10

1.2 The Elimination Method

18 In the elimination method, the ____________ of two equations results in an equation with one of the variables eliminated.

Answer: addition

19 In applying the elimination method it is sometimes first necessary to multiply one or both equations by nonzero ____________ in order to be able to eliminate a variable.

Answer: constants

In Problems 20–24, find the solution set of each system by the elimination method.

20 $\begin{cases} x + y = 3 & (1) \\ 2x - y = 3 & (2) \end{cases}$

Adding equations (1) and (2), we obtain

6, 2 | $3x =$ ____ or $x =$ ____

Substituting 2 for x in equation (1), we have

1 | $2 + y = 3$ or $y =$ ____

{(2, 1)} | The solution set is ________.

21 $\begin{cases} 2x + y = 4 & (1) \\ x + y = 1 & (2) \end{cases}$

Adding -1 times equation (2) to equation (1), we obtain

3 | $x =$ ____. Substituting 3 for x in equation (2), we have

-2 | $3 + y = 1$ or $y =$ ____

{(3, -2)} | The solution set is ________.

22 $\begin{cases} 2x + 3y = 2 & (1) \\ x + 2y = 0 & (2) \end{cases}$

Multiplying equation (2) by -2, we obtain the equivalent system

$\begin{cases} 2x + 3y \quad = 2 & (3) \\ -2x - \text{\_\_\_\_} = 0 & (4) \end{cases}$ | 4y

Adding equation (3) to equation (4), we have

2, -2 | $-y =$ ____ or $y =$ ____

Substituting -2 for y in equation (2), we obtain

-2, 4 | $x + 2($____$) = 0$ or $x =$ ____

{(4, -2)} | The solution set is ________.

23 $\begin{cases} 4x + 2y = 2 & (1) \\ 2x - 3y = 13 & (2) \end{cases}$

3, 2 | Multiplying equation (1) by ____ and equation (2) by ____, we obtain the equivalent system

$\begin{cases} 12x + 6y = 6 & (3) \\ 4x - 6y = 26 & (4) \end{cases}$

Adding equations (3) and (4), we have

32, 2 | $16x =$ ____ or $x =$ ____

2 | Substituting ____ for x in equation (1), we have

2 | $4($____$) + 2y = 2$, so

-6, -3 | $2y =$ ____ or $y =$ ____

{(2, -3)} | The solution set is ________.

Solution:

$$\begin{cases} 2x - y + z = 1 \\ 3x \quad + 2z = -1 \\ 4x + y + 2z = 2 \end{cases}$$

Multiply the second equation by $\frac{1}{3}$ to get

$$\begin{cases} 2x - y + z = 1 \\ x \quad + \frac{2}{3}z = -\frac{1}{3} \\ 4x + y + 2z = 2 \end{cases}$$

Add −2 times the first equation to the third equation to get

$$\begin{cases} 2x - y + z = 1 \\ x \quad + \frac{2}{3}z = -\frac{1}{3} \\ 3y \quad = 0 \end{cases}$$

Add −2 times the second equation to the first equation to get

$$\begin{cases} - y - \frac{1}{3}z = \frac{5}{3} \\ x \quad + \frac{2}{3}z = -\frac{1}{3} \\ 3y \quad = 0 \end{cases}$$

Interchange the first equation with the second equation, and the second equation with the third equation, to get

$$\begin{cases} x \quad + \frac{2}{3}z = -\frac{1}{3} \\ 3y \quad = 0 \\ -y - \frac{1}{3}z = \frac{5}{3} \end{cases}$$

From the second equation, $y = 0$, so that, after substituting in the third equation, $z = -5$, and, after substituting in the first equation, $x = 3$. Hence, the solution set is $\{(3, 0, -5)\}$.

24 $\begin{cases} 2x - y + z = 1 \\ 3x \quad + 2z = -1 \\ 4x + y + 2z = 2 \end{cases}$

2 Matrices and Row Reduction

Consider the system of linear equations (A).

$$\text{(A)} \begin{cases} a_{11}x_1 + a_{12}x_2 + a_{13}x_3 = b_1 \\ a_{21}x_1 + a_{22}x_2 + a_{23}x_3 = b_2 \\ a_{31}x_1 + a_{32}x_2 + a_{33}x_3 = b_3 \end{cases}$$

The matrix form of (A) is written as

$$A = \left[\begin{array}{ccc|c} a_{11} & a_{12} & a_{13} & b_1 \\ a_{21} & a_{22} & a_{23} & b_2 \\ a_{31} & a_{32} & a_{33} & b_3 \end{array}\right]$$

This is called an augmented matrix.

Notice that the system of linear equations (A) was replaced by an augmented matrix A whose elements are the coefficients and constants occurring in the equations. Notice also that we can work with the augmented matrix instead of the actual equations by performing the following operations on the augmented matrix:

1′ Interchange two rows of the matrix ($R_i \leftrightarrow R_j$).
2′ Multiply the elements in a row of the matrix by a nonzero number ($R_i \to kR_i$, where $k \neq 0$).
3′ Replace a row with the sum of a nonzero multiple of itself and a multiple of another row ($R_i \to kR_j + cR_i$, where $c \neq 0$ and $k \neq 0$).

The operations (1′), (2′), and (3′) are called the *elementary row operations*. A matrix is a *row-reduced echelon matrix* if all of the following conditions hold:

(i) The first nonzero entry in each row is 1; all other entries in that column are zeros.
(ii) Any rows that consist entirely of zeros are below all rows that contain a nonzero entry.
(iii) The first nonzero entry in each row is to the right of the first nonzero entry in the preceding row.

In Problems 25–29, find the solution set of the linear system of equations by using the elimination method. Show both the linear system and the corresponding matrix form in each step of the process.

25 $\begin{cases} x + y + z = 6 \\ 4x + 3z = 13 \\ -x - 4z = -13 \end{cases}$

Solution

$$\begin{cases} x + y + z = 6 \\ 4x + 3z = 13 \\ -x - 4z = -13 \end{cases} \qquad \left[\begin{array}{ccc|c} 1 & 1 & 1 & 6 \\ 4 & 0 & 3 & 13 \\ -1 & 0 & -4 & -13 \end{array}\right]$$

Add 4 times the last equation to the second equation and divide the result by -13. | Add 4 times the last row to the second row and divide the result by -13.

$$\begin{cases} x + y + z = 6 \\ z = 3 \\ -x - 4z = -13 \end{cases} \qquad \left[\begin{array}{ccc|c} 1 & 1 & 1 & 6 \\ 0 & 0 & 1 & 3 \\ -1 & 0 & -4 & -13 \end{array}\right]$$

Subtract the second equation from the first and add 4 times the second to the third, then multiply the third equation by -1. | Subtract the second row from the first and add 4 times the second to the third, then multiply the third row by -1.

$$\begin{cases} x + y = 3 \\ z = 3 \\ x = 1 \end{cases} \qquad \left[\begin{array}{ccc|c} 1 & 1 & 0 & 3 \\ 0 & 0 & 1 & 3 \\ 1 & 0 & 0 & 1 \end{array}\right]$$

Subtract the third equation from the first equation. | Subtract the third row from the first row.

$$\begin{cases} y = 2 \\ z = 3 \\ x = 1 \end{cases} \qquad \left[\begin{array}{ccc|c} 0 & 1 & 0 & 2 \\ 0 & 0 & 1 & 3 \\ 1 & 0 & 0 & 1 \end{array}\right]$$

Hence the solution set is $\{(1, 2, 3)\}$.

26 $\begin{cases} 2x + 3y + z = 6 \\ x - 2y + 3z = -3 \\ 3x + y - z = 8 \end{cases}$

Solution:

$$\begin{cases} 2x + 3y + z = 6 \\ x - 2y + 3z = -3 \\ 3x + y - z = 8 \end{cases} \qquad \left[\begin{array}{ccc|c} 2 & 3 & 1 & 6 \\ 1 & -2 & 3 & -3 \\ 3 & 1 & -1 & 8 \end{array}\right]$$

Subtract 3 times the second equation from the third, then multiply the second equation by -2 and add the first to it.

Subtract 3 times the second row from the third, then multiply the second row by -2 and add the first row to it.

$$\begin{cases} 2x + 3y + z = 6 \\ 7y - 5z = 12 \\ 7y - 10z = 17 \end{cases} \qquad \left[\begin{array}{ccc|c} 2 & 3 & 1 & 6 \\ 0 & 7 & -5 & 12 \\ 0 & 7 & -10 & 17 \end{array}\right]$$

Subtract the second equation from the third and divide the resulting third equation by -5.

Subtract the second row from the third and divide the resulting third row by -5.

$$\begin{cases} 2x + 3y + z = 6 \\ 7y - 5z = 12 \\ z = -1 \end{cases} \qquad \left[\begin{array}{ccc|c} 2 & 3 & 1 & 6 \\ 0 & 7 & -5 & 12 \\ 0 & 0 & 1 & -1 \end{array}\right]$$

Add 5 times the third equation to the second and subtract the third from the first. Then multiply the resulting second equation by $\frac{1}{7}$.

Add 5 times the third row to the second row and subtract the third row from the first row. Then multiply the resulting second row by $\frac{1}{7}$.

$$\begin{cases} 2x + 3y = 7 \\ y = 1 \\ z = -1 \end{cases} \qquad \left[\begin{array}{ccc|c} 2 & 3 & 0 & 7 \\ 0 & 1 & 0 & 1 \\ 0 & 0 & 1 & -1 \end{array}\right]$$

Subtract 3 times the second equation from the first; then divide the first by 2.

Subtract 3 times the second row from the first; then divide the first by 2.

$$\begin{cases} x = 2 \\ y = 1 \\ z = -1 \end{cases} \qquad \left[\begin{array}{ccc|c} 1 & 0 & 0 & 2 \\ 0 & 1 & 0 & 1 \\ 0 & 0 & 1 & -1 \end{array}\right]$$

The solution set is $\{(2, 1, -1)\}$.

27 $\begin{cases} x + y + 2z = 4 \\ x + y - 2z = 0 \\ x - y = 0 \end{cases}$

Solution:

$$\begin{cases} x + y + 2z = 4 \\ x + y - 2z = 0 \\ x - y = 0 \end{cases} \qquad \left[\begin{array}{ccc|c} 1 & 1 & 2 & 4 \\ 1 & 1 & -2 & 0 \\ 1 & -1 & 0 & 0 \end{array}\right]$$

Add the first equation to the second; then add 2 times the third equation to the second. Finally, divide the second equation by 4.

Add the first row to the second row; then add 2 times the third row to the second row. Finally, divide the second row by 4.

$$\begin{cases} x + y + 2z = 4 \\ x = 1 \\ x - y = 0 \end{cases} \qquad \left[\begin{array}{ccc|c} 1 & 1 & 2 & 4 \\ 1 & 0 & 0 & 1 \\ 1 & -1 & 0 & 0 \end{array}\right]$$

Subtract the second equation from the third, then multiply the third equation by -1.	Subtract the second row from the third, then multiply the third row by -1.
$\begin{cases} x+y+2z=4 \\ x \qquad\quad =1 \\ \quad y \qquad =1 \end{cases}$	$\left[\begin{array}{ccc\|c} 1 & 1 & 2 & 4 \\ 1 & 0 & 0 & 1 \\ 0 & 1 & 0 & 1 \end{array}\right]$
Subtract the second and third equations from the first, then divide the first by 2.	Subtract the second and third rows from the first, then divide the first by 2.
$\begin{cases} \qquad z=1 \\ x \qquad =1 \\ \quad y \quad =1 \end{cases}$	$\left[\begin{array}{ccc\|c} 0 & 0 & 1 & 1 \\ 1 & 0 & 0 & 1 \\ 0 & 1 & 0 & 1 \end{array}\right]$

Hence, the solution set is $\{(1, 1, 1)\}$.

28 $\begin{cases} x+y+\ z=4 \\ x-y+2z=8 \\ 2x+y-\ z=3 \end{cases}$

Solution:

$\begin{cases} x+y+\ z=4 \\ x-y+2z=8 \\ 2x+y-\ z=3 \end{cases}$	$\left[\begin{array}{ccc\|c} 1 & 1 & 1 & 4 \\ 1 & -1 & 2 & 8 \\ 2 & 1 & -1 & 3 \end{array}\right]$
Subtract 2 times the first equation from the third.	Subtract 2 times the first row from the third.
$\begin{cases} x+y+\ z=4 \\ x-y+2z=8 \\ \ -y-3z=-5 \end{cases}$	$\left[\begin{array}{ccc\|c} 1 & 1 & 1 & 4 \\ 1 & -1 & 2 & 8 \\ 0 & -1 & -3 & -5 \end{array}\right]$
Subtract the first equation from the second, then subtract 2 times the third equation from the second equation. Finally, divide the second equation by 7.	Subtract the first row from the second, then subtract 2 times the third row from the second row. Finally, divide the second row by 7.
$\begin{cases} x+y+\ z=4 \\ \qquad\quad z=2 \\ \ -y-3z=-5 \end{cases}$	$\left[\begin{array}{ccc\|c} 1 & 1 & 1 & 4 \\ 0 & 0 & 1 & 2 \\ 0 & -1 & -3 & -5 \end{array}\right]$
Add 3 times the second equation to the third; then multiply the third by -1.	Add 3 times the second row to the third; then multiply the third by -1.
$\begin{cases} x+y+z=4 \\ \qquad\ z=2 \\ \quad y \quad =-1 \end{cases}$	$\left[\begin{array}{ccc\|c} 1 & 1 & 1 & 4 \\ 0 & 0 & 1 & 2 \\ 0 & 1 & 0 & -1 \end{array}\right]$
Subtract the second and third equations from the first.	Subtract the second and third rows from the first.
$\begin{cases} x \qquad =3 \\ \qquad z=2 \\ \quad y \quad =-1 \end{cases}$	$\left[\begin{array}{ccc\|c} 1 & 0 & 0 & 3 \\ 0 & 0 & 1 & 2 \\ 0 & 1 & 0 & -1 \end{array}\right]$

The solution set is $\{(3, -1, 2)\}$.

Solution:

$$\begin{cases} 2x + y - z = 7 \\ -x + y \quad = 1 \\ -y + z = 1 \end{cases} \qquad \left[\begin{array}{ccc|c} 2 & 1 & -1 & 7 \\ -1 & 1 & 0 & 1 \\ 0 & -1 & 1 & 1 \end{array}\right]$$

Add the third equation to the first; then divide the first equation by 2. | Add the third row to the first; then divide the first by 2.

$$\begin{cases} x \quad = 4 \\ -x + y \quad = 1 \\ -y + z = 1 \end{cases} \qquad \left[\begin{array}{ccc|c} 1 & 0 & 0 & 4 \\ -1 & 1 & 0 & 1 \\ 0 & -1 & 1 & 1 \end{array}\right]$$

Add the first equation to the second; then add the second to the third. | Add the first row to the second; then add the second to the third.

$$\begin{cases} x \quad = 4 \\ y \quad = 5 \\ z = 6 \end{cases} \qquad \left[\begin{array}{ccc|c} 1 & 0 & 0 & 4 \\ 0 & 1 & 0 & 5 \\ 0 & 0 & 1 & 6 \end{array}\right]$$

The solution set is $\{(4, 5, 6)\}$.

29
$$\begin{cases} 2x + y - z = 7 \\ -x + y \quad = 1 \\ -y + z = 1 \end{cases}$$

3 Determinants

real number

30 The determinant is a function that associates a ____________ with each square matrix.

$a_{11}a_{22} - a_{21}a_{12}$

$a_{11}a_{22} - a_{21}a_{12}$

31 If A is a square matrix, a determinant of A is denoted by det A or by $|A|$. The determinant of the 2 × 2 matrix

$$A = \begin{bmatrix} a_{11} & a_{12} \\ a_{21} & a_{22} \end{bmatrix}$$

is defined to be the number ________________, and we write

$$|A| = \det A = \det \begin{bmatrix} a_{11} & a_{12} \\ a_{21} & a_{22} \end{bmatrix} = \begin{vmatrix} a_{11} & a_{12} \\ a_{21} & a_{22} \end{vmatrix} = \underline{\qquad\qquad}$$

In Problems 32–37, evaluate each of the given determinants.

3, -2, 10

32 $\begin{vmatrix} 1 & -2 \\ 3 & 4 \end{vmatrix} = 1(4) - (\_\_\_\_)(\_\_\_\_) = \_\_\_\_$

-4, 1, 13

33 $\begin{vmatrix} 3 & 1 \\ -4 & 3 \end{vmatrix} = 3(3) - (\_\_\_\_)(\_\_\_\_) = \_\_\_\_$

14, -29

34 $\begin{vmatrix} 5 & 7 \\ 2 & -3 \end{vmatrix} = -15 - \_\_\_\_ = \_\_\_\_$

48, -23

35 $\begin{vmatrix} 5 & 24 \\ 2 & 5 \end{vmatrix} = 25 - \_\_\_\_ = \_\_\_\_$

-72, 35, -107

36 $\begin{vmatrix} 24 & 7 \\ 5 & -3 \end{vmatrix} =$ ____ - ____ = ____

-9*mb*, 20*an*

37 $\begin{vmatrix} 9m & -4n \\ -5a & -b \end{vmatrix} =$ _____ - _____

38 We define the determinant of the 3 X 3 matrix

$$A = \begin{bmatrix} a_{11} & a_{12} & a_{13} \\ a_{21} & a_{22} & a_{23} \\ a_{31} & a_{32} & a_{33} \end{bmatrix}$$

by $|A| = a_{11}a_{22}a_{33} - a_{11}a_{32}a_{23} - a_{12}a_{21}a_{33}$

+ ________ + ________ - ________.

$a_{12}a_{31}a_{23}$, $a_{13}a_{21}a_{32}$, $a_{13}a_{31}a_{22}$

39 The determinant of A is given by the equation

$$|A| = \begin{vmatrix} a_{11} & a_{12} & a_{13} \\ a_{21} & a_{22} & a_{23} \\ a_{31} & a_{32} & a_{33} \end{vmatrix}$$

$$= \_\_\_\_ \begin{vmatrix} a_{22} & a_{23} \\ a_{32} & a_{33} \end{vmatrix} - \_\_\_\_ \begin{vmatrix} a_{21} & a_{23} \\ a_{31} & a_{33} \end{vmatrix} + \_\_\_\_ \begin{vmatrix} a_{21} & a_{22} \\ a_{31} & a_{32} \end{vmatrix}$$

a_{11}, a_{12}, a_{13}

In Problems 40–45, find the value of the given determinant.

$\begin{vmatrix} 9 & 4 \\ 6 & -2 \end{vmatrix}$, $\begin{vmatrix} 5 & 9 \\ 7 & 6 \end{vmatrix}$

40 $\begin{vmatrix} 2 & -1 & -3 \\ 5 & 9 & 4 \\ 7 & 6 & -2 \end{vmatrix} = 2$ ______ $- (-1) \begin{vmatrix} 5 & 4 \\ 7 & -2 \end{vmatrix} - 3$ ______

= ____ + (____) + ____

= ____

-84, -38, 99

-23

$\begin{vmatrix} 1 & -2 \\ 3 & 4 \end{vmatrix}$

41 $\begin{vmatrix} 2 & 3 & 1 \\ 1 & 5 & -2 \\ 3 & -4 & 4 \end{vmatrix} = 2 \begin{vmatrix} 5 & -2 \\ -4 & 4 \end{vmatrix} - 3$ ______ $+ 1 \begin{vmatrix} 1 & 5 \\ 3 & -4 \end{vmatrix}$

= ____ + (____) + (____)

= ____

24, -30, -19

-25

42 $\begin{vmatrix} 4 & 3 & 1 \\ -1 & 5 & -2 \\ -1 & -4 & 4 \end{vmatrix} =$ 4(____) - 3(____) + 1(____)

12, -6, 9

= ____

75

43 $\begin{vmatrix} 2 & 4 & 1 \\ 1 & -1 & -2 \\ 2 & 3 & 4 \end{vmatrix} =$ 2(____) - 4(____) + 1(____)

2, 8, 5

= ____

-23

44 $\begin{vmatrix} 2 & 3 & 4 \\ 1 & 5 & -1 \\ 3 & -4 & -1 \end{vmatrix} =$ 2(____) - 3(____) + 4(____)

-9, 2, -19

= _____

-100

Answers	Problems
0, 0, 0 0	**45** $\begin{vmatrix} 3 & 1 & 2 \\ 0 & 0 & 0 \\ 1 & 4 & -3 \end{vmatrix} = 3(\_\_\_) - 1(\_\_\_) + 2(\_\_\_)$ $= \_\_\_$

In Problems 46–48, solve for x.

Answers	Problems
$3x + 2x$ 15, 3	**46** $\begin{vmatrix} x & 2x \\ -1 & 3 \end{vmatrix} = 15$ The left side of the equation is ________, so that $5x = \_\_\_$ or $x = \_\_\_$.
$2x - x^2 - 20 = -7x$ $x^2 - 9x + 20$ $x - 5$, 4, 5	**47** $\begin{vmatrix} x & 5 \\ 4 & 2-x \end{vmatrix} = -7x$ Expanding the left side of the equation, we obtain ________________, so that the equation becomes ____________ = 0. By factoring, we get $(x - 4)(\_\_\_) = 0$. Therefore, $x = \_\_\_$ or $x = \_\_\_$.
1, 0, −5, 7 32	**48** $\begin{vmatrix} x & 4 & 5 \\ 0 & 1 & 0 \\ 5 & 2 & 1 \end{vmatrix} = 7$ Expanding the left side of the equation, we obtain $x(\_\_\_) - 4(\_\_\_) + 5(\_\_\_) = 7$, so that $x - 25 = \_\_\_$. Therefore, $x = \_\_\_$.

3.1 Properties of Determinants

Answers	Problems
factored k	**49** A common factor that appears in all entries in a row (or a column) of a determinant can be ________ out of the determinant. Expressed in symbols, we have $\begin{vmatrix} ka_{11} & ka_{12} \\ a_{21} & a_{22} \end{vmatrix} = \_\_\_ \begin{vmatrix} a_{11} & a_{12} \\ a_{21} & a_{22} \end{vmatrix}$
5	**50** $\begin{vmatrix} 5 & 25 \\ 3 & 4 \end{vmatrix} = \_\_\_ \begin{vmatrix} 1 & 5 \\ 3 & 4 \end{vmatrix}$
$\begin{vmatrix} 1 & 2 \\ 5 & 7 \end{vmatrix}$	**51** $\begin{vmatrix} 6 & 12 \\ 5 & 7 \end{vmatrix} = 6 \_\_\_\_\_$
15	**52** $\begin{vmatrix} 15 & 1 & 5 \\ 45 & 2 & 7 \\ 60 & 3 & -2 \end{vmatrix} = \_\_\_ \begin{vmatrix} 1 & 1 & 5 \\ 3 & 2 & 7 \\ 4 & 3 & -2 \end{vmatrix}$
$\begin{vmatrix} a_{21} & a_{22} \\ a_{11} & a_{12} \end{vmatrix}$	**53** If two rows of a square matrix are interchanged, the values of the determinants of the two matrices differ only in the algebraic sign. Expressed in symbols, we have $\begin{vmatrix} a_{11} & a_{12} \\ a_{21} & a_{22} \end{vmatrix} = - \_\_\_\_\_$

$\begin{vmatrix} 5 & 4 \\ 3 & 2 \end{vmatrix}$

54 $\begin{vmatrix} 3 & 2 \\ 5 & 4 \end{vmatrix} = -$ ________

14

55 If $\begin{vmatrix} 2 & 5 \\ 4 & 3 \end{vmatrix} = -14$, then $\begin{vmatrix} 4 & 3 \\ 2 & 5 \end{vmatrix} =$ ____.

-15

56 If $\begin{vmatrix} -3 & 1 \\ 0 & -5 \end{vmatrix} = 15$, then $\begin{vmatrix} 0 & -5 \\ -3 & 1 \end{vmatrix} =$ ____.

-2

57 If $\begin{vmatrix} 9 & 2 \\ -1 & 0 \end{vmatrix} = 2$, then $\begin{vmatrix} -1 & 0 \\ 9 & 2 \end{vmatrix} =$ ____.

-3

58 If $\begin{vmatrix} 1 & -1 & 0 \\ 3 & 0 & 4 \\ 2 & 1 & 5 \end{vmatrix} = 3$, then $\begin{vmatrix} 2 & 1 & 5 \\ 3 & 0 & 4 \\ 1 & -1 & 0 \end{vmatrix} =$ ____.

unchanged

59 If any nonzero multiple of one row is added to any other row of a square matrix, the value of the determinant is ____________.

60 Replace $\begin{vmatrix} 3 & 2 & 0 \\ 4 & 4 & 1 \\ 1 & -1 & 3 \end{vmatrix}$ by an equivalent 2 X 2 determinant and then evaluate.

8, -11

$-\frac{17}{8}$

$$\begin{vmatrix} 3 & 2 & 0 \\ 4 & 4 & 1 \\ 1 & -1 & 3 \end{vmatrix} = -\begin{vmatrix} 1 & -1 & 3 \\ 0 & \_\_ & \_\_ \\ 0 & 0 & \_\_ \end{vmatrix}$$ results from these operations:

(i) $R_1 \leftrightarrow R_3$

(ii) Add $-4R_1$ to R_2

(iii) Add $-3R_1$ to R_3

(iv) Add $-\frac{5}{8}R_2$ to R_3

1, $\begin{vmatrix} 8 & -11 \\ 0 & -\frac{17}{8} \end{vmatrix}$, 17

so that $\begin{vmatrix} 3 & 2 & 0 \\ 4 & 4 & 1 \\ 1 & -1 & 3 \end{vmatrix} = -(\_\_\_\_)$ ________ $=$ ____

3.2 Cramer's Rule

61 The solution of the system of equations

$$\begin{cases} a_{11}x_1 + a_{12}x_2 = c_1 \\ a_{21}x_1 + a_{22}x_2 = c_2 \end{cases}$$

by Cramer's rule is

$$x_1 = \frac{D_1}{D} \text{ and } x_2 = \frac{D_2}{D}$$

$\begin{vmatrix} a_{11} & a_{12} \\ a_{21} & a_{22} \end{vmatrix}$

where $D =$ ________ and $D \neq 0$

$\begin{vmatrix} c_1 & a_{12} \\ c_2 & a_{22} \end{vmatrix}$

$D_1 =$ ________

$\begin{vmatrix} a_{11} & c_1 \\ a_{21} & c_2 \end{vmatrix}$

$D_2 =$ ________

In Problems 62–65, use Cramer's rule to solve the given linear system.

62 $\begin{cases} 2x_1 - 3x_2 = 3 \\ x_1 + 4x_2 = 7 \end{cases}$

11 $D = \begin{vmatrix} 2 & -3 \\ 1 & 4 \end{vmatrix} = \_\_\_\_$

33 $D_1 = \begin{vmatrix} 3 & -3 \\ 7 & 4 \end{vmatrix} = \_\_\_\_$

11 $D_2 = \begin{vmatrix} 2 & 3 \\ 1 & 7 \end{vmatrix} = \_\_\_\_$

3, 1 $x_1 = \dfrac{D_1}{D} = \_\_\_\_$ and $x_2 = \dfrac{D_2}{D} = \_\_\_\_$

63 $\begin{cases} 5x + 7y = 24 \\ 2x - 3y = 5 \end{cases}$

-29 $D = \begin{vmatrix} 5 & 7 \\ 2 & -3 \end{vmatrix} = \_\_\_\_$

-107 $D_1 = \begin{vmatrix} 24 & 7 \\ 5 & -3 \end{vmatrix} = \_\_\_\_$

-23 $D_2 = \begin{vmatrix} 5 & 24 \\ 2 & 5 \end{vmatrix} = \_\_\_\_$

$\dfrac{107}{29}, \dfrac{23}{29}$ $x = \dfrac{D_1}{D} = \_\_\_\_$ and $y = \dfrac{D_2}{D} = \_\_\_\_$

64 $\begin{cases} 2x + 3y = -2 \\ x + 5y = 3 \end{cases}$

7 $D = \begin{vmatrix} 2 & 3 \\ 1 & 5 \end{vmatrix} = \_\_\_\_$

-19 $D_1 = \begin{vmatrix} -2 & 3 \\ 3 & 5 \end{vmatrix} = \_\_\_\_$

8 $D_2 = \begin{vmatrix} 2 & -2 \\ 1 & 3 \end{vmatrix} = \_\_\_\_$

$-\dfrac{19}{7}, \dfrac{8}{7}$ $x = \dfrac{D_1}{D} = \_\_\_\_$ and $y = \dfrac{D_2}{D} = \_\_\_\_$

65 $\begin{cases} 3x - y = 14 \\ \dfrac{x - 2y}{3} = 2 \end{cases}$

Rewrite the second equation without fractions to obtain the equivalent system:

$x - 2y = 6$ $\begin{cases} 3x - y = 14 \\ \_\_\_\_\_\_\_\_ \end{cases}$

-5 $D = \begin{vmatrix} 3 & -1 \\ 1 & -2 \end{vmatrix} = \_\_\_\_$

-22 $D_1 = \begin{vmatrix} 14 & -1 \\ 6 & -2 \end{vmatrix} = \_\_\_\_$

4

$$D_2 = \begin{vmatrix} 3 & 14 \\ 1 & 6 \end{vmatrix} = \_\_\_\_$$

$\frac{22}{5}, -\frac{4}{5}$

$x = \frac{D_1}{D} = \_\_\_\_$ and $y = \frac{D_2}{D} = \_\_\_\_$

66 The linear system in three unknowns

$$\begin{cases} a_{11}x_1 + a_{12}x_2 + a_{13}x_3 = c_1 \\ a_{21}x_1 + a_{22}x_2 + a_{23}x_3 = c_2 \\ a_{31}x_1 + a_{32}x_2 + a_{33}x_3 = c_3 \end{cases}$$

has the solution $x_1 = \frac{D_1}{D}$, $x_2 = \frac{D_2}{D}$, and $x_3 = \frac{D_3}{D}$

$\begin{vmatrix} a_{11} & a_{12} & a_{13} \\ a_{21} & a_{22} & a_{23} \\ a_{31} & a_{32} & a_{33} \end{vmatrix}$

where $D =$ ________________

$\begin{vmatrix} c_1 & a_{12} & a_{13} \\ c_2 & a_{22} & a_{23} \\ c_3 & a_{32} & a_{33} \end{vmatrix}$

$D_1 =$ ________________

$\begin{vmatrix} a_{11} & c_1 & a_{13} \\ a_{21} & c_2 & a_{23} \\ a_{31} & c_3 & a_{33} \end{vmatrix}$

$D_2 =$ ________________

$\begin{vmatrix} a_{11} & a_{12} & c_1 \\ a_{21} & a_{22} & c_2 \\ a_{31} & a_{32} & c_3 \end{vmatrix}$

$D_3 =$ ________________

In Problems 67–70, use Cramer's rule to solve the given linear system.

67 $$\begin{cases} 2x_1 + 3x_2 + x_3 = 4 \\ x_1 + 5x_2 - 2x_3 = -1 \\ 3x_1 - 4x_2 + 4x_3 = -1 \end{cases}$$

-25

$$D = \begin{vmatrix} 2 & 3 & 1 \\ 1 & 5 & -2 \\ 3 & -4 & 4 \end{vmatrix} = \_\_\_\_$$

75

$$D_1 = \begin{vmatrix} 4 & 3 & 1 \\ -1 & 5 & -2 \\ -1 & -4 & 4 \end{vmatrix} = \_\_\_\_$$

-50

$$D_2 = \begin{vmatrix} 2 & 4 & 1 \\ 1 & -1 & -2 \\ 3 & -1 & 4 \end{vmatrix} = \_\_\_\_$$

-100

$$D_3 = \begin{vmatrix} 2 & 3 & 4 \\ 1 & 5 & -1 \\ 3 & -4 & -1 \end{vmatrix} = \_\_\_\_$$

-3, 2, 4

$x_1 = \frac{D_1}{D} = \_\_\_\_$ $x_2 = \frac{D_2}{D} = \_\_\_\_$ $x_3 = \frac{D_3}{D} = \_\_\_\_$

68 $\begin{cases} 3x - y - 4z = 7 \\ 2x + 3y + 5z = 8 \\ 5x - 2y - 6z = 10 \end{cases}$

15 $D = \begin{vmatrix} 3 & -1 & -4 \\ 2 & 3 & 5 \\ 5 & -2 & -6 \end{vmatrix} = \underline{\qquad}$

30 $D_1 = \begin{vmatrix} 7 & -1 & -4 \\ 8 & 3 & 5 \\ 10 & -2 & -6 \end{vmatrix} = \underline{\qquad}$

45 $D_2 = \begin{vmatrix} 3 & 7 & -4 \\ 2 & 8 & 5 \\ 5 & 10 & -6 \end{vmatrix} = \underline{\qquad}$

-15 $D_3 = \begin{vmatrix} 3 & -1 & 7 \\ 2 & 3 & 8 \\ 5 & -2 & 10 \end{vmatrix} = \underline{\qquad}$

2, 3, -1 $x = \dfrac{D_1}{D} = \underline{\qquad}$ $y = \dfrac{D_2}{D} = \underline{\qquad}$ $z = \dfrac{D_3}{D} = \underline{\qquad}$

69 $\begin{cases} x + y + z = 2 \\ 2x - y + z = 0 \\ x + 2y - z = 4 \end{cases}$

7 $D = \begin{vmatrix} 1 & 1 & 1 \\ 2 & -1 & 1 \\ 1 & 2 & -1 \end{vmatrix} = \underline{\qquad}$

6 $D_1 = \begin{vmatrix} 2 & 1 & 1 \\ 0 & -1 & 1 \\ 4 & 2 & -1 \end{vmatrix} = \underline{\qquad}$

10 $D_2 = \begin{vmatrix} 1 & 2 & 1 \\ 2 & 0 & 1 \\ 1 & 4 & -1 \end{vmatrix} = \underline{\qquad}$

-2 $D_3 = \begin{vmatrix} 1 & 1 & 2 \\ 2 & -1 & 0 \\ 1 & 2 & 4 \end{vmatrix} = \underline{\qquad}$

$\frac{6}{7}, \frac{10}{7}, -\frac{2}{7}$ $x = \dfrac{D_1}{D} = \underline{\qquad}$ $y = \dfrac{D_2}{D} = \underline{\qquad}$ $z = \dfrac{D_3}{D} = \underline{\qquad}$

70 $\begin{cases} 2x + y - 3z = 0 \\ 3x - 2y + 4z = 5 \\ 4x - y - 2z = 1 \end{cases}$

23 $D = \begin{vmatrix} 2 & 1 & -3 \\ 3 & -2 & 4 \\ 4 & -1 & -2 \end{vmatrix} = \underline{\qquad}$

23 $D_1 = \begin{vmatrix} 0 & 1 & -3 \\ 5 & -2 & 4 \\ 1 & -1 & -2 \end{vmatrix} =$ ____

23 $D_2 = \begin{vmatrix} 2 & 0 & -3 \\ 3 & 5 & 4 \\ 4 & 1 & -2 \end{vmatrix} =$ ____

23 $D_3 = \begin{vmatrix} 2 & 1 & 0 \\ 3 & -2 & 5 \\ 4 & -1 & 1 \end{vmatrix} =$ ____

1, 1, 1 $x = \dfrac{D_1}{D} =$ ____ $y = \dfrac{D_2}{D} =$ ____ $z = \dfrac{D_3}{D} =$ ____

71 A man said to his son, "I am now three times as old as you, but in 10 years I shall be only twice as old as you." Find the age of each. Let x years be the man's age and y years be the son's age. The values of x and y are found by solving the system:

$x = 3y$
$x + 10 = 2(y + 10)$

$\begin{cases} \_\_\_\_\_\_\_\_ \\ \_\_\_\_\_\_\_\_ \end{cases}$

The system is written as:

$$\begin{cases} x - 3y = 0 \\ x - 2y = 10 \end{cases}$$

1 $D = \begin{vmatrix} 1 & -3 \\ 1 & -2 \end{vmatrix} =$ ____

30 $D_1 = \begin{vmatrix} 0 & -3 \\ 10 & -2 \end{vmatrix} =$ ____

10 $D_2 = \begin{vmatrix} 1 & 0 \\ 1 & 10 \end{vmatrix} =$ ____

30, 10 $x = \dfrac{D_1}{D} =$ ____ and $y = \dfrac{D_2}{D} =$ ____

30, 10 Thus, the man is ____ years old and the son is ____ years old.

72 The sum of two numbers is 12. If one of the numbers is multiplied by 5 and the other by 8, the sum of the products is 75. Find the numbers.
Let x be the first number and y be the second number. Then the values of x and y are found by solving the system:

$x + y = 12$
$5x + 8y = 75$

$\begin{cases} \_\_\_\_\_\_\_\_ \\ \_\_\_\_\_\_\_\_ \end{cases}$

3 $D = \begin{vmatrix} 1 & 1 \\ 5 & 8 \end{vmatrix} =$ ____

21 $D_1 = \begin{vmatrix} 12 & 1 \\ 75 & 8 \end{vmatrix} =$ ____

15 $D_2 = \begin{vmatrix} 1 & 12 \\ 5 & 75 \end{vmatrix} = \_\_\_\_$

7, 5 $x = \dfrac{D_1}{D} = \_\_\_\_$ and $y = \dfrac{D_2}{D} = \_\_\_\_$

7, 5 Therefore the numbers are ____ and ____.

73 A community organization distributed 500 pounds of mixed candy in its "Christmas cheer" boxes. Part of the candy was bought at 15 cents a pound and part at 20 cents a pound. How many pounds were bought at each price if the total cost was \$91.25?

Let x be the number of pounds bought at 15 cents per pound and y be the number of pounds bought at 20 cents per pound. The values of x and y are found by solving the following system:

$x + y = 500$
$15x + 20y = 9125$

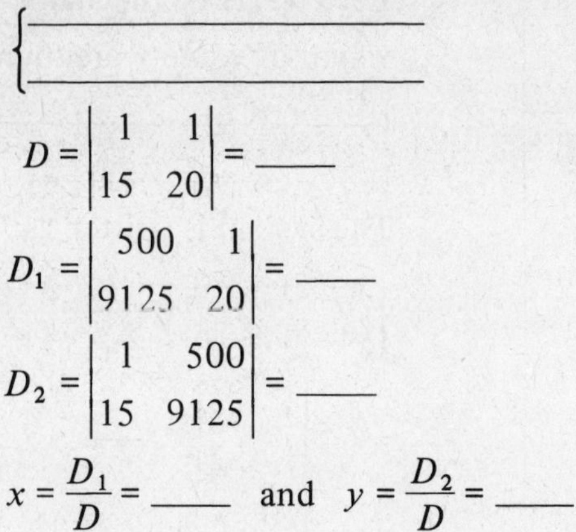

5 $D = \begin{vmatrix} 1 & 1 \\ 15 & 20 \end{vmatrix} = \_\_\_\_$

875 $D_1 = \begin{vmatrix} 500 & 1 \\ 9125 & 20 \end{vmatrix} = \_\_\_\_$

1625 $D_2 = \begin{vmatrix} 1 & 500 \\ 15 & 9125 \end{vmatrix} = \_\_\_\_$

175, 325 $x = \dfrac{D_1}{D} = \_\_\_\_$ and $y = \dfrac{D_2}{D} = \_\_\_\_$

15 cents Therefore, 175 pounds were bought at ________ a pound
20 cents and 325 pounds at ________ a pound.

74 The sum of the angles of any plane triangle is 180°. If one angle is twice another and the third is equal to one-fourth the sum of the first two, what are the angle measures?

Assume that the angles of the plane triangle are x, y, and z degrees. Then x, y, and z are found from solving the system:

$x + y + z = 180$
$x - 2y = 0$
$x + y - 4z = 0$

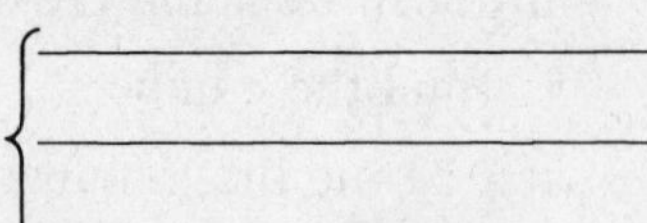

Using Cramer's rule,

15 $D = \begin{vmatrix} 1 & 1 & 1 \\ 1 & -2 & 0 \\ 1 & 1 & -4 \end{vmatrix} = \_\_\_\_$

1440 $D_1 = \begin{vmatrix} 180 & 1 & 1 \\ 0 & -2 & 0 \\ 0 & 1 & -4 \end{vmatrix} = \_\_\_\_$

720

$$D_2 = \begin{vmatrix} 1 & 180 & 1 \\ 1 & 0 & 0 \\ 1 & 0 & -4 \end{vmatrix} = \_\_\_\_$$

540

$$D_3 = \begin{vmatrix} 1 & 1 & 180 \\ 1 & -2 & 0 \\ 1 & 1 & 0 \end{vmatrix} = \_\_\_\_$$

96, 48, 36

$x = \dfrac{D_1}{D} = \_\_\_\_ \qquad y = \dfrac{D_2}{D} = \_\_\_\_ \qquad z = \dfrac{D_3}{D} = \_\_\_\_$

96°, 48°, 36°

Thus the angle measures of the triangle are ____, ____, and ____.

4 Systems of Linear Inequalities and Linear Programming

4.1 Systems of Linear Inequalities

half planes

75 The graph of a linear equation in two variables divides the plane into two disjoint ________________.

linear equation

76 The graph of a linear inequality in two variables in a plane consists of one of the two disjoint half planes (regions) determined by the graph of the associated ________________________.

one

77 To determine which region is the graph of a linear inequality, it is sufficient to test ____ point in either region.

In Problems 78–79, sketch the region containing the points that satisfy the given inequality.

satisfies
containing

78 $x > 3$

First, we sketch the graph of the associated linear equation $x = 3$. Next, we test one point, say, (4, 1), to determine the solution region. Since the point (4, 1) ____________ the inequality $x > 3$, the region ________________ the point (4, 1) is the solution region. The graph is

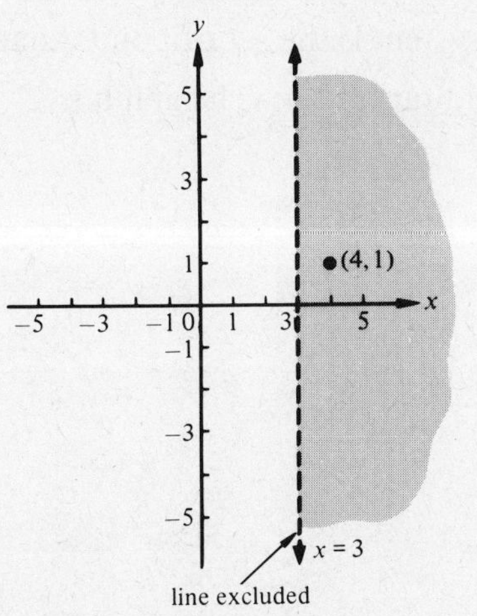

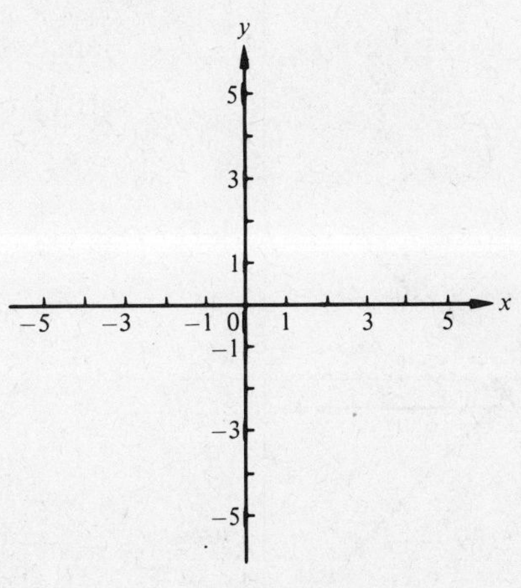

79 $y \geqslant x + 1$

First, we sketch the graph of the associated linear equation ____________. Next, we test one point not on the graph of the linear equation, say (4, 2), to determine the solution region. Since for $x = 4$ and $y = 2$, we have $2 \ngeqslant 4 + 1$, so that the solution region is the region ____________________ the point (4, 2). The graph is

$y = x + 1$

not containing

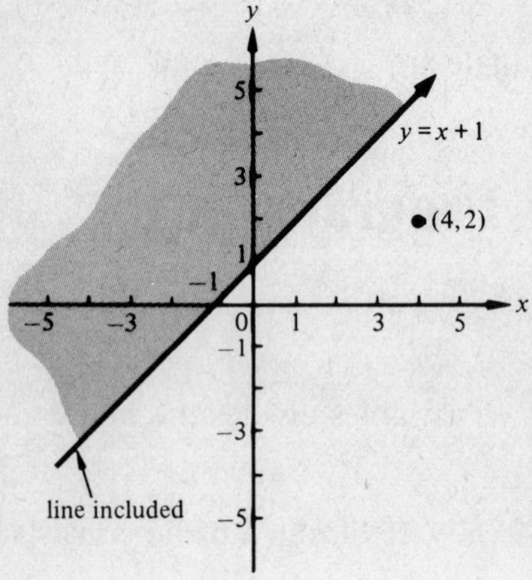

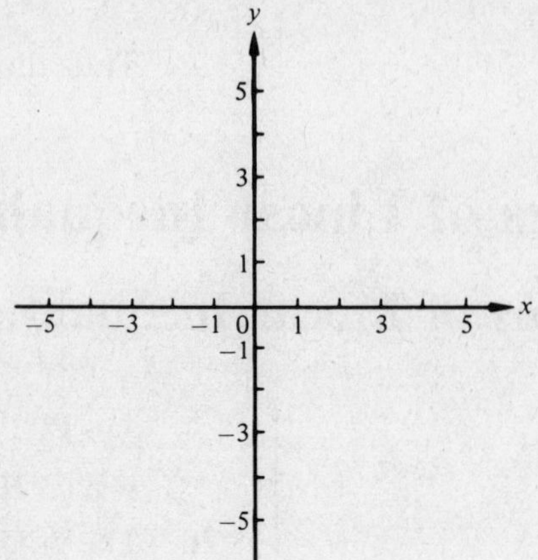

80 The region of points whose coordinates satisfy a system of linear inequalities can be found by finding the ____________________ of the regions that are the graphs of the respective inequalities in the system.

intersection

In Problems 81–84, sketch the region determined by the given system and find the corner points of the boundary of the region.

81 $\begin{cases} y > x - 2 \\ y > -x + 3 \end{cases}$

The solution set of $y > x - 2$ consists of all points that lie above the line ____________, and the solution set of $y > -x + 3$ consists of all points that lie above the line ________________. Hence, the solution set of the given system is the set of points that satisfy both __________________ simultaneously. The graph is

$y = x - 2$

$y = -x + 3$

inequalities

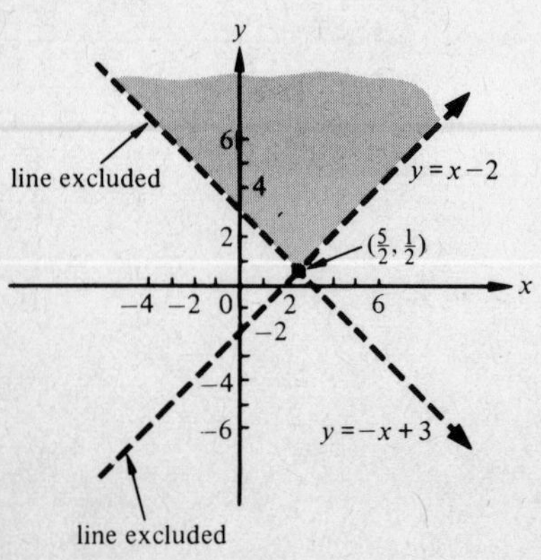

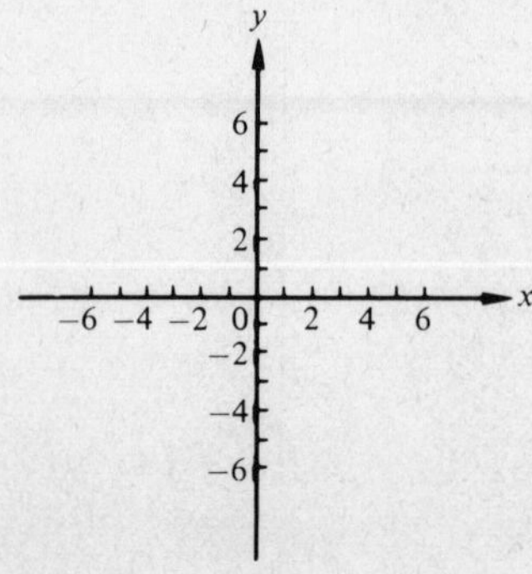

The corner point of the boundary can be found by solving the system

$$\begin{cases} y = x - 2 \\ y = -x + 3 \end{cases}$$

to get __________.

$(\frac{5}{2}, \frac{1}{2})$

82 $\begin{cases} y \leqslant 2x \\ y > -3 \end{cases}$

The solution set of $y \leqslant 2x$ consists of all points that lie below or coincide with the line ________ and the solution set of $y > -3$ consists of all points that lie ________ the line $y = -3$. Hence, the solution set of the given system consists of all points that satisfy both inequalities simultaneously. The graph is

$y = 2x$

above

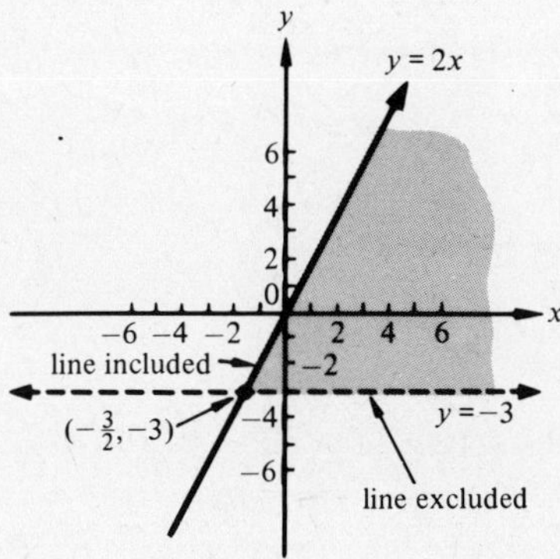

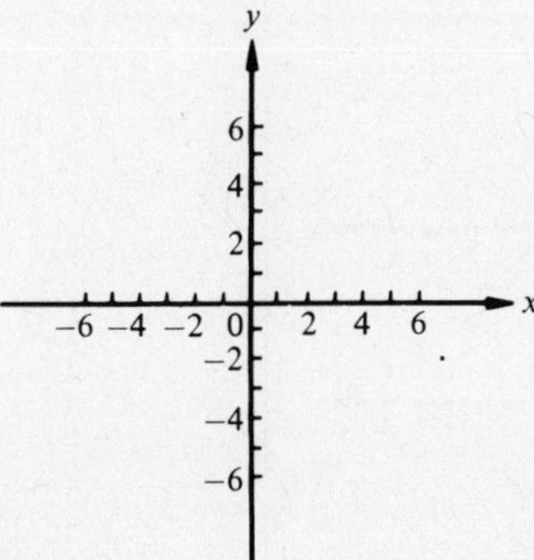

The corner point of the boundary can be found by solving the system

$$\begin{cases} y = 2x \\ y = -3 \end{cases}$$

to get __________.

$(-\frac{3}{2}, -3)$

83 $\begin{cases} 2x + \ \ y \geqslant 2 \\ -x + 3y \leqslant 4 \end{cases}$

The solution set of $2x + y \geqslant 2$ consists of all points that lie above or coincide with the line __________, and the solution set of $-x + 3y \leqslant 4$ consists of all points that lie below or coincide with the line __________. Hence, the solution set of the given system consists of all points that satisfy ________ inequalities simultaneously. The graph is

$2x + y = 2$

$-x + 3y = 4$

both

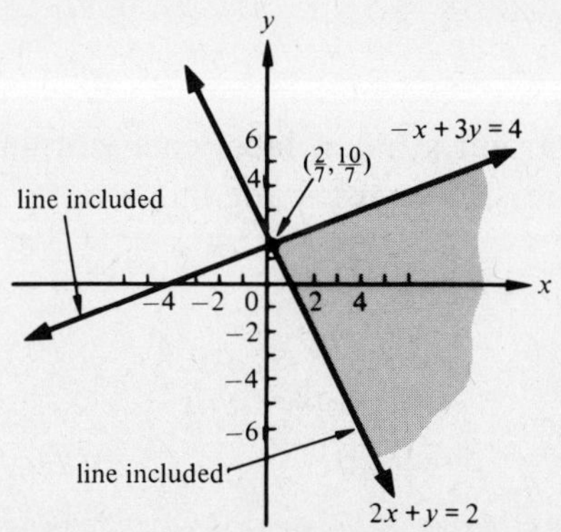

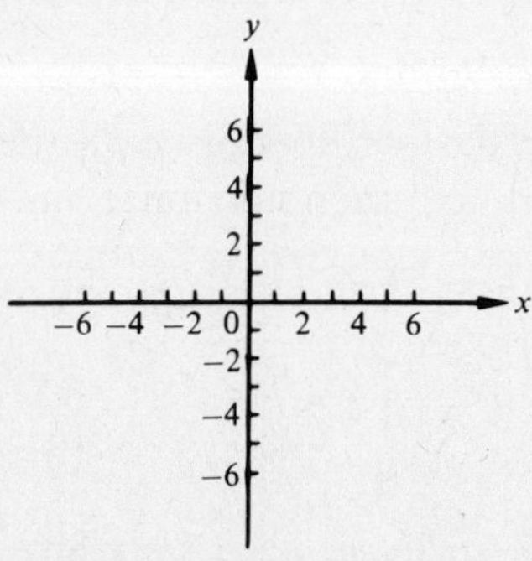

The corner point of the boundary can be found by solving the system

$$\begin{cases} 2x + y = 2 \\ -x + 3y = 4 \end{cases}$$

to get ______________.

$(\frac{2}{7}, \frac{10}{7})$

84 $\begin{cases} y < -6 \\ x < -3 \end{cases}$

The solution set of this system consists of all points that lie below the line ________ and to the left of the line ________. The graph is

$y = -6$, $x = -3$

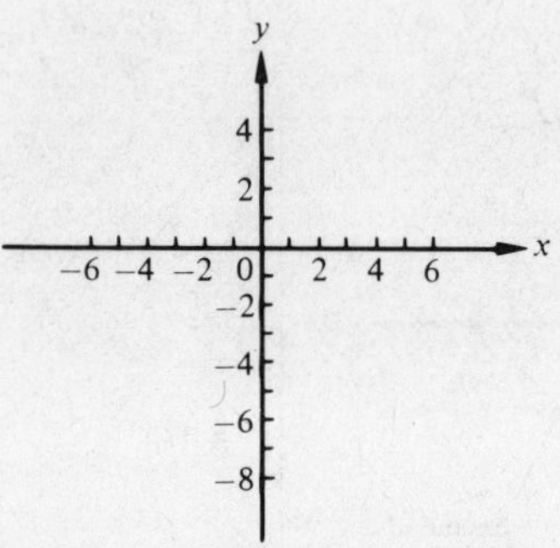

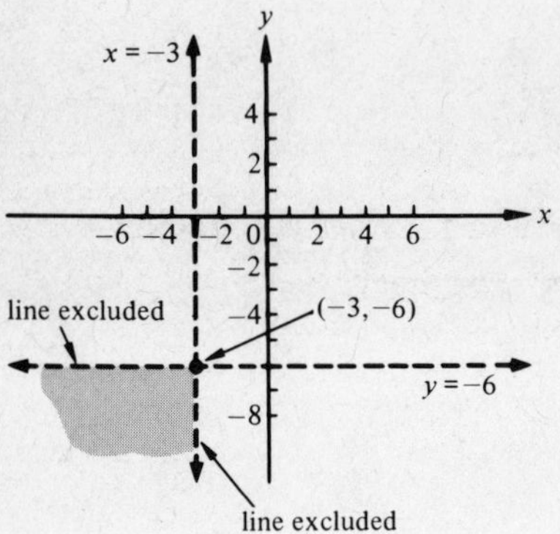

The corner point of the boundary can be found by solving the system

$$\begin{cases} y = -6 \\ x = -3 \end{cases}$$

to get ______________.

(-3, -6)

4.2 Linear Programming

85 The theory of linear programming establishes the fact that the maximum or minimum value of a linear expression occurs at a ________________ of the boundary of the constraint set.

corner point

In Problems 86–88, graph the region defined by each constraint system, label each corner point, and then maximize and minimize the given linear expression over the region.

86 $2x + y$

$$\begin{cases} 0 \leqslant x \leqslant 4 \\ 0 \leqslant y \leqslant 3 \end{cases}$$

The graph is

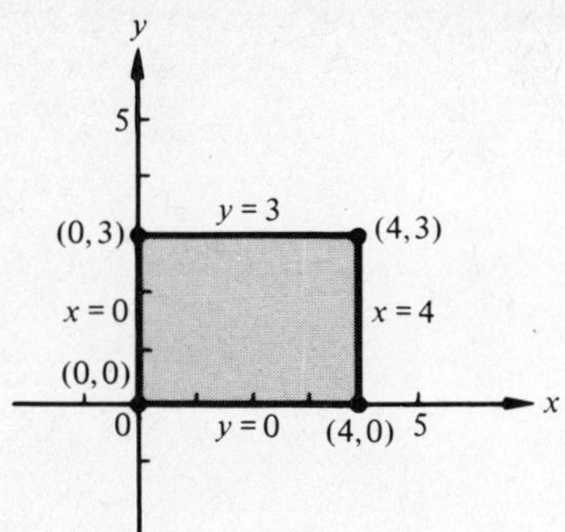

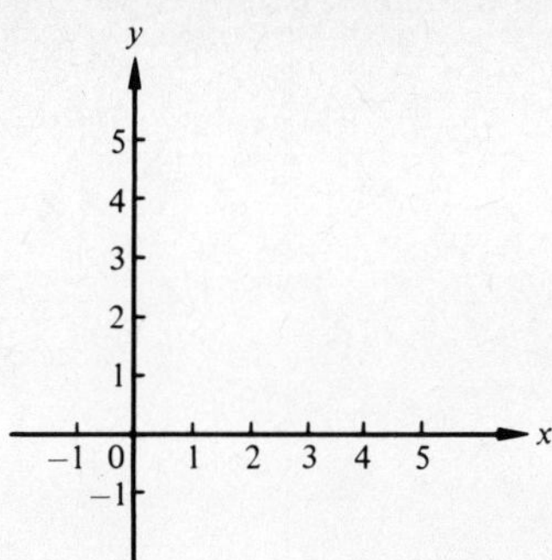

The following table indicates the value of $2x + y$ for each corner point.

corner point	value of $2x + y$
(0, 0)	$2(0) + 0 =$ ____
(0, 3)	$2(0) + 3 =$ ____
(4, 0)	$2(4) + 0 =$ ____
(4, 3)	$2(4) + 3 =$ ____

0
3
8
11

Thus, the maximum value of $2x + y$ is ____, and this occurs when $x =$ ____ and $y =$ ____. The minimum value of $2x + y$ is ____, which occurs when $x =$ ____ and $y =$ ____.

11
4, 3, 0
0, 0

87 $x + 3y$

$$\begin{cases} x \geqslant 0 \\ y \geqslant 0 \\ y \leqslant -\frac{1}{2}x + 2 \\ y \leqslant -2x + 4 \end{cases}$$

The graph is

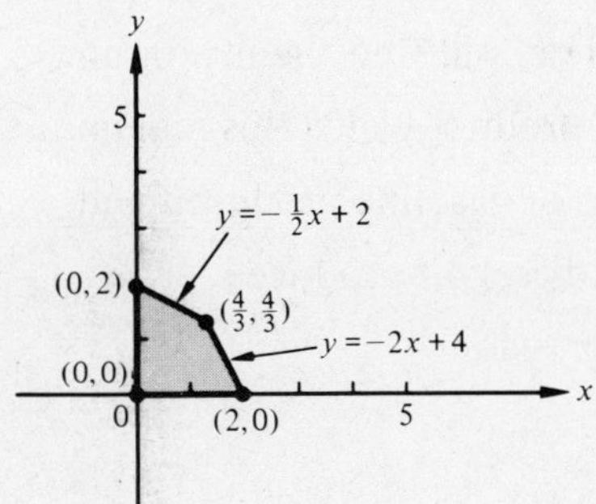

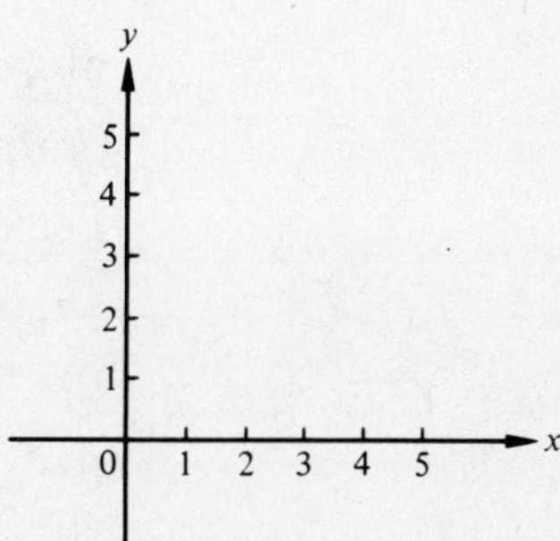

The following table indicates the value of $x + 3y$ for each corner point.

corner point	value of $x + 3y$
(0, 0)	$0 + 3(0) =$ ____
(0, 2)	$0 + 3(2) =$ ____
$(\frac{4}{3}, \frac{4}{3})$	$\frac{4}{3} + 3(\frac{4}{3}) =$ ____
(2, 0)	$2 + 3(0) =$ ____

0
6
$5\frac{1}{3}$
2

Thus, the maximum value of $x + 3y$ is ____, and this occurs when $x =$ ____ and $y =$ ____. The minimum value of $x + 3y$ is ____, which occurs when $x =$ ____ and $y =$ ____.

6
0, 2
0, 0, 0

88 $x + 2y$

$$\begin{cases} y \geqslant 0 \\ x + y \geqslant 1 \\ x + 2y \leqslant 2 \end{cases}$$

The graph is

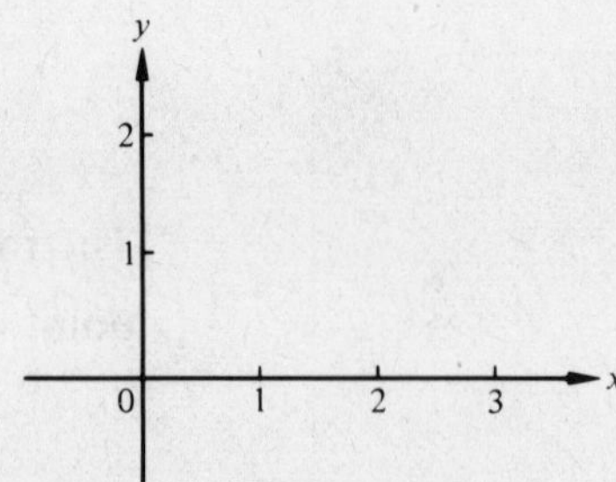

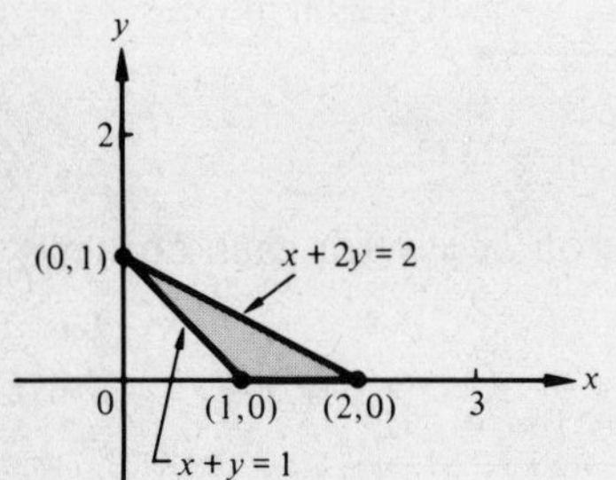

The following table indicates the value of $x + 2y$ for each corner point.

corner point	value of $x + 2y$
(0, 1)	$0 + 2(1) =$ ____
(1, 0)	$1 + 2(0) =$ ____
(2, 0)	$2 + 2(0) =$ ____

Answers: 2, 1, 2

Thus, the maximum value of $x + 2y$ is ____, and this occurs when $x = 0$ and $y = 1$ or $x =$ ____ and $y =$ ____. The minimum value of $x + 2y$ is ____, which occurs when $x =$ ____ and $y =$ ____.

Answers: 2; 2, 0; 1, 1, 0

89 A refinery produces a combined maximum of 25,000 barrels of gasoline and diesel oil per day. The refinery must produce at least 5000 barrels of diesel oil each day. If the profit is \$6.50 per barrel of gasoline and \$4.00 per barrel of diesel oil, find the maximum profit and how many barrels of each product yields this maximum. Let x represent the number of barrels of gasoline produced and y represent the number of barrels of diesel oil produced. Then, the profit P is given by the linear expression

$P =$ ________________

Answer: $6.50x + 4y$

The constraint set is

$$\begin{cases} 0 \leqslant x \leqslant 20{,}000 \\ 5{,}000 \leqslant y \leqslant 25{,}000 \\ x + y \leqslant \text{\_\_\_\_\_\_\_\_} \end{cases}$$

Answer: 25,000

The graph is

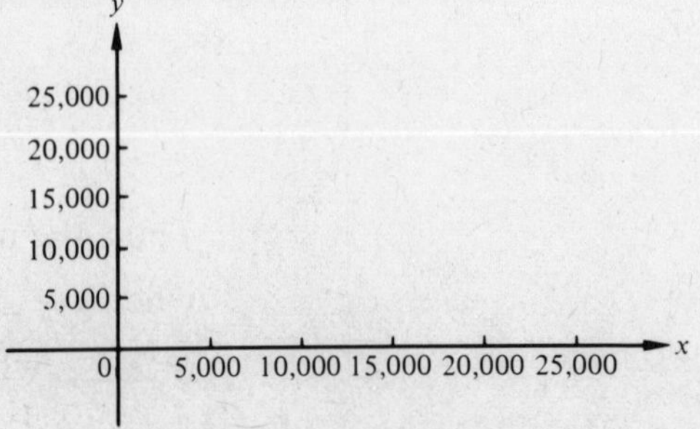

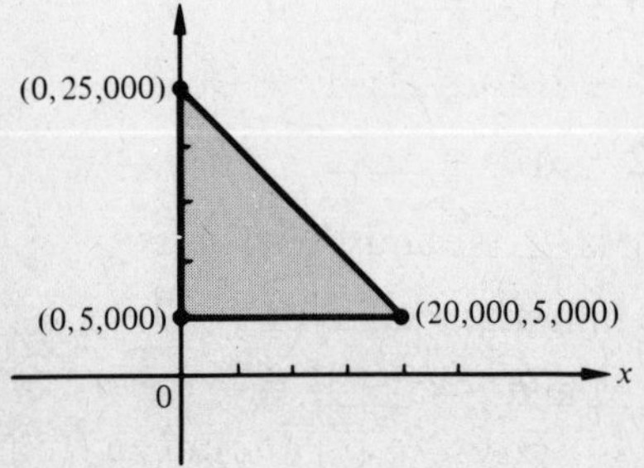

corner points	$P = 6.50x + 4y$
(0, 5000)	$P =$ ______
(0, 25000)	$P =$ ______
(20000, 5000)	$P =$ ______

Margin answers: 20,000; 100,000; 150,000

$150,000 — Thus, the maximum profit is ______, and this occurs

20,000, 5000 — when ______ barrels of gasoline and ______ barrels of diesel oil are produced per day.

90 Suppose that each serving of a special food contains 2 units of vitamin B and 5 units of iron and that each glass of a special drink contains 4 units of vitamin B and 2 units of iron. A minimum of 80 units of vitamin B and 60 units of iron must be provided each day. How much of the food and drink need be consumed in order to meet the daily requirements and to minimize costs if each serving of food costs $1.00 and each drink costs $0.80?

Let x represent number of servings of food and y represent number of drinks. Then the cost C is given by the linear expression

$x + 0.8y$ — $C =$ ______

The constraint set is

$$\begin{cases} x \geqslant 0 \\ y \geqslant 0 \\ 2x + 4y \geqslant 80 \\ 5x + 2y \geqslant 60 \end{cases}$$

The graph is

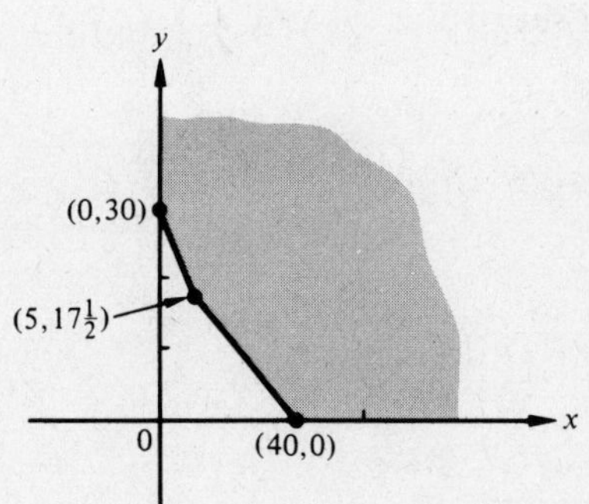

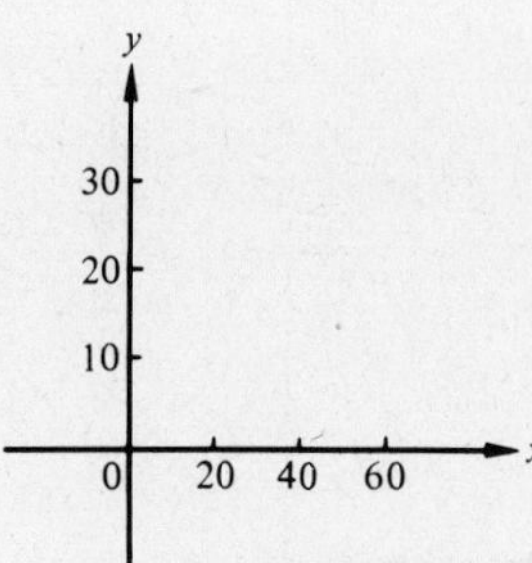

corner points	$C = x + 0.8y$
(0, 30)	$C =$ ____
$(5, 17\frac{1}{2})$	$C =$ ____
(40, 0)	$C =$ ____

Margin answers: 24; 14; 40

$14, 5 — Thus, the minimum daily cost is ____, and this occurs when ____

$17\frac{1}{2}$ — servings of the food and ____ glasses of the drink are consumed.

5 Systems with Quadratic Equations

91 Systems of equations in two variables can be solved by substitution, that is, by solving one of the equations explicitly for one variable in terms of the other and then substituting this solution into the other equation to obtain an equation that involves ____ variable.

one

92 Systems of equations involving two variables can be solved graphically by a method that produces only approximations to real __________, if they exist.

solutions

In Problems 93–97, solve each system of equations for real number solutions and verify the solution by sketching the graphs of the equations and noting the points of intersection.

93 $\begin{cases} x + y = 23 \\ x^2 + y^2 = 277 \end{cases}$

Solving the equation $x + y = 23$ explicitly for y in terms of x, we obtain __________. Substituting $23 - x$ for y in the equation $x^2 + y^2 = 277$, we obtain $x^2 + (\_\_\_\_\_\_)^2 = 277$.

$y = 23 - x$

$23 - x$

This equation is written in standard form as __________________. Then $x^2 - 23x + 126 = (\_\_\_\_\_\_)(\_\_\_\_\_\_) = 0$, so that ______ $= 0$ or ______ $= 0$

Therefore, $x =$ ____ or $x =$ ____.

$x^2 - 23x + 126 = 0$

$x - 9,\ x - 14$

$x - 9,\ x - 14$

9, 14

Replacing x by 9 or x by 14 in the equation $y = 23 - x$, we obtain $y =$ ____ or $y =$ ____. The solution set of the system is ______________. The graph is

14, 9

$\{(9, 14), (14, 9)\}$

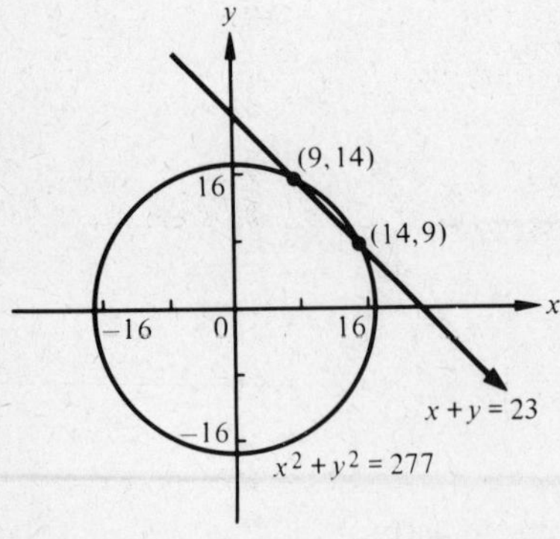

y
16
8
−16 −8 0 8 16
x
−8
−16

94 $\begin{cases} x - y = 7 \\ x^2 + y^2 = 169 \end{cases}$

Solve $x - y = 7$ explicitly for y to obtain __________.

$y = x - 7$

Substitute $x - 7$ for y in $x^2 + y^2 = 169$ to obtain $x^2 + (\_\_\_\_\_\_)^2 = 169$

$x - 7$

$2x^2 - 14x - 120 = 0$

$x - 12,\ x + 5$

$-5,\ 12$

$-12,\ 5$

$\{(-5, -12), (12, 5)\}$

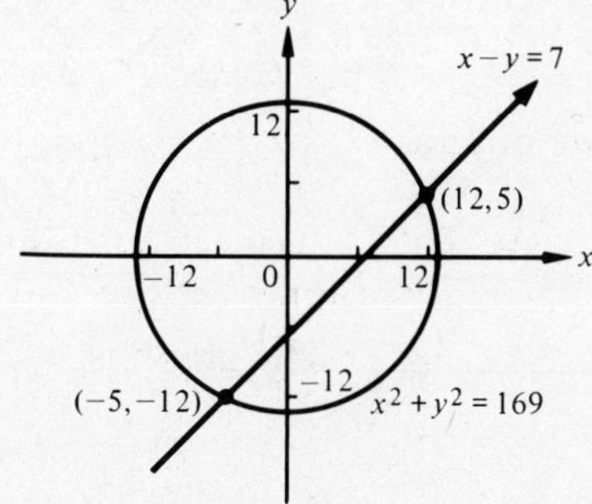

This equation is written in standard form as

__________________, so that

$x^2 - 7x - 60 = (\_\_\_\_\_\_)(\_\_\_\_\_\_) = 0$

That is, $x =$ ____ or $x =$ ____. Replacing x by -5 or by 12 in the equation $y = x - 7$, we obtain $y =$ ____ or $y =$ ____. The solution set of the system is __________________.

The graph is

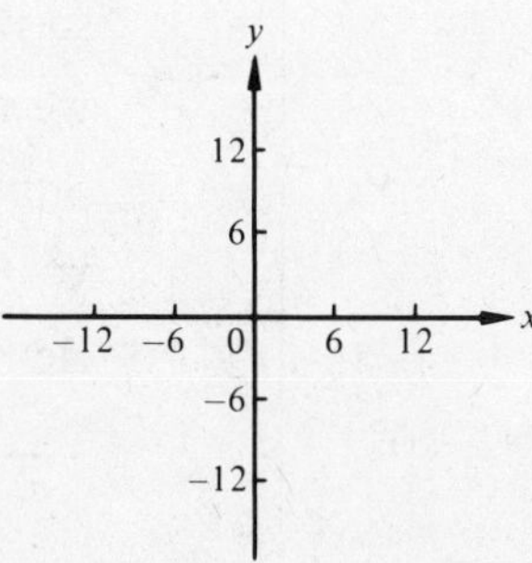

95 $\begin{cases} 4x^2 + 7y^2 = 32 & (1) \\ -3x^2 + 11y^2 = 41 & (2) \end{cases}$

Multiplying equation (1) by 3 and equation (2) by 4, we obtain

$\begin{cases} \_\_\_\_\_\_\_\_\_\_\_\_ \\ \_\_\_\_\_\_\_\_\_\_\_\_ \end{cases}$

$12x^2 + 21y^2 = 96$

$-12x^2 + 44y^2 = 164$

Adding these two equations, we obtain

__________ or $y^2 =$ ____

$65y^2 = 260,\ 4$

so that $y =$ ____ or $y =$ ____. When we substitute -2 for y in equation (1), we obtain $4x^2 + 7(\_\_\_\_)^2 = 32$, so $x =$ ____ or $x =$ ____

$-2,\ 2$

-2

$-1,\ 1$

When we substitute 2 for y in equation (1), we obtain $4x^2 + 7(\_\_\_\_)^2 = 32$, so $x =$ ____ or $x =$ ____.

$2,\ -1,\ 1$

The solution set of the system is

________________________. The graph is

$\{(1, 2), (-1, 2), (1, -2), (-1, -2)\}$

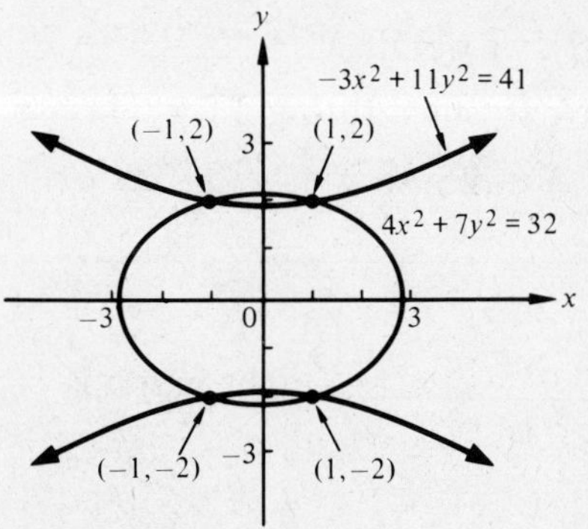

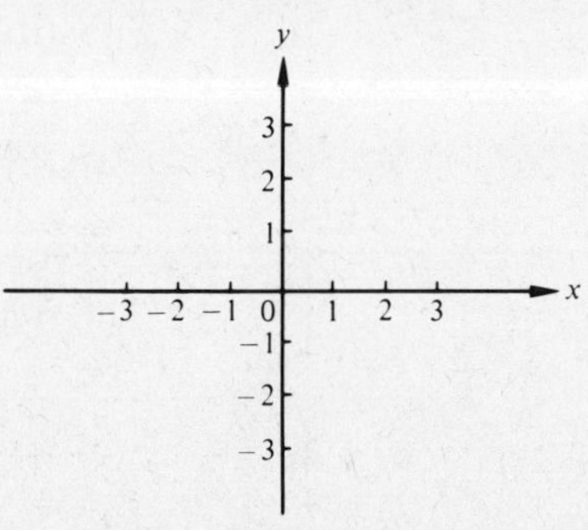

96 $\begin{cases} x^2 + 2y^2 = 22 & (1) \\ 2x^2 + y^2 = 17 & (2) \end{cases}$

Multiplying equation (1) by 2, we obtain

__________________ (3)

Solving equation (2) for $2x^2$ in terms of y, we obtain

____

Substituting $17 - y^2$ for $2x^2$ in equation (3), we obtain

________________________ or ________________

so $y =$ ____ or $y =$ ____. Substituting -3 for y in equation (1), we obtain $x^2 + 2($____$)^2 = 22$, so $x =$ ____ or $x =$ ____. Substituting 3 for y in equation (1), we obtain $x^2 + 2($____$)^2 = 22$, so $x =$ ____ or $x =$ ____.

Hence, the solution set is

____. The graph is

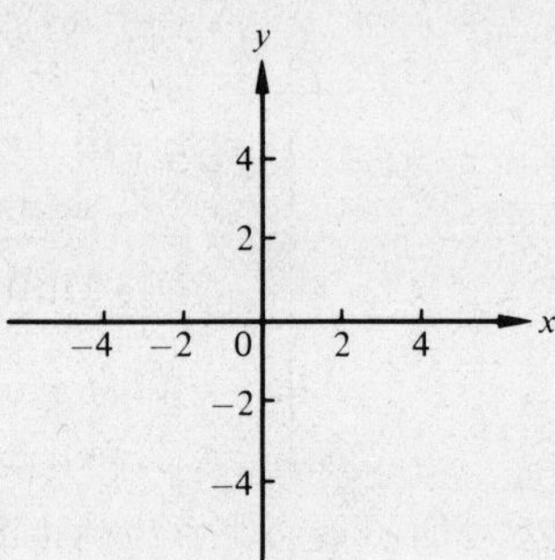

Answers:

$2x^2 + 4y^2 = 44$

$2x^2 = 17 - y^2$

$17 - y^2 + 4y^2 = 44,\ 3y^2 = 27$

$-3,\ 3$

$-3,\ -2,\ 2$

$3,\ -2,\ 2$

$\{(2, 3), (-2, 3), (2, -3), (-2, -3)\}$

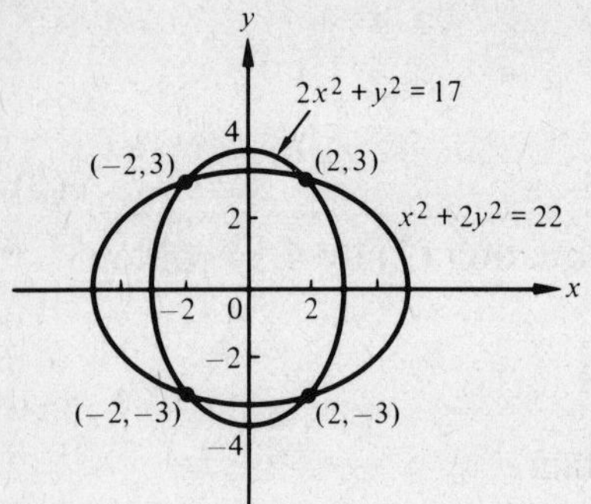

97 $\begin{cases} x^2 + 9y^2 = 57 & (1) \\ x^2 + \ \ y^2 = 25 & (2) \end{cases}$

Solving equation (1) explicitly for x^2, we obtain

____

Substituting $57 - 9y^2$ for x^2 in equation (2), we obtain

____

so $-8y^2 = -32$ or $y^2 = 4$. Therefore, $y =$ ____ or $y =$ ____. Substituting -2 for y in equation (2), we obtain $x^2 + ($____$)^2 = 25$, so $x =$ _______ or $x =$ ______. Next, substituting 2 for y in equation (2), we obtain $x^2 + ($____$)^2 = 25$, so $x =$ ________ or $x =$ ____.

The solution set is

____. The graph is

Answers:

$x^2 = 57 - 9y^2$

$57 - 9y^2 + y^2 = 25$

$-2,\ 2$

-2

$-\sqrt{21},\ \sqrt{21}$

$2,\ -\sqrt{21}$

$\sqrt{21}$

$\{(-\sqrt{21}, 2), (-\sqrt{21}, -2), (\sqrt{21}, 2), (\sqrt{21}, -2)\}$

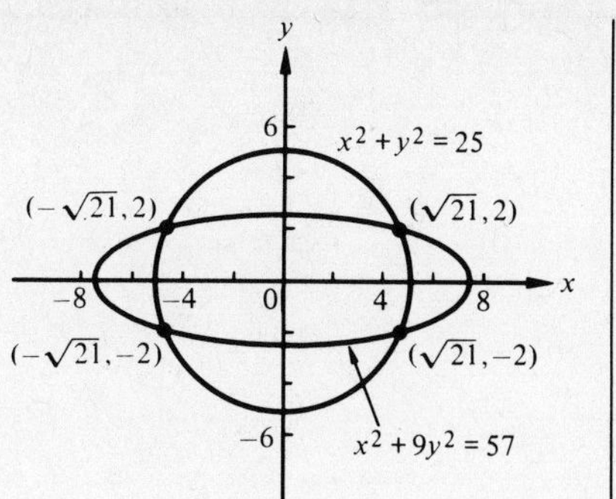

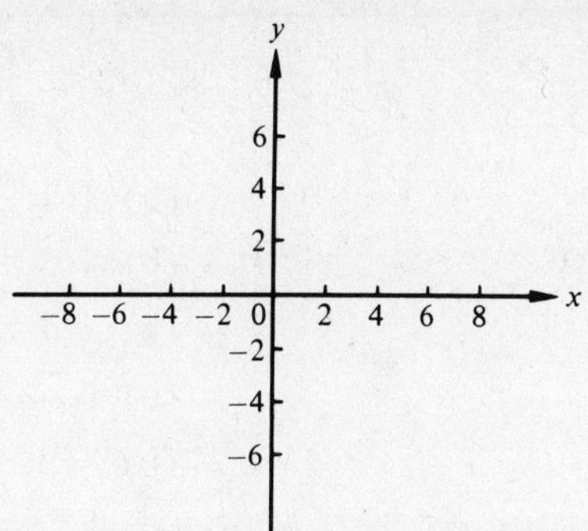

Chapter Test

1 Find the solution set of each given system by the substitution method.

(a) $\begin{cases} 2x + y = 4 \\ 3x - y = 1 \end{cases}$ (b) $\begin{cases} x + y + z = 6 \\ 2x - y + z = 3 \\ 3x + y - z = 2 \end{cases}$

2 Find the solution set of each given system by the elimination method.

(a) $\begin{cases} 3x + 2y = 8 \\ 2x - 3y = 14 \end{cases}$ (b) $\begin{cases} x + y + z = 6 \\ x - y + 2z = 12 \\ 2x + y - z = 1 \end{cases}$

3 Find the solution set of each system in Problem 1 by using matrices and row reduction.

4 Evaluate each of the given determinants.

(a) $\begin{vmatrix} 2 & -1 \\ 3 & 5 \end{vmatrix}$ (b) $\begin{vmatrix} 1 & 0 & 2 \\ 4 & 6 & -1 \\ -1 & 0 & -1 \end{vmatrix}$

5 Find the solution set of each system in Problem 2 by using Cramer's rule.

6 Sketch the region defined by the given system and find the corner points of the boundary of the region

$\begin{cases} x + y \geqslant 4 \\ x - y \leqslant 3 \end{cases}$

7 Find the maximum and minimum values of the linear expression $3x + 2y$ whose constraint set is

$\begin{cases} x \geqslant 0 \\ y \geqslant 0 \\ y \leqslant -2x + 3 \end{cases}$

8 Find the solution set of each of the following systems.

(a) $\begin{cases} x - y = 1 \\ x^2 + y^2 = 5 \end{cases}$ (b) $\begin{cases} 2x^2 - 3y^2 = 6 \\ 3x^2 + 2y^2 = 35 \end{cases}$

Answers

1 (a) $\{(1, 2)\}$ (b) $\{(1, 2, 3)\}$

2 (a) $\{(4, -2)\}$ (b) $\{(3, -1, 4)\}$

3 (a) $\{(1, 2)\}$ (b) $\{(1, 2, 3)\}$

4 (a) 13 (b) 6

5 (a) $\{(4, -2)\}$ (b) $\{(3, -1, 4)\}$

6

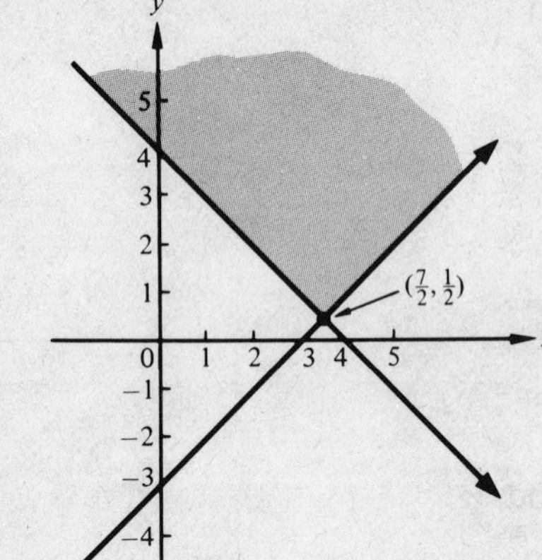

7 The maximum value is 6, minimum value is 0.

8 (a) $\{(-1, -2), (2, 1)\}$ (b) $\{(3, 2), (-3, 2), (3, -2), (-3, -2)\}$

Chapter 9 COMPLEX NUMBERS AND DISCRETE ALGEBRA

In this chapter we study complex numbers and their geometric representations. Other topics covered include the theory of equations, mathematical induction, the binomial theorem, and finite sums and geometric series. In working the problems of the chapter, the student will be able to:

1 Perform algebraic operations with complex numbers.
2 Represent complex numbers geometrically.
3 Find powers and roots of complex numbers.
4 Apply the theory of equations to find complex zeros of polynomial functions.
5 Apply the principle of mathematic induction and the binomial theorem.
6 Apply formulas for finding finite sums and sums of geometric series.

1 Complex Numbers

Answer	Frame
complex number	**1** An ordered pair of real numbers (a, b), which shall be denoted as $a + bi$, is called a ______.

In Problems 2–4, let $z_1 = a_1 + b_1 i$ and $z_2 = a_2 + b_2 i$ be complex numbers; then:

Answer	Frame
$a_2,\ b_2$	**2** Equality: $z_1 = z_2$ if and only if $a_1 =$ ______ and $b_1 =$ ______.
$a_1 + a_2,\ b_1 + b_2$	**3** Addition: $z_1 + z_2 =$ (______) + (______) i.
$a_1 a_2 - b_1 b_2,\ a_1 b_2 + a_2 b_1$	**4** Multiplication: $z_1 \cdot z_2 =$ (______) + (______) i.
real	**5** For $z = a + bi$, a is called the ______ part of z and b is called the
imaginary	______ part of z.
C	**6** The set of complex numbers is denoted by ______, so that
R	$C = \{a + bi \mid a, b \in$ ______ and $i^2 = -1\}$
	7 If the imaginary part of a complex number is 0, we shall consider
real	the number to be ______. In this sense, R can be considered to be
proper	a ______ subset of C.
6	**8** If $x + 3i = 6 + 3i$, then $x =$ ______.
4	**9** If $3x + 2i = 12 + 2i$, then $x =$ ______.
5	**10** If $5 + 2xi = 5 + 10i$, then $x =$ ______.
$7 - 4$	**11** $(13 + 7i) + (9 - 4i) = (13 + 9) +$ (______) i
$22 + 3i$	$=$ ______
$6 + 8$	**12** $(6 + 3i) + (8 + 5i) =$ (______) $+ (3 + 5)i$
$14 + 8i$	$=$ ______
$-2 + 2$	**13** $(11 - 2i) + (10 + 2i) = (11 + 10) +$ (______) i
$21 + 0i,\ 21$	$=$ ______ $=$ ______
$1 - 3i$	**14** $(1 + i) + (-4i) = 1 + (1 - 4)i =$ ______
$2,\ 20 + 2i$	**15** $(12 + 6i) + (8 - 4i) = 20 +$ (______) $i =$ ______
$4 + 6$	**16** $(4 - 2i)(-3 + i) = (-12 + 2) +$ (______) i
$-10 + 10i$	$=$ ______
$45 - 18$	**17** $(5 + 6i)(9 + 3i) =$ (______) $+ (15 + 54)i$
$27 + 69i$	$=$ ______
$6 + 4$	**18** $(3 - 4i)(2 + i) =$ (______) $+ (3 - 8)i$
$10 - 5i$	$=$ ______
$-1,\ 7 - i$	**19** $(1 + 2i)(1 - 3i) = (1 + 6) +$ (______) $i =$ ______
$23,\ 23 - 11i$	**20** $(7 + i)(3 - 2i) =$ (______) $+ (-11)i =$ ______
$0 + 1i$	**21** $i^2 = i \cdot i = (0 + 1i)$ (______)
$0 \cdot 1 + 1 \cdot 0$	$= (0 \cdot 0 - 1 \cdot 1) +$ (______) i
$0,\ -1$	$= -1 +$ ______ $i =$ ______
C	**22** The order properties of R do not hold for the set ______.

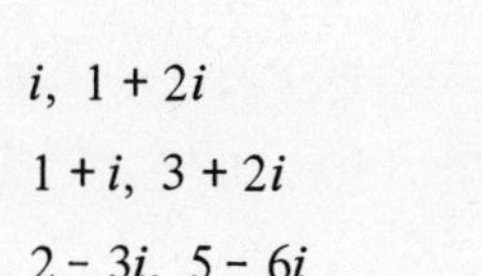

Answers	Problems
	23 The domain of functions can be extended to the set of complex numbers. Thus, if f is a function whose domain is the set of complex numbers, and if $f(z) = 2z + 1$, then
i, $1 + 2i$	$f(i) = 2(\_\_\_\_) + 1 = \_\_\_\_$
$1 + i$, $3 + 2i$	$f(1 + i) = 2(\_\_\_\_) + 1 = \_\_\_\_$
$2 - 3i$, $5 - 6i$	$f(2 - 3i) = 2(\_\_\_\_) + 1 = \_\_\_\_$

1.1 Properties of Addition and Multiplication

In Problems 24–27, assume that $z_1, z_2, z_3 \in C$; then the following properties hold.

Answers	Problems
	24 Closure of addition and multiplication:
C	(i) $z_1 + z_2 \in \_\_\_\_$
C	(ii) $z_1 \cdot z_2 \in \_\_\_\_$
	25 Commutativity of addition and multiplication:
$z_2 + z_1$	(i) $z_1 + z_2 = \_\_\_\_$
$z_2 \cdot z_1$	(ii) $z_1 \cdot z_2 = \_\_\_\_$
	26 Associativity of addition and multiplication:
$z_1 + z_2$	(i) $z_1 + (z_2 + z_3) = (\_\_\_\_) + z_3$
$z_1 \cdot z_2$	(ii) $z_1 \cdot (z_2 \cdot z_3) = (\_\_\_\_) \cdot z_3$
	27 Distributive properties:
$z_1 \cdot z_3$	(i) $z_1 \cdot (z_2 + z_3) = z_1 \cdot z_2 + \_\_\_\_$
$z_2 \cdot z_3$	(ii) $(z_1 + z_2) \cdot z_3 = z_1 \cdot z_3 + \_\_\_\_$
	28 Identities:
$0 + z$	(i) There exists $0 \in C$ such that $z + 0 = \_\_\_\_ = z$
C	for every $z \in \_\_\_\_$.
$z \cdot 1$	(ii) There exists $1 \in C$ such that $\_\_\_\_ = 1 \cdot z = z$
z	for every $\_\_\_\_ \in C$.
	29 Inverses:
$-z$	(i) If $z \in C$, then there exists $\_\_\_\_ \in C$ such that
0	$z + (-z) = (-z) + z = \_\_\_\_$
z^{-1}	(ii) If $z \in C$ and $z \neq 0$, then there exists $\_\_\_\_ \in C$ such that
1	$z \cdot z^{-1} = z^{-1} \cdot z = \_\_\_\_$

1.2 Subtraction of Complex Numbers

Answers	Problems
	30 If $z_1, z_2 \in C$, then $z_1 - z_2$, that is, the difference of z_1 and z_2, is
$-z_2$	defined as $z_1 - z_2 = z_1 + (\_\_\_\_)$.
$-5 + 4$, $-1 - i$	**31** $(4 - 5i) - (5 - 4i) = (4 - 5) + (\_\_\_\_)i = \_\_\_\_$
$4 - 7$, $-4 - 3i$	**32** $(3 + 4i) - (7 + 7i) = (3 - 7) + (\_\_\_\_)i = \_\_\_\_$
$-7 - 8$, $-15 - 9i$	**33** $(-7 - 6i) - (8 + 3i) = (\_\_\_\_) + (-6 - 3)i = \_\_\_\_$
$-5 + 4$, $1 - i$	**34** $(6 - 5i) - (5 - 4i) = (6 - 5) + (\_\_\_\_)i = \_\_\_\_$

1.3 Division of Complex Numbers

Answers	Frame
$a - bi,\ a^2 + b^2$	**35** The conjugate of a complex number $z = a + bi$, written $\bar{z}$, is defined as $\bar{z} =$ ________ and $z\bar{z} = (a + bi)(a - bi) =$ ________.
$z_1 \cdot \bar{z}_1$	**36** If $z_1 = a_1 + b_1 i$, for $z_1 \neq 0$, and $z_2 = a_2 + b_2 i$, then the quotient $z_2 \div z_1$ is given by $\dfrac{z_2}{z_1} = \dfrac{z_2 \cdot \bar{z}_1}{\_\_\_\_}$.
$1,\ -5i$	**37** $\dfrac{5}{i} = \dfrac{5(-i)}{i(-i)} = \dfrac{-5i}{\_\_\_} = \_\_\_$
$1 - 3i,\ 2 - 6i$	**38** $\dfrac{2}{1 + 3i} = \dfrac{2(1 - 3i)}{(1 + 3i)(\_\_\_\_)} = \dfrac{\_\_\_\_}{1 + 9}$
$10,\ \dfrac{6}{10},\ \dfrac{1}{5} - \dfrac{3}{5}i$	$= \dfrac{2 - 6i}{\_\_\_} = \left(\dfrac{2}{10}\right) - \left(\_\_\_\right)i = \_\_\_\_$
$3 - 4i,\ -8 - 9$	**39** $\dfrac{2 - 3i}{3 + 4i} = \dfrac{(2 - 3i)(\_\_\_\_)}{(3 + 4i)(3 - 4i)} = \dfrac{(6 - 12) + (\_\_\_\_)i}{3^2 + 4^2}$
$25,\ -\dfrac{6}{25},\ -\dfrac{17}{25},\ -\dfrac{6}{25} - \dfrac{17}{25}i$	$= \dfrac{-6 - 17i}{\_\_\_} = \left(\_\_\_\right) + \left(\_\_\_\right)i = \_\_\_\_$
$1 + i,\ 2,\ i$	**40** $\dfrac{1 + i}{1 - i} = \dfrac{(1 + i)(\_\_\_\_)}{(1 - i)(1 + i)} = \dfrac{2i}{\_\_\_} = \_\_\_$
$3 - 5i,\ 25$	**41** $\dfrac{1}{3 + 5i} = \dfrac{(3 - 5i)}{(3 + 5i)(\_\_\_\_)} = \dfrac{3 - 5i}{9 + \_\_\_}$
$\dfrac{3}{34},\ -\dfrac{5}{34},\ \dfrac{3}{34} - \dfrac{5}{34}i$	$= \left(\_\_\_\right) + \left(\_\_\_\right)i = \_\_\_\_$
$-i,\ 1,\ i$	**42** $i^{-3} = \dfrac{1}{i^3} = \dfrac{1}{i^2 i} = \dfrac{1}{\_\_\_} = \dfrac{i}{\_\_\_} = \_\_\_$
$2 - 2i,\ 6 - 4$	**43** $\dfrac{2 + 3i}{2 + 2i} = \dfrac{(2 + 3i)(2 - 2i)}{(2 + 2i)(\_\_\_\_)} = \dfrac{(4 + 6)(\_\_\_\_)i}{2^2 + 2^2}$
$10 + 2i,\ \dfrac{5}{4} + \dfrac{1}{4}i$	$= \dfrac{\_\_\_\_}{8} = \_\_\_\_$
$2,\ 5 - 12i$	**44** $\dfrac{3 - 2i}{3 + 2i} = \dfrac{(3 - 2i)(3 - 2i)}{(3)^2 + (\_\_\_)^2} = \dfrac{\_\_\_\_}{13}$
$\dfrac{5}{13},\ -\dfrac{12}{13},\ \dfrac{5}{13} - \dfrac{12}{13}i$	$= \left(\_\_\_\right) + \left(\_\_\_\right)i = \_\_\_\_$

2 Geometric Representation of Complex Numbers

Answers	Frame
$a + bi$	**45** Each ordered pair of real numbers (a, b) can be associated with the complex number $z =$ ________.
(a, b)	**46** Each complex number $z = a + bi$ can be associated with the ordered pair of real numbers ________.
point, (a, b)	**47** The complex number $z = a + bi$ can be represented graphically by the ______ in the plane associated with the ordered pair ______.

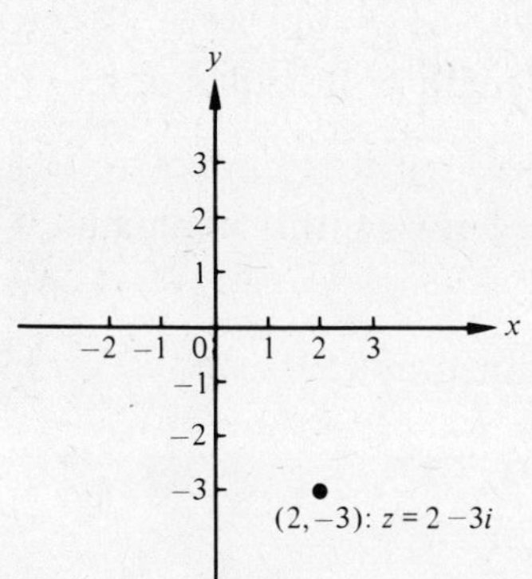

48 The graph of $z = 2 - 3i$ is

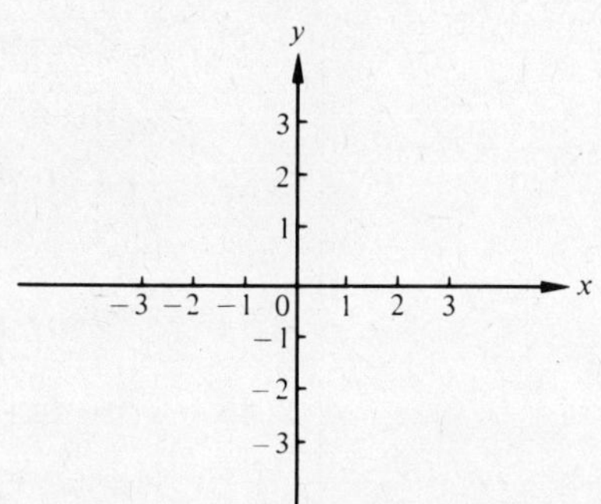

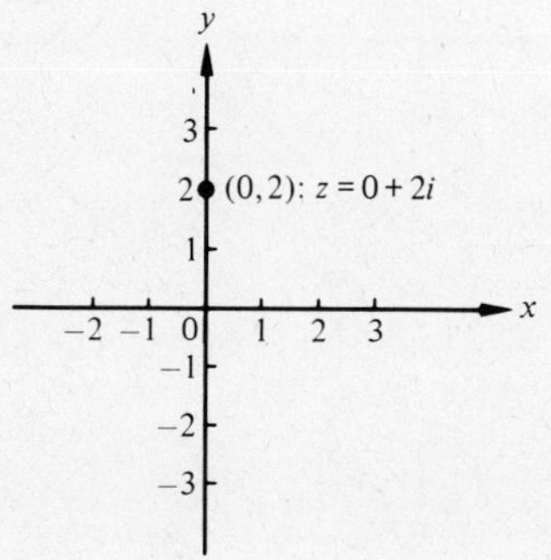

49 The graph of $z = 0 + 2i$ is

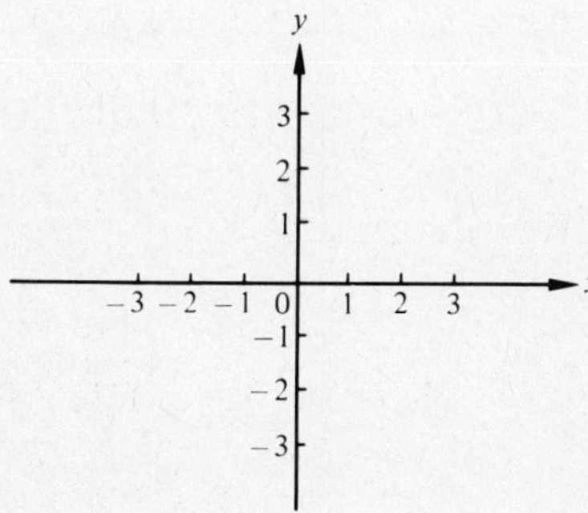

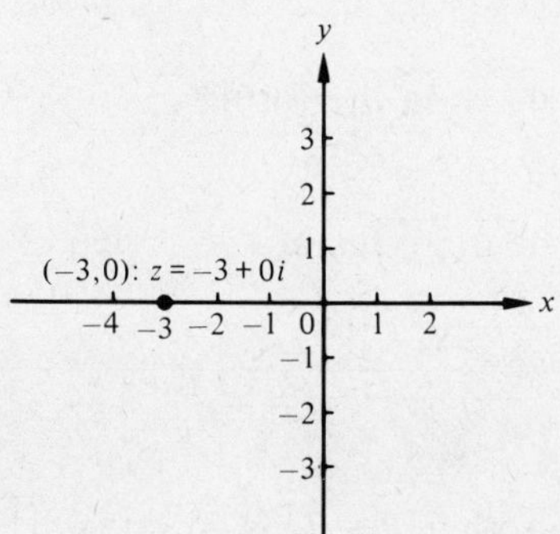

50 The graph of $z = -3 + 0i$ is

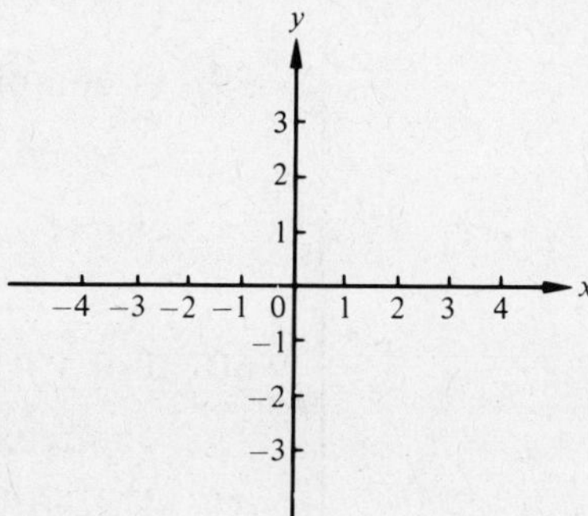

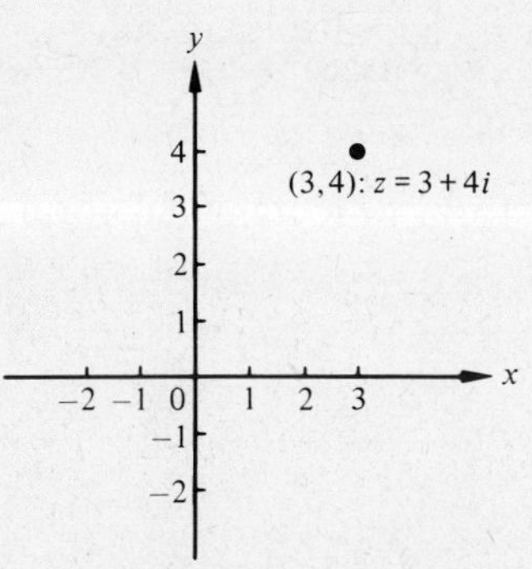

51 The graph of $z = 3 + 4i$ is

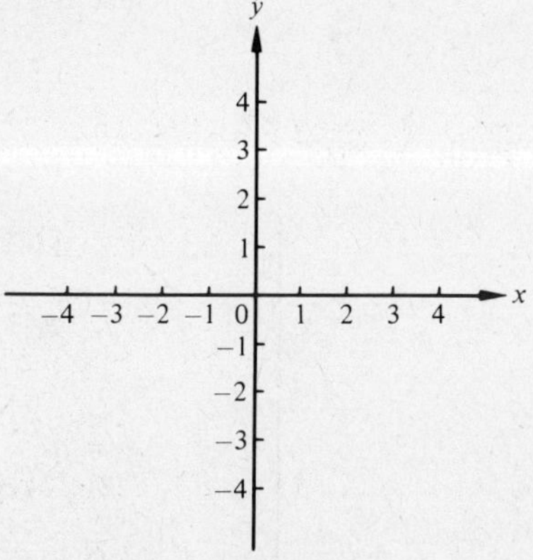

2.1 The Modulus of a Complex Number

Answers	Frames
$\sqrt{a^2+b^2}$	**52** If $z = a + bi$, then the absolute value or length or modulus of z, written $\lvert z \rvert$, is defined by $\lvert z \rvert =$ ________.
(a, b)	**53** The modulus of $z = a + bi$ is the distance between the origin and the point ______.
$\overline{z}$	**54** The modulus of $z = a + bi$ can be determined by $\lvert z \rvert = \sqrt{z \cdot (\_\_\_)}$.
	55 Find $\lvert z \rvert$ if $z = 2 - 3i$.
2, -3	Here $a =$ ____ and $b =$ ____, so that
2, -3, $\sqrt{13}$	$z = \sqrt{a^2 + b^2} = \sqrt{(\_\_\_)^2 + (\_\_\_)^2} =$ ____
	56 Find $\lvert z \rvert$ if $z = 3 + 4i$
3, 4	Here $a =$ ____ and $b =$ ____, so that
3, 4, 5	$z = \sqrt{(\_\_\_)^2 + (\_\_\_)^2} =$ ____
	57 Find $\lvert z \rvert$ if $z = 2 + 0i$.
$2 - 0i$	Here $\overline{z} =$ ____, so that
$2 - 0i$, 2	$z = \sqrt{z \cdot \overline{z}} = \sqrt{(2 + 0i)(\_\_\_\_)} =$ ____
	58 Find $\lvert z \rvert$ if $z = 0 - 3i$.
$0 + 3i$	Here $\overline{z} =$ ______, so that
$0 + 3i$, 9, 3	$z = \sqrt{z \cdot \overline{z}} = \sqrt{(0 - 3i)(\_\_\_\_)} = \sqrt{\_\_\_} =$ ____

2.2 Polar Form of Complex Numbers

Answers	Frames
	59 A complex number $z = x + yi$, when written in the form $z = r\cos\theta + ir\sin\theta = r(\cos\theta + i\sin\theta)$, where $x = r\cos\theta$, $y = r\sin\theta$, and r is the modulus of z, is called
polar, trigonometric	the ______ form or __________ form of z.
argument	**60** For $z = r(\cos\theta + i\sin\theta)$, θ is called an ________ of the complex number z.
	61 If $r(\cos\theta + i\sin\theta) = r(\cos\theta_1 + i\sin\theta_1)$, then $\theta - \theta_1$ is an integral
2π or 360°	multiple of ________, so that θ is not unique.
moduli	**62** Two complex numbers are equal if and only if their ______
arguments	are equal and their ________ differ by a multiple of 2π.

In Problems 63–68, change the given complex number from polar form to rectangular form.

Answers	Frames
	63 $z = 4(\cos 30° + i\sin 30°)$
$\frac{\sqrt{3}}{2}, \frac{1}{2}$	$\cos 30° = \frac{\quad}{\_\_\_}$ and $\sin 30° = \frac{\quad}{\_\_\_}$, so that
$\frac{\sqrt{3}}{2}, \frac{1}{2}$	$z = 4\left(\frac{\quad}{\_\_\_} + i\frac{\quad}{\_\_\_}\right)$
$2\sqrt{3}$, 2	$=$ ____ $+$ ____ i

64 $z = 6(\cos 60° + i \sin 60°)$

$\frac{1}{2}, \frac{\sqrt{3}}{2}$ — $\cos 60° =$ ____ and $\sin 60° =$ ____, so that

$\frac{1}{2}, \frac{\sqrt{3}}{2}$ — $z = 6\left(\_\_\_\_ + i\,\_\_\_\_\right)$

$3, 3\sqrt{3}$ — $= \_\_\_\_ + \_\_\_\_\,i$

65 $z = 3(\cos 270° + i \sin 270°)$

$0, -1$ — $\cos 270° =$ ____ and $\sin 270° =$ ____, so that

$0, -1$ — $z = 3[\_\_\_\_ + i\,(\_\_\_\_)]$

$0, -3$ — $= \_\_\_\_ + \_\_\_\_\,i$

66 $z = 5\left(\cos \frac{9\pi}{2} + i \sin \frac{9\pi}{2}\right)$

$0, 1$ — $\cos \frac{9\pi}{2} =$ ____ and $\sin \frac{9\pi}{2} =$ ____, so that

$0, 1$ — $z = 5(\_\_\_\_ + i\,\_\_\_\_)$

$0, 5$ — $= \_\_\_\_ + \_\_\_\_\,i$

67 $z = 8\left(\cos \frac{3\pi}{4} + i \sin \frac{3\pi}{4}\right)$

$-\frac{\sqrt{2}}{2}, \frac{\sqrt{2}}{2}$ — $\cos \frac{3\pi}{4} =$ ____ and $\sin \frac{3\pi}{4} =$ ____, so that

$-\frac{\sqrt{2}}{2}, \frac{\sqrt{2}}{2}$ — $z = 8\left[\left(\_\_\_\_\right) + i\left(\_\_\_\_\right)\right]$

$-4\sqrt{2}, 4\sqrt{2}$ — $= \_\_\_\_ + \_\_\_\_\,i$

68 $z = 2\left(\cos \frac{5\pi}{6} + i \sin \frac{5\pi}{6}\right)$

$-\frac{\sqrt{3}}{2}, \frac{1}{2}$ — $\cos \frac{5\pi}{6} =$ ____ and $\sin \frac{5\pi}{6} =$ ____, so that

$-\frac{\sqrt{3}}{2}, \frac{1}{2}$ — $z = 2\left[\left(\_\_\_\_\right) + i\left(\_\_\_\_\right)\right]$

$-\sqrt{3}, 1$ — $= \_\_\_\_ + \_\_\_\_\,i$

In Problems 69–73, change the given complex number from rectangular form to polar form.

69 $z = -1 + i$

$-1, 1$ — Here $x =$ ____ and $y =$ ____, so that

$-1, 1, \sqrt{2},$ — $r = \sqrt{x^2 + y^2} = \sqrt{(\_\_\_\_)^2 + (\_\_\_\_)^2} = \_\_\_\_$

-1, II — Also, $\tan \theta =$ ____. The point $(-1, 1)$ lies in quadrant ____, so

$\frac{3\pi}{4}$ — $\theta =$ ____. Since $z = r(\cos \theta + i \sin \theta)$, by substitution,

$\cos \frac{3\pi}{4}, \sin \frac{3\pi}{4}$ — $z = \sqrt{2}\left(\_\_\_\_ + i\,\_\_\_\_\right)$

70 $z = 1 + \sqrt{3}i$

Here $x =$ ____ and $y =$ ____, so that

$r = \sqrt{(\_\_\_)^2 + (\_\_\_)^2} =$ ____.

Also $\tan\theta =$ ____. The point $(1, \sqrt{3})$ lies in quadrant ____, so

$\theta =$ ____. Since $z = r(\cos\theta + i\sin\theta)$,

$z =$ ______________________

Answers: $1, \sqrt{3}$; $1, \sqrt{3}, 2$; $\sqrt{3}$, I; $60°$; $2(\cos 60° + i\sin 60°)$

71 $z = 0 + i$

Here $x =$ ____ and $y =$ ____, so that $r = \sqrt{\_\_\_ + \_\_\_} =$ ____.

Also, $\tan\theta =$ ________. Since (0, 1) lies on the ________ y axis, $\theta =$ ____. Hence, $z =$ ______________.

Answers: 0, 1, 0, 1, 1; undefined, positive; $\frac{\pi}{2}$, $\cos\frac{\pi}{2} + i\sin\frac{\pi}{2}$

72 $z = -\sqrt{2} - \sqrt{2}i$

Here $x =$ ____ and $y =$ ____, so that $r = \sqrt{\_\_\_ + \_\_\_} =$ ____.

Also, $\tan\theta =$ ____. Since $(-\sqrt{2}, -\sqrt{2})$ lies in quadrant ____,

$\theta =$ ____. Hence, $z =$ ______________________.

Answers: $-\sqrt{2}, -\sqrt{2}$, 2, 2, 2; 1, III; $\frac{5\pi}{4}$, $2\left(\cos\frac{5\pi}{4} + i\sin\frac{5\pi}{4}\right)$

73 $z = 3 + 0i$

Here $x =$ ____ and $y =$ ____, so that $r =$ ____. Also,

$\tan\theta =$ ____. Since (3, 0) lies on the positive x axis, $\theta =$ ____,

so that $z =$ ______________________.

Answers: 3, 0, 3; 0, 0; $3(\cos 0 + i\sin 0)$

2.3 Multiplication and Division of Complex Numbers in Polar Form

74 If $z_1 = r_1(\cos\theta_1 + i\sin\theta_1)$ and $z_2 = r_2(\cos\theta_2 + i\sin\theta_2)$, then

$z_1z_2 = (\_\_\_\_)\,[\cos(\theta_1 + \theta_2) + i\sin(\theta_1 + \theta_2)]$

Answer: r_1r_2

In Problems 75–78, given z_1 and z_2, find z_1z_2 in polar form.

75 $z_1 = 2(\cos 20° + i\sin 20°)$

$z_2 = 3(\cos 80° + i\sin 80°)$

$z_1z_2 = (\_\_\_\_)\,[\cos(\_\_\_\_) + i\sin(\_\_\_\_)]$

Answer: 6, 100°, 100°

76 $z_1 = 3(\cos 45° + i\sin 45°)$

$z_2 = 4(\cos 135° + i\sin 135°)$

$z_1z_2 = (\_\_\_\_)\,[\cos(\_\_\_\_) + i\sin(\_\_\_\_)]$

Answer: 12, 180°, 180°

77 $z_1 = 4\left(\cos\frac{\pi}{4} + i\sin\frac{\pi}{4}\right)$

$z_2 = 5\left(\cos\frac{\pi}{3} + i\sin\frac{\pi}{3}\right)$

$z_1z_2 = (\_\_\_\_)\left[\cos\left(\_\_\_\_\right) + i\sin\left(\_\_\_\_\right)\right]$

Answer: 20, $\frac{7\pi}{12}$, $\frac{7\pi}{12}$

12, $\frac{\pi}{4}$, $\frac{\pi}{4}$

78 $z_1 = 3(\cos \pi + i \sin \pi)$

$z_2 = 4\left[\cos\left(-\frac{3\pi}{4}\right) + i \sin\left(-\frac{3\pi}{4}\right)\right]$

$z_1 z_2 = (\_\_\_)\left[\cos\left(\_\_\_\right) + i \sin\left(\_\_\_\right)\right]$

$\frac{r_1}{r_2}$

79 If $z_1 = r_1(\cos \theta_1 + i \sin \theta_1)$ and $z_2 = r_2(\cos \theta_2 + i \sin \theta_2)$, then

$\frac{z_1}{z_2} = \left(\_\_\_\right)[\cos(\theta_1 - \theta_2) + i \sin(\theta_1 - \theta_2)]$ for $z_2 \neq 0$

In Problems 80–83, given z_1 and z_2, find z_1/z_2 in polar form.

4

2(cos 60° + *i* sin 60°)

80 $z_1 = 8(\cos 100° + i \sin 100°)$

$z_2 = 4(\cos 40° + i \sin 40°)$

$\frac{z_1}{z_2} = \frac{8}{\_\_\_}[\cos(100° - 40°) + i \sin(100° - 40°)]$

$= \_\_\_\_\_\_\_\_\_\_\_\_\_\_\_\_$

4, 30°, 30°

81 $z_1 = 20(\cos 120° + i \sin 120°)$

$z_2 = 5(\cos 90° + i \sin 90°)$

$\frac{z_1}{z_2} = \_\_\_(\cos \_\_\_ + i \sin \_\_\_)$

3, $\frac{2\pi}{3}$, $\frac{2\pi}{3}$

82 $z_1 = 12(\cos \pi + i \sin \pi)$

$z_2 = 4\left(\cos \frac{\pi}{3} + i \sin \frac{\pi}{3}\right)$

$\frac{z_1}{z_2} = \_\_\_\left(\cos \_\_\_ + i \sin \_\_\_\right)$

2, $-\frac{\pi}{12}$, $-\frac{\pi}{12}$

83 $z_1 = 10\left(\cos \frac{\pi}{4} + i \sin \frac{\pi}{4}\right)$

$z_2 = 5\left(\cos \frac{\pi}{3} + i \sin \frac{\pi}{3}\right)$

$\frac{z_1}{z_2} = \_\_\_\left[\cos\left(\_\_\_\right) + i \sin\left(\_\_\_\right)\right]$

3 Powers and Roots of Complex Numbers

integer

84 *DeMoivre's theorem:* If $z = r(\cos \theta + i \sin \theta)$, then $z^n = r^n(\cos n\theta + i \sin n\theta)$, for n a positive ______

20°, 20°

32(cos 100° + *i* sin 100°)

85 If $z = 2(\cos 20° + i \sin 20°)$, then

$z^5 = 2^5[\cos 5(\_\_\_) + i \sin 5(\_\_\_)]$

$= \_\_\_\_\_\_\_\_\_\_\_\_\_\_\_\_$

3, $\frac{2\pi}{3}$, $\frac{2\pi}{3}$

27(cos 2π + *i* sin 2π)

86 If $z = 3\left(\cos \frac{2\pi}{3} + i \sin \frac{2\pi}{3}\right)$, then

$z^3 = (\_\_\_)^3\left[\cos 3\left(\_\_\_\right) + i \sin 3\left(\_\_\_\right)\right]$

$= \_\_\_\_\_\_\_\_\_\_\_\_\_\_\_\_$

87 If $z = \sqrt{2}\left(\cos \frac{3\pi}{4} + i \sin \frac{3\pi}{4}\right)$, then

$z^4 = (\_\_\_)^4 (\cos \_\_\_ + i \sin \_\_\_)$

$= \_\_\_\_\_\_$

$\sqrt{2}$, 3π, 3π

$4(\cos 3\pi + i \sin 3\pi)$

88 If $z = \frac{1}{4}(\cos 20° + i \sin 20°)$ then

$z^4 = (\_\_\_)^4 (\cos \_\_\_ + i \sin \_\_\_)$

$= \_\_\_\_\_\_$

$\frac{1}{4}$, $80°$, $80°$

$\frac{1}{256}(\cos 80° + i \sin 80°)$

3.1 Roots

89 Assume $w = R(\cos \phi + i \sin \phi)$ and that $z = r(\cos \theta + i \sin \theta)$ is any root of $z^n = w$, where n is a positive integer. By ______,

$z^n = [r(\cos \theta + i \sin \theta)]^n = r^n(\cos n\theta + i \sin n\theta)$. Since $z^n = w$,

$r^n(\cos n\theta + i \sin n\theta) = R(\cos \phi + i \sin \phi)$.

DeMoivre's theorem

90 It follows from Problem 89 that $r^n = R$ or $r = \_\_\_$, where $r \geq 0$.

$\sqrt[n]{R}$

91 It also follows from Problem 89 that $n\theta = \phi + 2k\pi$ or $n\theta = \phi + 360°k$, so that $\theta = \_\_\_\_$ or $\theta = \_\_\_\_$, where $k = 0, \pm 1, \pm 2, \ldots$.

$\frac{\phi}{n} + \frac{2\pi k}{n}$, $\frac{\phi}{n} + \frac{360°k}{n}$

92 To find the arguments of all distinct n roots that exist, let k take on the integer values from 0 to ____.

$n - 1$

93 Find the cube roots of $27(\cos 180° + i \sin 180°)$.

Here $R = \_\_\_$ and $\phi = \_\_\_$, so that

$r = \sqrt[3]{\_\_\_} = \_\_\_$ and

$\theta = \frac{\_\_\_}{3} + \frac{(\_\_\_)k}{3}$

$= 60° + (\_\_\_)k$, where $k = 0, 1, 2$

The roots are

$z_0 = 3(\cos \_\_\_ + i \sin \_\_\_)$ for $k = 0$

$z_1 = 3(\cos \_\_\_ + i \sin \_\_\_)$ for $k = 1$

$z_2 = 3(\cos \_\_\_ + i \sin \_\_\_)$ for $k = 2$

27, 180°

27, 3

180°, 360°

120°

60°, 60°

180°, 180°

300°, 300°

94 Find the fourth roots of $16(\cos 3\pi + i \sin 3\pi)$.

Here $R = \_\_\_$ and $\phi = \_\_\_$, so that

$r = \sqrt[4]{\_\_\_} = \_\_\_$ and

$\theta = \frac{\_\_\_}{4} + \frac{(\_\_\_)k}{4}$, where $k = 0, 1, 2$, or 3

The roots are

$z_0 = 2\left(\cos \_\_\_ + i \sin \_\_\_\right)$ for $k = 0$

$z_1 = 2\left(\cos \_\_\_ + i \sin \_\_\_\right)$ for $k = 1$

16, 3π

16, 2

3π, 2π

$\frac{3\pi}{4}$, $\frac{3\pi}{4}$

$\frac{5\pi}{4}$, $\frac{5\pi}{4}$

$\frac{7\pi}{4}, \frac{7\pi}{4}$

$\frac{9\pi}{4}, \frac{9\pi}{4}$

243, 100°

243, 3

100°

20°

20°, 20°

92°, 92°

164°, 164°

236°, 236°

308°, 308°

unity

0°, 0°

1, 0°

1, 1

0°

0°, 0°

90°, 90°

180°, 180°

270°, 270°

$z_2 = 2\left(\cos \_\_\_\_ + i \sin \_\_\_\_\right)$ for $k = 2$

$z_3 = 2\left(\cos \_\_\_\_ + i \sin \_\_\_\_\right)$ for $k = 3$

95 Find the fifth roots of $243(\cos 100° + i \sin 100°)$.

Here, $R =$ ____ and $\phi =$ ____, so that

$r = \sqrt[5]{\_\_\_\_} =$ ____ and

$$\theta = \frac{\_\_\_\_}{5} + \frac{360°k}{5}$$

$= \_\_\_\_ + 72°k$, where $k = 0, 1, 2, 3, 4$

The roots are

$z_0 = 3(\cos \_\_\_\_ + i \sin \_\_\_\_)$ for $k = 0$

$z_1 = 3(\cos \_\_\_\_ + i \sin \_\_\_\_)$ for $k = 1$

$z_2 = 3(\cos \_\_\_\_ + i \sin \_\_\_\_)$ for $k = 2$

$z_3 = 3(\cos \_\_\_\_ + i \sin \_\_\_\_)$ for $k = 3$

$z_4 = 3(\cos \_\_\_\_ + i \sin \_\_\_\_)$ for $k = 4$

96 The solutions to $z^n = 1$ are called the nth roots of ______.

97 Find the fourth roots of unity.

First, the number 1 can be written in polar form as

$1 + 0i = 1(\cos \_\_\_\_ + i \sin \_\_\_\_)$.

Here $R =$ ____ and $\phi =$ ____, so that

$r = \sqrt[4]{\_\_\_\_} =$ ____ and

$$\theta = \frac{\_\_\_\_}{4} + \frac{360°k}{4}$$

$= 0° + 90°k$, where $k = 0, 1, 2, 3$

The roots are

$z_0 = 1(\cos \_\_\_\_ + i \sin \_\_\_\_)$ for $k = 0$

$z_1 = 1(\cos \_\_\_\_ + i \sin \_\_\_\_)$ for $k = 1$

$z_2 = 1(\cos \_\_\_\_ + i \sin \_\_\_\_)$ for $k = 2$

$z_3 = 1(\cos \_\_\_\_ + i \sin \_\_\_\_)$ for $k = 3$

4 Complex Zeros of Polynomial Functions

$<$

complex

quadratic, 1, 1

$-3, -\frac{1}{2} \pm \frac{\sqrt{3}}{2}i$

98 If the zeros of a quadratic polynomial function $f(x) = ax^2 + bx + c$, where $a \neq 0$, are not real numbers, that is, if $b^2 - 4ac$ ____ 0, the zeros are ______ numbers.

99 The zeros of $f(x) = x^2 + x + 1$ can be found by using the ______ formula. Thus, $x = \frac{-(\_\_\_\_) \pm \sqrt{1^2 - 4(1)(\_\_\_\_)}}{2(1)}$

or $x = \frac{-1 \pm \sqrt{\_\_\_\_}}{2}$, so that $x =$ ______.

complex

100 If $f(x)$ is a polynomial of degree $n \geqslant 1$ with complex coefficients, then there is a ________ number c such that $f(c) = 0$.

complex

101 *Factorization theorem:*
If $f(x) = a_n x^n + a_{n-1}x^{n-1} + a_{n-2}x^{n-2} + \cdots + a_1 x + a_0$, with $a_n \neq 0$ and n a positive integer, then $f(x) = a_n(x - c_1)(x - c_2) \cdots (x - c_n)$, where the numbers c_j are ________ numbers.

n

102 If f is a polynomial function of degree n, with $n \neq 0$, then $f(x) = 0$ has at most ____ roots. (Not all n roots are necessarily different.)

$\bar{z}$, $a - bi$

103 *Conjugate root theorem:*
If a polynomial of degree n, with $n \neq 0$, has real coefficients and $f(z_0) = 0$, where $z_0 = a + bi$, then $f(\_\_\_\_) = f(\_\_\_\_\_\_\_\_) = 0$.

$1 - i$

$1, -1$

$x^2 - 2x + 2$

$x^4 - 2x^3 + x^2 + 2x - 2$

104 Form a polynomial function f that has $1, -1$, and $1 + i$ as zeros.
Since $1 + i$ is a zero, then ______ is also a zero. Hence,

$f(x) = (x - \_\_\_\_)[x - (\_\_\_\_)][x - (1 + i)][x - (1 - i)]$

$= (x^2 - 1)(\_\_\_\_\_\_\_\_\_\_\_\_)$

$= \_\_\_\_\_\_\_\_\_\_\_\_\_\_\_\_\_\_\_\_$

$2 - i$

$3 + 2i$

$x^2 - 4x + 5$

$x^4 - 10x^3 + 42x^2 - 82x + 65$

105 Form a polynomial function f that has $2 + i$ and $3 - 2i$ as zeros.
Since $2 + i$ is a zero, then ______ is also a zero. Also, since $3 - 2i$ is a zero, ________ is a zero of $f(x)$. Hence,

$f(x) = [x - (2 + i)][x - (2 - i)][x - (3 + 2i)][x - (3 - 2i)]$

$= (\_\_\_\_\_\_\_\_\_\_\_\_)(x^2 - 6x + 13)$

$= \_\_\_\_\_\_\_\_\_\_\_\_\_\_\_\_\_\_\_\_$

106 Determine the multiplicity of the zeros of the polynomial function $f(x) = x^4 - 2x^3 + 2x^2 - 2x + 1$.
Using synthetic division for $c = 1$, we get

```
1 | 1  -2   2  -2   1
  |      1  -1   1  -1
  ---------------------
    1  -1   1  -1 | ____
```

0

Hence, 1 is a zero of $f(x)$ and $f(x) = (x - 1)(x^3 - x^2 + x - 1)$.
Repeating the process for $c = 1$, we get

```
1 | 1  -1   1  -1
  |      1   0   1
  -----------------
    1   0   1 | ____
```

0

Thus, $f(x) = (x - 1)(x - 1)(x^2 + 1)$.

i, i

Since $x^2 + 1 = (x + \_\_\_\_)(x - \_\_\_\_)$, then

$f(x) = (x - 1)^2(x + i)(x - i)$

2

1

Hence, 1 is a zero of f with multiplicity of ____; i and $-i$ are zeros, each of multiplicity ____.

107 Given that i is a zero of $f(x) = x^4 - 5x^3 + 7x^2 - 5x + 6$, find all other zeros of f.

First, since i is a zero of $f(x)$, ____ is also a zero. Thus, $(x - i)(x + i) =$ ________ is a ________ of $f(x)$.

Dividing $f(x)$ by $x^2 + 1$, another factor of $f(x)$ is ______________ $= (x - 2)($______$)$

Hence, $f(x) = (x - i)(x + i)(x - 2)($______$)$.

The zeros of f are ______________.

$-i$

$x^2 + 1$, factor

$x^2 - 5x + 6$, $x - 3$

$x - 3$

$i, -i, 2, 3$

5 Mathematical Induction

5.1 Principle of Mathematical Induction

108 Suppose that $S_1, S_2, S_3, \ldots$ is a sequence of assertions; that is, suppose that for each positive integer n we have a corresponding assertion S_n. Assume that the following two conditions hold:

(i) S_1 is true.

(ii) For each fixed positive integer k, the truth of S_k implies the truth of S_{k+1}.

Then it follows that every assertion ______________________; that is, S_n is true for all positive integers n.

$S_1, S_2, S_3, \ldots$ is true

In Problems 109–113, use mathematical induction to prove each of the following assertions. Assume that n represents a positive integer.

109 $1 + 3 + 5 + \cdots + (2n - 1) = n^2$ for all $n \geqslant 1$

Proof:

S_n is the equation
$1 + 3 + 5 + \cdots + (2n - 1) = n^2$

(i) S_1 becomes $1 = 1^2$, which is true.

(ii) Assume S_k is true, that is, assume that
$1 + 3 + 5 + \cdots + (2k - 1) = k^2$

Show that S_{k+1} is true. That is,
$1 + 3 + 5 + \cdots + (2k - 1) + (2k + 1) = (k + 1)^2$

Adding $2k + 1$ to both sides of S_k, we obtain
$1 + 3 + 5 + \cdots + (2k - 1) + (2k + 1) = k^2 + (2k + 1) = (k + 1)^2$

Hence, S_{k+1} is true. Thus, we conclude that
$1 + 3 + 5 + \cdots + (2n - 1) = n^2$

110 $2 + 2^2 + 2^3 + \cdots + 2^n = 2(2^n - 1)$ for $n \geqslant 1$

Proof:

S_n is the equation
$2 + 2^2 + 2^3 + \cdots + 2^n = 2(2^n - 1)$

(i) S_1 becomes $2 = 2(2 - 1)$, which is true.

(ii) Assume that S_k is true, that is, assume that
$2 + 2^2 + 2^3 + \cdots + 2^k = 2(2^k - 1)$

Show that S_{k+1} is true, that is,
$2 + 2^2 + 2^3 + \cdots + 2^k + 2^{k+1} = 2(2^{k+1} - 1)$
Adding 2^{k+1} to both sides of the equation S_k, we obtain
$2 + 2^2 + 2^3 + \cdots + 2^k + 2^{k+1}$
$= 2(2^k - 1) + 2^{k+1}$
$= 2 \cdot 2^k - 2 + 2^{k+1}$
$= 2^{k+1} - 2 + 2^{k+1}$
$= 2(2^{k+1} - 1)$
Hence, S_{k+1} is true. Thus, we conclude that
$2 + 2^2 + 2^3 + \cdots + 2^n = 2(2^n - 1)$

111 $2^n > 5n$ for all $n \geqslant 5$

Proof:

Let S_n be the inequality
$2^n > 5n$ for all $n \geqslant 5$

(i) S_5 is $2^5 > 5(5)$, which is true.

(ii) Assume that S_k is true, that is, $2^k > 5k$ for any $k > 5$.

Show that S_{k+1} is true, that is,
$2^{k+1} > 5(k + 1)$
$2^{k+1} > 2 \cdot 5k = 5k + 5k > 5(k + 1)$
Hence, S_{k+1} is true. Thus, we conclude that $2^n > 5n$.

112 If $0 < a < b$, then $a^n < b^n$ for all $n \geqslant 1$

Proof:

Let S_n be the inequality
$a^n < b^n$ for all $n \geqslant 1$

(i) S_1 will become $a^1 < b^1$.

(ii) Assume that S_k is true, that is, $a^k < b^k$ for $k \geqslant 1$

Show that S_{k+1} is true, that is, show that $a^{k+1} < b^{k+1}$. Multiply both sides of S_k by a, to get $a^{k+1} < b^k a$. Since $0 < a < b$, $b^k > 0$. Multiply both sides of $a < b$ by b^k to obtain $ab^k < b^{k+1}$.
Thus, $a^{k+1} < ab^k < b^{k+1}$
so that $a^{k+1} < b^{k+1}$.
Hence S_{k+1} is true. Thus we conclude that S_n is true.

113 $2 + 4 + 6 + \cdots + 2n = n(n + 1)$ for all $n \geqslant 1$

Proof:

Let S_n be the equation
$2 + 4 + 6 + \cdots + 2n = n(n + 1)$

(i) S_1 will become $2 = 1(1 + 1)$, which is true.

(ii) Assume that S_k is true, that is, assume that
$2 + 4 + 6 + \cdots + 2k = k(k + 1)$

Show that S_{k+1} is true, that is, show that
$2 + 4 + 6 + \cdots + 2k + (2k + 2) = (k + 1)(k + 2)$.

Add $2k + 2$ to both sides of the equation S_k to obtain

$$2 + 4 + 6 + \cdots + 2k + (2k + 2) = k(k + 1) + (2k + 2) = k(k + 1) + 2(k + 1) = (k + 1)(k + 2)$$

Hence, S_{k+1} is true, so S_n is true.

5.2 Binomial Expansions

114 If n is a positive integer, then $n(n - 1)(n - 2) \cdots 3 \cdot 2 \cdot 1$ is defined to be ____________, which is denoted by ____.

n factorial, $n!$

115 $5 \cdot 4 \cdot 3 \cdot 2 \cdot 1 =$ ____ $=$ ____

$5!$, 120

116 $6! =$ ______________________ $=$ ____

$6 \cdot 5 \cdot 4 \cdot 3 \cdot 2 \cdot 1$, 720

117 $(n - 1)! =$ ______________________

$(n - 1)(n - 2)(n - 3) \cdots 3 \cdot 2 \cdot 1$

118 $(n + 1)! =$ ______________________

$(n + 1)(n)(n - 1)(n - 2) \cdots 3 \cdot 2 \cdot 1$

119 $(n + 2)(n + 1)(n)(n - 1)(n - 2) \cdots 3 \cdot 2 \cdot 1 =$ __________

$(n + 2)!$

120 $2 \cdot 4 \cdot 6 \cdot 8 \cdots 2n = 2^n$ ____

$n!$

121 $(n + 1)n! =$ __________

$(n + 1)!$

122 $0! =$ ____

1

In Problems 123–135, simplify each of the given expressions.

123 $\dfrac{6!}{5!} = \dfrac{6 \cdot 5!}{\_\_\_\_} =$ ____

$5!$, 6

124 $\dfrac{7!}{5! \cdot 3!} = \dfrac{7 \cdot 6 \ \_\_\_\_}{5! \quad 3!} = \dfrac{7 \cdot 6}{\_\_\_\_\_\_\_\_} =$ ____

$5!$, $3 \cdot 2 \cdot 1$, 7

125 $\dfrac{8!}{4!6!} = \dfrac{8 \cdot 7 \cdot 6!}{6!(\_\_\_\_\_\_\_\_)} =$ ____

$4 \cdot 3 \cdot 2 \cdot 1$, $\frac{7}{3}$

126 $\dfrac{4!5!}{5! + 7!} = \dfrac{4!5!}{5!(\_\_\_\_\_\_\_\_)} =$ ____

$1 + 6 \cdot 7$, $\frac{24}{43}$

127 $\dfrac{3!6!}{7! - 6!} = \dfrac{3!6!}{6!(\_\_\_\_\_\_)} =$ ____

$7 - 1$, 1

128 $\dfrac{6! + 4!}{4! + 5!} = \dfrac{4!(\_\_\_\_\_\_\_\_)}{4!(1 + 5)} =$ ____

$6 \cdot 5 + 1$, $\frac{31}{6}$

24, 6, 1, 4, $\frac{5}{24}$

129 $\dfrac{1}{4!} + \dfrac{1}{3!} = \dfrac{1}{\_\_\_\_} + \dfrac{1}{\_\_\_\_} = \dfrac{\_\_\_\_ + \_\_\_\_}{24} = \_\_\_\_$

120, 24, 1, 5, $-\frac{1}{30}$

130 $\dfrac{1}{5!} - \dfrac{1}{4!} = \dfrac{1}{\_\_\_\_} - \dfrac{1}{\_\_\_\_} = \dfrac{\_\_\_\_ - \_\_\_\_}{120} = \_\_\_\_$

$n!$, $n + 1$

131 $\dfrac{(n+1)!}{n!} = \dfrac{(n+1)\ \_\_\_\_}{n!} = \_\_\_\_\_\_$

$(n-1)!$, $\dfrac{1}{(n+1)n}$

132 $\dfrac{(n-1)!}{(n+1)!} = \dfrac{(n-1)!}{(n+1)(n)\ \_\_\_\_\_\_} = \_\_\_\_\_\_$

$(2n)!$

$2n - 1$

$\dfrac{2n+1}{2n}$

133 $\dfrac{(2n+1)!(2n-1)!}{[(2n)!]^2} = \dfrac{(2n+1)!(2n-1)!}{(2n)!\ \_\_\_\_\_\_}$

$= \dfrac{(2n+1)(2n)!(2n-1)!}{(2n)!(2n)(\_\_\_\_\_\_)!}$

$= \dfrac{}{\_\_\_\_\_\_}$

$(n-k)!$

$(n-k-1)!$

134 $\dfrac{(n-k+1)!}{(n-k-1)!(n-k)!} = \dfrac{(n-k+1)\ \_\_\_\_\_\_}{(n-k-1)!(n-k)!}$

$= \dfrac{n-k+1}{\_\_\_\_\_\_}$

$n!$

$\frac{3}{2}$

$-\dfrac{2n}{3}$

135 $\dfrac{n! - (n+1)!}{n! + \frac{1}{2}n(n-1)!} = \dfrac{n! - (n+1)!}{n! + \frac{1}{2}(\_\_\_\_)}$

$= \dfrac{n!(1 - n - 1)}{n!(\_\_\_\_)}$

$= \_\_\_\_$

a^n

$1!$

$\dfrac{n(n-1)}{2!}$

$\dfrac{n(n-1)(n-2)}{3!}$

b^n

136 In expanding $(a + b)^n$, where n is a positive integer, the first term of the expansion is ____. The second term of the expansion is written as $\left(\dfrac{n}{\_\_\_\_}\right) a^{n-1}b$. The third term of the expansion is written as $\left[\dfrac{}{\_\_\_\_\_\_}\right] a^{n-2}b^2$. The fourth term of the expansion is written as $\left[\dfrac{}{\_\_\_\_\_\_\_\_}\right] a^{n-3}b^3$. The pattern of this expansion continues to the last term of the expansion, which is ____.

In Problems 137–140, expand each binomial.

137 $(2x + y)^4$

$2x$, $16x^4$

The first term of the expansion is $(\_\_\_\_)^4 = \_\_\_\_$.

$\dfrac{4}{1!}$, $32x^3y$

The second term is $\left(\dfrac{}{\_\_\_\_}\right)(2x)^3y = \_\_\_\_\_\_$.

$2!$, $24x^2y^2$

The third term is $\left(\dfrac{4 \cdot 3}{\_\_\_\_}\right)(2x)^2y^2 = \_\_\_\_\_\_$.

$3!$, $8xy^3$

The fourth term is $\left(\dfrac{4 \cdot 3 \cdot 2}{\_\_\_\_}\right)(2x)\,y^3 = \_\_\_\_\_\_$.

y^4

The fifth term is ____.

Therefore, the expansion is

$16x^4 + 32x^3y + 24x^2y^2 + 8xy^3 + y^4$

$(2x + y)^4 =$ ________________

138 $\left(2 + \frac{x}{3}\right)^5$

The first term of the expansion is

2, 32

$(\_\_\_)^5 =$ ____

The second term is

$\frac{80}{3}x$

$\left[\frac{5(2)^4}{1!}\right]\left(\frac{x}{3}\right) =$ ____

The third term is

4, $\frac{80}{9}x^2$

$\left[\frac{5(\_\_\_)(2)^3}{2!}\right]\left(\frac{x}{3}\right)^2 =$ ____

The fourth term is

3, $\frac{40}{27}x^3$

$\left[\frac{(5)(4)(\_\_\_)(2)^2}{3!}\right]\left(\frac{x}{3}\right)^3 =$ ____

The fifth term is

4, $\frac{10}{81}x^4$

$\left[\frac{(5)(4)(3)(2)(2)}{4!}\right]\left(\frac{x}{3}\right)^{\_\_\_} =$ ____

The sixth term is

$\frac{x}{3}, \frac{x^5}{243}$

$\left(\_\_\_\right)^5 =$ ____

Therefore, the expansion is

$\frac{40x^3}{27} + \frac{10x^4}{81} + \frac{x^5}{243}$

$\left(2 + \frac{x}{3}\right)^5 = 32 + \frac{80x}{3} + \frac{80x^2}{9} +$ ________________

139 $(3x^2 - y)^4$

$-y$

Write $(3x^2 - y)^4$ in the form of $(a + b)^n$ to obtain $[3x^2 + (\_\_\_)]^4$.

The first term of the expansion is

$3x^2$, $81x^8$

$(\_\_\_)^4 =$ ____

The second term is

$-y$, $-108x^6y$

$\left[\frac{4(3x^2)^3}{1!}\right](\_\_\_)^1 =$ ____

The third term is

3, $54x^4y^2$

$\left[\frac{4(\_\_\_)(3x^2)^2}{2!}\right](-y)^2 =$ ____

The fourth term is

3!, $-12x^2y^3$

$\left[\frac{4(3)(2)(3x^2)}{\_\_\_}\right](-y)^3 =$ ____

The fifth term is

4, y^4

$(-y)^{\_\_\_} =$ ____

Hence, the expansion is

$54x^4y^2 - 12x^2y^3 + y^4$

$(3x^2 - y)^4 = 81x^8 - 108x^6y +$ ________________

Answers	Problem
	140 $\left(x^2 - \frac{1}{2x}\right)^6$
$-\frac{1}{2x}$	Write $\left(x^2 - \frac{1}{2x}\right)^6$ in the form $(a+b)^n$ to obtain $\left[x^2 + (\_\_\_)\right]^6$.
	The first term of the expansion is
6, x^{12}	$(x^2)^{\_\_\_} = \_\_\_$
	The second term is
x^2, $-3x^9$	$\left[\frac{6(\_\_\_)^5}{1!}\right]\left(-\frac{1}{2x}\right) = \_\_\_$
	The third term is
x^2, $\frac{15}{4}x^6$	$\left[\frac{6(5)(\_\_\_)^4}{2!}\right]\left(-\frac{1}{2x}\right)^2 = \_\_\_$
	The fourth term is
4, $-\frac{5}{2}x^3$	$\left[\frac{6(5)(\_\_\_)(x^2)^3}{3!}\right]\left(-\frac{1}{2x}\right)^3 = \_\_\_$
	The fifth term is
3, $\frac{15}{16}$	$\left[\frac{6(5)(4)(\_\_\_)(x^2)^2}{4!}\right]\left(-\frac{1}{2x}\right)^4 = \_\_\_$
	The sixth term is
2, $-\frac{3}{16x^3}$	$\left[\frac{6(5)(4)(3)(\_\_\_)(x^2)}{5!}\right]\left(-\frac{1}{2x}\right)^5 = \_\_\_$
	The seventh term is
6, $\frac{1}{64x^6}$	$\left(-\frac{1}{2x}\right)^{\_\_\_} = \_\_\_$
	Hence, the expansion is
$\frac{15}{16} - \frac{3}{16x^3} + \frac{1}{64x^6}$	$\left(x^2 - \frac{1}{2x}\right)^6 = x^{12} - 3x^9 + \frac{15}{4}x^6 - \frac{5}{2}x^3 + \_\_\_$

In Problems 141–144, write the first four terms of the expansion. Do not simplify.

Answers	Problem
	141 $(x+y)^7$
x^7	The first term is ___.
7	The second term is $\frac{\_\_\_}{1!}x^6y$.
6	The third term is $\frac{7 \cdot \_\_\_}{2!}x^5y^2$.
$\frac{7 \cdot 6 \cdot 5}{3!}x^4y^3$	The fourth term is ___.
	142 $(1+2x)^{13}$
1	The first term is ___.
1!	The second term is $\frac{13(1)^{12}}{\_\_\_}(2x)$.

12 | The third term is $\frac{13(\_\_\_)(1)^{11}}{2!}(2x)^2$.

$\frac{(13)(12)(11)(1)^{10}}{3!}(2x)^3$ | The fourth term is ______________.

143 $(\sqrt{x} - 3y^2)^{11}$

Write $(\sqrt{x} - 3y^2)^{11}$ in the form of $(a + b)^n$ to obtain

$-3y^2$ | $[\sqrt{x} + (\_\_\_\_)]^{11}$

$(\sqrt{x})^{11}$ | The first term is ______.

$(\sqrt{x})^{10}$ | The second term is $\frac{11(\_\_\_\_)}{1!}(-3y^2)^1$.

$(\sqrt{x})^9$ | The third term is $\frac{11(10)(\_\_\_\_)}{2!}(-3y^2)^2$.

$\frac{(11)(10)(9)(\sqrt{x})^8}{3!}(-3y^2)^3$ | The fourth term is ______________.

144 $(1 + 0.01)^{10}$

1 | The first term is ____.

1! | The second term is $\frac{10(1)^9}{\_\_\_}(0.01)^1$.

9 | The third term is $\frac{10(\_\_\_)(1)^8}{2!}(0.01)^2$.

$\frac{10(9)(8)(1)^7(0.01)^3}{3!}$ | The fourth term is ______________.

145 The kth term of the expansion of $(a + b)^n$ is given by the formula

$(k - 1)!$ | $u_k = \frac{n(n-1)(n-2)\cdots(n-k+2)}{\_\_\_\_\_\_}a^{n-k+1}b^{k-1}$

146 The $(k + 1)$st term of the expansion of $(a + b)^n$ is given by the formula

$(n - k)!$ | $u_{k+1} = \frac{n!}{(\_\_\_\_\_)\cdot k!}a^{n-k}b^k$

$k!$ | $= \frac{n(n-1)(n-2)\cdots(n-k+1)}{\_\_\_}a^{n-k}b^k$

In Problems 147–152, find the indicated term for each expression and simplify.

147 The sixth term of $(x^2 + 2y)^{12}$

In the expansion, $n = 12$, and since the sixth term is being written

6 | $k =$ ____, use

$$u_k = \frac{n(n-1)(n-2)\cdots(n-k+2)}{(k-1)!}a^{n-k+1}b^{k-1}$$

x^2, $2y$ | where $a =$ ____ and $b =$ ____. Therefore,

$25{,}344x^{14}y^5$ | $u_6 = \frac{12\cdot 11\cdot 10\cdot 9\cdot 8}{(6-1)!}(x^2)^7(2y)^5 =$ ______________

148 The eighth term of $(x - y)^{12}$

12, 8, x, $-y$

In this expansion use $n =$ ____, $k =$ ____, $a =$ ____, and $b =$ ____.
Apply the formula

$$u_k = \frac{n(n-1)(n-2)\cdots(n-k+2)}{(k-1)!}a^{n-k+1}b^{k-1}$$

to obtain

7!, $-792x^5y^7$

$$u_8 = \frac{12\cdot 11\cdot 10\cdot 9\cdot 8\cdot 7\cdot 6}{(\_\_\_\_)}x^5(-y)^7 = \_\_\_\_\_\_\_\_$$

149 The sixth term of $(5x - \frac{1}{10}y)^{12}$

12, 6, $5x$, $-\frac{1}{10}y$

In this expansion, use $n =$ ____, $k =$ ____, $a =$ ____, and $b =$ ____.
Use the formula

$$u_k = \frac{n(n-1)(n-2)\cdots(n-k+2)}{(k-1)!}a^{n-k+1}b^{k-1}$$

to obtain

5!, $-\frac{2475}{4}x^7y^5$

$$u_6 = \frac{12\cdot 11\cdot 10\cdot 9\cdot 8}{\_\_\_\_}(5x)^7(-\tfrac{1}{10}y)^5 = \_\_\_\_\_\_\_\_$$

150 The term that involves x^4 of $(y^2 + 2x)^{12}$

12, 4, y^2, $2x$

In this expansion, use $n =$ ____, $k =$ ____, $a =$ ____, and $b =$ ____.
Apply the formula

$$u_{k+1} = \frac{n(n-1)(n-2)\cdots(n-k+1)}{k!}a^{n-k}b^k$$

to obtain

4!, $7920y^{16}x^4$

$$u_5 = \frac{12\cdot 11\cdot 10\cdot 9}{\_\_\_\_}(y^2)^8(2x)^4 = \_\_\_\_\_\_\_\_$$

151 The term that involves x^7 in the expansion of $(2 - x)^{12}$

12, 7, 2, $-x$

In this expansion, $n =$ ____, $k =$ ____, $a =$ ____, and $b =$ ____.
Use the formula

$$u_{k+1} = \frac{n!}{k!(n-k)!}a^{n-k}b^k$$

to obtain

$-32x^7$

$$u_8 = \frac{12!}{7!(12-7)!}(2)^5(-x)^7 = \frac{12\cdot 11\cdot 10\cdot 9\cdot 8}{5\cdot 4\cdot 3\cdot 2\cdot 1}(\_\_\_\_\_\_)$$

or

792; $-25{,}344x^7$

$$u_8 = (\_\_\_\_)(-32x^7) = \_\_\_\_\_\_\_\_$$

152 The fourth term of $\left(3x - \frac{y}{6}\right)^9$

In this expansion $n = 9$, and since the fourth term is being written,

4, $3x$, $-\frac{y}{6}$

$k =$ ____. Also, $a =$ ____ and $b =$ ____. Use

$$u_k = \frac{n(n-1)(n-2)\cdots(n-k+2)}{(k-1)!}a^{n-k+1}b^{k-1}$$

to obtain

3!, $-\frac{567}{2}x^6y^3$

$$u_4 = \frac{9\cdot 8\cdot 7}{\_\_\_\_}(3x)^6\left(-\frac{y}{6}\right)^3 = \_\_\_\_\_\_\_\_$$

6 Finite Sums and Series

6.1 Finite Sums

153 In considering sums of terms of a sequence we can employ a special symbol for a finite sum of the terms called the ________ notation or the ______ notation.

$\sum_{k=1}^{n} a_k$ is used to represent the sum ____________________.

The symbols above and below the Σ notation indicate that k is an integer running from 1 to n inclusive; k is called the ______ of summation. Notice that k is a dummy variable; hence it can be replaced by another letter.

summation, sigma

$a_1 + a_2, + \cdots + a_n$

index

In Problems 154–159, write each given summation in expanded form.

154 $\sum_{k=1}^{5} (k^2 + 2k)$

Replace k by 1 in $k^2 + 2k$ and obtain ____.

Replace k by 2 in $k^2 + 2k$ and obtain ____.

Replace k by 3 in $k^2 + 2k$ and obtain ____.

Replace k by 4 in $k^2 + 2k$ and obtain ____.

Replace k by 5 in $k^2 + 2k$ and obtain ____.

The expanded form is

$\sum_{k=1}^{5} (k^2 + 2k) =$ ____________________

3

8

15

24

35

$3 + 8 + 15 + 24 + 35$

155 $\sum_{k=1}^{4} (k^2 + 3)$

Replace k by 1 in $k^2 + 3$ and obtain ____.

Replace k by 2 in $k^2 + 3$ and obtain ____.

Replace k by 3 in $k^2 + 3$ and obtain ____.

Replace k by 4 in $k^2 + 3$ and obtain ____.

The expanded form is

$\sum_{k=1}^{4} (k^2 + 3) =$ ______________

4

7

12

19

$4 + 7 + 12 + 19$

156 $\sum_{k=3}^{7} \frac{1}{k}$

Replace k by 3 and obtain ____.

Replace k by 4 and obtain ____.

Replace k by 5 and obtain ____.

Replace k by 6 and obtain ____.

$\frac{1}{3}$

$\frac{1}{4}$

$\frac{1}{5}$

$\frac{1}{6}$

Answer	Problem
$\frac{1}{7}$	Replace k by 7 and obtain ____.
	The expanded form is
$\frac{1}{3}+\frac{1}{4}+\frac{1}{5}+\frac{1}{6}+\frac{1}{7}$	$\sum_{k=3}^{7} \frac{1}{k} =$ ________________
	157 $\sum_{k=2}^{5} \frac{k}{k-1}$
2	Replace k by 2 and obtain ____.
$\frac{3}{2}$	Replace k by 3 and obtain ____.
$\frac{4}{3}$	Replace k by 4 and obtain ____.
$\frac{5}{4}$	Replace k by 5 and obtain ____.
	The expanded form is
$2+\frac{3}{2}+\frac{4}{3}+\frac{5}{4}$	$\sum_{k=2}^{5} \frac{k}{k-1} =$ ________________
	158 $\sum_{k=1}^{3} \frac{(-1)^k}{(3k-1)^2}$
$-\frac{1}{4}$	Replace k by 1 and obtain ____.
$\frac{1}{25}$	Replace k by 2 and obtain ____.
$-\frac{1}{64}$	Replace k by 3 and obtain ____.
	The expanded form is
$-\frac{1}{4}+\frac{1}{25}-\frac{1}{64}$	$\sum_{k=1}^{3} \frac{(-1)^k}{(3k-1)^2} =$ ________________
	159 $\sum_{k=1}^{4} 2^{k+1}$
4	Replace k by 1 and obtain ____.
8	Replace k by 2 and obtain ____.
16	Replace k by 3 and obtain ____.
32	Replace k by 4 and obtain ____.
	The expanded form is
$4+8+16+32$	$\sum_{k=1}^{4} 2^{k+1} =$ ________________

In Problems 160–163, write the given finite sum in sigma notation, where the index k runs from 0 to 4.

Answer	Problem
	160 $1+\frac{1}{2}+\frac{1}{4}+\frac{1}{8}+\frac{1}{16}$
	The finite sum can be written as
$\frac{1}{2}$, $\frac{1}{2}$	$(\frac{1}{2})^0+(\frac{1}{2})^1+(\frac{1}{2})^2+(\_\_\_\_)^3+(\_\_\_\_)^4$
$(\frac{1}{2})^k$	The general term is ______. The required sigma notation is
$\sum_{k=0}^{4}\left(\frac{1}{2}\right)^k$	____________.

$\sqrt{2k+1}$

$\sum_{k=0}^{4} \sqrt{2k+1}$

161 $\sqrt{1}+\sqrt{3}+\sqrt{5}+\sqrt{7}+\sqrt{9}$

The general term of the finite sum is ________. The required sigma notation is ________.

$(-1)^k(k+1)^3$

$\sum_{k=0}^{4} (-1)^k(k+1)^3$

162 $1 - 8 + 27 - 64 + 125$

The general term of the finite sum is ________.

The required sigma notation is ________.

$\frac{2k+1}{3k+1}$

$\sum_{k=0}^{4} \frac{2k+1}{3k+1}$

163 $\frac{1}{1}+\frac{3}{4}+\frac{5}{7}+\frac{7}{10}+\frac{9}{13}$

The general term of the finite sum is ________. The required sigma notation is ________.

In Problems 164–167, determine whether the given statements are true or false.

False

164 $\sum_{k=1}^{16} k^2 = \left(\sum_{k=1}^{16} k\right)^2$ ________

True

165 $\sum_{k=1}^{20} (3k+6) = 3\left(\sum_{k=1}^{20} k\right) + 120$ ________

True

166 $\sum_{k=0}^{49} (k+1)^2 = \sum_{k=1}^{50} k^2$ ________

False

167 $\sum_{k=1}^{n} (k+1)^2 = \left(\sum_{k=1}^{n} k^2\right) + n$ ________

In Problems 168–172, find the numerical value of each of the given finite sums.

168 $\sum_{k=1}^{4} k$

The expanded form is

$1+2+3+4$

$\sum_{k=1}^{4} k =$ ________

10

The finite sum is equal to ____.

169 $\sum_{i=0}^{3} (\frac{1}{3})^i$

The expanded form is

$1+\frac{1}{3}+\frac{1}{9}+\frac{1}{27}$

$\sum_{i=0}^{3} (\frac{1}{3})^i =$ ________

$\frac{40}{27}$

The finite sum is equal to ____.

170 $\sum_{k=1}^{3}(4k^2 - 3k)$

The expanded form is

$\sum_{k=1}^{3}(4k^2 - 3k) = [4(1)^2 - 3(1)] + [4(2)^2 - 3(2)] +$ __________ $[4(3)^2 - 3(3)]$

$= 1 +$ ____ $+$ ____ 10, 27

$=$ ____ 38

171 $\sum_{i=3}^{6} i(i - 2)$

The expanded form is

$\sum_{i=3}^{6} i(i - 2) = 3(3 - 2) + 4(4 - 2) + 5(5 - 2) +$ __________ $6(6 - 2)$

$= 3 + 8 +$ ____ $+$ ____ 15, 24

$=$ ____ 50

172 $\sum_{k=2}^{5} \frac{k - 1}{k + 1}$

The expanded form is

$\sum_{k=2}^{5} \frac{k - 1}{k + 1} = \frac{2 - 1}{2 + 1} + \frac{3 - 1}{3 + 1} + \frac{4 - 1}{4 + 1} +$ ________ $\frac{5 - 1}{5 + 1}$

$=$ ____ $+$ ____ $+$ ____ $+$ ____ $\frac{1}{3}, \frac{2}{4}, \frac{3}{5}, \frac{4}{6}$

$=$ ____ $\frac{21}{10}$

6.2 Geometric Series

173 A series of the form $\sum_{k=1}^{\infty} ar^{k-1}$ or of the form $\sum_{k=0}^{\infty} ar^k$, where a is constant, is called a ____________________. geometric series

174 Given a geometric series $\sum_{k=1}^{\infty} ar^{k-1}$, if $|r| < 1$, then

$\sum_{k=1}^{\infty} ar^{k-1} =$ ________ or $\sum_{k=0}^{\infty} ar^k =$ ________ $\frac{a}{1 - r}, \frac{a}{1 - r}$

In Problems 175–179, find the sum of each geometric series.

175 $2 - \frac{4}{3} + \frac{8}{9} - \cdots + (-1)^n 2(\frac{2}{3})^n + \cdots$

The geometric series $2 - \frac{4}{3} + \frac{8}{9} - \cdots + (-1)^n 2(\frac{2}{3})^n + \cdots$

is written in summation notation as __________ $\sum_{k=0}^{\infty} 2(-1)^k(\frac{2}{3})^k$

Using $a =$ ____ and $r =$ ____, we have $2, -\frac{2}{3}$

$\sum_{k=0}^{\infty} 2(-1)^k(\frac{2}{3})^k = \frac{2}{1 - (-\frac{2}{3})} =$ ____ $\frac{6}{5}$

176 $1 + \frac{1}{2} + \frac{1}{4} + \frac{1}{8} + \cdots + (\frac{1}{2})^n + \cdots$

Using the summation notation, the series

$1 + \frac{1}{2} + \frac{1}{4} + \frac{1}{8} + \cdots + (\frac{1}{2})^n + \cdots$

is written as ______________. Using $a = 1$ and $r =$ _____, we have

$\sum_{k=0}^{\infty} (\frac{1}{2})^k = \frac{1}{1 - \frac{1}{2}} =$ _____

Answer: $\sum_{k=0}^{\infty} (\frac{1}{2})^k$, $\frac{1}{2}$; 2

177 $7 + \frac{7}{10} + \frac{7}{100} + \frac{7}{1000} + \cdots + 7(\frac{1}{10})^n + \cdots$

Using the summation notation, the series can be written as

______________. Using $a = 7$ and $r = \frac{1}{10}$, we obtain

$\sum_{k=0}^{\infty} 7\left(\frac{1}{10}\right)^k = \frac{7}{1 - \frac{1}{10}} =$ _____

Answer: $\sum_{k=0}^{\infty} 7(\frac{1}{10})^k$; $\frac{70}{9}$

178 $5 + \frac{5}{9} + \frac{5}{81} + \cdots + 5(\frac{1}{9})^n + \cdots$

Using the summation notation, the series can be written as

______________. Using $a = 5$ and $r = \frac{1}{9}$, we obtain

$\sum_{k=0}^{\infty} 5\left(\frac{1}{9}\right)^k = \frac{5}{1 - \frac{1}{9}} =$ _____

Answer: $\sum_{k=0}^{\infty} 5(\frac{1}{9})^k$; $\frac{45}{8}$

179 $3 + \frac{3}{7} + \frac{3}{49} + \cdots + 3(\frac{1}{7})^n + \cdots$

Using the summation notation, the series can be written as

______________. Using $a =$ _____ and $r =$ _____, we obtain

$\sum_{k=0}^{\infty} 3\left(\frac{1}{7}\right)^k = \frac{3}{1 - \frac{1}{7}} =$ _____

Answer: $\sum_{k=0}^{\infty} 3(\frac{1}{7})^k$, 3, $\frac{1}{7}$; $\frac{7}{2}$

In Problems 180–182, use geometric series to find the rational number that corresponds to the given decimal number.

180 $0.\overline{7}$

From the numerical expression $0.\overline{7}$, we obtain the geometric series

$$0.\overline{7} = \frac{7}{10} + \frac{7}{100} + \frac{7}{1000} + \frac{7}{10{,}000} + \cdots + \frac{7}{10^n} + \cdots$$

$=$ ______________________________

$=$ ______________

Using $a = \frac{7}{10}$ and $r = \frac{1}{10}$, we obtain

$0.\overline{7} = \frac{\frac{7}{10}}{1 - \frac{1}{10}} =$ _____

Answer: $\frac{7}{10} + \frac{7}{10}\left(\frac{1}{10}\right) + \frac{7}{10}\left(\frac{1}{10}\right)^2 + \cdots + \frac{7}{10}\left(\frac{1}{10}\right)^{n-1} + \cdots$; $\sum_{k-1}^{\infty} \frac{7}{10}\left(\frac{1}{10}\right)^{k-1}$; $\frac{7}{9}$

181 $0.\overline{32}$

$$0.\overline{32} = \frac{32}{100} + \frac{32}{10{,}000} + \frac{32}{1{,}000{,}000} + \cdots + \frac{32}{(100)^n} + \cdots$$

$= \underline{\qquad\qquad}$

$\dfrac{32}{100} + \dfrac{32}{100}\left(\dfrac{1}{100}\right) + \dfrac{32}{100}\left(\dfrac{1}{100}\right)^2 + \cdots + \left(\dfrac{32}{100}\right)\left(\dfrac{1}{100}\right)^{n-1} + \cdots$

$= \underline{\qquad\qquad}$

$\displaystyle\sum_{k=1}^{\infty} \frac{32}{100}\left(\frac{1}{100}\right)^{k-1}$

Using $a = \frac{32}{100}$ and $r = \frac{1}{100}$, we obtain

$$0.\overline{32} = \frac{\frac{32}{100}}{1 - \frac{1}{100}} = \underline{\qquad}$$

$\dfrac{32}{99}$

182 $0.\overline{534}$

$$0.\overline{534} = \frac{534}{1000} + \frac{534}{1{,}000{,}000} + \cdots + \frac{534}{(1000)^n} + \cdots$$

$= \underline{\qquad\qquad}$

$\displaystyle\sum_{k=1}^{\infty} \frac{534}{1000}\left(\frac{1}{1000}\right)^{k-1}$

Using $a = \frac{534}{1000}$ and $r = \frac{1}{1000}$, we have

$$0.\overline{534} = \frac{\frac{534}{1000}}{1 - \frac{1}{1000}} = \underline{\qquad}$$

$\dfrac{534}{999}$

Chapter Test

1 For $z_1 = 2 + 3i$ and $z_2 = 3 - i$, find

(a) $z_1 + z_2$ (b) $z_1 - z_2$ (c) $z_1 z_2$

(d) $\overline{z}_2$ (e) $\dfrac{z_1}{z_2}$ (f) $|z_1|$

2 Change the given complex number from polar form to rectangular form.

(a) $z = 4\left(\cos \dfrac{3\pi}{4} + i \sin \dfrac{3\pi}{4}\right)$

(b) $z = 6\left(\cos \dfrac{5\pi}{6} + i \sin \dfrac{5\pi}{6}\right)$

3 Change the given complex number from rectangular form to polar form.

(a) $z = 2\sqrt{3} + 2i$ (b) $z = \sqrt{2} - \sqrt{2}i$

4 If $z_1 = 3\left(\cos \dfrac{\pi}{2} + i \sin \dfrac{\pi}{2}\right)$ and $z_2 = 4\left(\cos \dfrac{\pi}{3} + i \sin \dfrac{\pi}{3}\right)$, find the following in polar form,

(a) $z_1 z_2$ (b) $z_1 \div z_2$ (c) z_1^4

5 Find the fourth roots of $z = 16(\cos 100° + i \sin 100°)$ in polar form.

6 Find the cube roots of $z = 2 - 2\sqrt{3}i$ in polar form.

7 Find the fifth roots of unity in polar form.

8 Find a fourth-degree polynomial function f that has -2, 2, and $1 - i$ as zeros.

9 Express each of the following finite sums in sigma notation.

(a) $1 + \frac{1}{4} + \frac{1}{9} + \frac{1}{16}$ (b) $\frac{1}{2} - \frac{1}{4} + \frac{1}{6} - \frac{1}{8}$

(c) $-\frac{1}{2^2} + \frac{1}{3^2} - \frac{1}{4^2} + \frac{1}{5^2}$ (d) $-\frac{1}{2} + \frac{2}{5} - \frac{3}{10} + \frac{4}{17}$

10 Find the numerical values of each of the following sums.

(a) $\sum_{k=0}^{3} \frac{1}{2^k}$ (b) $\sum_{k=1}^{4} 2k$ (c) $\sum_{k=1}^{3} (2k + 1)$

(d) $\sum_{k=0}^{3} \frac{2^k}{3k - 1}$ (e) $\sum_{k=0}^{\infty} (\frac{1}{7})^k$ (f) $\sum_{k=0}^{\infty} (\frac{3}{11})^k$

11 Simplify the following expressions.

(a) $\frac{3! + 5!}{6!}$ (b) $\frac{(n + 1)!}{n!}$ (c) $\frac{(n + 2)!}{(n - 1)!}$

12 Expand each of the following expressions by using the binomial theorem.

(a) $(x + y)^5$ (b) $(2x + y)^4$ (c) $(3x^2 - 2y)^3$ (d) $(x - 3y)^4$

13 Find the indicated term in each of the following binomial expansions and simplify your result.

(a) The seventh term of $(x - 2y)^{12}$

(b) The fifth term of $(x + y)^8$

(c) The term involving x^7 of $(2 - x)^{12}$

Answers

1 (a) $5 + 2i$ (b) $-1 + 4i$ (c) $9 + 7i$
(d) $3 + i$ (e) $\frac{3}{10} + \frac{11}{10}i$ (f) $\sqrt{13}$

2 (a) $-2\sqrt{2} + 2\sqrt{2}i$ (b) $-3\sqrt{3} + 3i$

3 (a) $4\left(\cos \frac{\pi}{6} + i \sin \frac{\pi}{6}\right)$ (b) $2\left(\cos \frac{7\pi}{4} + i \sin \frac{7\pi}{4}\right)$

4 (a) $12\left(\cos \frac{5\pi}{6} + i \sin \frac{5\pi}{6}\right)$ (b) $\frac{3}{4}\left(\cos \frac{\pi}{6} + i \sin \frac{\pi}{6}\right)$ (c) $81(\cos 2\pi + i \sin 2\pi)$

5 $2(\cos 25° + i \sin 25°)$, $2(\cos 115° + i \sin 115°)$,
$2(\cos 205° + i \sin 205°)$, $2(\cos 295° + i \sin 295°)$

6 $\sqrt[3]{4}(\cos 100° + i \sin 100°)$, $\sqrt[3]{4}(\cos 220° + i \sin 220°)$,
$\sqrt[3]{4}(\cos 340° + i \sin 340°)$

7 $\cos 0° + i \sin 0°$, $\cos 72° + i \sin 72°$, $\cos 144° + i \sin 144°$,
$\cos 216° + i \sin 216°$, $\cos 288° + i \sin 288°$

8 $x^4 - 2x^3 - 2x^2 + 8x - 8$

9 (a) $\sum_{k=1}^{4} \frac{1}{k^2}$ (b) $\sum_{k=1}^{4} \frac{(-1)^{k+1}}{2k}$

(c) $\sum_{k=1}^{4} \frac{(-1)^k}{(k + 1)^2}$ (d) $\sum_{k=1}^{4} \frac{(-1)^k k}{k^2 + 1}$

10 (a) $\frac{15}{8}$ (b) 20 (c) 15 (d) $\frac{9}{5}$ (e) $\frac{7}{6}$ (f) $\frac{11}{8}$

11 (a) $\frac{7}{40}$ (b) $(n+1)$ (c) $(n+2)(n+1)(n)$

12 (a) $x^5 + 5x^4y + 10x^3y^2 + 10x^2y^3 + 5xy^4 + y^5$ (b) $16x^4 + 32x^3y + 24x^2y^2 + 8xy^3 + y^4$
(c) $27x^6 - 54x^4y + 36x^2y^2 - 8y^3$ (d) $x^4 - 12x^3y + 54x^2y^2 - 108xy^3 + 81y^4$

13 (a) $59{,}136x^6y^6$ (b) $70x^4y^4$ (c) $-25{,}344x^7$

APPENDIX

APPENDIX

Tables

TABLE I COMMON LOGARITHMS

n	0.00	0.01	0.02	0.03	0.04	0.05	0.06	0.07	0.08	0.09
1.0	.0000	.0043	.0086	.0128	.0170	.0212	.0253	.0294	.0334	.0374
1.1	.0414	.0453	.0492	.0531	.0569	.0607	.0645	.0682	.0719	.0755
1.2	.0792	.0828	.0864	.0899	.0934	.0969	.1004	.1038	.1072	.1106
1.3	.1139	.1173	.1206	.1239	.1271	.1303	.1335	.1367	.1399	.1430
1.4	.1461	.1492	.1523	.1553	.1584	.1614	.1644	.1673	.1703	.1732
1.5	.1761	.1790	.1818	.1847	.1875	.1903	.1931	.1959	.1987	.2014
1.6	.2041	.2068	.2095	.2122	.2148	.2175	.2201	.2227	.2253	.2279
1.7	.2304	.2330	.2355	.2380	.2405	.2430	.2455	.2480	.2504	.2529
1.8	.2553	.2577	.2601	.2625	.2648	.2672	.2695	.2718	.2742	.2765
1.9	.2788	.2810	.2833	.2856	.2878	.2900	.2923	.2945	.2967	.2989
2.0	.3010	.3032	.3054	.3075	.3096	.3118	.3139	.3160	.3181	.3201
2.1	.3222	.3243	.3263	.3284	.3304	.3324	.3345	.3365	.3385	.3404
2.2	.3424	.3444	.3464	.3483	.3502	.3522	.3541	.3560	.3579	.3598
2.3	.3617	.3636	.3655	.3674	.3692	.3711	.3729	.3747	.3766	.3784
2.4	.3802	.3820	.3838	.3856	.3874	.3892	.3909	.3927	.3945	.3962
2.5	.3979	.3997	.4014	.4031	.4048	.4065	.4082	.4099	.4116	.4133
2.6	.4150	.4166	.4183	.4200	.4216	.4232	.4249	.4265	.4281	.4298
2.7	.4314	.4330	.4346	.4362	.4378	.4393	.4409	.4425	.4440	.4456
2.8	.4472	.4487	.4502	.4518	.4533	.4548	.4564	.4579	.4594	.4609
2.9	.4624	.4639	.4654	.4669	.4683	.4698	.4713	.4728	.4742	.4757
3.0	.4771	.4786	.4800	.4814	.4829	.4843	.4857	.4871	.4886	.4900
3.1	.4914	.4928	.4942	.4955	.4969	.4983	.4997	.5011	.5024	.5038
3.2	.5051	.5065	.5079	.5092	.5105	.5119	.5132	.5145	.5159	.5172
3.3	.5185	.5198	.5211	.5224	.5237	.5250	.5263	.5276	.5289	.5302
3.4	.5315	.5328	.5340	.5353	.5366	.5378	.5391	.5403	.5416	.5428
3.5	.5441	.5453	.5465	.5478	.5490	.5502	.5514	.5527	.5539	.5551
3.6	.5563	.5575	.5587	.5599	.5611	.5623	.5635	.5647	.5658	.5670
3.7	.5682	.5694	.5705	.5717	.5729	.5740	.5752	.5763	.5775	.5786
3.8	.5798	.5809	.5821	.5832	.5843	.5855	.5866	.5877	.5888	.5899
3.9	.5911	.5922	.5933	.5944	.5955	.5966	.5977	.5988	.5999	.6010
4.0	.6021	.6031	.6042	.6053	.6064	.6075	.6085	.6096	.6107	.6117
4.1	.6128	.6138	.6149	.6160	.6170	.6180	.6191	.6201	.6212	.6222
4.2	.6232	.6243	.6253	.6263	.6274	.6284	.6294	.6304	.6314	.6325
4.3	.6335	.6345	.6355	.6365	.6375	.6385	.6395	.6405	.6415	.6425
4.4	.6435	.6444	.6454	.6464	.6474	.6484	.6493	.6503	.6513	.6522
4.5	.6532	.6542	.6551	.6561	.6571	.6580	.6590	.6599	.6609	.6618
4.6	.6628	.6637	.6646	.6656	.6665	.6675	.6684	.6693	.6702	.6712
4.7	.6721	.6730	.6739	.6749	.6758	.6767	.6776	.6785	.6794	.6803
4.8	.6812	.6821	.6830	.6839	.6848	.6857	.6866	.6875	.6884	.6893
4.9	.6902	.6911	.6920	.6928	.6937	.6946	.6955	.6964	.6972	.6981

n	0.00	0.01	0.02	0.03	0.04	0.05	0.06	0.07	0.08	0.09
5.0	.6990	.6998	.7007	.7016	.7024	.7033	.7042	.7050	.7059	.7067
5.1	.7076	.7084	.7093	.7101	.7110	.7118	.7126	.7135	.7143	.7152
5.2	.7160	.7168	.7177	.7185	.7193	.7202	.7210	.7218	.7226	.7235
5.3	.7243	.7251	.7259	.7267	.7275	.7284	.7292	.7300	.7308	.7316
5.4	.7324	.7332	.7340	.7348	.7356	.7364	.7372	.7380	.7388	.7396
5.5	.7404	.7412	.7419	.7427	.7435	.7443	.7451	.7459	.7466	.7474
5.6	.7482	.7490	.7497	.7505	.7513	.7520	.7528	.7536	.7543	.7551
5.7	.7559	.7566	.7574	.7582	.7589	.7597	.7604	.7612	.7619	.7627
5.8	.7634	.7642	.7649	.7657	.7664	.7672	.7679	.7686	.7694	.7701
5.9	.7709	.7716	.7723	.7731	.7738	.7745	.7752	.7760	.7767	.7774
6.0	.7782	.7789	.7796	.7803	.7810	.7818	.7825	.7832	.7839	.7846
6.1	.7853	.7860	.7868	.7875	.7882	.7889	.7896	.7903	.7910	.7917
6.2	.7924	.7931	.7938	.7945	.7952	.7959	.7966	.7973	.7980	.7987
6.3	.7993	.8000	.8007	.8014	.8021	.8028	.8035	.8041	.8048	.8055
6.4	.8062	.8069	.8075	.8082	.8089	.8096	.8102	.8109	.8116	.8122
6.5	.8129	.8136	.8142	.8149	.8156	.8162	.8169	.8176	.8182	.8189
6.6	.8195	.8202	.8209	.8215	.8222	.8228	.8235	.8241	.8248	.8254
6.7	.8261	.8267	.8274	.8280	.8287	.8293	.8299	.8306	.8312	.8319
6.8	.8325	.8331	.8338	.8344	.8351	.8357	.8363	.8370	.8376	.8382
6.9	.8388	.8395	.8401	.8407	.8414	.8420	.8426	.8432	.8439	.8445
7.0	.8451	.8457	.8463	.8470	.8476	.8482	.8488	.8494	.8500	.8506
7.1	.8513	.8519	.8525	.8531	.8537	.8543	.8549	.8555	.8561	.8567
7.2	.8573	.8579	.8585	.8591	.8597	.8603	.8609	.8615	.8621	.8627
7.3	.8633	.8639	.8645	.8651	.8657	.8663	.8669	.8675	.8681	.8686
7.4	.8692	.8698	.8704	.8710	.8716	.8722	.8727	.8733	.8739	.8745
7.5	.8751	.8756	.8762	.8768	.8774	.8779	.8785	.8791	.8797	.8802
7.6	.8808	.8814	.8820	.8825	.8831	.8837	.8842	.8848	.8854	.8859
7.7	.8865	.8871	.8876	.8882	.8887	.8893	.8899	.8904	.8910	.8915
7.8	.8921	.8927	.8932	.8938	.8943	.8949	.8954	.8960	.8965	.8971
7.9	.8976	.8982	.8987	.8993	.8998	.9004	.9009	.9015	.9020	.9025
8.0	.9031	.9036	.9042	.9047	.9053	.9058	.9063	.9069	.9074	.9079
8.1	.9085	.9090	.9096	.9101	.9106	.9112	.9117	.9122	.9128	.9133
8.2	.9138	.9143	.9149	.9154	.9159	.9165	.9170	.9175	.9180	.9186
8.3	.9191	.9196	.9201	.9206	.9212	.9217	.9222	.9227	.9232	.9238
8.4	.9243	.9248	.9253	.9258	.9263	.9269	.9274	.9279	.9284	.9289
8.5	.9294	.9299	.9304	.9309	.9315	.9320	.9325	.9330	.9335	.9340
8.6	.9345	.9350	.9355	.9360	.9365	.9370	.9375	.9380	.9385	.9390
8.7	.9395	.9400	.9405	.9410	.9415	.9420	.9425	.9430	.9435	.9440
8.8	.9445	.9450	.9455	.9460	.9465	.9469	.9474	.9479	.9484	.9489
8.9	.9494	.9499	.9504	.9509	.9513	.9518	.9523	.9528	.9533	.9538
9.0	.9542	.9547	.9552	.9557	.9562	.9566	.9571	.9576	.9581	.9586
9.1	.9590	.9595	.9600	.9605	.9609	.9614	.9619	.9624	.9628	.9633
9.2	.9638	.9643	.9647	.9652	.9657	.9661	.9666	.9671	.9675	.9680
9.3	.9685	.9689	.9694	.9699	.9703	.9708	.9713	.9717	.9722	.9727
9.4	.9731	.9736	.9741	.9745	.9750	.9754	.9759	.9763	.9768	.9773
9.5	.9777	.9782	.9786	.9791	.9795	.9800	.9805	.9809	.9814	.9818
9.6	.9823	.9827	.9832	.9836	.9841	.9845	.9850	.9854	.9859	.9863
9.7	.9868	.9872	.9877	.9881	.9886	.9890	.9894	.9899	.9903	.9908
9.8	.9912	.9917	.9921	.9926	.9930	.9934	.9939	.9943	.9948	.9952
9.9	.9956	.9961	.9965	.9969	.9974	.9978	.9983	.9987	.9991	.9996

TABLE II NATURAL LOGARITHMS

t	0.00	0.01	0.02	0.03	0.04	0.05	0.06	0.07	0.08	0.09
1.0	0.0000	0.0100	0.0198	0.0296	0.0392	0.0488	0.0583	0.0677	0.0770	0.0862
1.1	0.0953	0.1044	0.1133	0.1222	0.1310	0.1398	0.1484	0.1570	0.1655	0.1740
1.2	0.1823	0.1906	0.1989	0.2070	0.2151	0.2231	0.2311	0.2390	0.2469	0.2546
1.3	0.2624	0.2700	0.2776	0.2852	0.2927	0.3001	0.3075	0.3148	0.3221	0.3293
1.4	0.3365	0.3436	0.3507	0.3577	0.3646	0.3716	0.3784	0.3853	0.3920	0.3988
1.5	0.4055	0.4121	0.4187	0.4253	0.4318	0.4383	0.4447	0.4511	0.4574	0.4637
1.6	0.4700	0.4762	0.4824	0.4886	0.4947	0.5008	0.5068	0.5128	0.5188	0.5247
1.7	0.5306	0.5365	0.5423	0.5481	0.5539	0.5596	0.5653	0.5710	0.5766	0.5822
1.8	0.5878	0.5933	0.5988	0.6043	0.6098	0.6152	0.6206	0.6259	0.6313	0.6366
1.9	0.6419	0.6471	0.6523	0.6575	0.6627	0.6678	0.6729	0.6780	0.6831	0.6881
2.0	0.6931	0.6981	0.7031	0.7080	0.7130	0.7178	0.7227	0.7275	0.7324	0.7372
2.1	0.7419	0.7467	0.7514	0.7561	0.7608	0.7655	0.7701	0.7747	0.7793	0.7839
2.2	0.7885	0.7930	0.7975	0.8020	0.8065	0.8109	0.8154	0.8198	0.8242	0.8286
2.3	0.8329	0.8372	0.8416	0.8459	0.8502	0.8544	0.8587	0.8629	0.8671	0.8713
2.4	0.8755	0.8796	0.8838	0.8879	0.8920	0.8961	0.9002	0.9042	0.9083	0.9123
2.5	0.9163	0.9203	0.9243	0.9282	0.9322	0.9361	0.9400	0.9439	0.9478	0.9517
2.6	0.9555	0.9594	0.9632	0.9670	0.9708	0.9746	0.9783	0.9821	0.9858	0.9895
2.7	0.9933	0.9969	1.0006	1.0043	1.0080	1.0116	1.0152	0.0188	1.0225	1.0260
2.8	1.0296	1.0332	1.0367	1.0403	1.0438	1.0473	1.0508	1.0543	1.0578	1.0613
2.9	1.0647	1.0682	1.0716	1.0750	1.0784	1.0818	1.0852	1.0886	1.0919	1.0953
3.0	1.0986	1.1019	1.1053	1.1086	1.1119	1.1151	1.1184	1.1217	1.1249	1.1282
3.1	1.1314	1.1346	1.1378	1.1410	1.1442	1.1474	1.1506	1.1537	1.1569	1.1600
3.2	1.1632	1.1663	1.1694	1.1725	1.1756	1.1787	1.1817	1.1848	1.1878	1.1909
3.3	1.1939	1.1970	1.2000	1.2030	1.2060	1.2090	1.2119	1.2149	1.2179	1.2208
3.4	1.2238	1.2267	1.2296	1.2326	1.2355	1.2384	1.2413	1.2442	1.2470	1.2499
3.5	1.2528	1.2556	1.2585	1.2613	1.2641	1.2669	1.2698	1.2726	1.2754	1.2782
3.6	1.2809	1.2837	1.2865	1.2892	1.2920	1.2947	1.2975	1.3002	1.3029	1.3056
3.7	1.3083	1.3110	1.3137	1.3164	1.3191	1.3218	1.3244	1.3271	1.3297	1.3324
3.8	1.3350	1.3376	1.3403	1.3429	1.3455	1.3481	1.3507	1.3533	1.3558	1.3584
3.9	1.3610	1.3635	1.3661	1.3686	1.3712	1.3737	1.3762	1.3788	1.3813	1.3838
4.0	1.3863	1.3888	1.3913	1.3938	1.3962	1.3987	1.4012	1.4036	1.4061	1.4085
4.1	1.4110	1.4134	1.4159	1.4183	1.4207	1.4231	1.4255	1.4279	1.4303	1.4327
4.2	1.4351	1.4375	1.4398	1.4422	1.4446	1.4469	1.4493	1.4516	1.4540	1.4563
4.3	1.4586	1.4609	1.4633	1.4656	1.4679	1.4702	1.4725	1.4748	1.4770	1.4793
4.4	1.4816	1.4839	1.4861	1.4884	1.4907	1.4929	1.4952	1.4974	1.4996	1.5019
4.5	1.5041	1.5063	1.5085	1.5107	1.5129	1.5151	1.5173	1.5195	1.5217	1.5239
4.6	1.5261	1.5282	1.5304	1.5326	1.5347	1.5369	1.5390	1.5412	1.5433	1.5454
4.7	1.5476	1.5497	1.5518	1.5539	1.5560	1.5581	1.5602	1.5623	1.5644	1.5665
4.8	1.5686	1.5707	1.5728	1.5748	1.5769	1.5790	1.5810	1.5831	1.5851	1.5872
4.9	1.5892	1.5913	1.5933	1.5953	1.5974	1.5994	1.6014	1.6034	1.6054	1.6074
5.0	1.6094	1.6114	1.6134	1.6154	1.6174	1.6194	1.6214	1.6233	1.6253	1.6273
5.1	1.6292	1.6312	1.6332	1.6351	1.6371	1.6390	1.6409	1.6429	1.6448	1.6467
5.2	1.6487	1.6506	1.6525	1.6544	1.6563	1.6582	1.6601	1.6620	1.6639	1.6658
5.3	1.6677	1.6696	1.6715	1.6734	1.6752	1.6771	1.6790	1.6808	1.6827	1.6845
5.4	1.6864	1.6882	1.6901	1.6919	1.6938	1.6956	1.6974	1.6993	1.7011	1.7029

t	0.00	0.01	0.02	0.03	0.04	0.05	0.06	0.07	0.08	0.09
5.5	1.7047	1.7066	1.7084	1.7102	1.7120	1.7138	1.7156	1.7174	1.7192	1.7210
5.6	1.7228	1.7246	1.7263	1.7281	1.7299	1.7317	1.7334	1.7352	1.7370	1.7387
5.7	1.7405	1.7422	1.7440	1.7457	1.7475	1.7492	1.7509	1.7527	1.7544	1.7561
5.8	1.7579	1.7596	1.7613	1.7630	1.7647	1.7664	1.7682	1.7699	1.7716	1.7733
5.9	1.7750	1.7766	1.7783	1.7800	1.7817	1.7834	1.7851	1.7867	1.7884	1.7901
6.0	1.7918	1.7934	1.7951	1.7967	1.7984	1.8001	1.8017	1.8034	1.8050	1.8066
6.1	1.8083	1.8099	1.8116	1.8132	1.8148	1.8165	1.8181	1.8197	1.8213	1.8229
6.2	1.8245	1.8262	1.8278	1.8294	1.8310	1.8326	1.8342	1.8358	1.8374	1.8390
6.3	1.8406	1.8421	1.8437	1.8453	1.8469	1.8485	1.8500	1.8516	1.8532	1.8547
6.4	1.8563	1.8579	1.8594	1.8610	1.8625	1.8641	1.8656	1.8672	1.8687	1.8703
6.5	1.8718	1.8733	1.8749	1.8764	1.8779	1.8795	1.8810	1.8825	1.8840	1.8856
6.6	1.8871	1.8886	1.8901	1.8916	1.8931	1.8946	1.8961	1.8976	1.8991	1.9006
6.7	1.9021	1.9036	1.9051	1.9066	1.9081	1.9095	1.9110	1.9125	1.9140	1.9155
6.8	1.9169	1.9184	1.9199	1.9213	1.9228	1.9242	1.9257	1.9272	1.9286	1.9301
6.9	1.9315	1.9330	1.9344	1.9359	1.9373	1.9387	1.9402	1.9416	1.9430	1.9445
7.0	1.9459	1.9473	1.9488	1.9502	1.9516	1.9530	1.9544	1.9559	1.9573	1.9587
7.1	1.9601	1.9615	1.9629	1.9643	1.9657	1.9671	1.9685	1.9699	1.9713	1.9727
7.2	1.9741	1.9755	1.9769	1.9782	1.9796	1.9810	1.9824	1.9838	1.9851	1.9865
7.3	1.9879	1.9892	1.9906	1.9920	1.9933	1.9947	1.9961	1.9974	1.9988	2.0001
7.4	2.0015	2.0028	2.0042	2.0055	2.0069	2.0082	2.0096	2.0109	2.0122	2.0136
7.5	2.0149	2.0162	2.0176	2.0189	2.0202	2.0215	2.0229	2.0242	2.0255	2.0268
7.6	2.0282	2.0295	2.0308	2.0321	2.0334	2.0347	2.0360	2.0373	2.0386	2.0399
7.7	2.0412	2.0425	2.0438	2.0451	2.0464	2.0477	2.0490	2.0503	2.0516	2.0528
7.8	2.0541	2.0554	2.0567	2.0580	2.0592	2.0605	2.0618	2.0631	2.0643	2.0665
7.9	2.0669	2.0681	2.0694	2.0707	2.0719	2.0732	2.0744	2.0757	2.0769	2.0782
8.0	2.0794	2.0807	2.0819	2.0832	2.0844	2.0857	2.0869	2.0882	2.0894	2.0906
8.1	2.0919	2.0931	2.0943	2.0956	2.0968	2.0980	2.0992	2.1005	2.1017	2.1029
8.2	2.1041	2.1054	2.1066	2.1078	2.1090	2.1102	2.1114	2.1126	2.1138	2.1150
8.3	2.1163	2.1175	2.1187	2.1199	2.1211	2.1223	2.1235	2.1247	2.1258	2.1270
8.4	2.1282	2.1294	2.1306	2.1318	2.1330	2.1342	2.1353	2.1365	2.1377	2.1389
8.5	2.1401	2.1412	2.1424	2.1436	2.1448	2.1459	2.1471	2.1483	2.1494	2.1506
8.6	2.1518	2.1529	2.1541	2.1552	2.1564	2.1576	2.1587	2.1599	2.1610	2.1622
8.7	2.1633	2.1645	2.1656	2.1668	2.1679	2.1691	2.1702	2.1713	2.1725	2.1736
8.8	2.1748	2.1759	2.1770	2.1782	2.1793	2.1804	2.1815	2.1827	2.1838	2.1849
8.9	2.1861	2.1872	2.1883	2.1894	2.1905	2.1917	2.1928	2.1939	2.1950	2.1961
9.0	2.1972	2.1983	2.1994	2.2006	2.2017	2.2028	2.2039	2.2050	2.2061	2.2072
9.1	2.2083	2.2094	2.2105	2.2116	2.2127	2.2138	2.2148	2.2159	2.2170	2.2181
9.2	2.2192	2.2203	2.2214	2.2225	2.2235	2.2246	2.2257	2.2268	2.2279	2.2289
9.3	2.2300	2.2311	2.2322	2.2332	2.2343	2.2354	2.2364	2.2375	2.2386	2.2396
9.4	2.2407	2.2418	2.2428	2.2439	2.2450	2.2460	2.2471	2.2481	2.2492	2.2502
9.5	2.2513	2.2523	2.2534	2.2544	2.2555	2.2565	2.2576	2.2586	2.2597	2.2607
9.6	2.2618	2.2628	2.2638	2.2649	2.2659	2.2670	2.2680	2.2690	2.2701	2.2711
9.7	2.2721	2.2732	2.2742	2.2752	2.2762	2.2773	2.2783	2.2793	2.2803	2.2814
9.8	2.2824	2.2834	2.2844	2.2854	2.2865	2.2875	2.2885	2.2895	2.2905	2.2915
9.9	2.2925	2.2935	2.2946	2.2956	2.2966	2.2976	2.2986	2.2996	2.3006	2.3016

TABLE III TRIGONOMETRIC FUNCTIONS—RADIAN MEASURE

t	sin t	cos t	tan t	cot t	sec t	csc t
.00	.0000	1.0000	.0000	—	1.000	—
.01	.0100	1.0000	.0100	99.997	1.000	100.00
.02	.0200	.9998	.0200	49.993	1.000	50.00
.03	.0300	.9996	.0300	33.323	1.000	33.34
.04	.0400	.9992	.0400	24.987	1.001	25.01
.05	.0500	.9988	.0500	19.983	1.001	20.01
.06	.0600	.9982	.0601	16.647	1.002	16.68
.07	.0699	.9976	.0701	14.262	1.002	14.30
.08	.0799	.9968	.0802	12.473	1.003	12.51
.09	.0899	.9960	.0902	11.081	1.004	11.13
.10	.0998	.9950	.1003	9.967	1.005	10.02
.11	.1098	.9940	.1104	9.054	1.006	9.109
.12	.1197	.9928	.1206	8.293	1.007	8.353
.13	.1296	.9916	.1307	7.649	1.009	7.714
.14	.1395	.9902	.1409	7.096	1.010	7.166
.15	.1494	.9888	.1511	6.617	1.011	6.692
.16	.1593	.9872	.1614	6.197	1.013	6.277
.17	.1692	.9856	.1717	5.826	1.015	5.911
.18	.1790	.9838	.1820	5.495	1.016	5.586
.19	.1889	.9820	.1923	5.200	1.018	5.295
.20	.1987	.9801	.2027	4.933	1.020	5.033
.21	.2085	.9780	.2131	4.692	1.022	4.797
.22	.2182	.9759	.2236	4.472	1.025	4.582
.23	.2280	.9737	.2341	4.271	1.027	4.386
.24	.2377	.9713	.2447	4.086	1.030	4.207
.25	.2474	.9689	.2553	3.916	1.032	4.042
.26	.2571	.9664	.2660	3.759	1.035	3.890
.27	.2667	.9638	.2768	3.613	1.038	3.749
.28	.2764	.9611	.2876	3.478	1.041	3.619
.29	.2860	.9582	.2984	3.351	1.044	3.497
.30	.2955	.9553	.3093	3.233	1.047	3.384
.31	.3051	.9523	.3203	3.122	1.050	3.278
.32	.3146	.9492	.3314	3.018	1.053	3.179
.33	.3240	.9460	.3425	2.920	1.057	3.086
.34	.3335	.9428	.3537	2.827	1.061	2.999
.35	.3429	.9394	.3650	2.740	1.065	2.916
.36	.3523	.9359	.3764	2.657	1.068	2.839
.37	.3616	.9323	.3879	2.578	1.073	2.765
.38	.3709	.9287	.3994	2.504	1.077	2.696
.39	.3802	.9249	.4111	2.433	1.081	2.630
.40	.3894	.9211	.4228	2.365	1.086	2.568
.41	.3986	.9171	.4346	2.301	1.090	2.509
.42	.4078	.9131	.4466	2.239	1.095	2.452
.43	.4169	.9090	.4586	2.180	1.100	2.399
.44	.4259	.9048	.4708	2.124	1.105	2.348

t	$\sin t$	$\cos t$	$\tan t$	$\cot t$	$\sec t$	$\csc t$
.45	.4350	.9004	.4831	2.070	1.111	2.299
.46	.4439	.8961	.4954	2.018	1.116	2.253
.47	.4529	.8916	.5080	1.969	1.122	2.208
.48	.4618	.8870	.5206	1.921	1.127	2.166
.49	.4706	.8823	.5334	1.875	1.133	2.125
.50	.4794	.8776	.5463	1.830	1.139	2.086
.51	.4882	.8727	.5594	1.788	1.146	2.048
.52	.4969	.8678	.5726	1.747	1.152	2.013
$\frac{\pi}{6}$	.5000	.8660	.5774	1.732	1.155	2.000
.53	.5055	.8628	.5859	1.707	1.159	1.978
.54	.5141	.8577	.5994	1.668	1.166	1.945
.55	.5227	.8525	.6131	1.631	1.173	1.913
.56	.5312	.8473	.6269	1.595	1.180	1.883
.57	.5396	.8419	.6410	1.560	1.188	1.853
.58	.5480	.8365	.6552	1.526	1.196	1.825
.59	.5564	.8309	.6696	1.494	1.203	1.797
.60	.5646	.8253	.6841	1.462	1.212	1.771
.61	.5729	.8196	.6989	1.431	1.220	1.746
.62	.5810	.8139	.7139	1.401	1.229	1.721
.63	.5891	.8080	.7291	1.372	1.238	1.697
.64	.5972	.8021	.7445	1.343	1.247	1.674
.65	.6052	.7961	.7602	1.315	1.256	1.652
.66	.6131	.7900	.7761	1.288	1.266	1.631
.67	.6210	.7838	.7923	1.262	1.276	1.610
.68	.6288	.7776	.8087	1.237	1.286	1.590
.69	.6365	.7712	.8253	1.212	1.297	1.571
.70	.6442	.7648	.8423	1.187	1.307	1.552
.71	.6518	.7584	.8595	1.163	1.319	1.534
.72	.6594	.7518	.8771	1.140	1.330	1.517
.73	.6669	.7452	.8949	1.117	1.342	1.500
.74	.6743	.7385	.9131	1.095	1.354	1.483
.75	.6816	.7317	.9316	1.073	1.367	1.467
.76	.6889	.7248	.9505	1.052	1.380	1.452
.77	.6961	.7179	.9697	1.031	1.393	1.437
.78	.7033	.7109	.9893	1.011	1.407	1.422
$\frac{\pi}{4}$	.7071	.7071	1.000	1.000	1.414	1.414
.79	.7104	.7038	1.009	.9908	1.421	1.408
.80	.7174	.6967	1.030	.9712	1.435	1.394
.81	.7243	.6895	1.050	.9520	1.450	1.381
.82	.7311	.6822	1.072	.9331	1.466	1.368
.83	.7379	.6749	1.093	.9146	1.482	1.355
.84	.7446	.6675	1.116	.8964	1.498	1.343

TABLE III TRIGONOMETRIC FUNCTIONS—RADIAN MEASURE

t	$\sin t$	$\cos t$	$\tan t$	$\cot t$	$\sec t$	$\csc t$
.85	.7513	.6600	1.138	.8785	1.515	1.331
.86	.7578	.6524	1.162	.8609	1.533	1.320
.87	.7643	.6448	1.185	.8437	1.551	1.308
.88	.7707	.6372	1.210	.8267	1.569	1.297
.89	.7771	.6294	1.235	.8100	1.589	1.287
.90	.7833	.6216	1.260	.7936	1.609	1.277
.91	.7895	.6137	1.286	.7774	1.629	1.267
.92	.7956	.6058	1.313	.7615	1.651	1.257
.93	.8016	.5978	1.341	.7458	1.673	1.247
.94	.8076	.5898	1.369	.7303	1.696	1.238
.95	.8134	.5817	1.398	.7151	1.719	1.229
.96	.8192	.5735	1.428	.7001	1.744	1.221
.97	.8249	.5653	1.459	.6853	1.769	1.212
.98	.8305	.5570	1.491	.6707	1.795	1.204
.99	.8360	.5487	1.524	.6563	1.823	1.196
1.00	.8415	.5403	1.557	.6421	1.851	1.188
1.01	.8468	.5319	1.592	.6281	1.880	1.181
1.02	.8521	.5234	1.628	.6142	1.911	1.174
1.03	.8573	.5148	1.665	.6005	1.942	1.166
1.04	.8624	.5062	1.704	.5870	1.975	1.160
$\frac{\pi}{3}$	.8660	.5000	1.732	.5774	2.000	1.155
1.05	.8674	.4976	1.743	.5736	2.010	1.153
1.06	.8724	.4889	1.784	.5604	2.046	1.146
1.07	.8772	.4801	1.827	.5473	2.083	1.140
1.08	.8820	.4713	1.871	.5344	2.122	1.134
1.09	.8866	.4625	1.917	.5216	2.162	1.128
1.10	.8912	.4536	1.965	.5090	2.205	1.122
1.11	.8957	.4447	2.014	.4964	2.249	1.116
1.12	.9001	.4357	2.066	.4840	2.295	1.111
1.13	.9044	.4267	2.120	.4718	2.344	1.106
1.14	.9086	.4176	2.176	.4596	2.395	1.101
1.15	.9128	.4085	2.234	.4475	2.448	1.096
1.16	.9168	.3993	2.296	.4356	2.504	1.091
1.17	.9208	.3902	2.360	.4237	2.563	1.086
1.18	.9246	.3809	2.427	.4120	2.625	1.082
1.19	.9284	.3717	2.498	.4003	2.691	1.077
1.20	.9320	.3624	2.572	.3888	2.760	1.073
1.21	.9356	.3530	2.650	.3773	2.833	1.069
1.22	.9391	.3436	2.733	.3659	2.910	1.065
1.23	.9425	.3342	2.820	.3546	2.992	1.061
1.24	.9458	.3248	2.912	.3434	3.079	1.057
1.25	.9490	.3153	3.010	.3323	3.171	1.054
1.26	.9521	.3058	3.113	.3212	3.270	1.050
1.27	.9551	.2963	3.224	.3102	3.375	1.047
1.28	.9580	.2867	3.341	.2993	3.488	1.044
1.29	.9608	.2771	3.467	.2884	3.609	1.041

t	sin t	cos t	tan t	cot t	sec t	csc t
1.30	.9636	.2675	3.602	.2776	3.738	1.038
1.31	.9662	.2579	3.747	.2669	3.878	1.035
1.32	.9687	.2482	3.903	.2562	4.029	1.032
1.33	.9711	.2385	4.072	.2456	4.193	1.030
1.34	.9735	.2288	4.256	.2350	4.372	1.027
1.35	.9757	.2190	4.455	.2245	4.566	1.025
1.36	.9779	.2092	4.673	.2140	4.779	1.023
1.37	.9799	.1994	4.913	.2035	5.014	1.021
1.38	.9819	.1896	5.177	.1931	5.273	1.018
1.39	.9837	.1798	5.471	.1828	5.561	1.017
1.40	.9854	.1700	5.798	.1725	5.883	1.015
1.41	.9871	.1601	6.165	.1622	6.246	1.013
1.42	.9887	.1502	6.581	.1519	6.657	1.011
1.43	.9901	.1403	7.055	.1417	7.126	1.010
1.44	.9915	.1304	7.602	.1315	7.667	1.009
1.45	.9927	.1205	8.238	.1214	8.299	1.007
1.46	.9939	.1106	8.989	.1113	9.044	1.006
1.47	.9949	.1006	9.887	.1011	9.938	1.005
1.48	.9959	.0907	10.983	.0910	11.029	1.004
1.49	.9967	.0807	12.350	.0810	12.390	1.003
1.50	.9975	.0707	14.101	.0709	14.137	1.003
1.51	.9982	.0608	16.428	.0609	16.458	1.002
1.52	.9987	.0508	19.670	.0508	19.695	1.001
1.53	.9992	.0408	24.498	.0408	24.519	1.001
1.54	.9995	.0308	32.461	.0308	32.476	1.000
1.55	.9998	.0208	48.078	.0208	48.089	1.000
1.56	.9999	.0108	92.620	.0108	92.626	1.000
1.57	1.0000	.0008	1255.8	.0008	1255.8	1.000
$\frac{\pi}{2}$	1.0000	.0000	—	.0000	—	1.000

TABLE IV TRIGONOMETRIC FUNCTIONS—DEGREE MEASURE

Degrees	Sin	Csc	Tan	Cot	Sec	Cos	
0° 0′	.0000	——	.0000	——	1.000	1.0000	90° 0′
10′	029	343.8	029	343.8	000	000	50′
20′	058	171.9	058	171.9	000	000	40′
30′	.0087	114.6	.0087	114.6	1.000	1.0000	30′
40′	116	85.95	116	85.94	000	0.9999	20′
50′	145	68.76	145	68.75	000	999	10′
1° 0′	.0175	57.30	.0175	57.29	1.000	.9998	89° 0′
10′	204	49.11	204	49.10	000	998	50′
20′	233	42.98	233	42.96	000	997	40′
30′	.0262	38.20	.0262	38.19	1.000	.9997	30′
40′	291	34.38	291	34.37	000	996	20′
50′	320	31.26	320	31.24	001	995	10′
2° 0′	.0349	28.65	.0349	28.64	1.001	.9994	88° 0′
10′	378	26.45	378	26.43	001	993	50′
20′	407	24.56	407	24.54	001	992	40′
30′	.0436	22.93	.0437	22.90	1.001	.9990	30′
40′	465	21.49	466	21.47	001	989	20′
50′	494	20.23	495	20.21	001	988	10′
3° 0′	.0523	19.11	.0524	19.08	1.001	.9986	87° 0′
10′	552	18.10	553	18.07	002	985	50′
20′	581	17.20	582	17.17	002	983	40′
30′	.0610	16.38	.0612	16.35	1.002	.9981	30′
40′	640	15.64	641	15.60	002	980	20′
50′	669	14.96	670	14.92	002	978	10′
4° 0′	.0698	14.34	.0699	14.30	1.002	.9976	86° 0′
10′	727	13.76	729	13.73	003	974	50′
20′	756	13.23	758	13.20	003	971	40′
30′	.0785	12.75	.0787	12.71	1.003	.9969	30′
40′	814	12.29	816	12.25	003	967	20′
50′	843	11.87	846	11.83	004	964	10′
5° 0′	.0872	11.47	.0875	11.43	1.004	.9962	85° 0′
10′	901	11.10	904	11.06	004	959	50′
20′	929	10.76	934	10.71	004	957	40′
30′	.0958	10.43	.0963	10.39	1.005	.9954	30′
40′	.0987	10.13	.0992	10.08	005	951	20′
50′	.1016	9.839	.1022	9.788	005	948	10′
6° 0′	.1045	9.567	.1051	9.514	1.006	.9945	84° 0′
10′	074	9.309	080	9.255	006	942	50′
20′	103	9.065	110	9.010	006	939	40′
30′	.1132	8.834	.1139	8.777	1.006	.9936	30′
40′	161	8.614	169	8.556	007	932	20′
50′	190	8.405	198	8.345	007	929	10′
7° 0′	.1219	8.206	.1228	8.144	1.008	.9925	83° 0′
10′	248	8.016	257	7.953	008	922	50′
20′	276	7.834	287	7.770	008	918	40′
30′	.1305	7.661	.1317	7.596	1.009	.9914	30′
40′	334	7.496	346	7.429	009	911	20′
50′	363	7.337	376	7.269	009	907	10′
8° 0′	.1392	7.185	.1405	7.115	1.010	.9903	82° 0′
	Cos	Sec	Cot	Tan	Csc	Sin	Degrees

Degrees	Sin	Csc	Tan	Cot	Sec	Cos	
8° 0′	.1392	7.185	.1405	7.115	1.010	.9903	82° 0′
10′	421	7.040	435	6.968	010	899	50′
20′	449	6.900	465	6.827	011	894	40′
30′	.1478	6.765	.1495	6.691	1.011	.8980	30′
40′	507	6.636	524	6.561	012	886	20′
50′	536	6.512	554	6.435	012	881	10′
9° 0′	.1564	6.392	.1584	6.314	1.012	.9877	81° 0′
10′	593	277	614	197	013	872	50′
20′	622	166	644	6.084	013	868	40′
30′	.1650	6.059	.1673	5.976	1.014	.9863	30′
40′	679	5.955	703	871	014	858	20′
50′	708	855	733	769	015	853	10′
10° 0′	.1736	5.759	.1763	5.671	1.015	.9848	80° 0′
10′	765	665	793	576	016	843	50′
20′	794	575	823	485	016	838	40′
30′	.1822	5.487	.1853	5.396	1.017	.9833	30′
40′	851	403	883	309	018	827	20′
50′	880	320	914	226	018	822	10′
11° 0′	.1908	5.241	.1944	5.145	1.019	.9816	79° 0′
10′	937	164	.1974	5.066	019	811	50′
20′	965	089	.2004	4.989	020	805	40′
30′	.1994	5.016	.2035	4.915	1.020	.9799	30′
40′	.2022	4.945	065	843	021	793	20′
50′	051	876	095	773	022	787	10′
12° 0′	.2079	4.810	.2126	4.705	1.022	.9781	78° 0′
10′	108	745	156	638	023	775	50′
20′	136	682	186	574	024	769	40′
30′	.2164	4.620	.2217	4.511	1.024	.9763	30′
40′	193	560	247	449	025	757	20′
50′	221	502	278	390	026	750	10′
13° 0′	.2250	4.445	.2309	4.331	1.026	.9744	77° 0′
10′	278	390	339	275	027	737	50′
20′	306	336	370	219	028	730	40′
30′	.2334	4.284	.2401	4.165	1.028	.9724	30′
40′	363	232	432	113	029	717	20′
50′	391	182	462	061	030	710	10′
14° 0′	.2419	4.134	.2493	4.011	1.031	.9703	76° 0′
10′	447	086	524	3.962	031	696	50′
20′	476	4.039	555	914	032	698	40′
30′	.2504	3.994	.2586	3.867	1.033	.9681	30′
40′	532	950	617	821	034	674	20′
50′	560	906	648	776	034	667	10′
15° 0′	.2588	3.864	.2679	3.732	1.035	.9659	75° 0′
10′	616	822	711	689	036	652	50′
20′	644	782	742	647	037	644	40′
30′	.2672	3.742	.2773	3.606	1.038	.9636	30′
40′	700	703	805	566	039	628	20′
50′	728	665	836	526	039	621	10′
16° 0′	.2756	3.628	.2867	3.487	1.040	.9613	74° 0′
	Cos	Sec	Cot	Tan	Csc	Sin	Degrees

TABLE IV TRIGONOMETRIC FUNCTIONS—DEGREE MEASURE

Degrees	Sin	Csc	Tan	Cot	Sec	Cos	
16° 0′	.2756	3.628	.2867	3.487	1.040	.9613	74° 0′
10′	784	592	899	450	041	605	50′
20′	812	556	931	412	042	596	40′
30′	.2840	3.521	.2962	3.376	1.043	.9588	30′
40′	868	487	.2994	340	044	580	20′
50′	896	453	3026	305	045	572	10′
17° 0′	.2924	3.420	.3057	3.271	1.046	.9563	73° 0′
10′	952	388	089	237	047	555	50′
20′	.2979	357	121	204	048	546	40′
30′	.3007	3.326	.3153	3.172	1.048	.9537	30′
40′	035	295	185	140	049	528	20′
50′	062	265	217	108	050	520	10′
18° 0′	.3090	3.236	.3249	3.078	1.051	.9511	72° 0′
10′	118	207	281	047	052	502	50′
20′	145	179	314	3.018	053	492	40′
30′	.3173	3.152	.3346	2.989	1.054	.9483	30′
40′	201	124	378	960	056	474	20′
50′	228	098	411	932	057	465	10′
19° 0′	.3256	3.072	.3443	2.904	1.058	.9455	71° 0′
10′	283	046	476	877	059	446	50′
20′	311	3.021	508	850	060	436	40′
30′	.3338	2.996	.3541	2.824	1.061	.9426	30′
40′	365	971	574	798	062	417	20′
50′	393	947	607	773	063	407	10′
20° 0′	.3420	2.924	.3640	2.747	1.064	.9397	70° 0′
10′	448	901	673	723	065	387	50′
20′	475	878	706	699	066	377	40′
30′	.3502	2.855	.3739	2.675	1.068	.9367	30′
40′	529	833	772	651	069	356	20′
50′	557	812	805	628	070	346	10′
21° 0′	.3584	2.790	.3839	2.605	1.071	.9336	69° 0′
10′	611	769	872	583	072	325	50′
20′	638	749	906	560	074	315	40′
30′	.3665	2.729	.3939	2.539	1.075	.9304	30′
40′	692	709	.3973	517	076	293	20′
50′	719	689	.4006	496	077	283	10′
22° 0′	.3746	2.669	.4040	2.475	1.079	.9272	68° 0′
10′	773	650	074	455	080	261	50′
20′	800	632	108	434	081	250	40′
30′	.3827	2.613	.4142	2.414	1.082	.9239	30′
40′	854	595	176	394	084	228	20′
50′	881	577	210	375	085	216	10′
23° 0′	.3907	2.559	.4245	2.356	1.086	.9205	67° 0′
10′	934	542	279	337	088	194	50′
20′	961	525	314	318	089	182	40′
30′	.3987	2.508	.4348	2.300	1.090	.9171	30′
40′	.4014	491	383	282	092	159	20′
50′	041	475	417	264	093	147	10′
24° 0′	.4067	2.459	.4452	2.246	1.095	.9135	66° 0′
	Cos	Sec	Cot	Tan	Csc	Sin	Degrees

Degrees	Sin	Csc	Tan	Cot	Sec	Cos	
24° 0′	.4067	2.459	.4452	2.246	1.095	.9135	66° 0′
10′	094	443	487	229	096	124	50′
20′	120	427	522	211	097	112	40′
30′	.4147	2.411	.4557	2.194	1.099	.9100	30′
40′	173	396	592	177	100	088	20′
50′	200	381	628	161	102	075	10′
25° 0′	.4226	2.366	.4663	2.145	1.103	.9063	65° 0′
10′	253	352	699	128	105	051	50′
20′	279	337	734	112	106	038	40′
30′	.4305	2.323	.4770	2.097	1.108	.9026	30′
40′	331	309	806	081	109	013	20′
50′	358	295	841	066	111	.9001	10′
26° 0′	.4384	2.281	.4877	2.050	1.113	.8988	64° 0′
10′	410	268	913	035	114	975	50′
20′	436	254	950	020	116	962	40′
30′	.4462	2.241	.4986	2.006	1.117	.8949	30′
40′	488	228	.5022	1.991	119	936	20′
50′	514	215	059	977	121	923	10′
27° 0′	.4540	2.203	.5095	1.963	1.122	.8910	63° 0′
10′	566	190	132	949	124	897	50′
20′	592	178	169	935	126	884	40′
30′	.4617	2.166	.5206	1.921	1.127	.8870	30′
40′	643	154	243	907	129	857	20′
50′	669	142	280	894	131	843	10′
28° 0′	.4695	2.130	.5317	1.881	1.133	.8829	62° 0′
10′	720	118	354	868	134	.816	50′
20′	746	107	392	855	136	802	40′
30′	.4772	2.096	.5430	1.842	1.138	.8788	30′
40′	797	085	467	829	140	774	20′
50′	823	074	505	816	142	760	10′
29° 0′	.4848	2.063	.5543	1.804	1.143	.8746	61° 0′
10′	874	052	581	792	145	732	50′
20′	899	041	619	780	147	718	40′
30′	.4924	2.031	.5658	1.767	1.149	.8704	30′
40′	950	020	696	756	151	689	20′
50′	.4975	010	735	744	153	675	10′
30° 0′	.5000	2.000	.5774	1.732	1.155	.8660	60° 0′
10′	025	1.990	812	720	157	646	50′
20′	050	980	851	709	159	631	40′
30′	.5075	1.970	.5890	1.698	1.161	.8616	30′
40′	100	961	930	686	163	601	20′
50′	125	951	.5969	675	165	587	10′
31° 0′	.5150	1.942	.6009	1.664	1.167	.8572	59° 0′
10′	175	932	048	653	169	557	50′
20′	200	923	088	643	171	542	40′
30′	.5225	1.914	.6128	1.632	1.173	.8526	30′
40′	250	905	168	621	175	511	20′
50′	275	896	208	611	177	496	10′
32° 0′	.5299	1.887	.6249	1.600	1.179	.8480	58° 0′
	Cos	Sec	Cot	Tan	Csc	Sin	Degrees

TABLE IV TRIGONOMETRIC FUNCTIONS—DEGREE MEASURE

Degrees	Sin	Csc	Tan	Cot	Sec	Cos	
32° 0′	.5299	1.887	.6249	1.600	1.179	.8480	58° 0′
10′	324	878	289	590	181	465	50′
20′	348	870	330	580	184	450	40′
30′	.5373	1.861	.6371	1.570	1.186	.8434	30′
40′	398	853	412	560	188	418	20′
50′	422	844	453	550	190	403	10′
33° 0′	.5446	1.836	.6494	1.540	1.192	.8387	57° 0′
10′	471	828	536	530	195	371	50′
20′	495	820	577	520	197	355	40′
30′	.5519	1.812	.6619	1.511	1.199	.8339	30′
40′	544	804	661	501	202	323	20′
50′	568	796	703	1.492	204	307	10′
34° 0′	.5592	1.788	.6745	1.483	1.206	.8290	56° 0′
10′	616	781	787	473	209	274	50′
20′	640	773	830	464	211	258	40′
30′	.5664	1.766	.6873	1.455	1.213	.8241	30′
40′	688	758	916	446	216	225	20′
50′	712	751	.6959	437	218	208	10′
35° 0′	.5736	1.743	.7002	1.428	1.221	.8192	55° 0′
10′	760	736	046	419	223	175	50′
20′	783	729	089	411	226	158	40′
30′	.5807	1.722	.7133	1.402	1.228	.8141	30′
40′	831	715	177	393	231	124	20′
50′	854	708	221	385	233	107	10′
36° 0′	.5878	1.701	.7265	1.376	1.236	.8090	54° 0′
10′	901	695	310	368	239	073	50′
20′	925	688	355	360	241	056	40′
30′	.5948	1.681	.7400	1.351	1.244	.8039	30′
40′	972	675	445	343	247	021	20′
50′	.5995	668	490	335	249	.8004	10′
37° 0′	.6018	1.662	.7536	1.327	1.252	.7986	53° 0′
10′	041	655	581	319	255	969	50′
20′	065	649	627	311	258	951	40′
30′	.6088	1.643	.7673	1.303	1.260	.7934	30′
40′	111	636	720	295	263	916	20′
50′	134	630	766	288	266	898	10′
38° 0′	.6157	1.624	.7813	1.280	1.269	.7880	52° 0′
10′	180	618	860	272	272	862	50′
20′	202	612	907	265	275	844	40′
30′	.6225	1.606	.7954	1.257	1.278	.7826	30′
40′	248	601	.8002	250	281	808	20′
50′	271	595	050	242	284	790	10′
39° 0′	.6293	1.589	.8098	1.235	1.287	.7771	51° 0′
10′	316	583	146	228	290	753	50′
20′	338	578	195	220	293	735	40′
30′	.6361	1.572	.8243	1.213	1.296	.7716	30′
40′	383	567	292	206	299	698	20′
50′	406	561	342	199	302	679	10′
40° 0′	.6428	1.556	.8391	1.192	1.305	.7660	50° 0′
	Cos	Sec	Cot	Tan	Csc	Sin	Degrees

Degrees	Sin	Csc	Tan	Cot	Sec	Cos	
40° 0′	.6428	1.556	.8391	1.192	1.305	.7660	50° 0′
10′	450	550	441	185	309	642	50′
20′	472	545	491	178	312	623	40′
30′	.6494	1.540	.8541	1.171	1.315	.7604	30′
40′	517	535	591	164	318	585	20′
50′	539	529	642	157	322	566	10′
41° 0′	.6561	1.524	.8693	1.150	1.325	.7547	49° 0′
10′	583	519	744	144	328	528	50′
20′	604	514	796	137	332	509	40′
30′	.6626	1.509	.8847	1.130	1.335	.7490	30′
40′	648	504	899	124	339	470	20′
50′	670	499	.8952	117	342	451	10′
42° 0′	.6691	1.494	.9004	1.111	1.346	.7431	48° 0′
10′	713	490	057	104	349	412	50′
20′	734	485	110	098	353	392	40′
30′	.6756	1.480	.9163	1.091	1.356	.7373	30′
40′	777	476	217	085	360	353	20′
50′	799	471	271	079	364	333	10′
43° 0′	.6820	1.466	.9325	1.072	1.367	.7314	47° 0′
10′	841	462	380	066	371	294	50′
20′	862	457	435	060	375	274	40′
30′	.6884	1.453	.9490	1.054	1.379	.7254	30′
40′	905	448	545	048	382	234	20′
50′	926	444	601	042	386	214	10′
44° 0′	.6947	1.440	.9657	1.036	1.390	.7193	46° 0′
10′	967	435	713	030	394	173	50′
20′	.6988	431	770	024	398	153	40′
30′	.7009	1.427	.9827	1.018	1.402	.7133	30′
40′	030	423	884	012	406	112	20′
50′	050	418	.9942	006	410	092	10′
45° 0′	.7071	1.414	1.000	1.000	1.414	.7071	45° 0′
	Cos	Sec	Cot	Tan	Csc	Sin	Degrees

TABLE V POWERS AND ROOTS

Number	Square	Square Root	Cube	Cube Root	Number	Square	Square Root	Cube	Cube Root
1	1	1.000	1	1.000	51	2,601	7.141	132,651	3.708
2	4	1.414	8	1.260	52	2,704	7.211	140,608	3.733
3	9	1.732	27	1.442	53	2,809	7.280	148,877	3.756
4	16	2.000	64	1.587	54	2,916	7.348	157,464	3.780
5	25	2.236	125	1.710	55	3,025	7.416	166,375	3.803
6	36	2.449	216	1.817	56	3,136	7.483	175,616	3.826
7	49	2.646	343	1.913	57	3,249	7.550	185,193	3.849
8	64	2.828	512	2.000	58	3,364	7.616	195,112	3.871
9	81	3.000	729	2.080	59	3,481	7.681	205,379	3.893
10	100	3.162	1,000	2.154	60	3,600	7.746	216,000	3.915
11	121	3.317	1,331	2.224	61	3,721	7.810	226,981	3.936
12	144	3.464	1,728	2.289	62	3,844	7.874	238,328	3.958
13	169	3.606	2,197	2.351	63	3,969	7.937	250,047	3.979
14	196	3.742	2,744	2.410	64	4,096	8.000	262,144	4.000
15	225	3.873	3,375	2.466	65	4,225	8.062	274,625	4.021
16	256	4.000	4,096	2.520	66	4,356	8.124	287,496	4.041
17	289	4.123	4,913	2.571	67	4,489	8.185	300,763	4.062
18	324	4.243	5,832	2.621	68	4,624	8.246	314,432	4.082
19	361	4.359	6,859	2.668	69	4,761	8.307	328,509	4.102
20	400	4.472	8,000	2.714	70	4,900	8.367	343,000	4.121
21	441	4.583	9,261	2.759	71	5,041	8.426	357,911	4.141
22	484	4.690	10,648	2.802	72	5,184	8.485	373,248	4.160
23	529	4.796	12,167	2.844	73	5,329	8.544	389,017	4.179
24	576	4.899	13,824	2.884	74	5,476	8.602	405,224	4.198
25	625	5.000	15,625	2.924	75	5,625	8.660	421,875	4.217
26	676	5.099	17,576	2.962	76	5,776	8.718	438,976	4.236
27	729	5.196	19,683	3.000	77	5,929	8.775	456,533	4.254
28	784	5.292	21,952	3.037	78	6,084	8.832	474,552	4.273
29	841	5.385	24,389	3.072	79	6,241	8.888	493,039	4.291
30	900	5.477	27,000	3.107	80	6,400	8.944	512,000	4.309
31	961	5.568	29,791	3.141	81	6,561	9.000	531,441	4.327
32	1,024	5.657	32,768	3.175	82	6,724	9.055	551,368	4.344
33	1,089	5.745	35,937	3.208	83	6,889	9.110	571,787	4.362
34	1,156	5.831	39,304	3.240	84	7,056	9.165	592,704	4.380
35	1,225	5.916	42,875	3.271	85	7,225	9.220	614,125	4.397
36	1,296	6.000	46,656	3.302	86	7,396	9.274	636,056	4.414
37	1,369	6.083	50,653	3.332	87	7,569	9.327	658,503	4.431
38	1,444	6.164	54,872	3.362	88	7,744	9.381	681,472	4.448
39	1,521	6.245	59,319	3.391	89	7,921	9.434	704,969	4.465
40	1,600	6.325	64,000	3.420	90	8,100	9.487	729,000	4.481
41	1,681	6.403	68,921	3.448	91	8,281	9.539	753,571	4.498
42	1,764	6.481	74,088	3.476	92	8,464	9.592	778,688	4.514
43	1,849	6.557	79,507	3.503	93	8,649	9.644	804,357	4.531
44	1,936	6.633	85,184	3.530	94	8,836	9.695	830,584	4.547
45	2,025	6.708	91,125	3.557	95	9,025	9.747	857,375	4.563
46	2,116	6.782	97,336	3.583	96	9,216	9.798	884,736	4.579
47	2,209	6.856	103,823	3.609	97	9,409	9.849	912,673	4.595
48	2,304	6.928	110,592	3.634	98	9,604	9.899	941,192	4.610
49	2,401	7.000	117,649	3.659	99	9,801	9.950	970,299	4.626
50	2,500	7.071	125,000	3.684	100	10,000	10.000	1,000,000	4.642